Optimization and Differentiation

MONOGRAPHS AND RESEARCH NOTES IN MATHEMATICS

Series Editors

John A. Burns
Thomas J. Tucker
Miklos Bona
Michael Ruzhansky

Published Titles

Actions and Invariants of Algebraic Groups, Second Edition, Walter Ferrer Santos
and Alvaro Rittatore

Analytical Methods for Kolmogorov Equations, Second Edition, Luca Lorenzi

Application of Fuzzy Logic to Social Choice Theory, John N. Mordeson, Davender S. Malik
and Terry D. Clark

*Blow-up Patterns for Higher-Order: Nonlinear Parabolic, Hyperbolic Dispersion and
Schrödinger Equations*, Victor A. Galaktionov, Enzo L. Mitidieri, and Stanislav Pohozaev

Bounds for Determinants of Linear Operators and Their Applications, Michael Gil'

Complex Analysis: Conformal Inequalities and the Bieberbach Conjecture, Prem K. Kythe

Computation with Linear Algebraic Groups, Willem Adriaan de Graaf

Computational Aspects of Polynomial Identities: Volume I, Kemer's Theorems, 2nd Edition
Alexei Kanel-Belov, Yakov Karasik, and Louis Halle Rowen

A Concise Introduction to Geometric Numerical Integration, Fernando Casas
and Sergio Blanes

Cremona Groups and Icosahedron, Ivan Cheltsov and Constantin Shramov

Delay Differential Evolutions Subjected to Nonlocal Initial Conditions
Monica-Dana Burlică, Mihai Necula, Daniela Roșu, and Ioan I. Vrabie

Diagram Genus, Generators, and Applications, Alexander Stoimenow

Difference Equations: Theory, Applications and Advanced Topics, Third Edition
Ronald E. Mickens

Dictionary of Inequalities, Second Edition, Peter Bullen

Elements of Quasigroup Theory and Applications, Victor Shcherbacov

Finite Element Methods for Eigenvalue Problems, Jiguang Sun and Aihui Zhou

Introduction to Abelian Model Structures and Gorenstein Homological Dimensions
Marco A. Pérez

Iterative Methods without Inversion, Anatoly Galperin

Iterative Optimization in Inverse Problems, Charles L. Byrne

Line Integral Methods for Conservative Problems, Luigi Brugnano and Felice Iavernaro

Lineability: The Search for Linearity in Mathematics, Richard M. Aron,
Luis Bernal González, Daniel M. Pellegrino, and Juan B. Seoane Sepúlveda

Modeling and Inverse Problems in the Presence of Uncertainty, H. T. Banks, Shuhua Hu,
and W. Clayton Thompson

Published Titles Continued

Monomial Algebras, Second Edition, Rafael H. Villarreal

Noncommutative Deformation Theory, Eivind Eriksen, Olav Arnfinn Laudal, and Arvid Siqveland

Nonlinear Functional Analysis in Banach Spaces and Banach Algebras: Fixed Point Theory Under Weak Topology for Nonlinear Operators and Block Operator Matrices with Applications, Aref Jeribi and Bilel Krichen

Optimization and Differentiation, Simon Serovajsky

Partial Differential Equations with Variable Exponents: Variational Methods and Qualitative Analysis, Vicenţiu D. Rădulescu and Dušan D. Repovš

A Practical Guide to Geometric Regulation for Distributed Parameter Systems Eugenio Aulisa and David Gilliam

Reconstruction from Integral Data, Victor Palamodov

Signal Processing: A Mathematical Approach, Second Edition, Charles L. Byrne

Sinusoids: Theory and Technological Applications, Prem K. Kythe

Special Integrals of Gradshteyn and Ryzhik: the Proofs – Volume I, Victor H. Moll

Special Integrals of Gradshteyn and Ryzhik: the Proofs – Volume II, Victor H. Moll

Spectral and Scattering Theory for Second-Order Partial Differential Operators, Kiyoshi Mochizuki

Stochastic Cauchy Problems in Infinite Dimensions: Generalized and Regularized Solutions, Irina V. Melnikova

Submanifolds and Holonomy, Second Edition, Jürgen Berndt, Sergio Console, and Carlos Enrique Olmos

Symmetry and Quantum Mechanics, Scott Corry

The Truth Value Algebra of Type-2 Fuzzy Sets: Order Convolutions of Functions on the Unit Interval, John Harding, Carol Walker, and Elbert Walker

Forthcoming Titles

Groups, Designs, and Linear Algebra, Donald L. Kreher

Handbook of the Tutte Polynomial, Joanna Anthony Ellis-Monaghan and Iain Moffat

Microlocal Analysis on R^n and on NonCompact Manifolds, Sandro Coriasco

Practical Guide to Geometric Regulation for Distributed Parameter Systems, Eugenio Aulisa and David S. Gilliam

MONOGRAPHS AND RESEARCH NOTES IN MATHEMATICS

Optimization and Differentiation

Simon Serovajsky

CRC Press is an imprint of the
Taylor & Francis Group, an **informa** business
A CHAPMAN & HALL BOOK

CRC Press
Taylor & Francis Group
6000 Broken Sound Parkway NW, Suite 300
Boca Raton, FL 33487-2742

© 2018 by Taylor & Francis Group, LLC
CRC Press is an imprint of Taylor & Francis Group, an Informa business

No claim to original U.S. Government works

Printed on acid-free paper
Version Date: 20170801

International Standard Book Number-13: 978-1-4987-5093-6 (Hardback)

This book contains information obtained from authentic and highly regarded sources. Reasonable efforts have been made to publish reliable data and information, but the author and publisher cannot assume responsibility for the validity of all materials or the consequences of their use. The authors and publishers have attempted to trace the copyright holders of all material reproduced in this publication and apologize to copyright holders if permission to publish in this form has not been obtained. If any copyright material has not been acknowledged please write and let us know so we may rectify in any future reprint.

Except as permitted under U.S. Copyright Law, no part of this book may be reprinted, reproduced, transmitted, or utilized in any form by any electronic, mechanical, or other means, now known or hereafter invented, including photocopying, microfilming, and recording, or in any information storage or retrieval system, without written permission from the publishers.

For permission to photocopy or use material electronically from this work, please access www.copyright.com (http://www.copyright.com/) or contact the Copyright Clearance Center, Inc. (CCC), 222 Rosewood Drive, Danvers, MA 01923, 978-750-8400. CCC is a not-for-profit organization that provides licenses and registration for a variety of users. For organizations that have been granted a photocopy license by the CCC, a separate system of payment has been arranged.

Trademark Notice: Product or corporate names may be trademarks or registered trademarks, and are used only for identification and explanation without intent to infringe.

Visit the Taylor & Francis Web site at
http://www.taylorandfrancis.com

and the CRC Press Web site at
http://www.crcpress.com

Printed and bound in Great Britain by
TJ International Ltd, Padstow, Cornwall

To Jacques-Louis Lions

Contents

Preface	**xv**
List of Figures	**xix**
List of Tables	**xxi**

I Minimization of functionals — 1

1 Necessary conditions of extremum for functionals — 3

1.1	Minimization of functions and stationary condition	4
1.2	Lagrange problem and Euler equation	11
1.3	Linear spaces	16
1.4	Linear normalized spaces	20
1.5	Directional differentiation	26
1.6	Gâteaux differentiation of functionals	27
1.7	Minimization of functionals and stationary condition	31
1.8	General functional spaces	33
1.9	Minimization of Dirichlet integral	38
1.10	Minimization of functionals on subspaces	39
1.11	Derivatives with respect to subspaces	42
1.12	Minimization of functionals on affine varieties	43
1.13	Minimization of functionals on convex sets	45
1.14	Comments	50

2 Minimization of functionals. Addition — 53

2.1	Sufficiency of extremum conditions	53
2.2	Existence of the function minimum	58
2.3	Weak convergence in Banach spaces	60
2.4	Existence of the functional minimum	63
2.5	Uniqueness of the functional minimum	66
2.6	Tihonov well-posedness of extremum problems	68
2.7	Hadamard well-posedness of extremum problems	70
2.8	Ekeland principle and the approximate condition of extremum	72
2.9	Non-smooth functionals and subdifferentiation	75

x   *Contents*

2.10 Fréchet differentiation of functionals   78
2.11 Approximate methods of functionals minimization   80
2.12 Comments   82

II   Stationary systems   87

3   Linear systems   89

3.1   Additional results of functional analysis   90
3.2   Abstract linear control systems   92
3.3   Dirichlet problem for the Poisson equation   98
3.4   Optimal control problem statement 102
3.5   Necessary conditions of optimality 104
3.6   Optimal control problems with local constraints 107
3.7   Lagrange multipliers method 110
3.8   Maximum principle 113
3.9   Penalty method 115
3.10 Additional properties of Sobolev spaces 117
3.11 Nonhomogeneous Dirichlet problem for the elliptic equation   121
3.12 Comments 125

4   Weakly nonlinear systems   129

4.1   Differentiation of operators 130
4.2   Inverse function theorem 135
4.3   Optimal control problems for weakly nonlinear systems . . 140
4.4   Equations with monotone operators 144
4.5   Additional results of the functional analysis 149
4.6   Nonlinear elliptic equation 152
4.7   Optimal control problems for nonlinear elliptic equations   . 157
4.8   Necessary conditions of optimality 159
4.9   Optimal control problems for general functionals 164
4.10 Optimal control problems for semilinear elliptic equations   . 168
4.11 Differentiation of the inverse operator 173
4.12 Comments 174

5   Strongly nonlinear systems   177

5.1   Introduction 178
5.2   Absence of the differentiability of the inverse operator . . . 180
5.3   Extended differentiation of operators 187
5.4   Necessary conditions of optimality for the strongly nonlinear
      system 195
5.5   Boundary control for Neumann problem 199
5.6   Extended differentiability of the inverse operator 206
5.7   Comments 209

Contents

xi

6 Coefficients optimization control problems   **211**

   6.1   Coefficients control problem   212
   6.2   Derivative with respect to a convex set   214
   6.3   Optimization control problem for bilinear systems   216
   6.4   Analysis of the coefficient optimization problem   220
   6.5   Nonlinear system with the control at the coefficients   224
   6.6   Differentiability with respect to a convex set for abstract
        systems   230
   6.7   Strongly nonlinear systems with the control at the coefficients   234
   6.8   Comments   240

7 Systems with nonlinear control   **243**

   7.1   Implicit function theorem   243
   7.2   Optimization control problems for abstract systems   246
   7.3   Weakly nonlinear control systems   248
   7.4   Extended differentiability of the implicit function   254
   7.5   Strongly nonlinear control systems   256
   7.6   Comments   260

III   Evolutional systems   261

8 Linear first-order evolution systems   **263**

   8.1   Abstract functions   264
   8.2   Ordinary differential equations   270
   8.3   Linear first order evolutional equations   271
   8.4   Optimization control problems for linear evolutional equations   278
   8.5   Optimization control problems for the heat equation   283
   8.6   Optimization control problems with the functional that is not
         dependent from the control   287
   8.7   Non-conditional optimization control problems   290
   8.8   Non-conditional optimization control problems for the heat
         equation   294
   8.9   Hamilton–Jacobi equation   297
   8.10   Bellman method for an optimization problem for the heat
         equation   301
   8.11   Comments   307

9 Nonlinear first order evolutional systems   **311**

   9.1   Nonlinear evolutional equations with monotone operators   .   312
   9.2   Optimization control problems for evolutional equations with
        monotone operator   321
   9.3   Optimization control problems for the nonlinear heat equation   323

xii   Contents

9.4   Necessary optimality conditions for the nonlinear heat
      equation   326
9.5   Optimization control problems with the functional that is not
      dependent upon the control   334
9.6   Sufficient optimality conditions for the nonlinear heat
      equation   337
9.7   Coefficient optimization problems for linear parabolic
      equations   339
9.8   Coefficient optimization problems for nonlinear parabolic
      equations   346
9.9   Initial optimization control problems for nonlinear parabolic
      equations   353
9.10 Optimization control problems for nonlinear parabolic
      equations with final functional   358
9.11 Comments   361

10 Second order evolutional systems   365

10.1   Linear second order evolutional equations   366
10.2   Optimization control problem for linear second order
       evolutional equations   372
10.3   Non-conditional optimization control problems   377
10.4   Optimization control problem for the wave equation   380
10.5   Nonlinear wave equation   383
10.6   Optimization control problem for the nonlinear wave
       equation   390
10.7   Optimality conditions for the optimization optimal problem
       for the nonlinear wave equation   392
10.8   Non-differentiability of the solution of the nonlinear wave
       equation with respect to the absolute term   397
10.9   Optimization control problem for the linear hyperbolic
       equation with coefficient control   398
10.10 Optimization control problem for the nonlinear wave equation
       with coefficient control   403
10.11 Comments   409

11 Navier–Stokes equations   411

11.1   Evolutional Navier–Stokes equations   411
11.2   Optimization control problems for the evolutional
       Navier–Stokes equations   419
11.3   Stationary Navier–Stokes equations   430
11.4   Optimization control problems for the stationary
       Navier–Stokes equations   434
11.5   System of Navier–Stokes and heat equations   440

Contents

11.6   Optimization control problems for Navier–Stokes and heat equations 448

11.7   Comments 456

IV   Addition   459

12 Functors of the differentiation   461

12.1   Elements of the categories theory 461

12.2   Differentiation functor and its application to the extremum theory 463

12.3   Additional properties of extended derivatives 466

12.4   Extended differentiation functor and its application to the extremum theory 469

12.5   Comments 471

Bibliography   473

Index   511

Preface

Extremum theory is one of the most beautiful mathematical areas. Its development is largely driven by a wealth of practical applications. However, these applications confront researchers with a wide complex of most difficult mathematical problems. This requires a serious development of the theory and significantly enriches the mathematics in general. It is no coincidence this direction has always attracted the greatest mathematicians: Férmat, Euler, Lagrange, Hamilton, Weierstrass, Hilbert, and others.

Already in the process of solving the simplest problems of minimization of function a deep connection was found between the theory of extremum and the differential calculus. Particularly, it was noticed that the velocity of change of a smooth function decreases to zero in the neighborhood of the point of extremum. The discovery of this law led to the first general result of the extremum theory, which dates back to Fermat. The derivative of the smooth function is equal to zero at the point of its extremum. Thus, the search of the function extremum reduces to the calculation of its derivative, finding the points at which this derivative is zero, and the study of the behavior of the function at these points.

The development of the classical calculus of variations associated with deep problems of mechanics and optics led to the need to find an extremum of functionals that is maps defined on the set of functions. Proposed by Lagrange the method of variation actually already contained a theory of functional differentiation. Particularly, the equality to the zero of the functional variation is the analog of the equality to the zero of the Gâteaux derivative of the functional at the point of extremum. This is a natural extension of the Fermat condition to the minimization problem for the functionals. Practical solving of the problems of variation calculus is reduced in reality to the determination to the derivative of the functional and the analysis of the points, where the derivative is equal to zero.

In the middle of the twentieth century optimization control problems began to be considered. The dependence of the minimized functional of unknown value (control) was of implicit character here. The functional depends upon the state function that satisfies the state equations that includes unknown control. The calculation of the derivative of this functional is related to the differentiation of control–state mapping determined by the state equation. Therefore, the transition from the simplest problems of function minimization and variational calculus to optimization control problems is actually reduced

xvi   *Preface*

to the more difficult problems of the differentiation of the general operators. The analysis of the optimization problems for nonlinear infinite-dimensional systems has shown clearly that the methods of differentiation operators are in reality the basis of the general extremum theory.

A considerable amount of literature is devoted to optimal control theory. A peculiarity of this book is its presentation as an application of the theory of differential operators. The first part of the book has an introductory nature. It is dedicated to minimization of the functionals with direct dependence from the unknown values. The determination of the extremum condition requires here the calculation of the functional derivative in one form or another.

The optimization control problems for the stationary and evolutional systems are considered in the second and third parts. These problems can be transformed to the previous problems, if control–state mapping has necessary properties. This dependence is Gâteaux differentiable for the linear systems and is easy enough for nonlinear systems too. Unfortunately, its Gâteaux derivative does not exist for the difficult enough systems. However, this dependence is differentiable in a weaker form. Particularly, we propose the extended derivative of an operator. This is sufficient for obtaining the necessary conditions of optimality.

The final part of the book is subsidiary. We interpret the differentiation by using the categories theory here as a special functor.

The book is written in accordance with special courses given by the author at the Mechanics and Mathematics Faculty of al-Farabi Kazakh National University. It includes the results of the author from different years. The original version of the book [486], [488], which differs significantly from the present one, was published in Russian.

I would like to thank the following persons with whom I discussed some issues with this book in varying degrees: S. Aisagaliev, F. Aliev, V. Amerbaev, A. Antipin, G. Bizhanova, V. Boltyansky, A. Butkovsky, M. Dzhenaliev, A. Egorov, R. Fedorenko, A. Fursikov, M. Goebel, S. Harin, A. Iskenderov, S. Kabanihin, O. Ladyzhenskaya, V. Litvinov, K. Lurie, V. Neronov, V. Osmolovsky, M. Otelbaev, U. Raitum, M. Ruzhansky, T. Sirazetdiniv, V. Shcherbak, Sh. Smagulov, U. Sultangazin, N. Temirgaliev, V. Tihomirov, N. Uraltseva, F. Vasiliev, and V. Yakubovich. I am also grateful to the students and staff of the Kazakh National University for their understanding and support. I want to especially thank Professor M. Ruzhansky, Imperial College London, for his great help in the publication of this book. I am grateful also to D. Nurseitov for his help in preparing the book for publication.

I dedicate this work to a remarkable mathematician, recognized as a classic in the field of optimal control theory, the equations of mathematical physics, and functional analysis, Jacques Louis Lions. I studied his excellent books. In addition, I am extremely grateful to him for the review, which he gave to my thesis (see Figure 1), and, in particular, to the concept of an extended derivative and its applications in optimal control theory, which is widely used in this book.

Preface

xvii

For those who wish to make comments, suggestions, and questions on the content of the book, I can be reached at serovajskys@mail.ru.

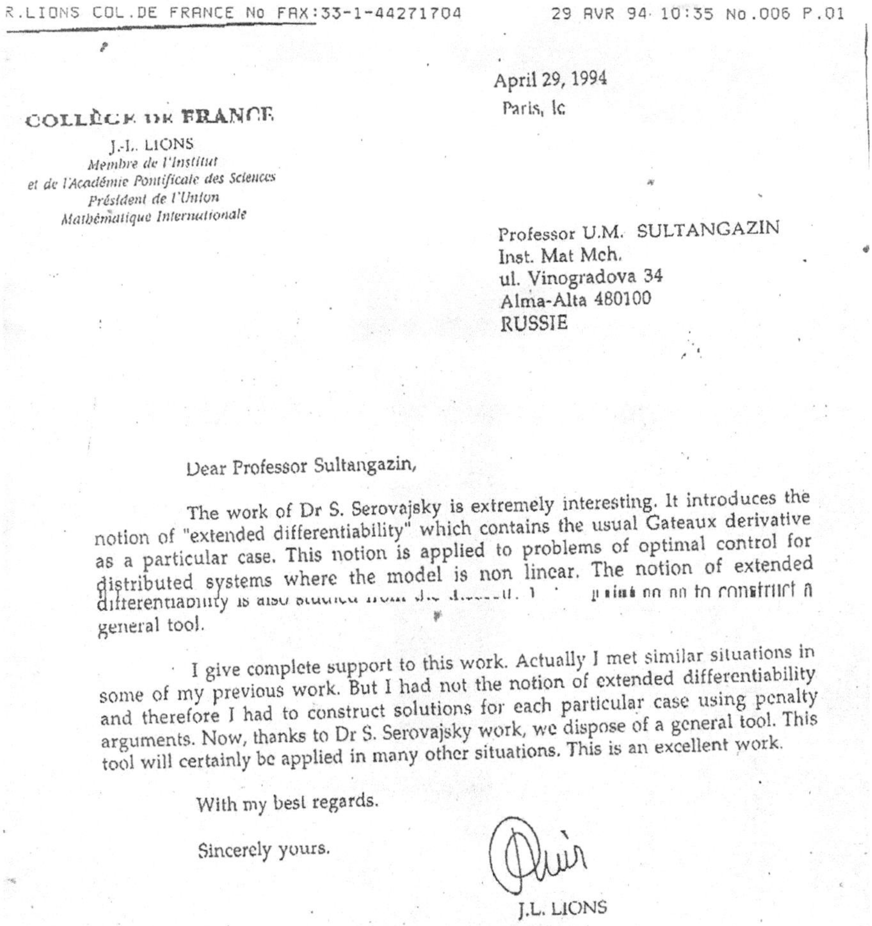

R.LIONS COL.DE FRANCE No FAX:33-1-44271704   29 AVR 94 10:35 No.006 P.01

April 29, 1994
Paris, le

COLLÈGE DE FRANCE

J.-L. LIONS
Membre de l'Institut
et de l'Académie Pontificale des Sciences
Président de l'Union
Mathématique Internationale

Professor U.M. SULTANGAZIN
Inst. Mat Mch.
ul. Vinogradova 34
Alma-Alta 480100
RUSSIE

Dear Professor Sultangazin,

The work of Dr S. Serovajsky is extremely interesting. It introduces the notion of "extended differentiability" which contains the usual Gateaux derivative as a particular case. This notion is applied to problems of optimal control for distributed systems where the model is non linear. The notion of extended differentiability is also studied from the different. 1 point on on to construct a general tool.

I give complete support to this work. Actually I met similar situations in some of my previous work. But I had not the notion of extended differentiability and therefore I had to construct solutions for each particular case using penalty arguments. Now, thanks to Dr S. Serovajsky work, we dispose of a general tool. This tool will certainly be applied in many other situations. This is an excellent work.

With my best regards.

Sincerely yours.

J.L. LIONS

FIGURE 1: Letter from J.L. Lions to U.M. Sultangazin

List of Figures

1	Letter from J.L. Lions to U.M. Sultangazin	xvii
1.1	Unique stationary point is the point of minimum.	5
1.2	Stationary points are three points of local extremum.	6
1.3	Classes of the functions' extremum.	7
1.4	Relations between the sets U_0 and U_*.	7
1.5	Function has two points of minimum.	8
1.6	Absence of stationary points for the insolvable problem.	9
1.7	Unique stationary point does not minimize the function.	9
1.8	Stationary condition is not applicable for the non-smooth case. .	10
1.9	Variation of a function u.	12
1.10	Function h of Lemma 1.2.	14
1.11	Different forms of closeness for continuous functions.	15
1.12	Euclid plane is the linear space.	17
1.13	The set $v + W$.	18
1.14	Subspace U and the affine variety $v + U$.	18
1.15	Convex set includes the segment connecting any of its points. . .	19
1.16	Convexity of functions.	19
1.17	Ball of a normalized space.	21
1.18	Relation between different classes of the functionals.	31
1.19	Annihilator $U^\perp$ of the subspace U.	42
1.20	Functional is the square of the distance from z to U.	48
1.21	Needle variation.	49
1.22	Solution u is the projection of the function z on the set U. . . .	50
2.1	Bounded sequence $\{\sin kx\}$ diverges in the space $L_2(0, \pi)$.	60
2.2	Semicontinuity of functions.	63
2.3	Smooth curve of the minimal length does not exist.	64
2.4	Strict convexity and the uniqueness of minimum.	67
2.5	Subgradient a of the function f at the point y.	76
2.6	Subgradient of the square function at zero.	76
2.7	Subgradients of the function $f(x) = |x|$ at zero.	77
2.8	Gradient method for a function of one variable.	81
3.1	Relation between weak and strong continuity.	92
3.2	Method of successive approximations.	95
3.3	Needle variation.	109

xix

List of Figures

3.4   Optimal control for the case of local constraints. 110

4.1   The Inverse function theorem. 135
4.2   If $f'(0) \neq 0$, then there exists the differentiable inverse function. 137
4.3   The non-differentiable inverse function exists everywhere. 137
4.4   For the case $f'(0) = 0$ the inverse function does not exist. 138
4.5   For the case $f'(0) = 0$ the inverse function does not exist. 138
4.6   There exist non-differentiable inverse functions. 139
4.7   Monotone functions. 145
4.8   The solvability of the linearized equation. 161

5.1   Unique stationary point is the point of minimum. 178
5.2   The absence of the differentiability of the inverse operator at points with heightened regularity. 181
5.3   The relations between the properties of the operator derivative and the differentiability of the inverse operator. 186
5.4   The spaces that characterize the linearized equation and the extended derivative. 189
5.5   Classical and extendedly differentiable operators. 195
5.6   The operators $G(z, \lambda)$, $\overline{G}(z, \lambda)$, and $\overline{G}(z, \lambda)^*$. 208

6.1   The structure of Chapter 6. 212

7.1   The implicit function exists at the neighborhood of the point M_0; this function does not exist at the neighborhood of the point M_1. 245

8.1   The solution of the program control problem. 293
8.2   The solution of the synthesis problem. 294
8.3   Bellman optimality principle. 304

9.1   The structure of Chapter 9. 312

10.1   The solution of the program control problem. 380
10.2   The solution of the synthesis problem. 381

List of Tables

1.1	Minimization problems and stationary conditions	11
1.2	Properties of stationary conditions	40
1.3	Sets of minimization and optimality conditions	47
4.1	Properties of inverse functions	139
4.2	Control systems	140
5.1	Relations between the properties of the operator derivative and the differentiability of the inverse operator	186

Part I

Minimization of functionals

Chapter 1

Necessary conditions of extremum for functionals

1.1	Minimization of functions and stationary condition	4
1.2	Lagrange problem and Euler equation	11
1.3	Linear spaces	16
1.4	Linear normalized spaces	20
1.5	Directional differentiation	26
1.6	Gâteaux differentiation of functionals	27
1.7	Minimization of functionals and stationary condition	31
1.8	General functional spaces	33
1.9	Minimization of Dirichlet integral	38
1.10	Minimization of functionals on subspaces	39
1.11	Derivatives with respect to subspaces	42
1.12	Minimization of functionals on affine varieties	43
1.13	Minimization of functionals on convex sets	45
1.14	Comments	50

The easiest result of extremum theory is the necessary condition of extremum for smooth functions of one variable. This is the equality to zero of its derivative at the point of extremum. Minimization problems for functionals are more difficult. The relevant stationary condition includes the derivative of the given functional. This result can be generalized to minimization problems on subspaces or on affine varieties. The variational inequality is the general necessary condition of minimum for functionals on convex sets. These results are the basis of the optimal control theory. The minimized functional depends on the state function that depends on the control here. The analysis of the minimization problems for functionals will be continued in the next chapter. We shall consider the sufficiency of extremum conditions, the existence of optimization problems, and some others there.

1.1 Minimization of functions and stationary condition

Consider an easy extremum problem. This is the problem of minimization for a function of one variable on the set of real numbers.

Problem 1.1 *Find the point that minimizes a function on the set of real numbers.*

We shall consider the classical necessary condition of minimum for differentiable functions only.

Theorem 1.1 *If the differentiable function $f = f(\sigma)$ has a minimum at the point τ, then it satisfies the equality*

$$f'(\tau) = 0. \tag{1.1}$$

Proof. Suppose τ is a point of minimum for the function f, then we have

$$f(\sigma) \geq f(\tau) \ \forall \sigma.$$

Hence, we get

$$f(\tau + h) \geq f(\tau) \ \forall h.$$

By Taylor formula, it follows that

$$f(\tau + h) = f(\tau) + f'(\tau)h + \eta(h),$$

where $\eta(h)/h \to 0$ as $h \to 0$. Hence, we transform the last inequality to

$$f'(\tau)h + \eta(h) \geq 0 \ \forall h. \tag{1.2}$$

If $h > 0$, then

$$f'(\tau) + \eta(h)/h \geq 0.$$

Passing to the limit as $h \to 0$, we get

$$f'(\tau) \geq 0. \tag{1.3}$$

Now suppose $h < 0$ in (1.2); then

$$f'(\tau) + \eta(h)/h \leq 0.$$

Passing to the limit as $h \to 0$, we obtain

$$f'(\tau) \leq 0. \tag{1.4}$$

It follows from the inequalities (1.3),(1.4) that the equality (1.2) holds. $\square$

Thus, we reduced Problem 1.1, i.e., finding the minimum for a differentiable function on the set of real numbers, to the analysis of the equality (1.1). This is a nonlinear algebraic equation with respect to the number τ.

Remark 1.1 Note the serious relation between extremum theory and equations theory. We shall find out the analogical relation for different extremum problems. Note also the application of the differentiation for optimization methods. Different forms of the differentiation will be used for the analysis of optimization problems. We shall base our future results on the common principles of the extremum, equations, and differentiation theories.

Definition 1.1 *The equation (1.1) is called the **stationary condition**; its solution is called a **stationary point** of the function f.*

Remark 1.2 Sometimes the solution of the equation (1.1) is called the *critical point* of the given function.

Consider examples.

Example 1.1 Let us have the function

$$f_1(\sigma) = \sigma^2.$$

It follows from the equality (1.1) the stationary condition $2\tau = 0$. The unique solution of this equation $\tau = 0$ is the point of minimum for the function f_1 (see Figure 1.1).

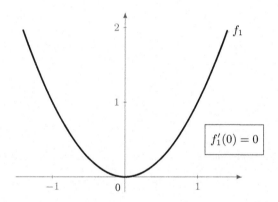

FIGURE 1.1: Unique stationary point is the point of minimum.

Remark 1.3 We do not have any difficulties in solving the stationary condition here. However, this is a nonlinear algebraic equation in the general case. Therefore, finding of stationary points can be a serious problem for difficult enough functions. We could find it by means of numerical algorithms only.

We considered the easiest example. However, the situation can be more difficult.

Example 1.2 Consider the function

$$f_2(\sigma) = 3\sigma^4 - 8\sigma^3 - 6\sigma^2 + 24\sigma.$$

From (1.1) the equation follows: $\tau^3 - 2\tau^2 - \tau + 2 = 0$. It has three solutions: $\tau_1 = -1$, $\tau_2 = 1$, $\tau_3 = 2$. Calculate the values $f_2(\tau_1) = -19$, $f_2(\tau_2) = 13$, $f_2(\tau_3) = 8$. It is obvious that the first stationary point only minimizes the function f_2 on the set of real numbers (see Figure 1.2). This is the point of its absolute minimum, i.e., the minimum on the domain of the given function. The third solution of the equation (1.1) is the point of local minimum for this function. We have the minimization of our function on the neighborhood of this point only. Finally, τ_2 is the point of the local maximum for the function f_2.

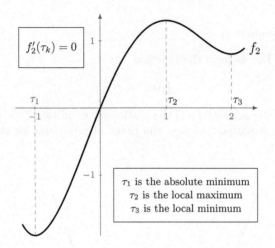

FIGURE 1.2: Stationary points are three points of local extremum.

Remark 1.4 We could exclude non-optimal stationary points of the function f_2 by means of the sign of the second derivative at these points. However, the calculation of second derivatives for difficult enough functions can be a serious problem.

Improve the classification of the points of extremum (see Figure 1.3).

Definition 1.2 *The function f has a **local minimum** (**local maximum**) at the point τ if there exists its neighborhood O such that the inequality $f(\tau) \leq f(\sigma)$ (respectively, $f(\tau) \geq f(\sigma)$) is true for all $\sigma \in O$. The minimum (and maximum too) is **strict** if the considered relation is true as the equality for $\sigma = \tau$ only. If these inequalities are true for all number σ, then τ is the point of the **absolute minimum** (maximum) for the function f.*

Consider an extremum problem P and a relation Q.

Definition 1.3 *The relation Q is called the **necessary condition of extremum** for the Problem P if each solution of P satisfies this relation. The relation Q is the **sufficient condition of extremum** for Problem P if each object that satisfies Q is a solution of Problem P.*

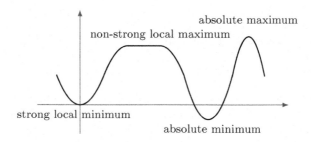

FIGURE 1.3: Classes of the functions' extremum.

Remark 1.5 If P is an optimal control problem, then Q is called the *necessary (sufficient) condition of optimality*.

If a relation Q is a necessary and sufficient condition of extremum, then it is equivalent to the initial Problem P (see Figure 1.4).

Thus, the relation (1.1) is the necessary condition of local minimum for the differentiable functions.

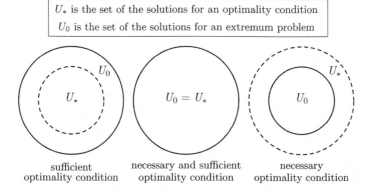

FIGURE 1.4: Relations between the sets U_0 and U_*.

Remark 1.6 Return to the proof of Theorem 1.1. If the considered point maximizes the function f, then we change the sign at the left-hand side of the inequality (1.2). Hence, the condition (1.3) will be replaced by (1.4). However, we have already obtained the inequality (1.4) from the condition (1.1) with negative value h. We get the inequality (1.3) too if τ is a point of maximum f. Therefore, the stationarity condition is true for the points of maximum too. If τ is a point of local minimum only, then the inequality (1.1) is true for small enough values h only. However, we can pass to the limit as $h \to 0$. Then the equality (1.1) is the condition of local extremum for the general case.

Remark 1.7 We shall return to the problem of sufficiency for an extremum condition in Chapter 2 (see Section 2.1).

Continue the analysis of properties of the stationary condition.

Example 1.3 Consider the function

$$f_3(\sigma) = \sigma^4 - 2\sigma^2.$$

The stationary condition for this function has three solutions: $\tau_1 = -1$, $\tau_2 = 0$, $\tau_3 = 1$. The second of them is the point of local maximum. Other points are the solutions of the minimization problem (see Figure 1.5). We do not have the sufficiency of the extremum condition and the uniqueness of the points of minimum for this problem.

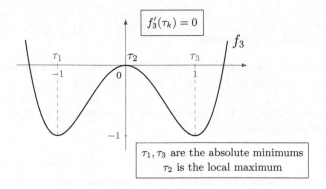

FIGURE 1.5: Function has two points of minimum.

Remark 1.8 If τ is a non-unique point of minimum for the function f, then the inequality $f(\tau) \geq f(\sigma)$ for all σ will be true as before.

Remark 1.9 The problem of uniqueness for extremum problems will be considered in Chapter 2 (see Section 2.4).

Example 1.4 Consider the function

$$f_4(\sigma) = \sigma.$$

Calculate the stationarity condition $1 = f_4'(\tau) = 0$. It is clear that the stationarity condition does not have any solutions. Certainly, the minimization problem for this function is insolvable (see Figure 1.6). However, the sets of the solutions of the extremum problem and stationarity condition are empty; we have the equality of these sets. Therefore, our necessary condition of extremum is sufficient too.

Remark 1.10 We start the proof of Theorem 1.1 from the supposition of the existence of the minimized point τ for the given function. If this supposition is false, then we cannot use any possibility to get the result. Therefore, the absence of the stationary points for the last example is normal.

Necessary conditions of extremum for functionals

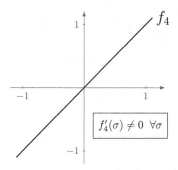

FIGURE 1.6: Absence of stationary points for the insolvable problem.

Remark 1.11 Note the independence between non-sufficiency of the stationary condition and the solvability of the problem. The minimization problem for Example 1.2 is solvable in the case of the non-sufficiency of the extremum condition; but we have the sufficiency for the insolvable problem for Example 1.4.

Example 1.5 Consider the function

$$f_5(\sigma) = \sigma^3.$$

The extremum condition has a unique solution $\tau = 0$ here. However, it does not minimize the function f_5 (see Figure 1.7). This minimization problem is insolvable. Thus, the equation (1.1) is a necessary but not sufficient condition of extremum.

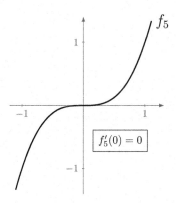

FIGURE 1.7: Unique stationary point does not minimize the function.

Remark 1.12 The supposition of the solvability of this minimization problem is false. Therefore, we cannot to substantiate the stationary condition. However, it does not guarantee the insolvability of the last relation.

Remark 1.13 The unique stationary point for the last example is not the point of the local extremum. This is a *point of inflection* for the given function. Thus, the properties of the stationary points can be different.

Remark 1.14 The second derivative of the function f_5 is zero at the stationary point. However, if the second derivative of a function is zero at the stationary point, then it can be the point of extremum too. It is true, for example, for the function $f(\sigma) = \sigma^4$.

Remark 1.15 Each stationary point can be the point of local extremum or inflection only for the functions with one variable. However, there are a lot of different forms of stationary points for the functions with many variables.

Remark 1.16 We understand the importance of the existence of solutions for extremum problems. We shall analyze the solvability of extremum problems in Chapter 2 (see Sections 2.2 and 2.4).

Example 1.6 Consider the function

$$f_6(\sigma) = |\sigma|.$$

This function is non-differentiable. Therefore, the extremum condition (1.1) is non-applicable here. We are in need of special methods for finding the unique solution $\tau = 0$ (see Figure 1.8) of this minimization problem.

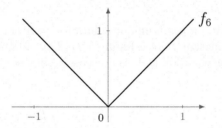

FIGURE 1.8: Stationary condition is not applicable for the non-smooth case.

Remark 1.17 If a minimized function is not differentiable, then we cannot obtain the inequality (1.2). However, the minimization problem makes sense. Therefore, minimization methods for non-smooth extremum problems are of a serious interest. We shall consider methods of non-smooth minimization in Chapter 2 (see also Section 1.5).

Remark 1.18 We will continue the analysis of these examples in Chapter 2.

By considered examples, we can declare the following problems of analysis of the stationary condition for Problem P (see also Table 1.1).

i) The *solvability of the extremum problem* (the necessary condition of extremum does not have any solutions or gives a false result for the insolvable problems);

Necessary conditions of extremum for functionals 11

TABLE 1.1: Minimization problems and stationary conditions

functions	*quantity of minimums*	*quantity of stationary points*	*property of stationary conditions*
f_1	1	1	sufficient
f_2	1	3	non-sufficient
f_3	2	3	non-sufficient
f_4	0	0	sufficient
f_5	0	1	non-sufficient
f_6	1	—	non-applicable

ii) The *uniqueness of the extremum problem* (the necessary condition of extremum has a non-unique solution in the case of the non-uniqueness of the extremum problem);

iii) The *sufficiency of the extremum condition* (the solution of the extremum condition can be non-optimal);

iv) *Practical solving of the extremum condition* (the stationary condition is a nonlinear algebraic equation; its practical solving is not obvious);

v) The *differentiability of the given function* (the stationary condition is non-applicable for non-smooth case).

Remark 1.19 Methods of practical solving for extremum problems will be considered in Chapter 2 (see Section 2.11).

Remark 1.20 The stationary condition is the *first order extremum condition* because it uses first derivative of the given function. There exist *high order extremum conditions*. However, we shall not consider them because the calculation of the high derivatives is an enormously difficult problem for serious optimal control problems.

1.2 Lagrange problem and Euler equation

Extend the previous results to more difficult extremum problems. We shall consider minimization problems for functionals on arbitrary sets. Analyze at first the easiest classical **Lagrange problem**. This is the problem of **variations calculus**. Determine the functional

$$I = I(v) = \int_{x_1}^{x_2} F[x, v(x), v'(x)] dx,$$

where F is a given function, $v = v(x)$ is an unknown function that satisfies the boundary conditions

$$v(x_1) = v_1, \quad v(x_2) = v_2, \tag{1.5}$$

and v_1 and v_2 are given numbers. Consider the following problem.

Problem 1.2 *Find the function v that satisfies the boundary conditions (1.5) and minimizes the functional I.*

Remark 1.21 We shall receive evidence soon that the set of functions satisfying the equalities (1.5) is an *affine variety*.

Try to use Theorem 1.1 for the analysis of our problem. Let u be a solution of the Lagrange problem. Determine the function

$$f = f(\sigma) = I(u + \sigma h),$$

where σ is a number, and h is a smooth enough function on the interval $[x_1, x_2]$ with homogeneous boundary conditions

$$h(x_1) = 0, \quad h(x_2) = 0. \tag{1.6}$$

Then the function $v_\sigma = u + \sigma h$ satisfies the boundary conditions (1.5) (see Figure 1.9). It is called the **variation of the function** u.

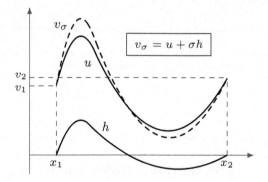

FIGURE 1.9: Variation of a function u.

Obviously, the functional I has a minimum at a point u whenever the number 0 is the point of minimum for the function f. By Theorem 1.1, the necessary condition of minimum (more exact, of local extremum) for a differentiable function at the given point is the equality to zero of its derivative at this point. Suppose the existence of the partial derivatives F_u and $F_{u'}$ of the function F with respect to the second and third arguments. Prove the following proposition.

Necessary conditions of extremum for functionals

Lemma 1.1 *Suppose the function F is smooth enough; then the function f has the derivative at the zero point that is equal to*

$$f'(0) = \int_{x_1}^{x_2} \left[F_u(x, u, u') - \frac{d}{dx} F_{u'}(x, u, u') \right] h \, dx. \tag{1.7}$$

Proof. Determine the value

$$f(\sigma) = \int_{x_1}^{x_2} F(x, u + \sigma h, u' + \sigma h') \, dx.$$

Using Taylor formula, we get

$$F(x, u + \sigma h, u' + \sigma h') = F(x, u, u') + \sigma F_u(x, u, u')h + \sigma F_{u'}(x, u, u')h' + \eta(\sigma),$$

where $\eta(\sigma) = O(\sigma)$, i.e., $\eta(\sigma)/\sigma \to 0$ as $\sigma \to 0$. Then we have the equality

$$f(\sigma) - f(0) = \sigma \int_{x_1}^{x_2} \left[F_u(x, u, u')h + F_{u'}(x, u, u')h' + \eta(\sigma)/\sigma \right] dx.$$

Dividing both sides by σ and passing to the limit as $\sigma \to 0$, we obtain

$$f'(0) = \int_{x_1}^{x_2} \left[F_u(x, u, u')h + F_{u'}(x, u, u')h' \right] dx.$$

Integrating by parts by using the boundary conditions, we get

$$\int_{x_1}^{x_2} F_{u'}(x, u, u')h' \, dx = - \int_{x_1}^{x_2} \frac{d}{dx} F_{u'}(x, u, u')h \, dx.$$

Finally, the antecedent equality can be transformed to (1.7). $\square$

The right side of the equality (1.7), i.e., the derivative of the function f at the zero point, is called the ***variation of the functional*** I at the point u with respect to the direction h and denoted by $\delta I(u, h)$.

Remark 1.22 This is called also the *derivative with respect to the direction* (see Section 1.5).

Thus, if u is a solution of the Lagrange problem, and there exists a variation of the functional I, then the equality

$$\delta I(u, h) = 0 \tag{1.8}$$

is true for all functions h that satisfy the boundary conditions (1.6) because of Theorem 1.1. Prove the following proposition, which is called the ***Euler–Lagrange lemma*** or the ***General lemma of variations calculus***.

Lemma 1.2 *Suppose g is a continuous function, as the following equality holds,*

$$\int_{x_1}^{x_2} g(x)h(x)dx = 0, \qquad (1.9)$$

for all continuous function h; then the function g is equal to the zero on the interval $[x_1, x_2]$.

Proof. Consider an arbitrary point $x_* \in [x_1, x_2]$ and a positive number ε such that

$$(x_* - \varepsilon, x_* + \varepsilon) \subset (x_1, x_2).$$

Choose the function h such that this satisfies the equality $h(x_*) = 1$ and this is zero outside the interval $(x_* - \varepsilon, x_* + \varepsilon)$ (see Figure 1.10). From equality (1.9) it follows that

$$\int_{x_*-\varepsilon}^{x_*+\varepsilon} g(x)h(x)dx = 0.$$

Using the Mean value theorem, we get

$$2\varepsilon g(x_\varepsilon)h(x_\varepsilon) = 0,$$

where $x_\varepsilon \in (x_* - \varepsilon, x_* + \varepsilon)$. Dividing by 2ε and passing to the limit as $\varepsilon \to 0$ by using the properties of the function h, we have the equality $g(x_*) = 0$. This completes the proof of the lemma because the point x_* is arbitrary. $\square$

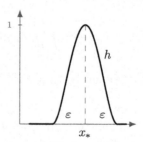

FIGURE 1.10: Function h of Lemma 1.2.

Remark 1.23 We shall use the analogical technique for the transformation of the integral extremum conditions to the pointwise form.

Substituting the value of the derivative from equality (1.7) in (1.8) and using Lemma 1.2, we get the following proposition.

Theorem 1.2 *If a smooth enough function u is a solution of the Lagrange problem, then it satisfies* **Euler equation**

$$F_u(x, u, u') - \frac{d}{dx} F_{u'}(x, u, u') = 0. \tag{1.10}$$

Remark 1.24 We shall not specify the sense of the smoothness, because now we have an interest in the transformation of the Lagrange problem to Problem 1.1 only.

Remark 1.25 The smooth solution of the Euler equation is called the *extremal*.

Remark 1.26 The Euler equation gives us the condition of the *weak minimum*. In this case our functional at its solution does not surpass the values of this functional for all functions that are close enough to u in the norm of the space of the continuously differentiable functions. It guarantees the smoothness of the functions but not its derivatives. If we have the smoothness in the sense of a larger class of the continuous functions, then we have the *strong minimum*. Each strong minimum is a weak minimum too. However, the inverse proposition is false for the general case, because the smoothness in the sense of the continuous functions is a stronger property (see Figure 1.11).

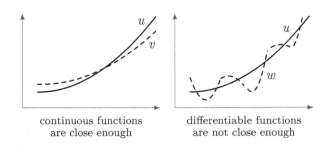

FIGURE 1.11: Different forms of closeness for continuous functions.

Remark 1.27 Variations calculus grants many conditions of extremum further to the Euler equation. However, we will not consider it, because its applicability for optimal control problems for nonlinear infinite dimensional systems is very limited.

By Theorem 1.2, the solution of Problem 1.2 satisfies the equalities (1.10), (1.5) for $v = u$. This is the *boundary problem* for the second order ordinary differential equation. Consider the following example.

Example 1.7 Let us minimize the functional

$$I(v) = \int_0^\pi \left[v(x)^2 + v'(x)^2 - 2v(x) \sin x - 2v'(x) \cos x \right] dx$$

on the set of twice continuously differentiable functions with homogeneous boundary conditions

$$v(0) = 0, \quad v(\pi) = 0.$$

16   *Optimization and Differentiation*

We have the equality

$$F_u(x, u, u') - \frac{d}{dx} F_{u'}(x, u, u') = 2(u - u'' - 2\sin x).$$

Then we get Euler equation

$$u - u'' = 2\sin x.$$

The obtained boundary problem has the unique solution $u = \sin x$ that minimizes our functional.

Remark 1.28 The given functional has the minimum at this solution in reality. Indeed, transform this functional. We have

$$I(v) = \int\limits_0^\pi \left\{ [v(x) - \sin x]^2 + [v'(x) - \cos x]^2 \right\} dx - 1.$$

Its value cannot have less than -1, as we have the equality to -1 here for $v = \sin x$ only.

Remark 1.29 We shall continue the analysis of this example in Section 1.12 and in Chapter 2 too (see Section 2.4).

Remark 1.30 We proved Theorem 1.2 by means of the stationary condition. Therefore, we can have known difficulties. Particularly, the Lagrange problem can have a unique or non-unique solution; furthermore, it can be insolvable. A solution of the problem (1.5), (1.6) can be non-optimal as it is necessary to substantiate the existence of the functional variation for each concrete example.

This used technique gives a hint for the method of the analysis for general extremum problems. We shall obtain analog of the stationary condition and the equality (1.8) for general functionals on the abstract mathematical spaces. However, we are in need of using properties of these spaces.

1.3   Linear spaces

We extended the method of the analysis of extremum functions to the Lagrange problem. Try to apply this method for abstract functional minimization problems. However, it is necessary to have some properties of the domains of these functionals.

Return to the definition of the function variation. We applied two properties. We used the possibility of summing of objects and of its multiplying by numbers. If we would like to use this technique for general functionals, then these properties should be true for its domains. Then we determine a special class of abstract mathematical spaces.

Definition 1.4 *A set V is called a **linear space** if for all its elements u, v and all real number a there exists the sum $u + v$ and the product av that are elements of this set, and the following conditions hold:*
i) $(u+v)+w = u+(v+w)$ $\forall u, v, w \in V$;
ii) $u+v = v+u$ $\forall u, v \in V$;
iii) *there exists an element $\theta \in V$ such that $\theta + v = v$ $\forall v \in V$;*
iv) *for all $v \in V$ there exists an element $-v$ of V such that $v + (-v) = \theta$;*
v) $a(u+v) = au + av$ $\forall u, v \in V, a \in \mathbb{R}$;
vi) $(a+b)v = av + bv$ $\forall v \in V, a, b \in \mathbb{R}$;
vii) $(ab)v = a(bv)$ $\forall v \in V, a, b \in \mathbb{R}$;
viii) $1v = v$ $\forall v \in V$.

Remark 1.31 The linear space is called sometimes the ***vector space***.

The natural example of the linear space is n-dimensional Euclid space $\mathbb{R}^n$ with standard operations of component-wise addition of vectors and multiplication of scalars by vectors. The sum of arbitrary vectors $u = (u_1, ..., u_n)$ and $v = (v_1, ..., v_n)$ is the vector $u + v = (u_1 + v_1, ..., u_n + v_n)$ and the product of a number a by a vector $v = (v_1, ..., v_n)$ is the vector $av = (av_1, ..., av_n)$ (see Figure 1.12). Another important example of linear space is the set of functions on a domain Ω with point-wise operations of the addition of functions and the multiplication of numbers by functions. The sum of functions u and v is here the function such that its value at an arbitrary point $x \in \Omega$ is $u(x) + v(x)$; and the product of a number a by a function v is the function av such that its value at the point x is equal to $av(x)$.

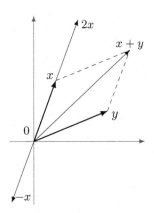

FIGURE 1.12: Euclid plane is the linear space.

If the domain of the functional is a linear space, then it is possible to determine the variation of its elements. We shall consider also conditional minimization problems. Therefore, we have an interest in subsets of linear spaces. Let W be a subset of a linear space V, and v is a point of V. The sum $v + W$ is the set of all sums $v + w$ (see Figure 1.13).

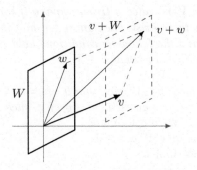

FIGURE 1.13: The set $v + W$.

Definition 1.5 *A subset U of a linear space V is called its **subspace** if this is the linear space; this is an **affine variety** if U is equal to $v + W$, where W is a subspace, $v \in V$; and U is a **convex set** if $\sigma u + (1 - \sigma)v \in U$ for all $\sigma \in [0, 1]$ and $u, v \in U$.*

Particularly, consider the Euclid plane, i.e., the space $\mathbb{R}^2$. Any line passing through the origin is a linear subspace, and any other line is an affine variety (see Figure 1.14). By geometry, a set is convex if for all its two points this set contains the interval connecting these points (see Figure 1.15). Each subspace is the affine variety, and each affine variety is convex.

Remark 1.32 We shall consider minimization problems for the functionals on subspaces, affine varieties, and convex sets in the Sections 1.10, 1.12, and 1.13. Note that the Lagrange problem is in reality the minimization problem for the functional on an affine variety. We shall analyze many optimization problems on convex sets.

Consider an important class of functionals on the convex sets.

Definition 1.6 *A functional I on a convex set U is called **convex** if*

$$I[\sigma u + (1 - \sigma)v] \leq \sigma I(u) + (1 - \sigma)I(v) \ \forall \sigma \in (0, 1), \ \forall u, v \in U.$$

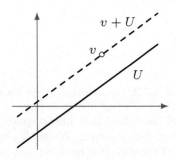

FIGURE 1.14: Subspace U and the affine variety $v + U$.

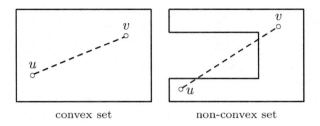

convex set   non-convex set

FIGURE 1.15: Convex set includes the segment connecting any of its points.

*The functional is said to be **strictly convex** if we have the equality here for $u = v$ only.*

Remark 1.33 The function is convex (strictly convex) if for all two points of its graph the interval connecting these points is located no less (above) than the corresponding arc of the curve (see Figure 1.16).

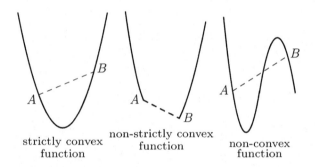

strictly convex function   non-strictly convex function   non-convex function

FIGURE 1.16: Convexity of functions.

Let us have a functional J on the set U such that

$$J[\sigma u + (1-\sigma)v] \leq \sigma J(u) + (1-\sigma)J(v) \ \forall \sigma \in [0,1], \ \forall u, v \in U.$$

Adding two last inequalities, we get

$$K[\sigma u + (1-\sigma)v] \leq \sigma K(u) + (1-\sigma)K(v) \ \forall \sigma \in [0,1], \ \forall u, v \in U,$$

where $K = I + J$. Therefore, the sum of convex functionals is convex. If one of these inequalities at least is strict for all $u \neq v$ then the last inequality is strict too. Hence, the sum of convex and strictly convex functionals is strictly convex. The multiplication of any functional by a positive constant does not change the properties of the convexity and strict convexity.

Remark 1.34 The convexity of functionals is used for the analysis of the extremum conditions and the solvability of optimization problems. We shall use the strong convexity for the proof of the uniqueness of optimization problems (see Chapter 2).

Optimization and Differentiation

Remark 1.35 We shall apply a stronger form of the convexity for the analysis of the well-posedness of extremum problems (see Chapter 2).

Determine the **dimension** of the linear spaces. The elements $v_1, ..., v_n$ of a linear space are called **linear dependent** if there exist numbers $a_1, ..., a_n$ that do not equal to zero together such that $a_1 v_1 + ... + a_n v_n = 0$. The linear space is called **n-dimensional** if there exist n linear independent elements, and all $n + 1$ elements are linear dependent.

Definition 1.7 *A space is **finite dimensional** if this is n-dimensional space for some n. A space is **infinite dimensional** if this is not a finite dimensional space.*

Particularly, the set of real numbers $\mathbb{R}$ is one-dimensional. Euclid space $\mathbb{R}^n$ is n-dimensional. We shall consider infinite dimensional spaces of functions and sequences.

Remark 1.36 We shall consider optimal control problems described by equations on infinite dimensional spaces. There are serious differences between topological properties of finite dimensional and infinite dimensional spaces (see Chapter 2, Section 2.3).

Now consider operators on linear spaces.

Definition 1.8 *An operator A on a linear space V is called **linear** if*

$$A(au + bv) = aAu + bAv \quad \forall a, b \in \mathbb{R}; \ \forall u, v \in V.$$

Particularly, a functional λ on a linear space V is linear if

$$\lambda(au + bv) = a\lambda u + b\lambda v \quad \forall a, b \in \mathbb{R}; \ \forall u, v \in V.$$

Denote the value of a linear functional λ at a point v by $\langle \lambda, v \rangle$.

An operator A is called **affine** if there exists a linear operator $B : V \to W$ and element $w \in W$ such that $Av = Bv + w$ for all $v \in V$. The operator A is affine if and only if

$$A[\sigma u + (1 - \sigma)v] = \sigma A(u) + (1 - \sigma)A(v) \ \forall \sigma \in [0, 1], \ u, v \in V.$$

We obtained the Euler equation by using a limit. Therefore, it will be necessary to determine the limit for general spaces.

1.4 Linear normalized spaces

Let a non-negative number $\|v\|$ be determined for all elements v of a linear space V. It is called the **norm** of v if the following properties hold:

i) $\|v\| = 0$ iff $v = 0$;
ii) $\|av\| = |a|\|v\| \ \forall a \in \mathbb{R}, v \in V$;
iii) $\|u+v\| \leq \|u\| + \|v\| \ \forall u, v \in V$.

If we would like to sign the norm of v for the concrete space V, then we use the denotation $\|v\|_V$.

Definition 1.9 *A linear space with a norm is called* **linear normalized**, *and its elements are called* **points**.

Let $v = (v_1, ..., v_n)$ be a vector of n-dimensional Euclid space $\mathbb{R}^n$. Determine its norm by the equality

$$\|v\| = \sum_{i=1}^{n} v_i^2.$$

Remark 1.37 Concrete linear normalized spaces of functions will be determined in Section 1.8.

The norm $\|u-v\|$ determines the distance between considered points. The set

$$B_\varepsilon(u) = \left\{ v \big| \ \|u-v\| \leq \varepsilon \right\}$$

is called the **ball** with the radius ε and the center u. This is the set of points such that its distance from u is not greater than ε (see Figure 1.17).

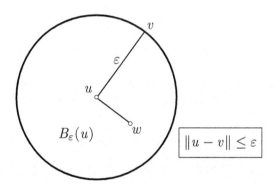

FIGURE 1.17: Ball of a normalized space.

Remark 1.38 It is convenient to choose the ball as a neighborhood of the point.

Determine the convergence by means of the norm.

Definition 1.10 *The sequence* $\{x_k\}$ *of a linear normalized space is called* **convergent** *(more exact,* **strongly convergent**) *to a point* v *(its* **limit**) *if*

$$\lim_{k \to \infty} \|v_k - v\| = 0.$$

22   *Optimization and Differentiation*

We denote this condition of the convergence by $v_k \to v$ in V or more completely $v_k \to v$ strongly in V.

Remark 1.39 The limit of the sum of sequences is equal to the sum of its limits because of the definition of the convergence. Besides, the limit of the product of the sequence elements by a number is equal to the product of the limit by this number. The algebraic and topological structures of the given set are accorded in this case. We shall use this property for passing to the limit in the variation $v_\sigma = u + \sigma h$ as $\sigma \to 0$ and obtain u as the limit. By the way, these results are true for the *linear topological spaces*. These are the sets, which are linear spaces and topological spaces, with accorded algebraic and topological structures.

The ***limit point*** of the sequence is the limit of some of its subsequence. If a sequence converges, then it has a unique limit point; it is equal to the limit of the sequence. If v is limit a point of all subsequences of the sequence, then this is the limit of the sequence.

Sometimes it is necessary to pass to the limit for a term, depending on a parameter.

Definition 1.11 *Suppose $v_k(\mu)$ are elements of a linear normalized space that depend on a parameter μ of a set M; then the sequence $\{v_k(\mu)\}$ is called **convergent** to a point $v(\mu)$ **uniformly** with respect to $\mu \in M$ if*

$$\lim_{k \to \infty} \sup_{\mu \in M} \left\| v_k(\mu) - v(\mu) \right\| = 0.$$

Determine an important class of operators.

Definition 1.12 *Operator $A : V \to W$ is called **continuous at a point** v if $Av_k \to Av$ in W whenever $v_k \to v$ in V. The operator is **continuous** if this is a continuous operator at the arbitrary point of its domain.*

Particularly, linear functional λ is continuous on the space V if $\langle \lambda, v_k \rangle \to \langle \lambda, v \rangle$ whenever $v_k \to v$ in V. The norm is the continuous nonlinear functional; we have $\|v_k\| \to \|v\|$ whenever $v_k \to v$ in V.

Remark 1.40 The continuity of the linear operator at the concrete point and on its domain is equivalent. However, this is false for the nonlinear case. For example, a function can be continuous at a point and discontinuous at another point.

Determine classes of subsets of linear normalized spaces. A subset U of a linear normalized space is called ***bounded*** if there exists a positive constant c such that $\|v\| \leq c$ for all $v \in U$. Particularly, a ***sequence*** $\{v_k\}$ is ***bounded*** if there exists a positive constant c such that $\|v_k\| \leq c$, $k = 1, 2, \dots$. A subset U is called ***closed*** if the limit of each convergent sequence of elements of this set belongs to U. The ***closure*** of a set U is the set of limits of all its convergent sequences. A set U is ***open*** in a space V if its complement $V \setminus U$ is closed. A set U is ***dense*** in a space V if its closure is equal to V, i.e., each element of V is the limit of some sequence of the set U. Particularly, the interval $[a, b]$ is closed, the interval (a, b) is open, and the set of all rational numbers is dense in the space of all real numbers.

Necessary conditions of extremum for functionals

Remark 1.41 The boundedness and the closeness of sets will be used for proving the solvability of extremum problems (see Chapter 2, Section 2.4). We shall use the density of sets in Section 1.10.

An operator $A : V \to W$ is called **bounded** if it takes the bounded sets of the space V to bounded sets of W. If a linear operator A is bounded, then there exists a positive constant c such that inequality $\|Av\| \leq \|v\|$ holds for all $v \in V$. The smallest of these numbers c is called the **norm of the operator** A. Therefore, we have the inequality $\|Av\| \leq \|A\|\|v\|$ for all $v \in V$. The properties of continuity and boundedness are equivalent for the linear operators.

The practical application of the convergence can be difficult enough because its definition uses the limit. However, we know usually the sequence only; the limit is unknown as a rule. Moreover, we do not believe often in the existence of a limit even. Therefore, it is used in the following notion. The sequence $\{v_k\}$ is called **fundamental** if

$$\lim_{k,n \to \infty} \|v_k - v_n\| = 0.$$

Any convergent sequence is fundamental because of the inequality

$$\|v_k - v_n\| \leq \|v_k - v\| + \|v - v_n\|.$$

However, the inverse proposition is false. Particularly, the sequence $\{1/k\}$ is fundamental on the set of positive numbers with absolute value as the norm. However, it does not have any positive number as its limit. The space is called **complete** if each of its fundamental sequences is convergent. Then we determine a very important class of mathematical spaces.

Definition 1.13 *The complete linear normalized space is called* ***Banach***.

The set of real numbers and Euclid space $\mathbb{R}^n$ are complete; therefore, these spaces are Banach spaces. The sets of positive or rational numbers are not Banach spaces.

Remark 1.42 The set of positive numbers is not even linear space. Particularly, multiplication of each of its elements by negative numbers does not belong to this set. However, it is a metric space with standard definition of the distance between points. The fundamental sequence can be definite for the metric spaces too. The set of positive numbers is a non-complete metric space. The set of rational numbers is a non-complete linear normalized space.

Remark 1.43 The completeness of the space will be used for the proof of the solvability of the extremum problems (see Chapter 2, Section 2.4) and boundary problems.

Definition 1.14 *The set of all linear continuous functionals on a linear normalized space V is called* ***adjoint*** *and is denoted by V'.*

24 *Optimization and Differentiation*

The adjoint space is linear with the following operations:

$$\langle \lambda + \mu, v \rangle = \langle \lambda, v \rangle + \langle \mu, v \rangle \ \forall \lambda, \mu \in V', \ v \in V;$$

$$\langle a\lambda, v \rangle = a\langle \lambda, v \rangle \ \forall \lambda, \mu \in V', \ a \in \mathbb{R}, \ v \in V.$$

Particularly, zero of the adjoint space is the functional, which takes each element of the given space to the number 0. Hence, if we have the equality $\langle \lambda, v \rangle = 0$ for all $\lambda \in V'$, $v \in V$, then $\lambda = 0$.

Remark 1.44 The symbol 0 of the equality $\lambda = 0$ is not a number. This is an element of the adjoint space.

Determine the norm of the adjoint space by the equality

$$\|\lambda\|_{V'} = \sup_{\|v\|_V = 1} |\langle \lambda, v \rangle|.$$

The dual equality

$$\|v\|_V = \sup_{\|\lambda\|_{V'} = 1} |\langle \lambda, v \rangle|$$

holds too. Note also the inequality

$$|\langle \lambda, v \rangle| \le \|\lambda\|_{V'} \|v\|_V \ \forall \lambda \in V', \ v \in V.$$

Remark 1.45 We shall consider problems of minimization of functionals on Banach spaces. The derivatives of these functionals are linear continuous functionals (see Section 1.6); then this belongs to the adjoint space.

Remark 1.46 The adjoint spaces are used for the definition of the weak convergence for linear normalized spaces (see Chapter 2, Section 2.3). We shall consider also the adjoint operator; it will be applied for obtaining optimality conditions (see Chapter 3).

Determine second adjoint space V'' as the adjoint space for V'. The space V is called **reflexive** if $V = V''$. Particularly, the Euclid space $\mathbb{R}^n$ is reflexive. This is even **self-adjoint**, i.e., this is equal to the adjoint space.

Remark 1.47 The reflexivity of the spaces will be used for the analysis of the weak convergence (see Section 2.3).

Definition 1.15 *Suppose $A : V \to W$ is a linear continuous operator, where V, W are Banach spaces; then the linear continuous operator $A^* : W' \to V'$ is called an* **adjoint operator** *if*

$$\langle p, Av \rangle = \langle A^*p, v \rangle \ \forall v \in V, \ p \in W'.$$

Example 1.8 *Matrix.* Determine $V = W = \mathbb{R}^n$. The linear continuous operator $A : V \to W$ is the square matrix of n degree $A = (a_{ij})$, $i, j = 1, ..., n$. Then

$$\langle p, Av \rangle = \sum_{i=1}^{n} \sum_{j=1}^{n} a_{ij} v_i p_j = \sum_{i=1}^{n} \sum_{j=1}^{n} a_{ji} v_j p_i = \langle A^*p, v \rangle \ \forall v, p \in \mathbb{R}^n.$$

Therefore, the adjoint operator is the adjoint matrix $A^* = (a_{ji})$, $i, j = 1, ..., n$ here.

Necessary conditions of extremum for functionals 25

Consider a linear continuous operator $A : V \to V'$, where the space V is reflexive. Then its adjoint operator has the same domain and codomain. The operator A is called **self-adjoint**, if $A = A^*$. For example, the matrix A from Example 1.8 is self-adjoint if $a_{ij} = a_{ji}$, i.e., the symmetric matrix is the self-adjoint operator.

Embedding of a Banach space V to a Banach space W is **continuous** if there exists a constant $c > 0$ such that

$$\|v\|_W \leq c\|v\|_V \; \forall v \in V.$$

Therefore, $v_k \to v$ in W whenever $v_k \to v$ in V. An operator that takes an arbitrary element of the space V to this element of space W is called an **embedding operator**. If embedding of spaces is continuous, then its embedding operator is continuous too.

Embedding of V to W is **dense** if the set V is dense in W. If embedding of a Banach space V to a Banach space W is continuous and dense, then the inclusion of the adjoint spaces $W' \subset V'$ holds, besides

$$\|\lambda\|_{V'} \leq c\|\lambda\|_{W'} \; \forall \lambda \in W'.$$

Therefore, this embedding is continuous. If the space V is reflexive additionally, then the set W' is dense in V'.

Remark 1.48 We shall consider also the compactness of embedding for Banach spaces.

There exists a very important class of the reflexive Banach spaces with extremely rich properties. Suppose for all elements u and v of linear spaces V there exists a number (u, v) such that

i) $(v, v) \geq 0 \;\; \forall v \in V$, $(v, v) = 0$ iff $v = 0$;

ii) $(u, v) = (v, u) \;\; \forall u, v \in V$;

iii) $(au + bv, w) = a(u, w) + b(u, w) \;\; \forall u, v, w \in V$.

This number is called the **scalar product** of the elements u and v. Determine the norm of the space with scalar product by the equality

$$\|v\| = \sqrt{(v, v)} \; \forall v \in V.$$

Definition 1.16 *A complete space with scalar product is called* ***Hilbert***.

Euclid space $\mathbb{R}^n$ is a Hilbert space with scalar product

$$(u, v) = \sum_{i=1}^{n} u_i v_i.$$

Each Hilbert space is a reflexive Banach space. By ***Riesz Theorem***, each linear continuous functional in a Hilbert space can be determined by the scalar product. Therefore, it is possible to identify each Hilbert space and its adjoint accurately within an isomorphism. Then there exists a bijection $\Lambda : V \to V'$,

26 *Optimization and Differentiation*

that is called the **canonic isomorphism** of Hilbert spaces V' and V, such that

$$\|\Lambda\lambda\|_V = \|\lambda\|_{V'}, \quad \langle \lambda, v \rangle = \langle \Lambda\lambda, v \rangle \; \forall \lambda \in V', \; v \in V.$$

We shall consider minimization problems for functionals that are determined on Hilbert spaces or Banach spaces.

Remark 1.49 We shall consider also some additional properties of Banach spaces.

1.5 Directional differentiation

Extend the method of the analysis of the Lagrange problem to general functionals. Let V be a Banach space V, and I is a functional on this space.

Problem 1.3 *Find a point of minimum for the functional I on the space V.*

Try to repeat the proof of Theorem 1.2 for this case. Determine the function

$$f = f(\sigma) = I(u + \sigma h),$$

where σ is a number and u and h are elements of the space V. The functional I has the minimum at the point u if and only if the number 0 is the point of the minimum for the function f. By Theorem 1.1, the necessary condition of minimum for the given function at this point is the equality of the function derivative at this point to zero. Of course, this proposition is true if this function is differentiable. Suppose the existence of the limit

$$\delta I(u, h) = \lim_{\sigma \to 0} \frac{I(u + \sigma h) - I(u)}{\sigma}.$$

Definition 1.17 *The value $\delta I(u, h)$ is called the **derivative** of the functional I at the point u **with respect to the direction h**.*

It is obvious that

$$\delta I(u, ah) = a\delta I(u, h) \;\; \forall a;$$

therefore, the dependence of the derivative from the direction is homogeneous. However, it is not linear (see following example).

Example 1.9 *Absolute value*. Consider the function $f(x) = |x|$. Of course, for all positive number x we can choose a number σ so small that the value $x + \sigma h$ is positive. Then we have the equality

$$|x + \sigma h| - |x| = x + \sigma h - x = \sigma h.$$

Therefore, the function f is differentiable at the arbitrary positive point with respect to the arbitrary direction h, as $\delta f(x, h) = h$.

If x is negative, then for all h the number σ can be chosen so small that the value $x + \sigma h$ is negative. Then we get

$$|x + \sigma h| - |x| = -(x + \sigma h) + x = -\sigma h.$$

Hence, the function f is differentiable at the arbitrary negative point with respect to the arbitrary direction h, as $\delta f(x, h) = -h$.

Finally, for $x = 0$ we obtain

$$\lim_{\sigma \to -0} \frac{|\sigma h|}{\sigma} = -|h|, \quad \lim_{\sigma \to +0} \frac{|\sigma h|}{\sigma} = |h|.$$

The result depends on the method of passing to the limit, because the left and the right limits are not equal. Therefore, our function is not differentiable at the zero point with respect to the arbitrary non-zero direction.

1.6   Gâteaux differentiation of functionals

Let u be a point of minimum for a functional I on a space V. Suppose the existence of its derivative at this point with respect to the arbitrary direction. Then we obtain the equality

$$\delta I(u, h) = 0 \quad \forall h \in V \tag{1.11}$$

that is an analog of the formula (1.8). If V is the set of real numbers, then the functional I is the classical function. If this is differentiable at the point u, then we have the equality $\delta I(u, h) = I'(u)h$. Therefore, the condition (1.11) can be transformed to the equality $I'(u)h = 0$ for all numbers h. Thus, we get $I'(u) = 0$. This is the standard stationarity condition (1.1).

The value $\delta I(u, h)$ is linear with respect to h for the differentiable case. This situation is true for general spaces too. If the map $h \to \delta I(u, h)$ is linear, then there exists a linear operator $I'(u) : V \to \mathbb{R}$, i.e., a linear functional such that $\delta I(u, h) = I'(u)h$ for all h.

Definition 1.18 *Suppose there exists a linear continuous functional $I'(u)$ on the space V such as*

$$\frac{I(u + \sigma h) - I(u)}{\sigma} \to I'(u)h$$

*for all $h \in V$ as $\sigma \to 0$; then the functional I is called **Gâteaux differentiable** at the point u. The object $I'(u)$ is called a **Gâteaux derivative** of this functional at the given point here.*

Optimization and Differentiation

Remark 1.50 If a functional is Gâteaux differentiable at some point, then there exists its derivative at this point with respect to the arbitrary direction as this derivative is linear with respect to the direction.

The derivative of the functional is a linear functional. Therefore, this is an element of the adjoint space V', i.e., $I'(u) \in V'$. Hence, it is natural to write $\langle I'(u), h \rangle$ in place of $I'(u)h$.

Consider examples of Gâteaux derivatives.

Example 1.10 *Functions of one variable*. If the function f is differentiable at a point x, then

$$\lim_{\sigma \to 0} \frac{f(x + \sigma h) - f(x)}{\sigma} = f'(x)h \; \forall h \in \mathbb{R}.$$

Thus, the Gâteaux derivative is equal to the standard derivative.

Example 1.11 *Functions of many variables*. Let V be Euclid space $\mathbb{R}^n$. Then the functional on the set V is a function f of n variables. Pass to the limit at the equality

$$\frac{f(x + \sigma h) - f(x)}{\sigma} = \frac{f(x_1 + \sigma h_1, ..., x_n + \sigma h_n) - f(x_1, ..., x_n)}{\sigma}$$

as $\sigma \to 0$ for all $h = (h_1, ..., h_n)$. If the function f is differentiable with respect to all its variables, then we get

$$\lim_{\sigma \to 0} \frac{f(x + \sigma h) - f(x)}{\sigma} = \sum_{i=1}^{n} \frac{\partial f(x)}{\partial x_i} h_i.$$

The right side of this equality is the scalar product of the vectors

$$\nabla f(x) = \left(\frac{\partial f(x)}{\partial x_1}, ..., \frac{\partial f(x)}{\partial x_n} \right)$$

and h. The Euclid space is Hilbert. Therefore, each linear continuous functional can be determined here by the scalar product because of the Riesz theorem. Then we have

$$\langle f'(x), h \rangle = \big(\nabla f(x), h \big) \; \forall h \in \mathbb{R}^n.$$

Hence, the Gâteaux derivative of the function of many variables is the vector with partial derivatives of this function as components, i.e., the **gradient**.

Remark 1.51 The derivative of functionals is called sometimes the gradient. Particularly, extremum problems can be solved by gradient methods (see Chapter 2). These algorithms are based on the determination of the functional derivatives.

Necessary conditions of extremum for functionals 29

Example 1.12 *Linear and affine functionals*. Let the functional I be linear. Then we have $I(u + \sigma h) - I(u) = \sigma I(h)$. Find $I'(u)h = I(h) = \langle I, h \rangle$. Hence, the derivative of the linear functional at the arbitrary point is equal to the given functional. Any affine functional J is determined by the equality $J(v) = I(v) + c$, where I is a linear functional, and c is a constant. Then we get $J'(u)h = I'(u)h = \langle I, h \rangle$. Therefore, the derivative of the affine functional at the arbitrary point is equal to the appropriate linear functional.

Example 1.13 *Square of the norm of a Hilbert space*. Determine the functional I on a Hilbert space V by the equality

$$I(v) = \|v\|^2.$$

Using the relation between the norm and the scalar product of Hilbert spaces, we get

$$I(u + \sigma h) = (u + \sigma h, u + \sigma h) = (u, u) + 2\sigma(u, h) + \sigma^2(h, h).$$

Divide the equality

$$I(u + \sigma h) - I(u) = 2\sigma(u, h) + \sigma^2 \|h\|^2$$

by σ. Passing to the limit as $\sigma \to 0$, we obtain

$$\langle I'(u), h \rangle = \sigma(2u, h) \ \forall h \in V.$$

By Riesz theorem, each linear continuous functional on Hilbert spaces can be determined by the scalar product. Then we have the equality

$$\langle \lambda, h \rangle = (\Lambda \lambda, h) \ \forall \lambda \in V', h \in V,$$

where Λ is a canonic isomorphism between the spaces V' and V. Hence, we get

$$(\Lambda I'(u), h) = (2u, h) \ \forall h \in V.$$

Therefore, $\Lambda I'(u) = 2u$. Finally, find $I'(u) = 2\Lambda^{-1}u$ as the derivative of the functional

$$J(v) = \|v - z\|^2,$$

where z is element of the space V, is determined by the equality

$$J'(u) = 2\Lambda^{-1}(u - z).$$

Remark 1.52 The derivative of the square of norm for a Hilbert space at the point u is equal to $2u$ with an accuracy to the isomorphism.

Example 1.14 *Quadratic form*. Let $A : V \to V'$ be a linear continuous self-adjoint operator, where V is a reflexive Banach space. Determine the

30 — *Optimization and Differentiation*

functional I by the equality $I(v) = \langle Av, v \rangle$. It is called the **quadratic form**. Find its Gâteaux derivative at the arbitrary point v. Divide the equality

$$I(u + \sigma h) - I(u) = \sigma \langle Au, h \rangle + \sigma \langle Ah, u \rangle + \sigma^2 \langle Ah, h \rangle$$

by σ. Passing to the limit as $\sigma \to 0$, we get

$$\langle I'(u), h \rangle = \langle Au, h \rangle + \langle Ah, u \rangle \ \forall h \in V.$$

Therefore, the derivative of our functional at the given point is $I'(u) = 2Au$, because the operator A is self-adjoint.

Example 1.15 *Discontinuous case*. Consider the function of two variables

$$f(x, y) = \begin{cases} y(x^2 + y^2)/x, & \text{if } x \neq 0, \\ 0, & \text{if } x = 0. \end{cases}$$

Find the difference

$$f(\sigma h, \sigma g) - f(0, 0) = \sigma^2 (h^2 + g^2) g/h.$$

Divide this equality by σ. Passing to the limit as $\sigma \to 0$, we obtain the equality of the derivative of the function f at the zero point to the zero (more exact, the second order zero vector). Now determine $x = t^3$, $y = t$; therefore, we have $f(x, y) = (t^4 + 1)$. Passing to the limit here as $t \to 0$, we have the convergence $(x, y) \to 0$ and $f(x, y) \to 1$. Hence the given function is discontinuous at the zero point.

By this example, the functional can be Gâteaux differentiable and discontinuous at the same point.

Remark 1.53 This property is false for the functions of one variable because Gâteaux differentiability is equivalent to the classical differentiability in this case.

Example 1.16 *Absolute value*. Consider the function $f(x) = |x|$. This function is not differentiable at the zero point (see also Example 1.9). Therefore, it does not have a Gâteaux derivative here.

Remark 1.54 We determined the Gâteaux differentiable functional, which is discontinuous. Now we considered the continuous functional, which is not Gâteaux differentiable. Hence, the relation between Gâteaux differentiability and continuity of the functionals is difficult enough (see Figure 1.18). We shall consider also Fréchet differentiable functionals that are always continuous and Gâteaux differentiable.

Remark 1.55 The Gâteaux derivative can be determined for general operators (see Chapter 4).

Now we return to the consideration of extremum problems.

Necessary conditions of extremum for functionals

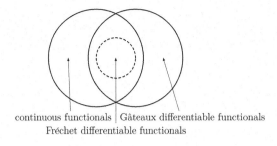
continuous functionals | Gâteaux differentiable functionals
Fréchet differentiable functionals

FIGURE 1.18: Relation between different classes of the functionals.

1.7 Minimization of functionals and stationary condition

Consider again Problem 1.3 of the minimization of a functional I on a Banach space V.

Theorem 1.3 *Suppose u is a point of minimum for the functional I on the space V, and there exists its Gâteaux derivative at this point; then the following equality holds*

$$I'(u) = 0. \qquad (1.12)$$

Proof. If the functional I is differentiable at the point u of its minimum, then the equality (1.11) is true, a.e. $\delta I(u, h) = 0$ for all $h \in V$. Transform this formula to

$$\langle I'(u), h \rangle = 0 \,\forall h \in V.$$

Then the Gâteaux derivative takes the arbitrary element h of the space V to the number zero. Therefore, $I'(u)$ is the zero element of the adjoint space V'. Using the standard symbol 0 for its denotation, we get the equality (1.12). □

We know that the Gâteaux derivative is equal to the classical derivative for the functions of one variable. Hence, the equality (1.12) is equivalent to (1.1) for $V = \mathbb{R}$. Therefore, this extremum condition is called the ***stationary condition*** too.

Remark 1.56 We do not use the completeness of the space for proving Theorem 1.3. This proposition is true for the general linear normalized spaces. Besides, the Gâteaux derivative can be extended to the larger class of linear topological spaces. The stationary condition can be obtained for this case too.

Problem 1.3 of functional extremum was transformed to the minimization problem for the function of one variable. We know that the equality (1.1) is a necessary condition of local extremum only. Therefore, the stationary condition (1.12) has the analogical properties. Then we extend Definition 1.2 to the general functionals.

32   *Optimization and Differentiation*

Definition 1.19 *A functional I has a **local minimum** (**local maximum**, respectively) on a set U at a point u if there exists a neighborhood O of this point on this set such that $I(u) \le I(v)$ ($I(u) \ge I(v)$, respectively) for all $v \in O$. If the equality can be true here only for $v = u$ then we have **strict local minimum** (maximum). If this inequality is true for all v from the set U, then u is the point of **absolute minimum** (maximum) of the functional I on the set U.*

Remark 1.57 We shall consider conditional extremum problems that are problems of minimization of functionals on some sets.

Example 1.17 *Function of many variables*. Consider the minimization problem for the function f of n variables. Using the value of its Gâteaux derivative (see Example 1.11), we have the following form of the stationary condition $\nabla f(x) = 0$. Then we obtain the equalities

$$\frac{\partial f(x_1, ..., x_n)}{\partial x_1} = 0, ..., \frac{\partial f(x_1, ..., x_n)}{\partial x_n} = 0.$$

Hence, we have the system of algebraic equations with respect to the vector $x = (x_1, ..., x_n)$.

Example 1.18 *Square of norm for Hilbert spaces*. Consider the minimization problem for the functional

$$I(v) = \left\| v \right\|^2$$

on a Hilbert space V. Substituting the known derivative of this functional (see Example 1.13) to the equality (1.12), we get $2\Lambda^{-1}u = 0$, where Λ is a canonic isomorphism between the spaces V' and V. Mapping the operator Λ to the last equality, we have $u = 0$. Hence, this is the unique stationary point.

Remark 1.58 Indeed, the norm is non-negative. It can be zero for the zero point only. Therefore, we found the unique solution of the given minimization problem.

Example 1.19 *Quadratic form*. Consider a linear continuous self-adjoint operator $A : V \to V'$ and a point $f \in V'$, where V is a reflexive Banach space. We have the minimization problem for functional

$$I(v) = \langle Av, v \rangle - 2\langle f, v \rangle.$$

This is the sum of the quadratic form from Example 1.14 and the linear functional. Its derivative at a point u is equal to $I'(u) = 2(Au - f)$. Hence, the stationary condition is the equation $Au = f$.

We would like to use Theorem 1.3 for the analysis of concrete variational problems. Therefore, it is necessary to determine functional spaces.

1.8 General functional spaces

Consider an open bounded set Ω of n-dimensional Euclid space $\mathbb{R}^n$. This is the set of points $x = (x_1, ..., x_n)$. Denote its closure by $\bar{\Omega}$.

Definition 1.20 *The **space** $C(\bar{\Omega})$ is the set of all continuous functions on the set $\bar{\Omega}$.*

Determine the pointwise addition of continuous functions and the pointwise multiplication of continuous functions by the numbers. Then we obtain the linear space. Determine its norm by the equality

$$\|v\| = \max_{x \in \bar{\Omega}} \left| v(x) \right|.$$

This linear normalized space is Banach.

Let $v = v(x)$ be a function on the set Ω. We will use the following notation for its partial derivatives

$$D^\alpha v = \frac{\partial^{|\alpha|} v}{\partial x_1^{\alpha_1} ... \partial x_n^{\alpha_n}},$$

where the vector $\alpha = (\alpha_1, ..., \alpha_n)$ is called the **multi-index**; and $\alpha_1, ..., \alpha_n$ are non-negative integer numbers, besides $|\alpha| = \alpha_1 + ... + \alpha_n$.

Definition 1.21 *The **space** $C^m(\bar{\Omega})$ is the set of all functions v on the set $\bar{\Omega}$ such that v has the continuous derivatives no greater than m degree on Ω, as each partial derivative $D^\alpha v$ with degree $|\alpha| \leq m$ has a continuous extension on the set $\bar{\Omega}$.*

The set $C^m(\bar{\Omega})$ is the Banach space with the norm

$$\|v\| = \sum_{|\alpha| \leq m} \max_{x \in \bar{\Omega}} \left| D^\alpha v(x) \right|.$$

The space $C(\bar{\Omega})$ is identical to $C^0(\bar{\Omega})$. The space $C^m(\bar{\Omega})$ is not adjoint to any Banach space for all non-negative m; this is non-reflexive space, of course. The embedding of the space $C^m(\bar{\Omega})$ to $C^r(\bar{\Omega})$ is continuous for all non-negative integer numbers $m > s$.

Determine classes of integrable functions.

Definition 1.22 *The **space** $L_p(\Omega)$, where $p \geq 1$, is the set of measurable p-degree Lebesgue integrable functions on Ω that is*

$$\int_\Omega \left| v(x) \right|^p dx < \infty.$$

34   *Optimization and Differentiation*

Remark 1.59 We do not determine the definitions of the measurability and Lebesgue integral. Note that the measurable function is determined up to the set of zero measure. Therefore, its change on the set of zero measure (for example, on a finite or countable sets) does give a new measurable function. Hence, the elements of these functional spaces are the equivalence classes of the functions. The elements of each equivalence class are equal almost everywhere (more succinctly, a.e.) on the set Ω, that is, it can be different on a set with zero measure. We will focus in the future on the fact that some pointwise property of the functions of $L_p(\Omega)$ or Sobolev spaces is true a.e. on the set Ω. We shall write succinctly that this property is true on Ω.

Determine the natural operations of the pointwise addition and multiplication by numbers on the set $L_p(\Omega)$. Then it is the linear space. Determine its norm by the equality

$$\|v\| = \left[\int_{\Omega} |v(x)|^p \, dx \right]^{1/p}.$$

The linear normalized space $L_p(\Omega)$ is Banach for all $p \geq 1$. Consider the most interesting case $p = 2$. The space $L_2(\Omega)$ with scalar product

$$(u, v) = \int_{\Omega} u(x)v(x)dx \; \forall u, v \in L_2(\Omega)$$

is Hilbert.

The measurable function v is called the ***essentially bounded function*** on the set Ω, if there exists a positive constant c such that the inequality $|v(x)| \leq c$ is true a.e. on Ω. The greatest lower bound of such constants c is denoted by $\text{vrai} \max_{x \in \Omega} |v(x)|$.

Definition 1.23 *The space $L_\infty(\Omega)$ is the set of all essentially bounded functions on Ω.*

The set $L_\infty(\Omega)$ is a Banach space with the norm

$$\|v\| = \text{vrai} \max_{x \in \Omega} |v(x)|.$$

The embedding of the space $C(\bar{\Omega})$ and $L_p(\Omega)$ to $L_q(\Omega)$ is continuous for all $p > q \geq 1$. The norm of the function v in the sense of the space $L_p(\Omega)$ will be denoted by $\|v\|_p$ for all numbers p such that $1 \leq p \leq \infty$. For all convergent sequences of the space $L_p(\Omega)$ there exists a subsequence, which converges a.e. on Ω.

Let x_* be a point of the set Ω. Consider a sequence of its neighborhood $\{O_k\}$ such that $m_k \to 0$ as $k \to \infty$, where m_k is the measure of the set O_k. By ***Lebesgue theorem***, the equality

$$\lim_{k \to \infty} \frac{1}{m_k} \int_{O_k} v(x)dx = v(x_*)$$

is true for almost all points $x_* \in \Omega$. These points are called ***Lebesgue points*** of the function v.

Necessary conditions of extremum for functionals

Remark 1.60 We can choose the ball $B_\varepsilon(x_*)$ with the center x_* and the radium $\varepsilon = \varepsilon(k)$ as a neighborhood O_k, where $\varepsilon(k) \to 0$ as $k \to \infty$. The ball is the interval for one-dimensional case; this is the circle for two-dimensional. The measure is the length, the area, and the volume for the interval, the circle, and the three-dimensional ball, respectively.

Remark 1.61 Lebesgue theorem will be applied for the transformation of integral extremum conditions to pointwise extremum conditions.

The space $L_p(\Omega)$ is reflexive for $p > 1$ as its adjoint space is (up an isomorphism) the space $L_{p'}(\Omega)$, where $1/p + 1/p' = 1$. The space $L_\infty(\Omega)$ is adjoint to $L_1(\Omega)$. However, the last space is not adjoint to any Banach space. Note the **Hölder inequality**

$$\left| \int_\Omega u(x)v(x)dx \right| \leq \|u\|_p \|v\|_{p'} \ \ \forall u \in L_p(\Omega), \ v \in L_{p'}(\Omega)$$

for all such that $1 \leq p \leq \infty$. If $u \in L_p(\Omega)$, $v \in L_q(\Omega)$ with $q > p'$, then the product uv is an element of the space $L_r(\Omega)$, where $1/r = 1/p + 1/q$.

Remark 1.62 The Hölder inequality is the partial case of the estimate for the value of the linear continuous functional (see Section 1.4).

The space $L_2(\Omega)$ is self-adjoint, that is, each linear continuous functional is not only isomorphic to an element of this space by the Riesz theorem. It can be even interpreted as an element of $L_2(\Omega)$. Therefore, we have the equality

$$\langle \lambda, v \rangle = (\lambda, v) \ \forall \lambda \in L_2(\Omega)', \ v \in L_2(\Omega).$$

We will consider optimal control problems for systems described by partial differential equations. Its solutions are elements of Sobolev spaces. These spaces are based on the distribution theory. Determine some notions of this theory.

The **support** of a continuous function on a bounded set Ω is the closure of all points of Ω, where this function has non-zero values. Let $D(\Omega)$ be the set of all infinite differentiable functions with closed bounded support. This is the linear space with natural operations of the pointwise addition and the multiplication by the number. Let the sequence $\{u_k\}$ of $D(\Omega)$ be convergent to zero in this space if there exists a subset M from Ω such that the supports of all functions u_k belong to M, and the sequence $\{D^\alpha u_k\}$ tends to zero uniformly on M for all multi-index α, i.e.,

$$\lim_{k \to \infty} \sup_{x \in M} \left| D^\alpha u_k(x) \right| = 0.$$

We have the convergence $u_k \to u$ in $D(\Omega)$ if the sequence $\{u_k - u\}$ tends to zero in this space.

Remark 1.63 This convergence does not accord even to any norm and metric. This is not important for us, because we have an interest in Sobolev spaces that are normalized. By the way, the set $D(\Omega)$ and its adjoint space (this is the distribution's space) are *local convex spaces*. However, we will consider narrower Banach spaces, which are associated with mathematical physics problems.

The linear functional v on $D(\Omega)$ is continuous if we have the convergence $\langle v, u_k \rangle \to \langle v, u \rangle$ whenever $u_k \to u$ in $D(\Omega)$.

Definition 1.24 *Linear continuous functionals on the space $D(\Omega)$ are called* **distributions**.

Hence, distribution is an element of the adjoint space. It is denoted by $D'(\Omega)$. For any distribution v we can determine its generalized derivative $D^\alpha v$ of the degree α by the equality

$$\langle D^\alpha v, u \rangle = (-1)^{|\alpha|} \langle v, D^\alpha u \rangle \quad \forall u \in D(\Omega).$$

Definition 1.25 **Sobolev space** $W_p^m(\Omega)$ *is the set of all distributions on Ω that belongs to the space $L_p(\Omega)$ along with all its generalized derivatives of degree α such that $|\alpha| \le m$, where $p \ge 1$, and m is an arbitrary natural number.*

Remark 1.64 The elements of Sobolev spaces are integrable functions. Therefore, the classical derivatives are non-applicable for this case. Hence, we use generalized derivatives that are determined by distribution theory.

The space $W_p^m(\Omega)$ is Banach with the norm

$$\|v\| = \left[\int_\Omega \sum_{|\alpha| \le m} |D^\alpha v(x)|^p dx \right]^{1/p}.$$

This is reflexive space for $p \ge 1$.

Definition 1.26 **Sobolev space** $\dot{W}_p^m(\Omega)$ *is the subspace of $W_p^m(\Omega)$ such that its elements are equal to the zero on the boundary of Ω with all its generalized derivatives with respect to the interior normal of degree less than m.*

The space $\dot{W}_p^m(\Omega)$ is Banach with the norm

$$\|v\| = \left[\int_\Omega \sum_{|\alpha| = m} |D^\alpha v(x)|^p dx \right]^{1/p}.$$

This space is reflexive if $p > 1$.

Consider the most important case $p = 2$. Denote by $H^m(\Omega)$ and $H_0^m(\Omega)$ Sobolev spaces $W_2^m(\Omega)$ and $\dot{W}_2^m(\Omega)$. The set $H^m(\Omega)$ is a Hilbert space with the scalar product

$$(u, v) = \sum_{|\alpha| \le m} \int_\Omega D^\alpha u(x) D^\alpha v(x) dx.$$

The set $H_0^m(\Omega)$ is a Hilbert space too; its scalar product can be determined by an easier formula,

$$(u, v) = \sum_{|\alpha| = m} \int_\Omega D^\alpha u(x) D^\alpha v(x) dx.$$

Necessary conditions of extremum for functionals 37

The adjoint space for $H_0^m(\Omega)$ is denoted by $H^{-m}(\Omega)$. This is a Hilbert space. Note the integral formula

$$\langle u, v \rangle = \int_\Omega u(x)v(x)dx \ \forall u \in H^{-m}(\Omega), \ v \in H_0^m(\Omega).$$

The space $L_2(\Omega)$ can be interpreted as $H^0(\Omega)$. Therefore, Sobolev space $H^m(\Omega)$ is determined for all integer numbers m. The generalized differentiation is the linear continuous operator from $H^m(\Omega)$ to $H^{m-1}(\Omega)$.

Remark 1.65 Sobolev spaces of non-integer degree have the sense too.

The most important Sobolev space is $H^1(\Omega)$. Determine the scalar product and the norm by the equalities

$$(u, v) = \int_\Omega \left[u(x)v(x) + \sum_{i=1}^n \frac{\partial u(x)}{\partial x_i} \frac{\partial v(x)}{\partial x_i} \right] dx,$$

$$\|v\| = \sqrt{\int_\Omega \left[|v(x)|^2 + \sum_{i=1}^n \left| \frac{\partial v(x)}{\partial x_i} \right|^2 \right] dx}.$$

We will consider often the space $H_0^1(\Omega)$ with the scalar product

$$(u, v) = \int_\Omega \sum_{i=1}^n \frac{\partial u(x)}{\partial x_i} \frac{\partial v(x)}{\partial x_i} dx$$

and the norm

$$\|v\| = \sqrt{\int_\Omega \sum_{i=1}^n \left| \frac{\partial v(x)}{\partial x_i} \right|^2 dx}.$$

All first derivatives of these equalities are generalized. Sometimes, we shall use the short denotation $\|v\|$ in place of $\|v\|_{H_0^1(\Omega)}$.

Remark 1.66 The generalized solutions for the mathematical physics problems for two degrees equations will be determined in the space $W_p^1(\Omega)$. It will be $H^1(\Omega)$ for the linear case.

Note the continuous embeddings of the space $W_p^m(\Omega)$ to $W_q^s(\Omega)$ and the space $H^m(\Omega)$ to $H^s(\Omega)$ for $p > q \geq 1$, $m > s$.

Remark 1.67 More exact results will be used in the next chapter.

The space $H^{-1}(\Omega)$ is adjoint to $H_0^1(\Omega)$; besides,

$$\langle u, v \rangle = \int_\Omega u(x)v(x)dx \ \forall u \in H^{-1}(\Omega), \ v \in H_0^1(\Omega).$$

We denote by $\|v\|_*$ the norm of v in the space $H^{-1}(\Omega)$. The embedding of the space $L_2(\Omega)$ to $H^{-1}(\Omega)$ is continuous.

Remark 1.68 We shall determine also Sobolev spaces with fractional degrees in Chapter 3.

1.9 Minimization of Dirichlet integral

Consider an example of the application of Theorem 1.3. Let Ω be an open bounded n-dimensional set of Euclid space $\mathbb{R}^n$ with boundary Γ. Determine the functional

$$I(v) = \int_\Omega \Big[\sum_{i=1}^n \Big|\frac{\partial v(x)}{\partial x_i}\Big|^2 + 2v(x)f(x)\Big]dx,$$

where f is a known function. This functional is called the **Dirichlet integral**.

Problem 1.4 *Find the function that minimizes the Dirichlet integral on the space $V = H_0^1(\Omega)$.*

Prove the following result.

Lemma 1.3 *The functional I is Gâteaux differentiable on the space V, as*

$$I'(u) = 2(f - \Delta u), \tag{1.13}$$

*where Δ is the **Laplace operator** (the sum of the second derivatives).*

Proof. For all functions $u, h \in V$ and the number σ determines the integral

$$I(u + \sigma h) = \int_\Omega \Big[\sum_{i=1}^n \Big|\frac{\partial u}{\partial x_i} + \sigma\frac{\partial h}{\partial x_i}\Big|^2 + 2(u + \sigma h)f(x)\Big]dx =$$

$$I(u) + 2\sigma\int_\Omega \Big(\sum_{i=1}^n \frac{\partial u}{\partial x_i}\frac{\partial h}{\partial x_i} + hf\Big)dx + \sigma^2\int_\Omega \sum_{i=1}^n \Big(\frac{\partial h}{\partial x_i}\Big)^2 dx.$$

Then we find

$$\big\langle I'(u), h\big\rangle = \lim_{\sigma \to 0}\frac{I(u + \sigma h) - I(u)}{\sigma} = 2\int_\Omega \sum_{i=1}^n \Big(\frac{\partial u}{\partial x_i}\frac{\partial h}{\partial x_i} + hf\Big)dx.$$

We use **Green's formula**

$$\int_\Omega \sum_{i=1}^n \frac{\partial u}{\partial x_i}\frac{\partial h}{\partial x_i}dx = -\int_\Omega h\Delta u\,dx + \int_\Gamma h\frac{\partial u}{\partial n}dx,$$

where n is the interior normal to the boundary Γ of the set Ω. We get

$$\big\langle I'(u), h\big\rangle = 2\int_\Omega (f - \Delta u)dx\ \forall h \in V$$

Necessary conditions of extremum for functionals 39

because the function h is zero on the boundary Γ. Using the determination of the linear continuous functional on the space $H_0^1(\Omega)$, we have the equality (1.13). $\square$

Using Theorem 1.3 and Lemma 1.3, we obtain the following result.

Theorem 1.4 *The solution u of the minimization problem for the Dirichlet integral on the space $H_0^1(\Omega)$ satisfies the equality*

$$\Delta u = f, \, x \in \Omega. \tag{1.14}$$

This equality is called the **Poisson equation**. Note that each element of the space $H_0^1(\Omega)$ is equal to the zero on the boundary of the given set. Therefore, we have the boundary condition

$$u = 0, \, x \in \Gamma. \tag{1.15}$$

The system (1.14), (1.15) is called the **Dirichlet problem**. This is a well-known result of theory of mathematical physics equations: the minimization of the Dirichlet integral on the Sobolev space is transformed to the Dirichlet problem for the Poisson equation.

Remark 1.69 We shall continue to analyze the problem of minimization for the Dirichlet integral in Chapter 2 (see Sections 2.1 and 2.4).

Remark 1.70 We shall consider an optimization problem for the system described by the Dirichlet problem for the Poisson equation. Therefore, it will be necessary to know the properties of this boundary problem.

Remark 1.71 Consider the properties of the stationary conditions for the different functionals (see Table 1.2). The Gâteaux derivative, a.e. the term at the left-hand side of the stationary condition (1.12), is a point of the adjoint space for the general case. The value at its right-hand side is the zero point in the adjoint space. If the domain of the minimized functional is $\mathbb{R}$, a.e. we minimize a function of one variable, then the derivative of the functional at the point of minimum is a number, the zero of the equality (1.12) is the zero number, and the stationary condition is an algebraic equation. If the domain of the functional is a Euclid space, a.e. we have a function of many variables, then the derivative of the functional is a vector, the zero is the zero vector, and the stationary condition is a system of algebraic equations. We find the function of one variable for the Lagrange problem. The derivative of the functional is a function of one variable too here, the zero of the equality (1.12) is the zero function, and the stationary condition is an ordinary differential equation. We find the function of many variables for the problem of minimization of the Dirichlet integral. The derivative of the functional is a function of many variables too here, the zero is the zero function of many variables, and the stationary condition is a partial differential equation.

1.10   Minimization of functionals on subspaces

We considered before unconditional extremum problems only. However, the conditional optimization problems are more interesting. These problems

40 *Optimization and Differentiation*

TABLE 1.2: Properties of stationary conditions

object of minimization	Gâteaux derivative	zero point	stationary condition
general functional	point of adjoint space	zero point of adjoint space	operator equation
function of one variable	number	zero number	algebraic equation
function of many variables	vector	zero vector	system of algebraic equations
Lagrange functional	function of one variable	zero function of one variable	ordinary differential equation
Dirichlet integral	function of many variables	zero function of many variables	partial differential equation

are more difficult. Return to the proof of Theorem 1.3. We had the inequality

$$I(v) \geq I(u) \; \forall v, \tag{1.16}$$

where u is a solution of the given problem. Then chosen v equals to $v_\sigma = u + \sigma h$, where h is an arbitrary element of the space V, and σ is a number. Therefore, we can determine the function of one variable $f(\sigma) = I(u + \sigma h)$. The relation (1.16) is transformed to the inequality $f(\sigma) \geq f(0)$ for all σ; and we can use Problem 1.1.

If we minimize the functional on a subset U of the given space, then the inequality (1.16) is true for the element v of this set only. However, we cannot guarantee the inclusion $v_\sigma \in U$. The inequality $I(v_\sigma) \geq I(u)$ can be false for this case. Thus, the previous technique is not applicable. The unconditional extremum problems with arbitrary constraints are impossible for the analysis. However, we can solve these problems for some special classes of the set U. Consider the case where the set U is a subspace.

Problem 1.5 *Find the element that minimizes a functional I on a subspace U of a Banach space V.*

Theorem 1.5 *If u is a solution of Problem 1.5, and there exists a Gâteaux derivative of the functional I at this point, then*

$$\langle I'(u), h \rangle = 0 \; \forall h \in U. \tag{1.17}$$

Indeed, we determine the function $f(\sigma) = I(u + \sigma h)$, where h is an arbitrary element of the subspace U. Then we equate to the zero its derivative at the zero point.

Remark 1.72 The Gâteaux derivative of the functional is an element of the adjoint space V'. Therefore, this is a linear continuous functional on the initial space. Its value at the arbitrary element h of the set U is zero by the equality (1.17). The set of all elements with this property is called the **annihilator** of set U. Denote it by $U^\perp$. Then the relation (1.17) can be transformed to the inclusion $I'(u) \in U^\perp$.

Necessary conditions of extremum for functionals 41

Example 1.20 *Dense subspace*. Let a subspace U be dense in V. Then for any element $v \in V$ there exists a sequence $\{v_k\}$ of the set U such that $v_k \to v$ in V. Determine $h = u_k$ in the formula (1.17); we get

$$\langle I'(u), v_k \rangle = 0, \ k = 1, 2, \dots.$$

Using the continuity of the map $h \to \langle I'(u), h \rangle$, after passing to the limit we obtain

$$\langle I'(u), v \rangle = 0.$$

The element $v \in V$ is arbitrary here. Therefore, we have the equality $I'(u) = 0$ that is the stationary condition (1.12). Thus, the minimization problems on the space and on its dense subspace have the same necessary condition of extremum.

Remark 1.73 We noted that the condition (1.17) is equal to the inclusion $I'(u) \in U^\perp$. By the last example, the annihilator of the dense subspace has a unique element. This is the zero.

Example 1.21 *Function of two variables*. Consider the minimization problem of the function of two variables $f(x, y) = x^2 + y^2$ on the subspace

$$U = \{(x, y) | y = 2x\}$$

of the Euclid plan (see Figure 1.19). The Gâteaux derivative of the function f at the point $u = (x, y)$ is the gradient of this function at the given point, i.e., the vector $f'(u) = (2x, 2y)$. By the equality (1.17), its scalar product with a vector $h = (\varphi, \psi)$ of the set U is the zero. Then we have the equality $\psi - 2\varphi$ because of $h \in U$. Thus, the condition (1.17) is transformed to

$$2x\varphi + 4y\varphi = 0 \ \forall \varphi \in \mathbb{R}.$$

Therefore, the set of points (x, y) that satisfies the equality (1.17) is

$$\{(x, y) | x + 2y = 0\}.$$

This is the annihilator of the set U (see Figure 1.19). By Theorem 1.5, the Gâteaux derivative of the function f at the point $u = (x, y)$ of its minimum on the set U is an element of $U^\perp$. Therefore, we have $2x + 4y = 0$. However, this point belongs to the set U that is $y = 2x$. Using the two last inequalities, we determine the point $u = (0, 0)$, which is the solution of the given problem. Indeed, our function is non-negative; and it is equal to the zero for the point u that is the element of the set U.

Remark 1.74 The annihilator of the set U is the subspace of the Euclid plan, particularly the line.

Remark 1.75 We shall continue the consideration of this example in the last section.

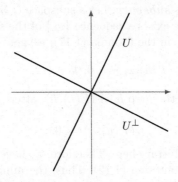

FIGURE 1.19: Annihilator $U^\perp$ of the subspace U.

1.11 Derivatives with respect to subspaces

We used the directions h from the given subspace only for the determination of the relation (1.17). Therefore, we cannot suppose the existence of the Gâteaux derivative for solving Problem 1.5. Thus, we obtain a new form of the functional derivatives. Let U be a subspace of the space V, and let I be the functional on V.

Definition 1.27 *Suppose the convergence*

$$\frac{I(u+\sigma h) - I(u)}{\sigma} \to I'_U(u)h \quad \forall h \in U$$

as $\sigma \to 0$, as the map $h \to I'_U(u)h$ is linear and continuous. Then the element $I'_U(u)h$ is called the **derivative** *of the functional I at the point u* **with respect to the subspace** *U.*

The derivative with respect to the subspace is the linear continuous functional on the space U. Therefore, this is an element of the adjoint space U'. Thus we can denote the value $I'_U(u)h$ by $\langle I'_U(u), h \rangle$. If U is the given space V, then the derivative with respect to the subspace is equal to the standard Gâteaux derivative.

Remark 1.76 The derivative with respect to the subspace is an extension of the Gâteaux derivative. It uses all directions of the space, whereas now we use the directions from the given subspace only. We shall consider with the lapse of time a larger notion of the *derivative with respect to the convex set*. We shall define also the *extended derivative* (see Chapter 5) that is an extension of the derivative with respect to the subspace.

The following result is proved as an analog of Theorem 1.5.

Necessary conditions of extremum for functionals 43

Theorem 1.6 *If u is a solution of Problem 1.5, and there exists the derivative with respect to the subspace U of the functional I at this point, then*

$$I'_U(u) = 0. \tag{1.18}$$

Indeed, it is sufficient to determine again the function $f(\sigma) = I(u + \sigma h)$ with $h \in U$. The condition (1.18) is the corollary of the equality to the zero of the derivative of f at the zero point.

Remark 1.77 The symbol 0 at the right side of the equality (1.18) is the zero element of the adjoint space U'.

Example 1.22 *Function of two variables*. Consider again the problem of minimization of the function $f(x, y) = x^2 + y^2$ on the subspace

$$U = \{(x, y) \,|\, y = 2x\}$$

of the Euclid plan (see Example 1.21). Find the derivative of this function with respect to this subspace at a point $u = (x, y)$. Each vector $h \in U$ can be characterized by the equality $h = (\varphi, 2\varphi)$. Then we have

$$f(u + \sigma h) = (x + \sigma\varphi)^2 + (y + 2\sigma\varphi)^2 = f(u) + 2\sigma\varphi(x + 2y) = 5\sigma^2\varphi^2.$$

Determine the derivative of the function f with respect to this subspace at a point u by the equality

$$f'_U(u)h = f'_U(u)(\varphi, 2\varphi) = 2(x + 2y)\varphi \quad \forall \varphi \in \mathbb{R}.$$

Thus, the derivative $f'_U(u)$ takes an arbitrary vector $h = (\varphi, 2\varphi)$ on the plan (i.e., is an element of the set U) to a number $2(x + 2y)\varphi$. By the condition (1.18), this derivative is the zero element set U' that is the zero linear continuous functional on U. Therefore, it takes an arbitrary element $\varphi \in U$ to the zero. Then we transform the formula (1.18) to the equality $2(x + 2y)\varphi = 0$ for all $\varphi \in U$. Thus, we get the equality $x + 2y = 0$. However, the point $u = (x, y)$ is an element of the set U. Then we have the equality $y = 2x$. Thus, there exists the unique point $u = 0$ that satisfies the condition (1.18). This is the origin of coordinates. We obtained before this result by using Theorem 1.5 (see Example 1.21).

1.12 Minimization of functionals on affine varieties

Let a set U be an affine variety of a space V, i.e., there exists a subspace W of V such that $U = u_0 + W$, where a point u_0 does not belong to W. The elements of the set U are the points $v = u_0 + w$, where $w \in W$.

44 *Optimization and Differentiation*

Problem 1.6 *Find the element that minimizes a functional I on the affine variety U of the Banach space V.*

Theorem 1.7 *If the functional I is differentiable with respect to the subspace W at the solution u of Problem 1.6, then the following equality holds*

$$I'_W(u) = 0. \tag{1.19}$$

Proof. The solution u of Problem 1.6 belongs to U. Therefore, there exists an element $w_0 \in W$ such that $u = u_0 + w_0$. Then the point $u + \sigma h = u_0 + (w_0 + \sigma h)$ is an element of the set U for all $h \in W$ and the constant σ. Determine the function $f(\sigma) = I(u + \sigma h)$. It has the minimum at the point $\sigma = 0$. Its derivative at the zero point is zero. Using the definition of the derivative with respect to the subspace, we obtain the equality (1.19). $\square$

Each subspace is an affine variety. Therefore, Theorem 1.7 characterizes the solution of Problem 1.5 too. Thus, the equality (1.19) can be transformed to (1.18) for $U = W$.

Remark 1.78 If the functional I is Gâteaux differentiable, then we transform the equality (1.19) to $\langle I'(u), h \rangle = 0$ for all $h \in W$ that is the analog of (1.17).

Return to Problem 1.2, i.e., the Lagrange problem. We would like to minimize the functional

$$I = I(v) = \int_{x_1}^{x_2} F[x, v(x), v'(x)] dx$$

on the set of all functions $v = v(x)$ that satisfy the boundary conditions

$$v(x_1) = v_1, \ v(x_2) = v_2.$$

We cannot transform the Lagrange problem to the minimization problem on a linear space because of the boundary conditions. Particularly, if two functions satisfy these equalities, then it will be false for its sum. However, we can interpret Problem 1.2 as the minimization problem of the given functional on an affine variety. Suppose the function u is equal to the values v_1 and v_2 at the points x_1 and x_2. Determine the space $V = C^2[x_1, x_2]$ of twice continuously differentiable functions on this interval and its subspace W of functions such that it is zero at the points x_1 and x_2. Choose the affine variety $U = u_0 + W$.

Lemma 1.4 *Suppose the function F is continuously differentiable; then the functional I is differentiable with respect to W, as*

$$I'_W(u) = F_u(x, u, u') - \frac{d}{dx} F'_u(x, u, u').$$

Necessary conditions of extremum for functionals 45

Proof. Determine the difference

$$I(u + \sigma h) - I(u) = \int\limits_{x_1}^{x_2} \Big[F\big(x, u + \sigma h, u' + \sigma h'\big) - F(x, u, u')\Big]\,dx,$$

where $h \in W$. Using the standard technique (see the proof of Lemma 1.1), we get

$$I(u + \sigma h) - I(u) = \sigma \int\limits_{x_1}^{x_2} \Big[F_u(x, u, u')h + F_{u'}(x, u, u')h' + \eta(\sigma)/\sigma\Big]\,dx,$$

where $\eta(\sigma) = O(\sigma)$. Dividing by σ and passing to the limit as $\sigma \to 0$, we obtain

$$I'_W(u) = \int\limits_{x_1}^{x_2} \Big[F_u(x, u, u')h + F_{u'}(x, u, u')h'\Big]\,dx.$$

After integration by parts by using the boundary values of the function h we have

$$I'_W(u) = \int\limits_{x_1}^{x_2} \Big[F_u(x, u, u') - \frac{d}{dx}F'_u(x, u, u')\Big]h\,dx \ \forall h \in V.$$

Then the formula for the derivative with respect to the subspace follows from the Euler–Lagrange lemma. $\square$

Using the equality (1.19), we have Euler equation

$$I'_W(u) = F_u(x, u, u') - \frac{d}{dx}F'_u(x, u, u').$$

Therefore, Theorem 1.2 is the corollary of Theorem 11.7.

Remark 1.79 The assertions of Lemma 1.4 are true for a larger class of functions F. However, now we have the interest to the reduction Lagrange problem to Problem 1.6 and the relation between the Euler equation and Theorem 1.7 only.

Remark 1.80 We could apply Theorem 1.6 for the analysis of the Dirichlet problem with nonhomogeneous boundary conditions.

1.13 Minimization of functionals on convex sets

Consider a Banach space V, its convex subset U, and a functional I on the space V.

46   *Optimization and Differentiation*

Problem 1.7 *Find the element that minimizes the functional I on the convex set U of the Banach space V.*

Theorem 1.8 *Suppose u is a solution of Problem 1.7. If the functional I is Gâteaux differentiable at the point u, then*

$$\langle I'(u), v - u \rangle \geq 0 \ \forall v \in U. \tag{1.20}$$

Proof. Try to use the standard technique. Determine the function $f(\sigma) = I(u + \sigma h)$, where σ is a number, and h is an element of the space V. However, we cannot guarantee the inclusion of the point $u + \sigma h$ to the set U; and the comparison of the values $f(0)$ and $f(\sigma)$ is senseless. But the point $u + \sigma(v - u)$ belongs to U for all $v \in V$ and number $\sigma \in [0, 1]$ because of the convexity of the set U. Therefore, determine the function $f(\sigma) = I[u + \sigma(v - u)]$. Then the point u is a solution of Problem 1.7 if and only if the function f has a minimum on the interval $[0, 1]$ at the point $\sigma = 0$. Hence, we have the inequality $f(\sigma) - f(0) \geq 0$ for all small enough positive values σ. Dividing this inequality by σ and passing to the limit as $\sigma \to 0$, we get $f'(0) \geq 0$. Using the definition of Gâteaux derivative, we find $f'(0) = \langle I'(u), v - u \rangle$. Then the condition (1.20) holds by the last inequality. $\square$

Definition 1.28 *The condition (1.20) is called the **variational inequality**.*

Thus, the variational inequality is the necessary condition of minimum for the differentiable functional on a convex set.

Remark 1.81 We will not use the density of the space here. The variational inequality can be obtained for linear topological spaces too.

Consider examples.

Example 1.23 *Non-conditional extremum problem*. Suppose the set U is equal to the space V. Then the variational inequality (1.20) is transformed to

$$\langle I'(u), v - u \rangle \geq 0 \ \forall v \in V.$$

Determine v equal to $u + h$ and $u - h$ here, where h is an arbitrary element of the space V; we get the inequalities

$$\langle I'(u), h \rangle \geq 0, \ \langle I'(u), h \rangle \leq 0.$$

Then we have

$$\langle I'(u), h \rangle = 0 \ \forall h \in V.$$

Therefore, $I'(u) = 0$. Thus, the variational inequality (1.20) for the non-conditional extremum problem can be transformed to the stationary condition (1.12).

Necessary conditions of extremum for functionals

TABLE 1.3: Sets of minimization and optimality conditions

sets of minimization	optimality conditions
space V	$I'(u) = 0$
affine variety $u_0 + W$	$I'_W(u) = 0$
convex set U	$\langle I'(u), v - u \rangle \geq 0 \; \forall v \in U$

Remark 1.82 Non-conditional extremum Problem 1.3 is the partial case of Problem 1.7. We proved that the stationary condition (1.12) is the corollary of the variational inequality (1.20) for this case. Hence, the variational inequality is the natural extension of the stationary condition to the problem of the functional minimization on the convex set.

Remark 1.83 If the set U is a subspace of V, then we can determine $v = u + h$ and $v = u - h$ at the inequality (1.20), where h is an arbitrary element of the subspace U. Then we get the inequalities $\langle I'(u), h \rangle \geq 0$ and $\langle I'(u), h \rangle \leq 0$, and the equality (1.17) holds. Thus, Theorem 1.8 characterizes the solution of Problem 1.5 too. If U is an affine variety with subspace W, then we choose $v = h$ and $v = -h$ at the inequality (1.20) with arbitrary value $h \in W$. Hence, we get the equality $\langle I'(u), h \rangle = 0$ for all $h \in W$. Then we have the necessary condition of extremum for Problem 1.6 (see also Table 1.3).

Example 1.24 *Minimization of the distance.* Consider a Hilbert space V, its convex closed subset U, and the point $z \in V$. Determine functional

$$I(v) = \left\| v - z \right\|^2.$$

This is the square of the distance between the points v and z (see Figure 1.20). Hence, we have the problem of finding the nearest point of the set U to the given point z. Using the relation between scalar products and norms of Hilbert spaces, determine the value

$$I(u + \sigma h) = \left(u + \sigma h, u + \sigma h\right) = I(u) + 2\sigma(u - z, h) + \sigma^2 \|h\|^2.$$

Thus, the derivative of the functional satisfies the equality

$$\langle I'(u), h \rangle = 2\left(u - z, h\right) \; \forall h \in V.$$

Now transform the variational inequality (1.20); we get

$$\left(u - z, v - z\right) \geq 0 \;\; \forall v \in U. \tag{1.21}$$

Determine the **projector** $P_U : V \to U$, which takes an arbitrary point of the space V to its projection on the set U (see Figure 1.20). Therefore, the point u satisfies the variational inequality (1.21). We can obtain a more exact result after the concretization of the problem statement (see next example).

Remark 1.84 We shall continue the analysis of Example 1.16 in Chapter 2 (see Sections 2.1 and 2.4).

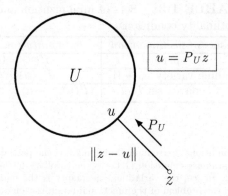

FIGURE 1.20: Functional is the square of the distance from z to U.

Example 1.25 *Minimization of the distance for the space $L_2(\Omega)$*. Consider the space $V = L_2(\Omega)$, the function $z \in V$, the set

$$U = \{v \in V \mid a \le v(x) \le b \text{ a.e. on } \Omega\},$$

and the functional

$$I(v) = \int_\Omega (v-z)^2 dx,$$

where a,b are given constants, as $a < b$. Then determine the variational inequality (1.21):

$$\int_\Omega (u-z)(v-u)dx \ge 0 \ \forall v \in U. \qquad (1.22)$$

Let x_* be an arbitrary interior point of the set Ω, $v_* \in [a,b]$. Consider the ball $B_\varepsilon(x_*)$, where the positive number ε is small enough such that this ball is the subset of Ω. Determine the function (see Figure 1.21)

$$v_\varepsilon = \begin{cases} u(x), & \text{if } x \in B_\varepsilon(x_*), \\ v_*, & \text{if } x \notin B_\varepsilon(x_*). \end{cases}$$

It is called the **needle variation** of the function u. For all $x_* \in \Omega$, $v_* \in [a,b]$ we have the inclusion $v_\varepsilon \in U$. Choose $v = v_\varepsilon$ at the inequality (1.22). Dividing by the measure m_ε of the ball $B_\varepsilon(x_*)$, we get

$$\frac{1}{m_\varepsilon} \int_{B_\varepsilon(x_*)} (u-z)(v_*-u)dx \ge 0.$$

Passing to the limit as $\varepsilon \to 0$ by using the Lebesgue Theorem (see Section 1.8), we obtain

$$[u(x_*) - z(x_*)][v_* - u(x_*)] \ge 0 \ \forall v_* \in [a,b]. \qquad (1.23)$$

This formula is true for almost everywhere points $x_* \in \Omega$.

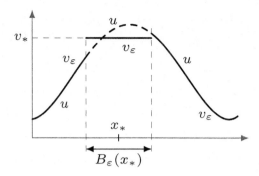

FIGURE 1.21: Needle variation.

The first multiplier of the left side does not depend on the arbitrary parameter v_* here. Therefore, it has a concrete sign. Suppose $u(x_*) > z(x_*)$. Dividing the inequality (1.23) by the first multiplier, we have $v_* \geq u(x_*)$ for all $v_* \in [a,b]$. Thus, the value $u(x_*)$ is not greater than arbitrary number $v_* \in [a,b]$. Hence, it does not exceed the most minimal number of this interval, i.e., a. However, $u(x_*)$ is not less than a by definition of the set U. Therefore, $u(x_*) = a$. Thus, if the number $z(x_*)$ is less than $u(x_*)$ that is equal to a, then $u(x_*) = a$.

Suppose now the inequality $u(x_*) < z(x_*)$. Then we have $v_* \leq u(x_*)$ for all $v_* \in [a,b]$ by the condition (1.23). Hence, the value $u(x_*)$ is not less than arbitrary number $v_* \in [a,b]$ that is $u(x_*) \geq b$. However, it cannot be greater than b by definition of the set U. Therefore, $u(x_*) = b$. Thus, if the number $z(x_*)$ is greater than $u(x_*)$ that is equal to b, then $u(x_*) = b$.

We consider also the case $u(x_*) = z(x_*)$. However, the function u belongs to U. Hence, this value has the sense for the inclusion $z(x_*) \in [a,b]$ only.

Thus, our function u at the arbitrary point x of the set Ω (more exact, a.e. on Ω, but integrable functions have the sense up the set of zero measure, in principle) is determined by the formula

$$u(x) = \begin{cases} a, & \text{if } z(x) < a, \\ z(x), & \text{if } a \leq z(x) \leq b, \\ b, & \text{if } z(x) > b. \end{cases}$$

The value $u(x)$ is the projection of the number $z(x)$ on the interval $[a,b]$ (see Figure 1.22), i.e., $u = P_U z$. Thus, the variational inequality (1.20) characterizes the projection on the subset U of the space $L_2(\Omega)$.

Remark 1.85 We shall apply the needle variation and consider the technique at Chapter 3 for obtaining the necessary optimality condition in the form of the maximum principle.

Remark 1.86 We could consider the case, where the functional has the sense on the convex set U only. We cannot any possibility to determine its Gâteaux derivative, because the

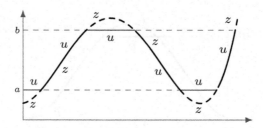

FIGURE 1.22: Solution u is the projection of the function z on the set U.

value $I(u + \sigma h)$ is not determined for the arbitrary $h \in V$. Then we could obtain necessary conditions of optimality by using the derivative with respect to the convex set that is an extension of the derivative with respect to the subspace (see Chapter 6).

1.14 Comments

The stationary condition as the necessary condition of extremum was obtained by Fermat for polynomials. G.W. Leibniz extended this result for the general functions. However, N. Oresme and I. Kepler saw that the velocity of change of the function decreases at the neighborhood of the extremum.

The properties of the stationary (critical) points are considered by **Morse theory** [260] (see also J.P. Aubin and I. Ekeland [31], J. Milnor [357], H. Seifert and W. Threllfall [454], M.M. Postnikov [409]), that is, the *calculus of variations on the large*.

Euler equation was obtained by Euler by using the approximation method and by Lagrange by using the variational method (see the proof of Theorem 1.2). The solutions of the Euler equation (extremals) can be analyzed by the conditions of Legendre, Weierstrass, Jacobi, and others. There exist many books with complete consideration of the calculus of variations (see, for example, N.I. Ahiezer [7], G.A. Bliss [75], O. Bolza [82], G. Buttazzo, G. Mariano and S. Hildebrandt [104], E. Elsgolts [164], I.M. Gelfand and S.V. Fomin [205], M.R. Hestens [241], M.A.Lavrentiev and L.A. Lusternik [300], G. Leitmann [307], L. Young [563], and L. Zlaf [570]).

The history of the calculus of variations can be found in G.A. Bliss [75] and E.R. Smolyakov [506] (see also L. Markus [340] and H.J. Pesch and R. Bulirsch [400]).

The general problems of the optimal control theory are considered by S.A. Aisagaliev [13], V.M. Alekseev, V.M. Tihomirov, and S.V. Fomin [14], M. Aoki [20], A. Balakrishnan [41], A. Balakrishnan and L. Neustadt [42], R. Bellman [57], L.D. Berkovitz [65], V.G. Boltyansky [78], V.V. Dikussar and A.A. Milutin [141], W.H. Fleming and R.W. Rishell [178], R. Fletcher [179], A.V. Fursikov [197], R. Gabasov and F.M. Kirillova [198], R.V. Gamkrelidze [202], P.E. Gill, W. Murrey, and M.H. Wright [212], A.D. Ioffe and V.M. Tihomirov [253], V.F. Krotov [285], V.F. Krotov, V.Z. Bukreev, and L.I. Gurman [286], V.F. Krotov and L.I. Gurman [287], G. Leitmann [307], X. Li and J. Yong [313], J.L. Lions [315], A.S. Matveev and V.A. Yakubovich [343], A.A. Milutin [359], N.N. Moiseev [365], L. Neustadt [386], L.S. Pontryagin, V.G. Boltyansky, R.V. Gamkrelidze, and E.F. Mishchenko [408], B.N. Pshenichny [416], V.M. Tihomirov [522], F.P. Vasiliev [538], [539], J. Warga [553], L. Young [563], J. Zabkzyk [565], and others.

G. Minkowski and J. Jensen obtain the first results of the convex analysis. This area has many applications in the extremum theory (see W. Fenchel [174], J.J. Moreau [370], and

Necessary conditions of extremum for functionals

R.T. Rockafellar [432]). The complete enough consideration of the convex analysis and its applications is given by E. Asplund [29], J.P. Aubin and I. Ekeland [31], V.G. Boltyansky and P.S. Soltan [81], B. Grunbaum [224], V.F. Demyanov and A.M. Rubinov [137], I. Ekeland and R. Temam [162], and B.N. Pshenichny [415].

The differentiation of the operators is the direction of the nonlinear functional analysis. Note the following books of the nonlinear functional analysis: J.P. Aubin and I. Ekeland [31], M.C. Joshi and R.K. Bose [263], M.A. Krasnoselsky and P.P. Zabreiko [282], L. Nirenberg [387], T.L. Saaty and J. Bram [441], and E. Zarantonello [566]. The general theory of the operator differentiation in the linear normalized space is given, for example, by Yu.G. Borisovich, V.G. Zvyagin, and Yu.I. Sapronov [86], J. Dieudonné [140], N.V. Ivanov [261], L.V. Kantorovich and G.P. Akilov [267], A.N. Kolmogorov and S.V. Fomin [276] and M.A. Krasnoselsky and P.P. Zabreiko [282]. By D. Priess [412], there exists a continuous functional on a Hilbert space that is Gâteaux differentiable everywhere and nondifferentiable in the sense of Fréchet on the essential set of this space.

V. Volterra and J. Hadamard determine the derivative of the functional. However, Weierstrass has, in reality, the idea of Fréchet derivative. There are many forms of operator derivatives on the linear topological spaces (see V.I. Averbuh and O.G. Smolyanov [36], [37], A. Frölicher and W.J. Bucher [190], Gil de Lamadrid [209], H.H. Keller [268], J. Sebastiãoe e Silva [451] and C. Ursescu [534]). R. Gâteaux determined the derivative of the functional with respect to the direction. The derivative with respect to the subspace is given by V.I. Averbuh and O.G. Smolyanov [36], [37]. Example 1.15 of the discontinuous Gâteaux differentiable functional is considered by J.P. Aubin and I. Ekeland [31].

The spaces of continuous and integrable functions and the theory of the measurable functions and Lebesgue integration are described in detail in many books on functions theory and the functional analysis. The theory of distributions can be found, for example, in the following books: H. Bremermann [90], F.G. Friedlander [185], I.M. Gelfand and G.E. Shilov [206], J.I. Richards and H.K. Joun [430], L. Schwartz [448], S.L. Sobolev [508], and V.S. Vladimirov [542]. Note also the distributions theory that is based on the completeness technique (see P. Antosic, J. Mikusinski, and R. Sikorski [18], J.F. Colombeau [129], and Yu.V. Egorov [157]). The different Sobolev spaces are considered by R. Adams [4], O.V. Besov, V.P. Ilyin, and S.M. Nikolsky [67], H. Gajewski, K. Gröger, and K. Zacharias [200], J.L. Lions and E. Magenes [319], and S.L. Sobolev [508].

The minimization problem for the Dirichlet integral was analyzed by K.F. Gauss, W. Thomson, P.L. Dirichlet, B. Riemann, and others. D. Hilbert substantiated the variational method for mathematical physics problems. The applications of the variational method for the different mathematical physics problems are given by V.S. Vladimirov [543], K. Lanczos [295], and S.G. Mihlin [355], [355].

The variational inequality was determined by G. Stampacchia. Its first physical application was given by G. Fichera [176] (see also G. Duvaut and J.L. Lions [149]). The general theory of the variational inequalities is described by G. Duvaut and J.L. Lions [149] and D. Kinderlehrer and G. Stampacchia [270]. The variational inequalities as the form of the necessary conditions of optimality became widely known after the release of the classic monograph of J.L. Lions [315]. The well-known Theorem 1.8 is described, for example, by I. Ekeland and R.Temam [162], J.L. Lions [315], J. Séa [450], and others. D.E. Ward and G.M. Lee [551] determines the relation between the variational inequalities and Pareto optimality for the multi-criterion problems. The optimal control problems for the systems described by variational inequalities are considered by I. Bock and J. Lovisek [76], Q. Chen, D. Chu, and R.C.E. Tan [123], F. Mignot and J.P. Puel [354], F. Patrone [397], S. Saguez [444], and S.Ya. Serovajsky [496], [497]. G. Knowles [272], and J.L. Lions [315] analyze minimization problems on the non-convex set.

The needle variation was determined by K. Weierstrass. The different variations are considered by L. Graves, G. Kelly, R.E. Kopp, and A.G. Moyer (see, for example, R. Gabasov and F.M. Kirillova [199]). High order extremum conditions are determined, for example, by E.R. Avakov [34], J.F. Bonnans [83], R. Brockett [92], E. Casas [110], R. Gabasov and F.M. Kirillova [199], U. Ledzewicz and H. Schättler [303], E.S. Levitin, A.A. Milutin, and N.P. Osmolovsky [311], and Z. Páles and V. Zeidan [391].

52   *Optimization and Differentiation*

Lusternik theorem on the approximation of the smooth variety [334] has a great influence on the extremum theory with general constraints (see the survey of A.V. Dmitruk, A.A. Milutin, and N.P. Osmolovsky [142]). The conditions of extremum for the problems with difficult enough constraints are given by E.R. Avakov [34], E.R. Avakov and A.V. Arutunov [35], V.M. Alekseev, V.M. Tihomirov, and S.V. Fomin [14], A.V. Arutunov [25], [26], V.G. Boltyansky [80], A.V. Dmitruk, A.A. Milutin, and N.P. Osmolovsky [142], A.V. Dubovitsky and A.A. Milutin [147], R.V. Gamkrelidze [202], X. Guo and J. Ma [228], C.-W. Ha [230], H. Halkin [235], A.D. Ioffe and V.M. Tihomirov [253], U. Ledzewicz and H. Schättler [303], E.S. Levitin, A.A. Milutin, and N.P. Osmolovsky [311], A.S. Matveev and V.A. Yakubovich [343], A.A. Milutin [360], L. Neustadt [385], [386], B.N. Pshenichny [414], [415], S.Ya. Serovajsky [471], V.M. Tihomirov [522], V.A. Yakubovich [558], J.J. Ye [560], and others.

Chapter 2

Minimization of functionals. Addition

2.1	Sufficiency of extremum conditions	53
2.2	Existence of the function minimum	58
2.3	Weak convergence in Banach spaces	60
2.4	Existence of the functional minimum	63
2.5	Uniqueness of the functional minimum	66
2.6	Tihonov well-posedness of extremum problems	68
2.7	Hadamard well-posedness of extremum problems	70
2.8	Ekeland principle and the approximate condition of extremum .	72
2.9	Non-smooth functionals and subdifferentiation	75
2.10	Fréchet differentiation of functionals	78
2.11	Approximate methods of functionals minimization	80
2.12	Comments	82

We considered before the necessary conditions of extremum for the functionals. Now we analyze additional problems of the functionals' extremum theory. This is the sufficiency of extremum conditions, the solvability of extremum problems, the minimization of non-convex and non-smooth functionals, approximate methods of extremum theory, and some others.

2.1 Sufficiency of extremum conditions

We considered before the stationary condition and the variational inequality. These are the necessary conditions of extremum. Therefore, its solutions can be non-optimal. However, these conditions of extremum and the initial extremum problem are equivalent under some additional suppositions. Return to Problem 1.7 of the minimization of a functional I on a convex subset U of a Banach space. Its solution u satisfies the variational inequality

$$\langle I'(u), v - u \rangle \geq 0 \ \ \forall v \in U \qquad (2.1)$$

because of Theorem 1.8.

53

54   *Optimization and Differentiation*

Theorem 2.1 *If the functional I is convex and Gâteaux differentiable, and the point $u \in U$ satisfies the variational inequality (2.1), then it minimizes this functional on the set U.*

Proof. Using the convexity of the functional, we obtain the inequality

$$I[(1-\sigma)u + \sigma v] \le (1-\sigma)I(u) + \sigma I(v) \ \ \forall u, v \in U, \ \sigma \in (0,1).$$

Then

$$I[u + \sigma(v-u)] - I(u) \le \sigma[I(v) - I(u)] \ \forall u, v \in U, \ \sigma \in (0,1).$$

Dividing this formula by σ and passing to the limit as $\sigma \to 0$ by using the differentiability of the functional, we get

$$\langle I'(u), v - u \rangle \le I(v) - I(u) \ \ \forall u, v \in U.$$

If the point u satisfies the variational inequality (2.1), then the term in the left-hand side of this inequality is non-negative. Then we have the inequality $0 \le I(v) - I(u)$ for all $v \in U$. This completes the proof of Theorem 2.1. $\square$

Remark 2.1 We know that the variational inequality can be transformed to the stationary condition for the problem of unconditional extremum. Therefore, from Theorem 2.1 it follow that the stationary condition is the necessary and sufficient condition of minimum for a convex differentiable functional on a Banach space. We can obtain analogical results for the problems of the functional minimization on subspaces and affine manifolds.

Use Theorem 2.1 for the additional analysis of the previous examples.

Example 2.1 *Function of one variable*. We consider the minimization of the function $f = f_1(x) = x^2$ (see Example 1.1). For any $u, v \in U, \ \sigma \in (0,1)$ we have

$$f[(1-\sigma)u + \sigma v] - (1-\sigma)f(u) - \sigma f(v) =$$
$$[(1-\sigma)^2 - (1-\sigma)]u^2 + 2(1-\sigma)\sigma uv + (\sigma^2 - \sigma)v^2 =$$
$$-(1-\sigma)\sigma(u^2 - 2uv + v^2) = -(1-\sigma)\sigma(u-v)^2 \le 0.$$

Then we obtain the inequality

$$f[(1-\sigma)u + \sigma v] \le (1-\sigma)f(u) + \sigma f(v) \ \forall u, v \in R, \ \sigma \in (0,1).$$

We have the equality here for $u = v$ only. Hence, the square function is strictly convex (see Figure 1.1). Then the solution of the stationary condition is the point of minimum by Theorem 2.1 (see Example 1.1). The functions $f_2(x) = 3x^4 - 8x^3 - 6x^2 + 24x$, $f_3(x) = x^4 - 2x^2$, and $f_5(x) = x^3$ of Examples 1.2, 1.3, 1.5 are non-convex (see Figure 1.2, 1.5, and 1.7). The solutions of the suitable stationary conditions do not minimize the considered functions. The linear function $f_4(x) = x$ from Example 1.4 is convex but not strictly convex (see Figure 1.6). Hence, Theorem 2.1 is applicable, and stationary conditions

Minimization of functionals. Addition 55

are necessary and sufficient conditions of extremum. This function does not have the minimum on the set of real numbers. But this does not contradict Theorem 2.1 because the stationary condition has the form $0 = f_4'(x) = 1$. This is an insolvable equation, of course. Therefore, the set of the points of minimum for this function is equal to the set of the solutions of the stationary condition, because both sets are empty. Finally, the function $f_6(x) = |x|$ of Example 1.6 is convex but non-differentiable (see Figure 1.8). Hence, Theorem 1.7 is not applicable.

Example 2.2 *Lagrange problem*. The properties of the integral functional for the Lagrange problem are determined by the function under the integral. Consider the functional

$$I(v) = \int_0^\pi \left[v(x)^2 + v'(x)^2 - 2v(x)\sin x - 2v'(x)\cos x \right] dx.$$

We transform it to the equality

$$I(v) = \int_0^\pi \left[v(x)^2 - 2v(x)\sin x + \sin^2 x \right] dx +$$

$$\int_0^\pi \left[v'(x)^2 - 2v'(x)\cos x + \cos^2 x \right] dx - \pi = J(v) + K(v) - \pi,$$

where

$$J(v) = \int_0^\pi \left[v(x) - \sin x \right]^2 dx, \quad K(v) = \int_0^\pi \left[v'(x) - \cos x \right]^2 dx.$$

We proved before the strict convexity of the square function $f(v) = v^2$. The function $f(v) = (v - a)^2$ has the same property for all a. Then we have

$$\left[(1 - \sigma)u(x) + \sigma v(x) - a \right]^2 \leq (1 - \sigma)\left[u(x) - a \right]^2 + \sigma[v(x) - a]^2,$$

$$\left[(1 - \sigma)u'(x) + \sigma v'(x) - a \right]^2 \leq (1 - \sigma)\left[u'(x) - a \right]^2 + \sigma[v'(x) - a]^2,$$

where $a = sinx$. Integrating in x, we prove the convexity of the functionals J and K:

$$J\left[(1 - \sigma)u + \sigma v \right] \leq (1 - \sigma)J(u) + \sigma J(v) \; \forall u, v \in U, \sigma \in (0, 1),$$

$$K\left[(1 - \sigma)u + \sigma v \right] \leq (1 - \sigma)K(u) + \sigma K(v) \; \forall u, v \in U, \sigma \in (0, 1).$$

Adding these inequalities, we obtain the strict convexity of the functional I. Hence, the stationary condition for this problem is necessary and a sufficient condition of extremum.

56 *Optimization and Differentiation*

Remark 2.2 Of course, the functional for the general Lagrange problem is not convex (and differentiable too). Therefore, the Euler equation is not a sufficient condition of the extremum for the general case.

Example 2.3 *Dirichlet integral*. Consider the functional

$$I(v) = \int_\Omega \Big[\sum_{i=1}^n \Big(\frac{\partial v(x)}{\partial x_i} \Big)^2 + 2v(x)f(x) \Big] dx,$$

where the function f is given. Transform it to the sum

$$I(v) = \sum_{i=0}^n J_i(v),$$

where

$$J_0(v) = 2 \int_\Omega v(x)f(x)dx, \quad J_i(v) = 2 \int_\Omega \Big[\frac{\partial v(x)}{\partial x_i} \Big]^2 dx, \; i = 1, ..., n.$$

The functional J_0 is linear. Each linear functional is non-strictly convex. Therefore, it satisfies the equality

$$J_0\big[(1-\sigma)u + \sigma v\big] = (1-\sigma)J_0(u) + \sigma J_0(v) \; \forall u,v \in U, \; \sigma \in (0,1).$$

Using the convexity of the square function, we get

$$\Big[(1-\sigma)\frac{\partial u(x)}{\partial x_i} + \sigma \frac{\partial v(x)}{\partial x_i}\Big]^2 \le (1-\sigma)\Big[\frac{\partial u(x)}{\partial x_i}\Big]^2 + \sigma\Big[\frac{\partial v(x)}{\partial x_i}\Big]^2.$$

After integration we obtain the convexity (strict convexity, in reality) of the functional J_i, i.e.,

$$J_i\big[(1-\sigma)u + \sigma v\big] \le (1-\sigma)J_i(u) + \sigma J_i(v) \; \forall u,v \in U, \; \sigma \in (0,1).$$

After the addition of these inequalities we prove the convexity Dirichlet integral. Moreover, it is strictly convex. Note that the sum of the convex and strict convex functionals is strictly convex. Therefore, the stationary condition for our case is necessary and a sufficient condition of extremum. The equivalence of the Dirichlet problem for the Poisson equation to the minimization problem for the Dirichlet integral is a well-known result of the mathematical physics equations theory.

Example 2.4 *Quadratic form*. Consider the problem of minimization for the functional

$$I(v) = \langle Av, v \rangle, \; -2\langle f, v \rangle$$

on a reflexive Banach space V, where $A : V \to V'$ is a linear continuous self-adjoint operator, and $f \in V'$. We have the equalities

Minimization of functionals. Addition 57

$$I\big[(1-\sigma)u + \sigma v\big] - (1-\sigma)I(u) - \sigma)I(v) = \big[(1-\sigma)^2 - (1-\sigma)\big]\langle Au, u\rangle +$$

$$2(1-\sigma)\sigma\langle Au, v\rangle + (\sigma^2 - \sigma)\langle Av, v\rangle =$$

$$-(1-\sigma)\sigma\langle Au - v, Au - v\rangle \ \forall u, v \in V; \ \sigma \in (0,1).$$

Suppose the operator A is positive, i.e., the following inequality holds:

$$\langle Av, v\rangle \geq 0 \ \forall v \in V.$$

Therefore, our functional is convex. We proved before (see Example 1.19) that this functional is Gâteaux differentiable, and the stationary condition is transformed to the equality $Au = f$. Using Theorem 2.1, we prove that this extremum condition is necessary and sufficient, i.e., minimization problem for the given functional and the last operator equation are equivalent.

Example 2.5 *Minimization of the distance.* Consider the problem of finding the nearest point to the convex closed subset U of a Hilbert space V from a point $z \in V$. This is the minimization problem of the functional

$$I(v) = \|v - z\|^2$$

on the given set. Using the relation between the norm of the Hilbert space and its scalar product, we obtain

$$I\big[(1-\sigma)u + \sigma v\big] - (1-\sigma)I(u) - \sigma I(v) = \big[(1-\sigma)^2 - (1-\sigma)\big]\|u - z\|^2 +$$

$$2(1-\sigma)\sigma\big(u - z, v - z\big) + (\sigma^2 - \sigma)\|v - z\|^2 =$$

$$-(1-\sigma)\sigma\Big[\|u - z\|^2 - 2\big(u - z, v - z\big) + \|v - z\|^2\Big] =$$

$$-(1-\sigma)\sigma\|u - v\|^2 \leq 0 \ \forall u, v \in V; \ \sigma \in (0,1).$$

We have the equality here for $u = v$ only. Therefore, this functional is strictly convex. Then the variational inequality (2.1) is necessary and a sufficient condition of minimum for this functional; and it characterizes the projector on the convex subset of the Hilbert space.

Example 2.6 *Function of two variables.* Consider the minimization problem for the function $f(x, y) = x^2 + y^2$ on the subspace $U = \big\{(x, y)|y = 2x\big\}$ of Euclid plan (see Example 1.21). The square function is convex. Therefore, the solution of the extremum condition, i.e., the origin of the coordinates, minimizes in reality this function on the given subspace.

Remark 2.3 We could prove the sufficiency of the extremum condition from properties of the second derivative of the given functionals. However, our general subject is optimal control problems for nonlinear infinite dimensional systems. The calculation of second derivatives for these functionals is a very difficult problem. Therefore, the optimization methods for these problems that use second derivatives are non-effective.

Our next step is the solvability of extremum problems. This is not a very serious problem for functions. However, it is necessary to use additional properties of Banach spaces for extending these results to the general functionals.

2.2 Existence of the function minimum

The easiest result of the solvability for the extremum problem theory is the **Weierstrass theorem** of the existence of the minimum for continuous functions.

Theorem 2.2 *The continuous function has a minimum on a bounded closed set.*

Proof. Consider a continuous function $f = f(x)$ on a bounded closed set U. Using lower boundedness of the numerical set $f(U)$, prove the existence of its lower bound $\inf f(U)$. Therefore, there exists a sequence $\{x_k\}$ of the set U such that $f(x_k) \to \inf f(U)$. This sequence is called **minimizing**. It is a bounded sequence because of the boundedness of the set U. Then there exists a constant $c > 0$ such that $|x_k| \leq c$, $k = 1, 2, \dots$. By **Bolzano–Weierstrass theorem** for any uniformly bounded sequence we can extract a convergent subsequence. Hence, there exists a subsequence of $\{x_k\}$ (we save the initial denotation for it) such that $x_k \to x$. Using the closeness of the set U, we get the inclusion $x \in U$. Then we have the convergence $f(x_k) \to f(x)$ because of the continuity of the function f. However, the limit of the sequence $\{f(x_k)\}$ is $\inf f(U)$. Therefore, we obtain the equality $f(x) = \inf f(U)$. Thus, there exists a point of the set U that minimizes the function f. $\square$

Remark 2.4 The Weierstrass theorem and other results of the solvability of the extremum problem are non-constructive. It does not contain any methods of finding extremum points. This is true because the Bolzano--Weierstrass theorem is not constructive too. It guarantees the existence of the convergent subsequence, and does not give the method of its extracting.

We could not use this result for proving the existence of minimum for the functions of Chapter 1 because we had non-conditional extremum problems. Therefore, we had the minimization problem on the unbounded sets. However, we can prove the solvability of these problems by using a special property of minimizing functions.

Definition 2.1 *The functional I on a linear normalized space V is* **coercive** *if $I(v_k) \to +\infty$ whenever $\|v_k\| \to \infty$.*

Theorem 2.3 *The continuous coercive function has a minimum on a closed numerical set.*

Proof. Suppose the minimizing sequence $\{x_k\}$ is unbounded, i.e., $|x_k| \to \infty$. Using the coercive property, we obtain $f(x_k) \to +\infty$. However, we have the convergence $f(x_k) \to \inf f(U)$. Therefore, our supposition is false; and the sequence $\{x_k\}$ is bounded. Hence, we can repeat the previous proof. $\square$

Minimization of functionals. Addition

Example 2.7 *Functions of one variable*. We considered before the continuous functions

$$f_1(x) = x^2, \ f_2(x) = 3x^4 - 8x^3 - 6x^2 + 24x, \ f_3(x) = x^4 - 2x^2,$$

$$f_4(x) = x, \ f_5(x) = x^3, \ f_6(x) = |x|.$$

Its domain R is closed and unbounded. Therefore, we can analyze the existence problems of its minimum by using Theorem 2.3. The functions f_1, f_2, f_3, and f_6 are coercive. We prove the solvability of the problem of its minimization on the set of real numbers by Theorem 2.3 (see also Figures 1.1, 1.2, 1.5, and 1.8). However the functions f_4 and f_5 are not coercive, because the condition $|x_k| \to \infty$ can be true for $x_k \to -\infty$. Therefore, Theorem 2.3 is not applicable here. The problems of its minimization on the set of real numbers are unsolvable (see Figure 1.6 and 1.7).

Remark 2.5 The results of Theorem 2.2 and 2.3 are true for the functions of many variables. Therefore, using coercivity of the two variables function of Example 2.6, we prove the solvability of the problem of its minimization on the given subspace.

We consider minimization problems for the functionals that are determined on general sets. We could try to use Theorem 2.2 or Theorem 2.3 for proving the solvability of the Lagrange problem and the existence of the minimum of the Dirichlet integral. The boundedness of the minimizing sequences is not serious problem. However, we cannot guarantee the existence of the convergent subsequences for it.

Example 2.8 Consider sequence of the functions (see Figure 2.1)

$$v_k(x) = \sin kx, \ k = 1, 2, \dots$$

and the space $L_2(0, \pi)$. We have the equality

$$\|v_k\|^2 = \int_0^\pi \sin^2 kx \, dx = \frac{\pi}{2}, \ k = 1, 2, \dots.$$

Therefore, this sequence is bounded. Suppose there exists a function v such that $v_k \to v$ in $L_2(0, \pi)$. By Fourier series theory the function $v \in L_2(0, \pi)$ can be determined as the convergent series

$$v = \sum_{j=1}^{\infty} a_j \sin jx.$$

Besides, the **Parceval equality** holds that

$$\|v\|^2 = \sum_{j=1}^{\infty} |a_j|^2.$$

Then we can find

$$\|v_k - v\|^2 = \sum_{j=1, j \neq k}^{\infty} |a_j|^2 + |1 - a_k|^2.$$

We have $v_k \to v$ in $L_2(0,\pi)$. Therefore, $a_j = 0$, $j = 1,2,...$, $j \neq k$; and $a_k \to 1$. It is obvious that these properties are impossible. Then the limit of the sequence $\{v_k\}$ does not exist (see also Figure 2.1). We obtain the analogical result for the supposition of the existence of the convergent subsequence. Thus, the boundedness of the sequence for the space $L_2(0,\pi)$ does not guarantee the existence of its convergent subsequence, i.e., Bolzano–Weierstrass theorem is not realized here.

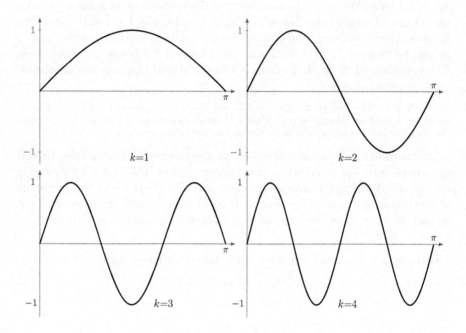

FIGURE 2.1: Bounded sequence $\{\sin kx\}$ diverges in the space $L_2(0,\pi)$.

Bolzano–Weierstrass theorem is realized for linear normalized space whenever this space is finite dimensional. We shall consider, as a rule, infinite dimensional spaces. However, there exists an extension of the Bolzano–Weierstrass theorem to an important class of Banach spaces. But it uses a weaker form of the convergence.

2.3 Weak convergence in Banach spaces

Determine the weak form of the convergence for linear normalized spaces.

Definition 2.2 *A sequence $\{v_k\}$ of a linear normalized space V tends to v **weakly** if $\langle u, v_k \rangle \to \langle u, v \rangle$ for all $u \in V'$.*

Minimization of functionals. Addition   61

We denote this property by $v_k \to v$ weakly in V. If the space V is Hilbert the convergence $v_k \to v$ weakly in V is true whenever $(u, v_k) \to (u, v)$ for all $u \in V$ because of Rietz theorem.

We have

$$\left|\langle u, v_k \rangle - \langle u, v \rangle\right| = \left|\langle u, v_k - v \rangle\right| \le \|u\|\|v_k - v\|.$$

Then from the condition $v_k \to v$ it follows that $v_k \to v$ weakly in V, i.e., the weak convergence is the corollary of the strong one. However, the inverse assertion can be false.

Example 2.9 Consider the sequence $v_k = \sin kx$, $k = 1, 2, \ldots$ of the space $L_2(0, \pi)$ (see Example 2.8). This space is self-adjoint. For any function $u \in L_2(0, \pi)$ we have the equality

$$u = \sum_{k=1}^{\infty} a_k \sin kx.$$

Determine its *Fourier coefficients*

$$a_k = \frac{2}{\pi} \int_0^\pi u(x) \sin kx dx, \ k = 1, 2, \ldots.$$

The integral in the right side of this equality is $\langle u, v_k \rangle$. The sequence of the coefficients $\{a_k\}$ converges to zero because of the convergence of the Fourier series. Therefore, $\langle u, v_k \rangle \to 0$ for all functions $u \in L_2(0, \pi)$. It is obvious, that the zero element 0 of the space $L_2(0, \pi)$ satisfies the equality $0 = \langle u, 0 \rangle$ for all $u \in L_2(0, \pi)$. Hence, we have the convergence $\langle u, v_k \rangle \to \langle u, 0 \rangle$; and $v_k \to 0$ weakly in $L_2(0, \pi)$. Determine the norm of the difference

$$\|v_k - 0\|^2 = \int_0^\pi \sin^2 kx dx = \frac{\pi}{2}.$$

Then the strong convergence $v_k \to 0$ in $L_2(0, \pi)$ is false. Thus, the weak convergence is weaker than the strong one.

Using Rietz theorem, we have the definition of the weak convergence for Hilbert spaces. A sequence $\{v_k\}$ of a Hilbert space V converges to v weakly if $(u, v_k) \to (u, v)$ for all $u \in V$.

We consider also an additional class of the convergence. Let the space V be adjoint to a linear normalized space W, i.e., $W' = V$.

Definition 2.3 *A sequence $\{v_k\}$ of V converges to v *-**weakly** if $\langle v_k, w \rangle \to \langle v, w \rangle$ for all $w \in W$.*

62   *Optimization and Differentiation*

Remark 2.6 If the space V is reflexive, i.e., it is equal to its adjoint space, then $W = V'$. Therefore, the weak convergence is equal to $*$-weak one. The $*$-weak convergence follows from weak convergence for the non-reflexive case.

Remark 2.7 The convergence is determined by **topological properties**. The **strong topology** of the linear normalized space characterizes the convergence that is determined by the norm. The weak and $*$-weak convergence are related to the **weak** and **$*$-weak topologies** of the space.

If we have the convergence $u_k \to u$ strongly in V' and $v_k \to v$ weakly in V, where V is a reflexive Banach space, then $\langle u_k, v_k \rangle \to \langle u, v \rangle$. Particularly, if we have the convergence $u_k \to u$ strongly in V and $v_k \to v$ weakly in V, where V is a Hilbert space, then $(u_k, v_k) \to (u, v)$. If we have the convergence $w_k \to w$ strongly in W and $v_k \to v$ $*$-weakly in $V = W'$, where W is a Banach space, then $\langle v_k, w_k \rangle \to \langle v, w \rangle$.

Now we consider an extension of the Bolzano–Weierstrass theorem. By the **Banach–Alaoglu theorem** for all bounded sequences of a reflexive Banach space (space that is adjoint to a Banach space) there exists a weak ($*$-weak) convergent subsequence.

Remember that the spaces $L_p(\Omega)$ of integrable functions and Sobolev spaces $W_p^m(\Omega)$, $\dot{W}_p^m(\Omega)$, $H^m(\Omega)$, and $H_0^m(\Omega)$ for $1 < p < \infty$ are reflexive. The space $L_\infty(\Omega)$ is not reflexive; this space is adjoint to Banach space $L_1(\Omega)$. The spaces of continuous, continuously differentiable functions, and $L_1(\Omega)$ are not reflexive; moreover, these spaces are not adjoint to Banach spaces. The Banach–Alaoglu theorem is not applicable for it.

This result is used for proving the solvability of extremum problems for infinite dimensional systems. Now we determine the weak analog of closeness. A subset U of a linear normalized space V is called **weak** (**$*$-weak**) **closed** if for all sequence $\{v_k\}$ of U such that $v_k \to v$ weakly ($*$-weakly) in V we have $v \in U$.

Remark 2.8 The weak and $*$-weak closeness are equivalent for the reflexive spaces. However, $*$-weak closeness is weaker for the general case. The weak closeness is equal to the natural, i.e., strong, closeness for the finite dimensional case.

The quantity of the weakly convergent (especially, $*$-weakly convergent) sequences of the infinite dimensional spaces is greater than the quantity of the strongly convergent. Therefore, each weakly and $*$-weakly closed sets are strongly closed, but the inverse proposition is false, as a rule. However, each convex closed set of reflexive Banach spaces is weakly closed by the **Mazur lemma**. Besides, each convex closed set of space that is adjoint to a Banach is $*$-weakly closed.

We use also additional properties of the functionals. These are semicontinuity properties. The **lower limit** $\inf \lim\limits_{k \to \infty} x_k$ of the numerical sequence $\{x_k\}$ is the lower bound of the limits of all subsequences from $\{x_k\}$.

Definition 2.4 *The functional I on a linear normalized space V is called* **lower semicontinuous** (**weakly lower semicontinuous**, **$*$-weakly lower**

semicontinuous) inf $\lim_{k\to\infty} I(v_k) \geq I(v)$ *whenever* $v_k \to v$ *strongly (weakly, ∗-weakly) in V*.

Of course, each continuous functional is semicontinuous, but the inverse proposition is false (see Figure 2.2). Each convex continuous functional on a reflexive Banach space is weakly lower semicontinuous because of the Mazur lemma. Each convex continuous functional on a space that is adjoint to a Banach space is ∗-weakly lower semicontinuous. The norm is the weakly lower semicontinuous functional, i.e., inf $\lim_{k\to\infty} \|v_k\| \geq \|v\|$ whenever $v_k \to v$ weakly.

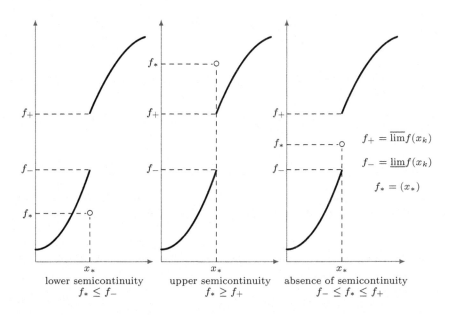

FIGURE 2.2: Semicontinuity of functions.

2.4 Existence of the functional minimum

Consider the following example.

Example 2.10 *The smooth curve of the minimal length*. Find the curve of the minimal length that passes through the given points A,B,C, where C is not a point of the interval AB. The broken line ACB is not admissible because this is not a smooth curve. But this broken line can be approximated by a smooth curve (see Figure 2.3). Therefore, the lower bound of our functional (the length of the curve) is equal to the sum of the length of the interval AC and CB. However, the problem of finding a smooth curve of minimal length that passes through three points is insolvable.

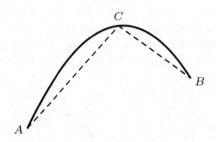

FIGURE 2.3: Smooth curve of the minimal length does not exist.

We shall consider the functionals on the extended numerical line

$$\overline{\mathbb{R}} = \mathbb{R} \cup \{-\infty\} \cup \{+\infty\}.$$

However, serious results can be obtained for the **proper functionals** that are equal to $-\infty$ nowhere and equal to $+\infty$ identically.

Theorem 2.4 *Each proper $*$-weak lower semicontinuous functional on a nonempty convex closed bounded subset of a space that is adjoint to a Banach space has a minimum. If the functional is coercive, then the boundedness of the set is not obligatory.*

Proof. Consider a functional I on a subset U of a space V. Let $\{v_k\}$ be a minimizing sequence, i.e., $v_k \in U$, $k = 1, 2, \ldots$, $I(v_k) \to \inf I(U)$. We do not know for a while yet the inequality of the value $\inf I(U)$ to $-\infty$. Using the boundedness of the set U or the coercivity of the functional I (see the proof of Theorem 2.3), we prove the boundedness of the sequence $\{v_k\}$. Then we use the Banach–Alaoglu theorem and extract a subsequence (we save the initial denotation for it) such that $v_k \to v$ $*$-weakly in V. Using the Mazur lemma, we have the $*$-weak closeness of the set U, and the inclusion $v \in U$. Therefore, we get $\inf \lim\limits_{k \to \infty} I(v_k) \geq I(v)$ because of the *-weak lower semicontinuity of the functional I. The left side of this inequality is equal to $\inf I(U)$. Hence, we have $I(v) \leq \inf I(U)$. The lower bound of the functional I on the set U cannot be greater than the value of its value at the arbitrary point of this set. Then we have the equality $I(v) = \inf I(U)$. Our functional is proper. Therefore, we obtain the inequality $\inf I(U) > -\infty$. Thus, there exists a point of the set U such that the value of the functional at this point is equal to its lower bound.
□

Minimization of functionals. Addition 65

Remark 2.9 If the space V is reflexive, then we change all $*$-weak notions by its weak analogs.

Remark 2.10 We use here the completeness of the space that guarantee the applicability of the Banach–Alaoglu Theorem.

Apply Theorem 2.4 for proving the solvability of the considered extremum problems.

Example 2.11 *Dirichlet integral.* Consider the minimization problem for the functional

$$I(v) = \int_\Omega \left[\sum_{i=1}^n \left(\frac{\partial v(x)}{\partial x_i} \right)^2 + 2v(x)f(x) \right] dx$$

on the space $H_0^1(\Omega)$. This functional is the sum of the square of the norm of this space and the linear continuous functional determined by the second term under the integral. We proved before its continuity, coercivity, and convexity. Therefore, this problem has a solution because of Theorem 2.4.

Example 2.12 *Square form.* Consider the minimization problem for the functional

$$I(v) = \langle Av, v \rangle - 2\langle f, v \rangle$$

on a Hilbert space V. The point $f \in V'$ is given here, and $A : V \to V'$ is a linear continuous self-adjoint operator such that $\langle Av, v \rangle \geq \alpha \|v\|^2$ for all $v \in V$, where α is a positive constant. Using the inequality

$$|\langle f, v \rangle| \leq \|f\| \|v\| \ \ \forall f \in V', v \in V,$$

we get

$$I(v) \geq \|v\| \big(\alpha \|v\| - \|f\| \big) \ \ \forall v \in V.$$

If $\|v\| \to \infty$, then $I(v) \to \infty$, i.e., our functional is coercive. Then all conditions of Theorem 2.4 are true. Thus, this minimization problem is solvable.

Remark 2.11 We know that the last minimization problem is equivalent to the operator equation $Au = f$. Hence, for all $f \in V'$ this equation is solvable on the space V.

Example 2.13 *Minimization of the distance.* Consider the minimization problem for the functional $I(v) = \|v - z\|^2$ on a convex subset U of a Hilbert space V, where $z \in V$. This functional is convex, continuous, and coercive. Therefore, this problem is solvable because of Theorem 2.4.

Example 2.14 *Lagrange Problem.* Minimize the functional

$$I(v) = \int_{x_1}^{x_2} F\big[x, v(x), v'(x)\big] dx$$

66   *Optimization and Differentiation*

on a set of functions that satisfy the boundary condition $v(x_1) = v_1$, $v(x_2) = v_2$. This functional is not convex for the general case. However, we considered before the functional

$$I(v) = \int_0^\pi \left[v(x)^2 + v'(x)^2 - 2v(x)\sin x - 2v'(x)\cos x \right] dx$$

with homogeneous boundary conditions $v(0) = 0$, $v(\pi) = 0$. Determine the space $V = H_0^1(\Omega)$. Transform this functional (see Example 2.3) to

$$I(v) = \int_0^\pi [v(x) - \sin x]^2 dx + \int_0^\pi [v('x) - \cos x]^2 dx - \pi.$$

This is the sum of two square and one constant functionals. Therefore, Theorem 2.4 is applicable and this problem has a solution.

2.5   Uniqueness of the functional minimum

Consider an easy uniqueness theorem for the solution of the extremum problem.

Theorem 2.5 *Under the conditions of Theorem 2.4 the minimization problem has a unique solution whenever the functional is strictly convex.*

Proof. Suppose there exist two solutions u and v of the problem, i.e.,

$$I(u) = I(v) = \min I(U).$$

Then the point $w = (u + v)/2$ is the element of the set U by the convexity of this set. Using the strict convexity of the functional, we get

$$I(w) = I\left(\frac{1}{2}u + \frac{1}{2}v\right) < \frac{1}{2}I(u) + \frac{1}{2}I(v) = \min I(U).$$

Therefore, the value of the functional I at the point w of the set U is less than the minimum of this functional on the given set. This contradiction completes the proof of Theorem 2.5. $\square$

Remark 2.12 If the functional is convex only, it can have a non-unique minimum (see, for example, Figure 2.4 a). It can have an infinite set of minimum points even. Indeed, if we have $I(u) = I(v) = \min I(U)$ and for $\sigma \in [0, 1]$ the equality $I[(1-\sigma)u+\sigma v] = (1-\sigma)I(u)+\sigma I(v)$ holds, then the element $(1 - \sigma)u + \sigma v$ minimizes this functional too.

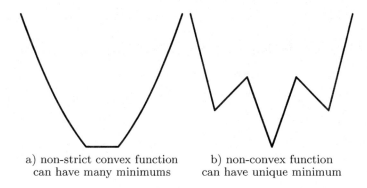

a) non-strict convex function can have many minimums
b) non-convex function can have unique minimum

FIGURE 2.4: Strict convexity and the uniqueness of minimum.

Remark 2.13 The non-convex functional can have a unique point of minimum (see Figure 2.4 b). Note that the function of this figure is strictly convex on a neighborhood of the point of minimum.

Apply these results for the analysis of the uniqueness of the solution of extremum problems.

Example 2.15 *Minimization of functions.* We proved the existence of minimum of the functions

$$f_1(x) = x^2, \ f_2(x) = 3x^4 - 8x^3 - 6x^2 + 24x, \ f_3(x) = x^4 - 2x^2, \ f_6(x) = |x|$$

on the set of real numbers by using Theorem 2.4. Determine additional results. The functions f_1 and f_6 are strictly convex; the uniqueness of its minimum is the corollary of Theorem 2.5. The functions f_2 and f_3 are not convex. Therefore, this theorem is not applicable for it. In reality the function f_3 has two point of minimum (see Figure 1.5). We have the breach of the condition of the uniqueness theorem and its assertion. However, the function f_2 has a unique point of minimum (see Figure 1.2). Hence, a non-convex function can have a unique point of minimum.

Remark 2.14 Theorem 2.5 gives only sufficient conditions of uniqueness of the solution of the extremum problem.

The function of two variables of Example 2.6 is strictly convex. Therefore, the problem of its minimization on the given subspace has a unique solution. We can obtain analogical results for the minimization problem of Example 2.11–2.13.

Remark 2.15 The minimization problem for the square functional and linear operator equation $Au = f$ is equivalent. Therefore, for any $f \in V'$ this equation has a unique solution $u \in V$. We shall consider optimal control problems for systems described by linear operator equations (see next chapter). We determine its single-valued solvability by using this result.

2.6 Tihonov well-posedness of extremum problems

Single-valued solvable extremum problems can have strange enough properties.

Example 2.16 *Ill-posed problem.*Consider the minimization problem for the functional

$$I(v) = \int_0^\pi v^2 dx$$

on the space $V = H_0^1(0, \pi)$. It has a unique solution. This is the function v_0 that is equal to zero as, $I(v_0) = 0$. Determine the sequence

$$v_k(x) = k^{-1} \sin kx, \ k = 1, 2, \dots .$$

We have the convergence $I(v_k) \to 0$. Therefore, this is the minimizing sequence. However, we have the equality

$$\|v_k - v_0\|^2 = \int_0^\pi \left(\frac{dv_k}{dx}\right)^2 dx = \int_0^\pi \cos^2 kx dx = \frac{\pi}{2}.$$

Thus, the minimizing sequence does not converge to the optimal control. This is the property of ill-posed problems.

Definition 2.5 *The extremum problem is called **Tihonov well-posed** if it has a unique solution, and each minimizing sequence converges to this solution.*

The optimization problem of Example 2.16 is **ill-posed**. Tihinov well-posedness is important for characterizing the approximate solution of the problem. An element u_* of the set U is called the **approximate solution** of the minimization problem for the functional I on this set if it is sufficiently close to the solution of this problem. However, we could define another form of the approximate solution. This is an element u_* of the set U such that the value $I(u_*)$ is close enough to the lower bound of this functional on the given set. If the functional is continuous, then values of the functional are sufficiently close whenever this property is true for its arguments. Therefore, the second form of the approximate solution is a corollary of its first form. If the problem is Tihonov well-posed, then the inverse property is true too; and these forms are equivalent. However, this is false for the ill-posed problems. Particularly, the values $I(v_k)$ and $I(v_k)$ are sufficiently close for close enough numbers k. But the functions v_k and v_0 are not near in the sense of the norm of the space V. Hence, finding of the functional minimum and the points of its minimum is not equivalent for ill-posed problems.

Minimization of functionals. Addition

Remark 2.16 We shall consider approximate solutions of extremum problems in the next section (see also final section of this chapter).

Remark 2.17 Ill-posed problems are weakly sensitive with respect to the argument of the functional. The small change of the functional can be the corollary of the small change of its argument.

Remark 2.18 If the extremum problem does not have any solutions, then the first form of the approximate solution is senseless. However, the lower bound of the functional can exist even for the case of the absence of the minimum point. Therefore, the second form of the approximate solution is admissible for the insolvable problems too. This is a basis of the extension methods of the extremum problems.

Remark 2.19 We could determine a weaker form of the approximate solution of the minimization problem for a functional I on a set U. The point u_* could be an approximate solution of this problem if it is sufficiently close to an element of the set U, and the value $I(u_*)$ is sufficiently close to the lower bound of this functional on the given set. The previous forms of the approximate solution satisfy exactly the given constraints. Now the small enough breach of this condition is admissible.

Determine the sufficient condition of the Tihonov well-posedness problem.

Definition 2.6 *A functional I on a convex set U of a linear normalized space is called* **strictly uniformly convex** *if there exists a continuous function $\delta = \delta(\tau)$ such that $\delta(0) = 0$, $\delta(\tau) > 0$ for $\tau > 0$, and $\delta(\tau) \to \infty$ as $\tau \to \infty$, and for all $u, v \in U$, $\sigma \in [0, 1]$ the following inequality holds:*

$$I[(1 - \sigma)u + \sigma v] \leq (1 - \sigma)I(u) + \sigma I(v) - \sigma(1 - \sigma)\delta(\|u - v\|).$$

Each strict uniformly convex functional is strictly convex. The sum of the convex and strictly uniformly convex functional is strictly uniformly convex.

Theorem 2.6 *Under the conditions of Theorem 2.4 the extremum problem is the Tihonov well-posedness problem whenever the minimizing functional is strictly uniformly convex.*

Proof. The considered problem has a unique solution u by Theorem 2.5. Consider an arbitrary minimizing sequence $\{u_k\}$. We have

$$I[(1 - \sigma)u + \sigma u_k] \leq (1 - \sigma)I(u) + \sigma I(u_k) - \sigma(1 - \sigma)\delta(\|u - u_k\|).$$

Then

$$I[(1 - \sigma)u + \sigma u_k] - I(u) \leq \sigma[I(u_k) - I(u)] - \sigma(1 - \sigma)\delta(\|u_k - u\|).$$

The left-hand side of this inequality is not negative because of the optimality of the control u. Therefore, we get

$$(1 - \sigma)\delta(\|u_k - u\|) \leq \sigma[I(u_k) - I(u)].$$

Passing to the limit as $\sigma \to 0$, we obtain the inequality

$$0 \leq \delta(\|u_k - u\|) \leq \sigma[I(u_k) - I(u)].$$

The sequence $\{u_k\}$ is minimizing. Hence, we have the convergence

$$\delta(\|u_k - u\|) \to 0.$$

70   *Optimization and Differentiation*

The function δ of the non-negative argument can be equal to zero for the zero value of its argument only. Then $\|u_k - u\| \to 0$ and $u_k \to u$. Therefore, each minimizing sequence converges to the solution of the problem. Hence, this problem is Tihonov well-posed. $\square$

Remark 2.20 The functional of Example 2.16 is strictly convex. However, this is not uniformly strictly convex.

Remark 2.21 Theorem 2.6 gives the sufficient condition of the well-posedness. The problem can be Tihonov well-posed with the absence of the uniform strict convexity of the minimizing functional.

Example 2.17 *Well-posed problem.* Consider the problem of the minimization for the functional

$$I(v) = \int\limits_0^\pi \left(\frac{dv}{dx}\right)^2 dx$$

on the space $V = H_0^1(0, \pi)$. We have the equality

$$I[\sigma u + (1 - \sigma)v] = \|\sigma u + (1 - \sigma)v\|^2 =$$

$$\sigma^2\|u\|^2 + 2\sigma(1 - \sigma)(u, v) + (1 - \sigma)^2\|v\|^2 \quad \forall u, v \in U, \sigma \in (0, 1)$$

with norm and scalar product of the space V. Then we get the equality

$$I[\sigma u + (1 - \sigma)v] = \sigma I(u) + (1 - \sigma)I(v) - \sigma(1 - \sigma)\|u - v\|^2.$$

Thus, we have the condition of the strict uniform convexity in the form of the equality with function $\delta(\tau) = \tau^2$. Therefore, the well-posedness of the problem follows from Theorem 2.6.

Remark 2.22 We shall apply Theorem 2.6 for proving of the Tihonov well-posedness of optimal control problems for linear systems (see next section and Section 2.8).

Remark 2.23 The analysis of the ill-posed problem can be based on the *Tihonov regularization method.* Consider the functional

$$I_\varepsilon(v) = I(v) + \varepsilon\|v\|^2,$$

where $\varepsilon > 0$ is a regularization parameter. The regularization functional is strictly uniformly convex for our example because this is the sum of the convex functional I and strictly uniformly convex second term (see Example 2.17). Besides, the initial and regularization functionals are sufficiently nearest for the small enough value of the parameter ε. Thus, we could analyze the initial ill-posed problem by using a nearest well-posed problem.

2.7   Hadamard well-posedness of extremum problems

Consider another class of the well-posedness of the extremum problems.

$$\textit{Minimization of functionals. Addition} \qquad 71$$

Example 2.18 *Ill-posed problem.* Consider the minimization problem for the functional

$$I_\mu(v) = \int\limits_0^\pi [v - \mu \sin(x/\mu)]^2 dx$$

on the space $V = H_0^1(0,\pi)$, where $\mu = \mu_k = 1/k$ is a parameter, $k = 1,2,\dots$. This problem has a unique solution $u(\mu) = \mu \sin(x/\mu)$. We have the convergence $\mu_k \to \mu_\infty$ as $k \to \infty$, where $\mu_\infty = 0$. Determine $\mu = \mu_k$ and $k \to \infty$. For any $v \in V$ we have the convergence $I_\mu(v) \to I_\infty(v)$, where

$$I_\infty(v) = \int\limits_0^\pi v^2 dx.$$

The minimization problem of the functional I_∞ on the space V has a unique solution $u(\mu_\infty) = 0$. Thus, the parameters μ_k and μ_∞ are sufficiently nearest for the large enough value k. However, the solutions $u(\mu_k)$ and $u(\mu_\infty)$ are not close enough in the sense of the space V. Therefore, this extremum problem is not well-posed.

Definition 2.7 *The minimization problem for the functional I_μ on a set U with parameter μ of a set M is called **Hadamard well-posed** if for all $\mu \in M$ it has a unique solution $u(\mu)$, as $\mu \to u(\mu)$ is the continuous mapping from M to U.*

Remark 2.24 The extremum problem is Hadamard well-posed if it has a unique solution that is continuous with respect to the parameter of the problem. This is an analog of the well-known notion of well-posedness for mathematical physics problems. The small error of the parameter can be the cause of the large error of the solution for ill-posed problems. Therefore, the absence of the well-posedness can be a very serious difficulty for practical extremum problems.

Determine relations between Tihonov and Hadamard well-posedness.

Theorem 2.7 *If for all $\mu \in M$ the minimization problem for the functional I_μ on the set U is Tihonov well-posed, and the map $\mu \to I_\mu(u)$ is continuous uniformly with respect to $u \in U$, then this problem is Hadamard well-posed.*

Proof. Let us have the convergence $\mu_k \to \mu_\infty$ in M as $k \to \infty$. Denote by I^k the value of the functional I_μ for $\mu = \mu_k$ and u_k is its point of minimum on the set U. It exists because of the Tihonov well-posedness of the problem. We have

$$I^\infty(u_k) - I^\infty(u_\infty) = \left[I^\infty(u_k) - I^k(u_k)\right] + \left[I^k(u_k) - I^k(u_\infty)\right] + \left[I^k(u_\infty) - I^\infty(u_\infty)\right].$$

The point u_k minimizes the functional I^k on the set U. Then we have the inequality

$$0 \leq I^\infty(u_k) - I^\infty(u_\infty), \quad I^k(u_k) - I^k(u_\infty) < 0.$$

72   *Optimization and Differentiation*

We get

$$0 \leq I^\infty(u_k) - I^\infty(u_\infty) \leq \left| I^\infty(u_k) - I^k(u_k) \right| +$$

$$\left| I^k(u_\infty) - I^\infty(u_\infty) \right| \leq 2 \sup_{u \in U} \left| I^k(u) - I^\infty(u) \right|.$$

Using the convergence $I^k(u) \to I^\infty(u)$ uniformly with respect to $u \in U$ we determine that the right side of the last inequality tends to zero. Then we have the convergence $I^\infty(u_k) \to I^\infty(u_\infty)$, and the sequence $\{u_k\}$ is minimizing for the functional I^∞. Using Tihonov well-posedness of the problem, we get $u_k \to u_\infty$. Thus, we have the convergence of the problem solution whenever the convergence of the parameters sequence is true. Therefore, the extremum problem is Hadamard well-posed. $\square$

Example 2.19 *Well-posed problem*. Consider the minimization problem for the functional

$$I_\mu(v) = \int\limits_0^\pi \left(\frac{dv}{dx} - \frac{d\mu}{dx} \right)^2 dx$$

on a convex closed bounded set U of the space $V = H_0^1(0, \pi)$, where $\mu \in V$ is the parameter of the problem. We prove the Tihonov well-posedness of the minimization problem for functional I_μ on the set U (see Example 2.17). Let the parameters μ and ν be close in the sense of the norm of the space V. We have

$$\sup_{u \in U} \left| I_\mu(u) - I_\nu(u) \right| = \sup_{u \in U} \left| \|u - \mu\|^2 - |u - \nu|^2 \right| \leq \sup_{u \in U} \|2u - \mu - \nu\| \|\mu - \nu\|.$$

Using the closeness of the parameters μ and ν, and boundedness of the set U, we prove the smallness of the left side of the last inequality. Therefore, the given functional is continuous with respect to the parameter uniformly with respect to $u \in U$ Hadamard well-posedness of the problem that follows from Theorem 2.7.

Remark 2.25 We shall prove Hadamard well-posedness of an optimal control problem for a linear system.

Remark 2.26 Using the relation between Tihonov and Hadamard well-posedness, we can apply the Tihonov regularization method for Hadamard ill-posed problems too.

2.8   Ekeland principle and the approximate condition of extremum

Consider again Problem 1.3 of minimization of the functional I on a Banach space V. If this functional is lower bounded, then for any $\varepsilon > 0$ there

Minimization of functionals. Addition 73

exists a point $u_\varepsilon \in V$ such that

$$I(u_\varepsilon) \leq \inf I(U) + \varepsilon. \tag{2.2}$$

The point u_ε can be interpreted as an **approximate solution** of Problem 1.3. Its exactness depends on the smallness of the number ε.

Theorem 2.8 (**Ekeland principle**). *Let V be a space that is adjoint to a Banach, I be a lower semicontinuous lower bounded functional, and the point $u_\varepsilon \in V$ satisfies the inequality (2.2) for a constant $\varepsilon > 0$. Then for all $\alpha > 0$ there exists a point $v_\alpha \in V$ such that the following inequalities hold:*

$$\|v_\alpha - u_\varepsilon\| \leq \alpha, \tag{2.3}$$

$$I(v_\alpha) \leq I(u_\varepsilon), \tag{2.4}$$

$$I(v_\alpha) \leq I(v) + (\varepsilon/\alpha)\|v - v_\alpha\| \ \ \forall v \in V. \tag{2.5}$$

Proof. For all constant $\alpha > 0$ determine the functional

$$J_\alpha(v) = I(v) + (\varepsilon/\alpha)\|v - u_\varepsilon\|,$$

which is coercive. Then the minimization problem for the functional J_α on a space V has a solution v_α because of Theorem 2.4. Then we have the inequality

$$J_\alpha(v_\alpha) \leq J_\alpha(v) \ \forall v \in V.$$

Therefore, we get

$$\Big[I(v_\alpha) + (\varepsilon/\alpha)\|v_\alpha - u_\varepsilon\|\Big] \leq \Big[I(v) + (\varepsilon/\alpha)\|v - u_\varepsilon\|\Big] \ \forall v \in V.$$

Transform this inequality to

$$I(v_\alpha) \leq I(v) + (\varepsilon/\alpha)\Big[\|v - u_\varepsilon\| - \|v_\alpha - u_\varepsilon\|\Big] \ \forall v \in V. \tag{2.6}$$

Determine $v = v_\varepsilon$; we obtain

$$I(v_\alpha) + (\varepsilon/\alpha)\|v_\alpha - u_\varepsilon\| \leq I(u_\varepsilon).$$

Estimate the first term of the left-hand side of this inequality. We have the inequality (2.4). From (2.2) it follows also that

$$(\varepsilon/\alpha)\|v_\alpha - u_\varepsilon\| \leq I(u_\varepsilon) - I(v_\alpha) \leq \Big[\inf I(V) - I(v_\alpha)\Big] + \varepsilon.$$

The term under the square brackets is not positive here. Then we obtain the inequality (2.3). Using the estimate inequality

$$\|v - u_\varepsilon\| \leq \|v - v_\alpha\| + \|v_\alpha - u_\varepsilon\|$$

and (2.6), we get inequality (2.5). $\square$

By Theorem 2.5, if a point u_ε is an approximate solution of Problem 1.3, i.e., the inequality (2.2) holds, then for all its neighborhood there exists v_α, where the value of the given functional is not greater. Moreover, the inequality (2.5) is true. We can use it as an approximate solution of the problem. The Ekeland principle has many important corollaries. Note the following result.

74 *Optimization and Differentiation*

Theorem 2.9 *Under the conditions of Theorem 2.8 suppose the functional I is Gâteaux differentiable, and the point u_ε satisfies the inequality (2.2). Then there exists a point v_ε such that $\|u_\varepsilon - v_\varepsilon\| \le \sqrt{\varepsilon}$, $I(v_\varepsilon) \le I(u_\varepsilon)$, and the following inequality holds:*

$$\|I'(v_\varepsilon)\| \le \sqrt{\varepsilon}. \tag{2.7}$$

Proof. Consider the conditions (2.3)–(2.5) for $\alpha = \sqrt{\varepsilon}$. Denote the value v_α by v_ε; this point is close enough to u_ε and the value of the functional at this point is less than $I(u_\varepsilon)$. Return to the inequality (2.5). Determine $v = v_\varepsilon + \sigma h v_\alpha$, where $\sigma > 0$, $h \in V$. Therefore, we have the inequality

$$I(v_\varepsilon + \sigma h) - I(v_\varepsilon) \ge -\sqrt{\varepsilon}\|h\|.$$

Dividing by σ and passing to the limit, we get

$$\langle I'(v_\varepsilon), h \rangle \ge -\sqrt{\varepsilon}\|h\|.$$

Replace here the point h by $-h$. We have

$$-\langle I'(v_\varepsilon), h \rangle \ge -\sqrt{\varepsilon}\|h\|.$$

Therefore, we have the estimate

$$|\langle I'(v_\varepsilon), h \rangle| \le \sqrt{\varepsilon}\|h\|.$$

Using the definition of the norm for the adjoint space, we have

$$\|I'(v_\varepsilon)\| = \sup_{\|h\|=1} |\langle I'(v_\varepsilon), h \rangle| \le \sqrt{\varepsilon}.$$

Thus, the inequality (2.7) is true. $\square$

The inequality (2.7) can be interpreted as an approximate form of the stationary condition (2.2). By this result there exists a point v_ε such that the value of the functional here is close enough to the minimum of the functional, and the stationary condition holds with small enough error. If $\varepsilon \to 0$, then we have the convergence $I(\varepsilon) \to \inf I(V)$ and $I'(v_\varepsilon) \to 0$ in V'. Note that these results do not use the convexity of the functional. Hence these assertions can be true for the case of insolvability of the extremum problem.

Remark 2.27 The lower bound of the minimizing functional for the insolvable case can exist. For example, the linear function $f(x) = x$ on the set U of all positive numbers does not have any minimum. However, there exists a small enough positive number, i.e., an element of the set U such that the value of the function f at this point is close enough to its lower bound on this set, i.e., zero. Therefore, insolvable extremum problems have a sense. We could try to find its approximate solution that can be characterized by the inequality (2.2). The Ekeland principle could give the method of finding it.

Theorem 2.8 does not use the differentiability of the minimizing functional. Therefore, it can be applied for the minimization problems with non-smooth functionals. However, the analogical results can be obtained by using an extension of Gâteaux derivatives.

2.9 Non-smooth functionals and subdifferentiation

We use Gâteaux differentiability of the functionals for obtaining the stationary condition and the variational inequality. However, sometimes functionals are non-smooth; and Gâteaux derivatives are non-applicable. This is true, for example, for the function $f(x) = |x|$. However, there exist extensions of the functional derivatives.

Consider a Banach space V and a functional $I : V \to \overline{\mathbb{R}}$.

Definition 2.8 *The functional I is called **subdifferentiable** at a point $u \in V$ if there exists an affine continuous functional l on the space V such that*

$$l(v) \leq I(v) \ \forall v \in V, \ \ l(u) = I(u). \tag{2.8}$$

Each affine functional is the sum of linear and constant functionals. The previous formulas are true if and only if there exists an element $\lambda \in V'$ such that

$$l(v) = \langle \lambda, v - u \rangle + I(u) \ \forall v \in V.$$

This element is called the **subgradient** of the functional I at the point u, and the set $\partial I(u)$ of all subgradients of the given functional at this point is called the **subdifferential** of the functional I at the point u.

Particularly, the function f is subdifferentiable at a point y, if there exists an affine continuous functional l such that

$$l(x) \leq f(x) \ \forall x \in \mathbb{R}, \ \ l(y) = f(y). \tag{2.9}$$

Each affine functional l on a set of real numbers is the sum of linear and constant functionals; therefore, we have $l(x) = ax + c$, i.e., we have the line. By first condition (2.9), the graph of this function is not lower than the line l; additionally, the value and the value of the function f and this linear function at the point y are equal by its second condition (see Figure 2.5). The constant that is the angle coefficient of this line is the subgradient of f at the point y.

If the convex functional $I : V \to \overline{\mathbb{R}}$ is bounded and continuous at a point then it is subdifferentiable at this point. Each convex Gâteaux differentiable functional $I : V \to \overline{\mathbb{R}}$ at point $u \in V$ is subdifferentiable at this point, as its subdifferential has a unique point $I'(u)$ i.e., $\partial U(U) = \{I'(u)\}$. However, if the functional $I : V \to \overline{\mathbb{R}}$ is continuously bounded, and has a unique subgradient at a point then there exists its Gâteaux derivative at this point that is equal to this subgradient. Therefore, the differentiable convex functional has a unique subgradient. Consider examples.

Example 2.20 *Smooth function.* Let us have the function $f(x) = x^2$. It is convex; then this is everywhere a subdifferentiable function. Find its subgradiont. Second relation (2.9) is the equality. Then $=0$, i.e., the considered line

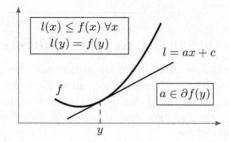

FIGURE 2.5: Subgradient a of the function f at the point y.

passes through the origin of the coordinates. Therefore, we have an interest in the lines that pass through the origin of the coordinates and do not place higher than the given parabola. These properties are true for the unique line $l(x) = 0$ that determine $a = 0$ (see Figure 2.6). Then we have the equality $\partial f(0) = \{0\}$. Note that $f'(0) = 0$. Thus, the subdifferential of the smooth function f has a unique point, in reality. This is the derivative of this function.

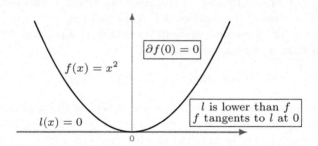

FIGURE 2.6: Subgradient of the square function at zero.

Example 2.21 *Non-smooth function.* Determine the subgradient of the function $f(x) = |x|$ at the zero point. Consider the line $l(x) = ax + c$ that passes through the coordinate's origin and places non-higher than the graph of the function f. The condition $ax \leq |x|$ for all $x \in R$ is true for all line $l(x) = ax$ with $a \in [0, 1]$ (see Figure 2.7). Therefore, the considered function is subdifferentiable at zero; besides, $\partial f(0) = [-1, 1]$. Note that the absence of subgradient uniqueness is realized because of the non-differentiability of this function at the given point.

Determine a condition of minimum for the subdifferentiable functionals.

Minimization of functionals. Addition 77

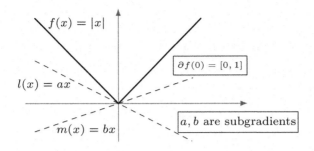

FIGURE 2.7: Subgradients of the function $f(x) = |x|$ at zero.

Theorem 2.10 *If the functional I is subdifferentiable at a point u, then it has the minimum at this point if and only if the following condition holds:*

$$0 \in \partial I(u). \tag{2.10}$$

Proof. If the functional has the minimum in the space V at the point u, then $I(u) \leq I(v)$ for all $v \in V$. Determine the affine functional l by the equality $l(v) = \langle 0, v - u \rangle + I(u)$. Therefore, we obtain the relations (2.8), and 0 is the subgradient of this functional at the point u. Now suppose the condition (2.10) is true. Then, the affine functional l that determines the subgradient is constant and equal to $I(u)$. By first condition (2.8) we get the inequality $l(v) \leq I(v)$ for all $v \in V$. It can be transformed to $I(u) \leq I(v)$ for all $v \in V$. Hence, the functional I has the minimum at the point u. We know that the subgradient of the convex Gâteaux differentiable functional is unique and equal to the Gâteaux derivative. Then the condition (2.10) can be transformed to the equality $I'(u) = 0$ that is the stationary condition. This is the necessary and sufficient condition of minimum of convex Gâteaux differentiable functionals on the space V by Theorem 2.8. □

Thus, we can apply Theorem 2.10 for finding the extremum of non-smooth functionals.

Example 2.22 *Non-smooth function.* Consider the minimization problem for the non-smooth convex function $f(x) = |x|$. It has the derivative $f'(x) = -1$ for $x < 0$ and $f'(x) = 1$ for $x > 0$. Our function is not differentiable at the zero point. However, this subdifferentiable function is everywhere. Its subdifferential $\partial f(x)$ is a one-element set -1 for negative x, 1 for positive x, and the interval $[-1, 1]$ for $x = 0$. By the condition (2.10) zero belongs to the subdifferential at the point of minimum. It can be true for $x = 0$ only. Therefore, this is the unique solution of the given problem (see Figure 2.7).

The final result of this section is practical solving of extremum problems. It uses another form of functional derivative.

2.10 Fréchet differentiation of functionals

Consider another definition of the function derivative. The derivative of a function f at a point x is a number a such that

$$f(x + h) - f(x) = ah + \eta(x, h),$$

where $\eta(x, h)/h \to 0$ as $h \to 0$. Use the existence of the one-to-one correspondence between the set of real numbers a and the set of linear functions on $\mathbb{R}$. Therefore, a function f is differentiable at a point x if and only if there exists a linear mapping $h \to f'(x)h$ on $\mathbb{R}$ such that the difference $\eta(x, h)$ between the increment $f(x + h) - f(x)$ of the given function and the value of the linear function $f'(x)h$ at the point h has the higher order of smallness than h. Use this property as a basis of another form of the functional's derivatives. Let I be a functional on a Banach space V.

Definition 2.9 *An element $I'(u)$ is called the **Fréchet derivative** of the functional I at a point u if the following equality holds:*

$$I(u + h) = I(u) + \langle I'(u), h \rangle + \eta(u, h),$$

*where $\|h\|^{-1}|\eta(u, h)| \to 0$ as $\|h\| \to 0$. The functional I is called **Fréchet differentiable** at the point u here.*

Suppose there exists a Fréchet derivative of a functional I at a point u. Then we have the equality

$$I(u + \sigma h) - I(u) = \langle I'(u), \sigma h \rangle + \eta(u, \sigma h).$$

Dividing by σ and passing to the limit as $\sigma \to$, we have

$$\lim_{\sigma \to} \left[I(u + \sigma h) - I(u) \right]/\sigma = \langle I'(u), h \rangle.$$

Then $I'(u)$ is a Gâteaux derivative of this functional at the given point. Therefore, if there exists a Fréchet derivative of a functional at a concrete point, then there exists its Gâteaux derivative that is equal to a Fréchet derivative. The inverse proposition can be false. Particularly, a Fréchet differentiable functional is continuous. But, a Gâteaux differentiable functional can be discontinuous (see Example 1.13). However, if there exists the Gâteaux derivative $I'(v)$ of the functional I for all points v from a neighborhood of a point u, and mapping $v \to I'(v)$ is continuous at u, then there exists a Fréchet derivative of this functional at a point u that is equal to $I'(u)$.

Remark 2.28 We would like to determine the Gâteaux derivative for the concrete functional. We can calculate the relevant limit and check its linearity with respect to h. However, the definition of a Fréchet derivative is not constructive. We could check only if a concrete linear continuous functional is a Fréchet derivative of the given functional or not. Therefore, sometimes we determine a Gâteaux derivative and analyze its properties. A Gâteaux derivative with additional properties is a Fréchet derivative.

Minimization of functionals. Addition

Consider examples.

Example 2.23 *Dirichlet integral.* Consider Dirichlet integral

$$I(v) = \int_\Omega \Big[\sum_{i=1}^n \Big(\frac{\partial v(x)}{\partial x_i} \Big)^2 + 2v(x)f(x) \Big] dx$$

on the Sobolev space $H_0^1(\Omega)$, where Ω is an open bounded set of n-dimensional Euclid space R^n. It has Gâteaux derivative $I'(u) = 2(f - \Delta u)$. The Laplace operator is the continuous mapping $\Delta : H_0^1(\Omega) \to H^{-1}(\Omega)$. Therefore, we obtain the continuity of mapping $v \to I'(v)$. Hence, the given functional is Fréchet differentiable with equality Gâteaux and Fréchet derivatives. Indeed, we have

$$I(u+h) - I(u) - \langle I'(u), h \rangle = \|u+h\|^2 = 2\langle f, u+h \rangle - \|u\|^2 -$$
$$2\langle f, u \rangle = 2(u, h) + \|h\|^2 + 2\langle f, h \rangle.$$

We use the norm and the scalar product of the space $H_0^1(\Omega)$ here. Using Green's formula

$$(u, h) = -\langle \Delta u, h \rangle \ \forall u, h \in H_0^1(\Omega),$$

transform the previous equality to the formula

$$I(u+h) - I(u) - \langle I'(u), h \rangle = \|h\|^2.$$

Thus, $I'(u)$ is a Fréchet derivative, in reality.

Remark 2.29 We shall use these properties of the Laplace operator in the next section. We shall consider control systems that are determined by the Laplace operator.

Example 2.24 *Function of two variables.* Determine the function of two variables by the equalities

$$f(x, y) = \frac{y(x^2 + y^2)}{x}, \ x \neq 0; \ f(0, y) = 0.$$

We know (see Example 1.13) that mapping $f : R^2 \to R$ is Gâteaux differentiable at the origin of the coordinates, and its Gâteaux derivative is equal to zero. Suppose there exists its Fréchet derivative at this point. Then it is equal to the Gâteaux derivative. Therefore, we get the equality

$$\eta(0, h) = f(h) - f(0) - \langle f'(0), h \rangle = f(h),$$

where

$$f(h) \equiv f(h_1, h_2) = h_2\big[(h_1)^2 + (h_2)^2\big]/h_1, \ h_1 \neq 0; \ f(0, h_2) = 0.$$

Determine $h_1 = t^2$, $h_2 = t$. We have $\|h\| \to 0$ as $t \to 0$ as we get

$$\eta(0, h)\|h\|^{-1} = h_2\sqrt{(h_1)^2 + (h_2)^2}/h_1 = \sqrt{t^2 + 1}.$$

The value in the right side of this equality does not tend to zero. Therefore, the Gâteaux derivative is not a Fréchet derivative. Thus, our supposition about Fréchet differentiability of the given function at the zero point is false.

80 *Optimization and Differentiation*

Remark 2.30 This is natural, because our function is discontinuous at the zero point.

Apply Fréchet derivatives for approximate methods of solving extremum problems.

2.11 Approximate methods of functionals minimization

Consider Problem 1.3 of minimization of the functional I on the space V. We try to determine a sequence $\{v_k\}$ on the given space. Let us recall the previous iteration v_k. Determine the next iteration by the formula

$$v_{k+1} = v_k + \sigma_k h_k, \qquad (2.11)$$

where a positive number σ_k and an element $h_k \in V$ are chosen such that they guarantee the minimization of the functional.

Suppose the functional I is Fréchet differentiable, and V is a Hilbert space. The derivative of the functional is an element of the given space; it is determined by the equality

$$I(u + h) = I(u) + \big(I'(u), h\big) + \eta(u, h).$$

Using the equality (2.11), we have

$$I(v_{k+1}) = I\big(v_k + \sigma_k h_k\big) = I(v_k) + \big(I'(v_k), \sigma_k h_k\big) + \eta(v_k, \sigma_k h_k). \qquad (2.12)$$

Determine the value

$$h_k = -I'(v_k)\big\|I'(v_k)\big\|^{-1}. \qquad (2.13)$$

Put it in the formula (2.11); we get (see Figure 2.8)

$$v_{k+1} = v_k - \sigma_k\big\|I'(v_k)\big\|^{-1} I'(v_k).$$

Remark 2.31 If $I'(v_k) = 0$, then the point v_k satisfies the stationary condition. Maybe, this is the solution of the problem. Probably, the norm in the formula (2.13) is not equal to zero; this formula makes sense for this case.

Suppose the sequence of positive numbers $\{\sigma_k\}$ tends to zero. By the formula (2.13) the norm of y_k is equal to 1. Therefore, the sequence of the norms $\{\|\sigma_k h_k\|\}$ tends to zero. Hence, we have the convergence

$$\eta(v_k, \sigma_k h_k)/\sigma_k \to 0. \qquad (2.14)$$

Using the definition of the norm of a Hilbert space, we have

$$\big(I'(v_k), \sigma_k h_k\big) = -\sigma_k\Big(I'(v_k), I'(v_k)\|I'(v_k)\|^{-1}\Big) = -\sigma_k\|I'(v_k)\|.$$

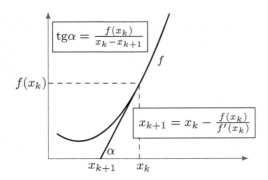

FIGURE 2.8: Gradient method for a function of one variable.

Then the second term of the right-hand side of the equality (2.12) is negative. By the condition (2.14) the third term of the equality (2.12) has the higher order of smallness than σ_k. Therefore, the number σ_k can be chosen so small that
$$\bigl(I'(v_k), \sigma_k h_k\bigr) + \eta(v_k, \sigma_k h_k) < 0.$$
Then $I(v_{k+1}) < I(v_k)$ because of the equality (2.12), i.e., the sequence of the functionals $\{I(v_k)\}$ decreases.

Remark 2.32 We cannot even guarantee the convergence of the algorithm to the solution of the extremum problem and decrease of the functional's sequence $\{I(v_k)\}$.

Considered algorithms differ by the choice of the sequence $\{\sigma_k\}$. These are **gradient methods** because they use the functional derivative that is also called a gradient.

Remark 2.33 Gradient methods make sense for the case of Gâteaux differentiability of the functional. We use the properties of Fréchet derivatives for proving the increase of the functionals only.

Remark 2.34 We could consider gradient methods if the space is not Hilbert. The derivative of the functional is an element of the adjoint space. Therefore, we cannot use the formula (2.13) directly. However, this problem is solvable.

Consider a conditional extremum problem. Try to minimize the functional I on a set U. We cannot use the formula (2.11) for this case because we do not guarantee that v_{k+1} is an element of the set U. However, we could apply the **projection gradient method** that is determined by the equality
$$v_{k+1} = P_U\bigl(v_k + \sigma_k h_k\bigr),$$
where P_U is the projector on the set U. The number σ_k and the element h_k are chosen such that they guarantee the inequality $I(v_{k+1}) < I(v_k)$. We can use, for example, the formula
$$v_{k+1} = P_U\bigl(v_k + \sigma_k I'(v_k)\bigr).$$

82 *Optimization and Differentiation*

Remark 2.35 The projection gradient method is effective enough for the minimization problem of the smooth functional on a convex subset of a Hilbert space.

We consider the general method of minimization of functionals. Then we analyze optimal control problems, where the minimizing functional depends on a control not only directly. It depends from state function that satisfies a state equation and depends on control.

2.12 Comments

The classic sufficient conditions of extremum of variations calculus were obtained by A.M. Legendre, K. Weierstrass, and others (see, for example, N.I. Ahiezer [7], G.A. Bliss [75], O. Bolza [82], G. Buttazzo, G. Mariano, and S. M. Hildebrandt [104], E. Elsgolts [164], I.M. Gelfand and S.V. Fomin [205], M.R. Hestens [241], M.A. Lavrentiev and L.A. Lusternik [300], G. Leitmann [307], L. Zlaf [570], and L. Young [563]. *Krotov method* gives the sufficient conditions of optimality for the optimal control problems (see [285], [286], and [287]). Sufficient conditions of optimality for the concrete problem are considered also by V.I. Blagodatskikh [74], [73], V.G. Boltyansky [77], V.N. Brandin [88], E. Casas and F. Tröltzsch [115], Yu.P. Krivenkov [284], Z. Nehari [382], S. Russak [440], S.Ya. Serovajsky [482], Z.S. Wu and K.L. Teo [557], and V.A. Yakubovich [559] (see also the review of D. Peterson and J.H. Zalkind [401]).

The weak convergence in Banach spaces is considered in the literature on functional analysis (see, for example, B. Baeuzamy [38], H. Brésis [91], J. Dieudonné [140], A. Friedmann [187], A. Gleason [213], P. Halmos [236], [236], V. Hutson and J.S. Pym [247], K. Iosida [256], L.V. Kantorovich and G.P. Akilov [267], A.N. Kolmogorov and S.V. Fomin [276], S.G. Krein and others [283], S.S. Kutateladze [291], L.A. Lusternik and V.I. Sobolev [334], M. Reed and B. Simon [426], W. Rudin [439], S.L. Sobolev [508], and V.A. Trenogin [530]).

The existence of the lower bound of the bounded numerical set was proved by B. Bolzano. It is well known as the classical *Weierstrass theorem* of the existence of the continuous function on the closed bounded set. The first example of the non-solvable variational problem was given by K. Weierstrass. The analogical examples were considered also by V.M. Alekseev, V.M. Tihomirov and S.V. Fomin [14], L. Cesari [116], [117], I. Ekeland and R. Temam [162], W.H. Fleming and R.W. Rishell [178], I.M. Gelfand and S.V. Fomin [205], C. Olech [389], L. Tonelli [527], and J. Warga [552]. The non-solvable optimal control problems described by ordinary differential equations were considered by A. Balakrishnan [41], J. Warga [552], S.Ya. Serovajsky [482], and A.D. Ioffe and V.M. Tihomirov [253]. Optimization problems for the systems with distributed parameters without the optimal control are analyzed by J. Baranger [47], M.F. Bidaut [70], J.L. Lions [315], K.A. Lurie [333], F. Murat [373], [374], M.B. Suryanarayana [514], and L. Tartar [515].

The first existence theorem for calculus of variations was proved by D. Hilbert. The existence Theorem 2.4 and uniqueness Theorem 2.5 are the classic results of the extremum theory (see, for example, A. Balakrishnan [41], I. Ekeland and R. Temam [162], A.D. Ioffe and V.M. Tihomirov [253], J.L. Lions [315], and F.P. Vasiliev [538], [539]). The solvability of the abstract extremum problems is considered also by I. Ekeland and R. Temam [162], J.L. Lions [315], J. Séa [450], F.P. Vasiliev [538], [539], and others. The solvability of the problem of the calculus of variations is considered by V.M. Alekseev, V.M. Tihomirov, and S.V. Fomin [14], M.F. Bidaut [70], F. Flores-Bazán [180], A.D. Ioffe and V.M. Tihomirov [253], L. Levaggi [310], B.T. Polyak [407], J.P. Raymond [424], L. Tonelli [527], F.P. Vasiliev [538], [539], J. Warga [552], [553], J. Yong [562], L. Young [563], and others.

Minimization of functionals. Addition 83

The existence theorems for the optimization problems for ordinary differential equations are proved by V.M. Alekseev, V.M. Tihomirov, and S.V. Fomin [14], A. Balakrishnan [41], L. Cesari [116], [118], [119], A.F. Filippov [177], R.V. Gamkrelidze [202], J.P. Gossez [220], A.D. Ioffe [250], C. Olech [389], A.M. Steinberg and H.L. Stalford [512], F.P. Vasiliev [538], [539], J. Warga [552], [553], J. Yong [562], J. Zabkzyk [565], and others. The solvability of the optimization problems for distributed parameter systems is considered by N.V. Andreev and V.S. Melnik [17], M.F. Bidaut [70], J. Baranger [47], J.P. Cesari [119], A.V. Fursikov [197], [194], [195], J.L. Lions [314], [315], [317], F. Murat [373], [374], V.I. Plotnikov [404], T. Seidman and H.X. Zhou [453], S.Ya. Serovajsky [482], [486], M.B. Suryanarayana [514], L. Tartar [515], T. Zolezzi [572], [574], and others. The solvability of the optimization problems for the integral equations is considered, for example, by E.J. Balder [43], D. Cowles [131], and C. Simionescu [502]. The existence of the solutions for the minimax problems is proved by A. Balakrishnan [40], I. Ekeland and R. Temam [162], and I.K. Gogodze [216]. The solvability of the multiextremal problems is analyzed by J.L. Lions [318] and R.T. Vescan [541]. A problem of the solvability for non-convex variational problems is considered by J.P. Raymond [424].

The existence theorems for the extremum problems use the semicontinuity of the functionals (see, for example, E.J. Balder [44], L. Cesari [117], B. Dacorogna [136], I. Ekeland and R. Temam [162], A.D. Ioffe [251], A.D. Ioffe and V.M. Tihomirov [253], S.F. Morozov and V.I. Plotnikov [371], M. Nacinovich and R. Schianchi [379], B.T. Polyak [407], C. Trombetti [533], and X.P. Zhou [568]). H. Lebesgue and J. Hadamard proved the first semicontinuity theorems for functionals.

The analysis of the non-solvable extremum problems can be realized by using *extension methods*. The idea of the extension methods was proposed by D. Hilbert. L. Young [563] and E. McShane [346] realized it for one-dimensional variational problems. The extension of the results to the optimal control problems is given by A.F. Filippov [177], R.V. Gamkrelidze [202], V.F. Krotov [285], [286], [287], J. Warga [552], [553], and others. Problems of the extension theory are considered also by I. Ekeland [160], I. Ekeland and R. Temam [162], P. Marcellini and C. Sbordone [339], T. Roubiček [434], J.E. Rubio and [438], C. Trombetti [533] for variational calculus problems; A.V. Arutunov [25], L. Cesari [117], [118], [119], V.I. Gurman [229], A.D. Ioffe and V.M. Tihomirov [252], T. Roubiček [434], E. Roxin [436], and others for the ordinary differential equations; N. Arada and J.P. Raymond [21], J. Baranger [47], M.F. Bidaut [69], I. Ekeland [158], I. Ekeland and R. Temam [162], K.A. Lurie [333], U. Raitums [422], S.Ya. Serovajsky [479], M.I. Sumin [513], L. Tartar [515], and other for the elliptic equations; Yu.G. Borisovich and V.V. Obuhovsky [85] for the parabolic equations; M.V. Suryanarayana [514] for the hyperbolic equations; T. Roubiček [435] for the integral equations.

Examples of non-unique solutions of the optimal control problems are given by R. Gabasov and F.M. Kirillova [199], J.L. Lions [315], T. Seidman and H.X. Zhou [453], and S.Ya. Serovajsky [482]. The uniqueness theorems for nonlinear extremum problems are proved by A.V. Fursikov [195], R. Lucchetti and F. Patrone [332], V.I. Plotnikov [404], and J.P. Raymond [424].

E. Asplund [29], J. Baranger [47], J. Baranger and R. Temam [48], M.F. Bidaut [69], L. Cesari [118], M. Edelstein [151], I. Ekeland [159], I. Ekeland and R. Temam [162], F. Flores-Bazán [180], C. J. Goh and X.Q. Yang [217], B.S. Mordukhovich and B. Wang [369], J.P. Raymond [424], and V. Zizler [569] consider optimal control problems with non-convex functionals.

The well-posedness of the extremum problems are determined by A.N. Tihonov [523], though there exists a relation between this well-posedness and the difference between the weak and the strong extremum that was determined by K. Weiershtrass. Example 2.16 and Theorem 2.6 are considered by V.P. Vasiliev [538] (see also S.Ya. Serovajsky [482]), and Theorem 2.7 is considered by S.Ya. Serovajsky [482]. Other examples of ill-posed extremum problems are given by W.H. Fleming and R.W. Rishell [178], and R.P. Fedorenko [172]. T. Zolezzi determines the criterion of the well-posedness of the optimization problems for the systems described by ordinary differential equations [572]. Tihonov well-posedness is considered also by I. Ekeland and R. Temam [162], G. Le Bourg [301], F.V. Lubyshev and M.E. Pairuzov [330], R. Lucchetti and F. Patrone [332], S.Ya. Serovajsky [482], A.N.

84 *Optimization and Differentiation*

Tihonov [523], A.N. Tihonov and V.Ya. Arsenin [524], A.N. Tihonov and V.P. Vasiliev [525], V.P. Vasiliev [538], [539], and T. Zolezzi [573], [574]. Hadamard well-posedness is analyzed by E. Asplund [29], J. Baranger [47], J. Baranger and R. Temam [48], M.L. Bennati [60], M.F. Bidaut [69], A. Cheng and K. Morris [124], I. Ekeland and R. Temam [162], G. Le Bourg [301], R. Lucchetti [331], R. Lucchetti and F. Patrone [332], N. S. Papageorgiou [395], T. Seidman and H.X. Zhou [453], C. Yu and J. Yu [564], and T. Zolezzi [572], [573]. X.X. Huang [246], A.D. Ioffe [250], J.P. Revalski and N.V. Zhivkov [427], and T. Zolezzi [574] consider other forms of well-posedness for the extremum problems. The relations between different forms of well-posedness are obtained by B. Lemaire, C.O.A. Salem, and J.P. Revalski [308], R. Lucchetti and F. Patrone [332], J.P. Penot [399], J.P. Revalski and N.V. Zhivkov [427], S.Ya. Serovajsky [482], and T. Zolezzi [572].

The regularization methods for extremum problems are proposed by A.N. Tihonov [523] (see also J. Baranger [47], M. Edelstein [151], H.W. Engl [166], R.P. Fedorenko [172], T. Kobayashi [275], J.L. Lions [315], [317], F.V. Lubyshev and M.E. Fairuzov [330], M.M. Potapov [410], S.Ya. Serovajsky [456], [464], A.N. Tihonov and V.Ya. Arsenin [524], A.N. Tihonov and F.P. Vasiliev [525], and F.P. Vasiliev [537], [538], [539]). Well-posedness of the extremum problems with non-convex functional is considered by E. Asplund [29], J. Baranger [47], J. Baranger and R. Temam [48], M.F. Bidaut [69], M. Edelstein [151], I. Ekeland and R. Temam [162], and V. Zizler [569]. X.X. Huang [246] analyzes the well-posedness for multiextremal problems.

Theorem 2.8 and 2.9 are proved by I. Ekeland [159]. Ekeland principle is considered also by A.V. Arutyunov, N. Bobylev, and S. Korovin [28], H. Attouch and G. Beer [30], J.P. Aubin and I. Ekeland [31], V.F. Demyanov and A.M. Rubinov [137], I. Ekeland [159], [161], I. Ekeland and R. Temam [162], J. Fu [191], T.X.D. Ha [233], and A.D. Ioffe and V.M. Tihomirov [254], P. Loridan [326], B.S. Mordukhovich and B. Wang [369], and J.H. Qiu [419]. D. Dentcheva and S. Helbig [138], A.D. Ioffe and A.J. Zaslavski [255], and A.D. Ioffe and V.M. Tihomirov [254] give modifications of Ekeland principle. B.S. Mordukhovich and B. Wang [369] describe a subdifferential variational principle. J.F. Bonnans and E. Casas [84], E. Casas and L.A. Fernández [112], and I. Ekeland [158] use Ekeland principle for the analysis of the optimal control problems for nonlinear infinite dimensional systems. X.X. Huang [246] applies a variational principle for the analysis of the well-posedness of vector optimization problems. The different forms of the approximate conditions of optimality are considered by J.P. Aubin and I. Ekeland [31], V.F. Demyanov and A.M. Rubinov [137], I. Ekeland and R. Temam [162], P. Loridan [326], B.S. Mordukhovich [366], and S.Ya. Serovajsky [458].

Non-smooth variational problems were considered by D. Bernulli. K. Weierstrass analyzed the non-smooth solutions of the variational problems. H. Minkowski gave the basis of the subdifferential calculus. The general theory of the subdifferentiation was obtained by J.J. Moreau [370], R.T. Rockafellar [432], and others. B.S. Mordukhovich and J.V. Outrata [367] determined the second order subdifferential. The application of the subdifferential calculus to the non-smooth extremum problems is considered by J.P. Aubin and I. Ekeland [31], V.F. Demyanov and A.M. Rubinov [137], I. Ekeland and R. Temam [162], B.N. Pshenichny [415], and R. Rockafellar [432]. Theorem 2.10 is given in the book by I. Ekeland and R. Temam [162].

Note the very important extension of the subgradient. This is the *Clarke derivative* [128]. The methods of non-smooth analysis and its applications in the extremum problems are considered also by E. Asplund [29], H. Frankowska [181], I.D. Georgieva [208], J.R. Giles and S. Sciffer [211], W. Huyer and A. Neumaier [248], W. Johnson, J. Lindenstrauss, D. Priess, and G. Schechtman [262], B.S. Mordukhovich and J.V. Outrata [367], A.A. Milutin [360], B.Sh. Mordukhovich [366], L. Moreau [370], A. Myslinski and J. Sokolowski [376], A. Nedic and D.P. Bertsekas [381], J.V. Outrata and W. Römisch [390], B.N. Pshenichny [415], S.Ya. Serovajsky [481], [485], [478], [480], J.S. Treiman [528], J.J. Ye [560], and others.

The approximation methods for optimization problems are described in the reviews of H.A. Spang [510], G. Zoutendijk [575], and F.L. Chernousko and V.V. Kolmanovsky [126]. These problems are considered in the books of A. Auslender [33], A.E. Bryson and Ho Yu-Chi [95], F.L. Chernousko and N.V. Banichuk [125], R.P. Fedorenko [172], R. Fletcher [179], N.N. Moiseev [365], E. Polak [406], J. Séa [450], F.P. Vasiliev [538], [539] (see also

Minimization of functionals. Addition 85

A. Balakrishnan [40], [41], P. Burgmeier and H. Jasinski [99], J. Cullum [133], C. Cuvelier [134], G. Fraser-Andrews [184], P.E. Gill, W. Murrey and M.H. Wright [212], V.K. Isaev and V.V. Sonin [257], H.J. Kelley, R.E. Kopp, and H.G. Moyer [269], I.A. Krylov and F.L. Chernousko [288], R. McGill [345], C. Meyer and A. Rösch [352], B.T. Polyak [407], S.Ya. Serovajsky [476], and M. Sidar [501]). J.W. He and R. Glowinski [238], A. Sage and P. Chaudhuri [443], S.Ya. Serovajsky [472], and Z.S. Wu and K.L. Teo [556], [557] use gradient methods for the analysis of the infinite dimensional optimal control problems.

Part II

Stationary systems

Chapter 3

Linear systems

3.1	Additional results of functional analysis	90
3.2	Abstract linear control systems	92
3.3	Dirichlet problem for the Poisson equation	98
3.4	Optimal control problem statement	102
3.5	Necessary conditions of optimality	104
3.6	Optimal control problems with local constraints	107
3.7	Lagrange multipliers method	110
3.8	Maximum principle	113
3.9	Penalty method	115
3.10	Additional properties of Sobolev spaces	117
3.11	Nonhomogeneous Dirichlet problem for the elliptic equation ...	121
3.12	Comments	125

We considered at first the problem of minimization of a function on one variable $f = f(\sigma)$ on the set of real numbers. Suppose this function is differentiable. If τ is a point of its minimum, then $f'(\tau) = 0$. Then we had the minimization problem for a functional I on a Banach space V. The point u is its solution whenever the function of one variable $f = f(\sigma) = I(u + \sigma h)$ has the minimum at the zero point for all $h \in V$. If the functional I is Gâteaux differentiable at a point u, then we have the *stationary condition*

$$I'(u) = 0 \tag{3.1}$$

because of the necessary condition of minimum for the function f.

Consider the problem of minimization of the functional I on a convex subset U of a Banach space V. If u is a solution of this problem, then we have the inequality $I[(u + \sigma(v - u)] - I(u) \geq 0$ for all $v \in U$. Dividing this inequality by the number σ and passing to the limit as $\sigma \to 0$, we have the *variational inequality*

$$\langle I'(u), v - u \rangle \geq 0 \ \forall v \in U. \tag{3.2}$$

However, the minimizing functional depends, as a rule, on the unknown function, i.e., control, non-directly. We have the dependence $I = I(v, y[v])$, where state function $y[v]$ satisfies a state equation with a control v as a parameter. We could apply here the previous technique. However, we have some difficulties. This is the calculation of the functional derivative. The general difficulty is the analysis of mapping $v \to y[v]$ that is determined by the state

89

90   *Optimization and Differentiation*

equation. However, there exists a class equation such that the desired result can be obtained without serious difficulties. These are linear state equations; the dependence $y = y[v]$ is affine for it. The necessary conditions of optimality are the corollaries of the results of Chapter 1 here.

The general subject of this section is optimal control problems for linear systems in infinite dimensional spaces. We obtain necessary and sufficient optimality conditions. The control system described by the Poisson equation is an application of these results. This is extended to the general elliptic equation and the nonhomogeneous boundary problem. The nonlinear control systems are the subject of the next chapter.

3.1   Additional results of functional analysis

Let us have Banach spaces V and Y. The set $V \times Y$ of pairs

$$\{(v,y)|\ v \in V,\ y \in Y\}$$

is called the **product** of the sets V and Y. Particularly, Euclid plane $\mathbb{R}^2$ is the product $\mathbb{R} \times \mathbb{R}$, and n-dimensional Euclid space $\mathbb{R}^n$ is the n degree product of the set of real number $\mathbb{R}$. The set $V \times Y$ is the linear space with natural operations

$$(v,y) + (v',y') = (v+v', y+y')\ \ \forall v, v' \in V,\ y, y' \in Y;$$

$$a(v,y) = (av, ay)\ \ \forall v \in V,\ y \in Y\ a \in \mathbb{R}.$$

This is a Banach space with norm

$$\big\|(v,y)\big\|_{V \times Y} = \|v\| + \|y\|.$$

This space is reflexive whenever V and Y are reflexive spaces. Suppose V and Y are Hilbert spaces; then its product is a Hilbert space with scalar product

$$\big((v,y),(v',y')\big)_{V \times Y} = (v,v')_V + (y,y')_Y.$$

Remark 3.1 We shall consider the pair "control–state" as a single object. This is an element of the product of corresponding spaces (see, particularly, Section 3.9).

The space Y is called **separable** if there exists a dense countable set here. All considered spaces except $L_\infty(\Omega)$ are separable.

Remark 3.2 If the space is separable, then each point can be approximated by elements of a countable set that can be interpreted as a sequence. We use this property for proving of existence theorems for the boundary problems that describe control systems.

Linear systems

Consider the **series** $\sum_{m=1}^{\infty} y_m$ with elements of Banach space Y, and the sums $z_k = \sum_{m=1}^{k} y_m$. The series **converges** to an element $y \in Y$ that is called the **sum** of the series if $z_k \to y$ in Y.

Remark 3.3 We prove the existence of the solution of boundary problems that describe control systems by the Galerkin method. This solution is the sum of a series.

The set $\{\mu_m\}$ is called **complete** in a space Y, if for all k the elements $\mu_1, ..., \mu_k$ are linear independent, and linear combinations of these elements are dense in Y. If the space Y is separable, then there exists a complete set $\{\mu_m\}$. Then the arbitrary element $y \in Y$ is the sum of a series

$$y = \sum_{m=1}^{\infty} a_m \mu_m,$$

where a_m are numbers, $m = 1, 2, ...$.

Consider linear continuous operator $A : Y \to Z$, where Y and Z are Banach spaces. By **Banach Inverse Operator Theorem**, if this operator is invertible, i.e., this is bijection, then the inverse operator is continuous. If the operator A is surjection, i.e., the equation $Ay = z$ has a solution of the space Y for all $z \in Z$, and there exists a constant $c > 0$ such that

$$\|Ay\|_Z \geq c\|y\|_Y \ \forall y \in Y,$$

then the operator is a continuous inverse operator with a norm that is not greater than $1/c$.

The existence of the continuous inverse operator A^{-1} is equivalent to the continuous inverse operator $(A^*)^{-1}$, as $(A^{-1})^* = (A^*)^{-1}$.

Remark 3.4 Thus, the operations of transformation to adjoint and inverse operators are commutated.

By the last equality, the equation $Ay = z$ has a unique solution $y \in Y$ for all $z \in Z$ whenever the **adjoint equation** $A^*p = \mu$ has the unique solution $p \in Z'$ for all $\mu \in Y'$. An operator A is a surjection whenever there exists a constant $c > 0$ such that

$$\|A^*p\|_{Y'} \geq c\|p\|_{Z'} \ \forall p \in Z'.$$

Remark 3.5 We shall consider the adjoint operator and adjoint equation in the next chapter. The simultaneous unique solvability of the initial and adjoint equations will be important enough there.

An operator $A : Y \to Z$ is called **weakly continuous** at a point if $Ay_k \to Ay$ weakly in Z whenever $y_k \to y$ weakly Y. If is linear continuous operator, then

$$\left|\langle p, Ay_k - Ay \rangle\right| = \left|\langle A^*p, y_k - y \rangle\right| \ \forall p \in Z'.$$

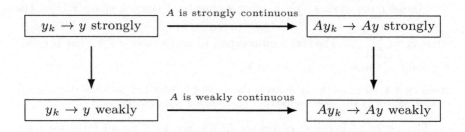

FIGURE 3.1: Relation between weak and strong continuity.

Therefore, if $y_k \to y$ weakly in Y, then $Ay_k \to Ay$ weakly in Z. Thus, each linear continuous operator is weakly continuous.

Remark 3.6 This property can be false for a nonlinear operator. For a weakly continuous operator the weaker condition (weak convergence of the argument) has the weaker corollary (weak convergence of the value). Therefore, the classes of the continuous and weakly continuous operators do not belong and have a difficult relation (see Figure 3.1). However, the values of the functionals are numbers; the weak and strong convergence is equivalent on the set of real numbers. Hence, the weak continuity and semicontinuity of functionals follow from its strong continuity and semicontinuity.

Remark 3.7 We shall use weak continuity of the dependence of the state function from the control for proving the solvability of optimal control problems.

Remark 3.8 An operator A is *-weakly continuous* if $Ay_k \to Ay$ *-weakly whenever $y_k \to y$ *-weakly.

3.2 Abstract linear control systems

Let us have a mathematical model of a phenomenon. It can be a *state equation* of the *control system*. Suppose the system is described by the equation
$$Ay = Bv + f, \tag{3.3}$$
where y is the *state* of the system, v is a *control*, A is a *state operator*, B is a *control operator*, and f is an absolute term. Consider also a *control space* V, a *state space* Y, and a *value space* Z. Suppose there are Banach spaces. Let A be a linear mapping from Y to Z, B be a linear mapping from Y to Z, and f belongs to Z. Suppose for all control $v \in V$ there exists a unique solution $y = y[v]$ of the equation (3.3) that is an element set Y. If the operator is invertible, then we find $y[v] = A^{-1}(Bv + f)$.

Linear systems 93

Definition 3.1 *The dependence $y = y[v]$ is called* **control--state mapping** *of the system.*

Remark 3.9 We shall substantiate this supposition for each concrete case.

Remark 3.10 There exists a possibility to analyze extremum problems without supposition of unique solvability of the state equation for all controls. We can interpret the state equation as a constraint. Then we minimize the functional on the set of pairs "control–state" that satisfy the given equation. The result can be obtained here by use of the penalty method (see Section 3.9).

Remark 3.11 We shall consider also a larger class of the state equations. Particularly, these are systems with control in the coefficients of the state operator (see Chapter 6) or general equation $A(v, y[v]) = 0$ (see Chapter 7).

We choose the control from a convex subset U of the space V, which is called the **set of admissible controls**. Thus, for all admissible control v that is an element of the set U there exists a state that satisfies the equation (3.3).

Remark 3.12 We could consider also state constraints. However, this is not our aim because this is another form of difficulties.

Determine also the **cost functional**

$$I(v) = J(v) + K(y[v]),$$

where $J : V \to \mathbb{R}$, $K : Y \to \mathbb{R}$.

Remark 3.13 We shall consider also a larger form of the functional $I = I(v, y[v])$.

Thus, we have the following **optimal control problem**.

Problem 3.1 *Find a control $u \in U$ that minimize the functional I on the set U.*

Try to use the known technique for solving this problem. Determine the value

$$I(u + \sigma h) - I(u) = \left[J(u + \sigma h) - J(u)\right] + \left\{K(y[u + \sigma h]) - K(y[u])\right\}.$$

Then we would like to divide this equality by σ and pass to the limit as $\sigma \to 0$ for all $h \in V$. The linearity of the considered operators is important here.

Remark 3.14 Our general aim is optimal control problems for nonlinear infinite dimensional systems. We consider the linear equation to determine the difficulties of nonlinear systems only.

Lemma 3.1 *Suppose the operators A and B are continuous, there exists the inverse operator A^{-1}, and the functionals J and K are Gâteaux differentiable at the points $u \in V$ and $y = y[u]$; then the functional I has Gâteaux derivative*

$$I'(u) = J'(u) - B^* p \tag{3.4}$$

at the points u, where p is the solution of the adjoint equation

$$A^* p = -K'(y). \tag{3.5}$$

94   *Optimization and Differentiation*

Proof. Find the difference

$$I(v_\sigma) - I(u) = \big[J(v_\sigma) - J(u)\big] + \big[K(y_\sigma) - K(y)\big],$$

where $v_\sigma = u + \sigma h$, $y_\sigma = y[v_\sigma]$, σ is a constant, and h is an arbitrary element of the space V. By equalities

$$Ay = Bu + f, \ Ay_\sigma = Bv_\sigma + f$$

we get $A(y_\sigma - y) = \sigma Bh$. Using the invertibility of the operator A, we find $y_\sigma = y + \sigma A^{-1} Bh$. Therefore, we get

$$I(v_\sigma) - I(u) = \big[J(v_\sigma) - J(u)\big] + \big[K(y + \sigma A^{-1} Bh) - K(y)\big].$$

The operator A^{-1} is continuous because of the Banach inverse operator theorem. Dividing by σ and passing to the limit as $\sigma \to 0$, we have

$$\langle I'(u), h \rangle = \langle J'(u), h \rangle + \langle K'(y), A^{-1} Bh \rangle \ \forall h \in V.$$

Transform the second term in the right-hand side of this equality by using the definition of the adjoint operator.

$$\langle K'(y), A^{-1} Bh \rangle = \langle (A^{-1})^* K'(y), Bh \rangle = \langle B^* (A^{-1})^* K'(y), h \rangle.$$

Using the equality $(A^{-1})^* = (A^*)^{-1}$, we have

$$\langle K'(y), A^{-1} Bh \rangle = \langle B^* (A^*)^{-1} K'(y), h \rangle = -\langle B^* p, h \rangle,$$

where $p = -(A^*)^{-1} K'(y)$. Therefore this is a solution of the adjoint equation (3.5). Finally, we have

$$\langle I'(u), h \rangle = \langle J'(u) - B^* p, h \rangle \ \forall h \in V,$$

and the formula (3.4) is true. $\square$

Determine the necessary condition of optimality for Problem 3.1.

Theorem 3.1 *Under the conditions of Lemma 3.1 the solution u of Problem 3.1 satisfies the variational inequality*

$$\langle J'(u) - B^* p, v - u \rangle \geq 0 \ \forall v \in U. \tag{3.6}$$

Indeed, put the derivative of the functional from formula (3.4) in to (3.2); we get the inequality (3.6).

Remark 3.15 The variational inequality (3.6) was obtained by the transformation of our optimal control problem to an abstract problem of minimization of the differentiable functional on the convex subset of a Banach space. Analogical result could be obtained by using the *Lagrange multipliers method* (see Section 3.7) or the *penalty method* (see Section 3.9).

Linear systems

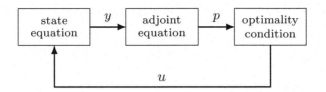

FIGURE 3.2: Method of successive approximations.

Thus, we have the state equation (3.3) for $v = u$, the adjoint equation (3.5), and the variational inequality (3.6) for finding three unknown values u, y, p.

Remark 3.16 We can use the *method of successive approximations* for solving this system (see Figure 3.2). Suppose initial iteration of the control is given. Then we determine the state function from the state equation (3.3). Solving adjoint equation (3.5) with known function y, we find the adjoint state p. The new iteration of the control can be determined by the variational inequality (3.6).

Remark 3.17 We could apply also the projection gradient method (see Chapter 2) for finding an approximate solution of Problem 3.1. Suppose the space V is Hilbert. Then the next approximation of the control can be determined by the formula

$$u_{k+1} = P_U[u_k - \sigma_k I'(u_k)],$$

where P_U is the projection on the set U, and the functional derivative is known.

Suppose we have the problem of minimization of the functional I on the space V. Then we use stationary condition (3.1). We get

$$J'(u) - B^*p = 0. \qquad (3.7)$$

Remark 3.18 We can use the subdifferential form of the optimality conditions if our functionals are non-smooth.

We know that the relations (3.1) and (3.2) are necessary optimality conditions only for the general case. Therefore, this is true for the relations (3.6) and (3.7) too. However, we can use Theorem 2.1 for obtaining sufficiency of the optimality conditions under additional suppositions. Particularly, we would like to guarantee the convexity of the functional.

Lemma 3.2 *Suppose the operators A and B are continuous, there exists the inverse operator A^{-1}, and the functionals J and K are convex. Then the functional I is convex.*

Proof. The sum of convex functionals is convex. Therefore, it is sufficient to prove the convexity of the map $v \to K(y[v])$. For all functions $u, v \in U$ and constant $\sigma \in (0, 1)$ consider the control $v_\sigma = (1 - \sigma)u + \sigma v$. We have the equalities

$$Ay[u] = Bu + f, \quad Ay[v] = Bv + f.$$

96 *Optimization and Differentiation*

Multiply the first of them by $(1 - \sigma)$, and the second equality by σ. After addition we get

$$A\{(1 - \sigma)y[u] + \sigma y[v]\} = (1 - \sigma)(Bu + f) + \sigma(Bv + f) = Bv_\sigma + f.$$

Therefore, we have the inequality

$$K\{y[(1 - \sigma)u + \sigma v]\} \leq (1 - \sigma)K(y[u]) + \sigma K(y[v])$$

for all $u, v \in U$, $\sigma \in (0, 1)$ because of the convexity of the functional K. Thus, the map $v \to K(y[v])$ is convex. This completes the proof of Lemma 3.2.

Using Theorem 2.1, we prove the following result.

Theorem 3.2 *Suppose the operators A and B are continuous, there exists the inverse operator A^{-1}, and the functionals J and K are convex and Gâteaux differentiable at the points $u \in V$ and $y = y[u]$. Then the relation (3.7) is a necessary and sufficient optimality condition.*

Remark 3.19 The equality (3.7) is a necessary and sufficient optimality condition for the unconditional optimization problem under the supposition of Theorem 3.2.

Remark 3.20 We proved that the mapping $v \to y[v]$ is affine. By this property we proved the convexity of the functional I for Problem 3.1. However, the control-state mapping is not affine for the nonlinear systems. Therefore, we cannot extend Theorem 3.2 to the nonlinear case.

Remark 3.21 Another form of necessary optimality condition for Problem 3.1 will be obtained in Section 3.7.

Theorem 3.1 uses the supposition of the solvability of Problem 3.1. It is necessary to substantiate this assumption. By Theorem 2.4, a convex lower bounded lower semicontinuous functional on a non-empty convex closed bounded subset of a reflexive Banach space has the minimum; besides, the boundedness of the set is not necessary if the functional is coercive. The convexity of the minimized functional for Problem 3.1 is guaranteed by Lemma 3.2. Prove the lower semicontinuity of the functional.

Lemma 3.3 *If the spaces V and Y are reflexive, and operators A and B are continuous, there exists the inverse operator A^{-1}, and if the functionals J and K are lower semicontinuous, then the functional I is lower semicontinuous.*

Proof. Let us have the convergence $v_k \to v$ weakly in V. The equation (3.3) has a unique solution $y[v] = A^{-1}(Bv + f)$ for all $v \in V$. It is obvious that $A^{-1}B$ is a linear continuous operator. Therefore, this is a weakly continuous operator. Then we have the convergence $y[v_k] \to y[v]$ weakly in Y. Using the semicontinuity of the functionals J and K, we get

$$\inf \lim_{k \to \infty} J(v_k) \geq J(v); \quad \inf \lim_{k \to \infty} K(y[v_k]) \geq K(y[v]).$$

Linear systems 97

Therefore, we have

$$\inf \lim_{k \to \infty} I(v_k) \geq I(v).$$

Thus this inequality follows from the convergence $v_k \to v$ weakly in V. $\square$

Using Theorem 2.4 with Lemmas 3.2 and 3.3, we have

Theorem 3.3 *Suppose the spaces V and Y are reflexive, the operators A and B are continuous, there exists the inverse operator A^{-1}, and the functionals J and K are convex, lower bounded and lower are semicontinuous. If the set U is bounded or the functional J is coercive, then Problem 3.1 has a solution.*

Remark 3.22 The considered spaces can be adjoint to Banach spaces.

Remark 3.23 There exist differences between the conditions of Theorem 3.1 and Theorem 3.3. We do not use the convexity of the functionals for obtaining the optimality conditions. However, its differentiability is not necessary for proving the solvability of the optimal control problem. Note that we use the convexity of the functionals in Theorem 3.2 for the justification of the sufficiency for the optimality conditions. We shall obtain also necessary optimality conditions that apply the differentiability of the functionals without its convexity (see Section 3.7).

Now prove the uniqueness of the optimal control.

Theorem 3.4 *Under conditions of Theorem 3.3 suppose one of the functionals J and K at least is strictly convex. Then the optimal control for Problem 3.1 is unique.*

Proof. By Theorem 2.5 the optimal control is unique under the strict convexity of the minimized functional. This is the sum of the functional J and mapping $v \to K\big(y[v]\big)$. The first of them is convex by the conditions of the theorem, and the convexity of the second functional was proved before (see Lemma 3.2). The dependence $v \to K\big(y[v]\big)$ is the superposition of K and affine mapping $v \to y[v]$ (see Lemma 3.2). Then this is strictly convex functional whenever K is strictly convex. Therefore, the functional I is strictly convex because this is the sum of the convex and strictly convex functionals. Thus, the uniqueness of the solution is guaranteed by Theorem 2.5. $\square$

Remark 3.24 We cannot prove analogs of Lemma 3.2 for nonlinear systems. However, the convexity of the functional is a means there of the justification of its weak or *-weak lower semicontinuous. We shall prove this property for nonlinear optimal control problems directly. Therefore, the solvability of these problems can be proved without the convexity of the functional.

Remark 3.25 If the convexity is absent, then we can use the Ekeland principle for proving the existence of the approximate solution of the problem and obtaining approximate conditions of optimality (see Theorem 2.6 and Theorem 2.7).

We supposed before the invertibility of the operator A. However, we considered before the unique solvability of a linear operator equation. Let us

98 *Optimization and Differentiation*

have a linear continuous self-adjoint operator $A : Y \to Y'$ that satisfies the condition

$$\langle Ay, y \rangle \geq \alpha \|y\|^2 \; \forall y \in Y, \tag{3.8}$$

where $\alpha > 0$, Y is a Hilbert space. Then for all $f \in Y'$ the equation $Ay = f$ has a unique solution $y \in Y$ i.e., the operator A is invertible. Using the previous theorems, we have the following result.

Corollary 3.1 *Suppose the space Y is Hilbert, the operator $A : Y \to Y'$ is self-adjoint and satisfies the condition (3.8), the operator B is continuous, and the functionals J and K are convex, Gâteaux differentiable, lower bounded, and lower semicontinuous as the set U is bounded or the functional J is coercive. Then Problem 3.1 is solvable, and the variational inequality (3.6) is the necessary sufficient optimality condition. If one of the functionals at least is strictly convex, then the solution of the problem is unique.*

Now consider an example.

3.3 Dirichlet problem for the Poisson equation

Let us have an open bounded set Ω of Euclid space R^n with boundary Γ. Consider the equation

$$-\Delta y(x) = v(x) + f(x), \; x \in \Omega \tag{3.9}$$

with the boundary condition

$$y(x) = 0, \; x \in \Gamma, \tag{3.10}$$

where Δ is the Laplace operator, v is a control, y is the state function, and f is a given function.

The equation

$$\Delta y = z \tag{3.11}$$

is called a **Poisson equation**. This is the *linear elliptic equation*. The equality (3.9) can be transformed to (3.11) with $z = -(v + f)$. The boundary problem (3.9), (3.10) is called the **Dirichlet problem**.

Consider at first the boundary problem (3.10), (3.11). Transform it to the operator equation (3.3). Multiply the equality (3.9) by the function y; after formal integration we have

$$\int_\Omega y \Delta y \, dx = \int_\Omega y z \, dx,$$

where $x = (x_1, ..., x_n)$ is a point of the set Ω. Using Green's formula, we get the equality

$$\int_\Omega y \Delta p \, dx = - \int_\Omega \sum_{i=1}^n \frac{\partial y}{\partial x_i} \frac{\partial p}{\partial x_i} dx + \int_\Gamma y \frac{\partial p}{\partial n} dx$$

for all smooth enough functions y and p, where n is exterior normal to the boundary Γ. Using the boundary condition (3.10), we get

$$\int_\Omega \sum_{i=1}^n \left| \frac{\partial y}{\partial x_i} \right|^2 dx = - \int_\Omega y z \, dx.$$

The value in the left-hand side of this equality is the square of norm for the function of the Sobolev space $H_0^1(\Omega)$. The right-hand side is the value of the linear continuous functional $-z$ at the point y of this space. It is known (see Section 3.1) that it is the adjoint space $H^{-1}(\Omega)$ to $H_0^1(\Omega)$, i.e., the set of all linear continuous functionals that are determined on this space. For any linear normalized space we have the inequality

$$|\langle \lambda, y \rangle| \le \|\lambda\|_{Y'} \|y\|_Y \ \forall \lambda \in Y', y \in Y.$$

Particularly, the following inequality holds:

$$\left| \int_\Omega y z \, dx \right| \le \|y\| \|z\|_* \ \forall y \in H_0^1(\Omega), z \in H^{-1}(\Omega),$$

where $\|y\|$ and $\|z\|_*$ are the norms of the given elements with respect to the spaces $H_0^1(\Omega)$ and $H^{-1}(\Omega)$. Therefore, we get

$$\|y\|^2 \le \|y\| \|z\|_*.$$

Thus, the following estimate is true

$$\|y\| \le \|z\|_*. \tag{3.12}$$

This inequality is called **a priori estimate** of the solution of the boundary problem (3.10), (3.11).

Our transformations were done formally because we cannot guarantee the existence of the considered integrals. However, we can use it for choosing the functional spaces. Try to determine the solution of the boundary problem as an element of the Sobolev space $H_0^1(\Omega)$, besides let us choose the absolute term of the equation (3.11) from the set $H^{-1}(\Omega)$.

Remark 3.26 Note we considered these spaces for the analysis of the Dirichlet integral, which has a relation with the Dirichlet problem for the Poisson equation.

We have the equality

$$\langle Ay, y \rangle = - \int_\Omega y \Delta y dx = \int_\Omega \sum_{i=1}^n \left| \frac{\partial y}{\partial x_i} \right|^2 dx = \|y\|^2 \ \forall y \in H_0^1(\Omega),$$

i.e., the condition (3.8) is true as an equality with the constant $\alpha = 1$. Therefore, it is possible to determine the solvability of the problem (3.10), (3.11) by using the known technique. However, we give another method of proving this result.

Theorem 3.5 *For all functions* $z \in H^{-1}(\Omega)$ *the problem* (3.10), (3.11) *has a unique solution* $y \in H_0^1(\Omega)$.

Proof. 1. We use the **Galerkin method** for proving the existence of the solution of the boundary problem. The space $H_0^1(\Omega)$ is separable; therefore, there exists a complete set of linear independent elements $\{\mu_m\}$ here. Determine the sequence $\{y_k\}$ on the space $H_0^1(\Omega)$ by the equalities

$$y_k = \sum_{m=1}^k \xi_{mk} \mu_m, \ k = 1, 2, ..., \tag{3.13}$$

where the numbers $\xi_{1k}, ..., \xi_{kk}$ satisfy the equalities

$$\int_\Omega \sum_{i=1}^n \frac{\partial y_k}{\partial x_i} \frac{\partial \mu_j}{\partial x_i} dx = \int_\Omega z \mu_j dx, \ j = 1, ..., k. \tag{3.14}$$

Using (3.13), transform it to the equalities

$$\sum_{m=1}^k \chi_{jm} \xi_{mk} = \beta_j, \ j = 1, ...k,$$

where

$$\chi_{jm} = \int_\Omega \sum_{i=1}^n \frac{\partial \mu_m}{\partial x_i} \frac{\partial \mu_j}{\partial x_i} dx, \ \beta_j = \int_\Omega z \mu_j dx, \ j, m = 1, ..., k.$$

We have the system of the linear algebraic equations (3.14). Its determinant is not equal to zero because of the linear independence of the elements $\mu_1, ..., \mu_k$. Therefore, this problem has a unique solution $\xi_{1k}, ..., \xi_{kk}$, which determine the element y_k of the space $H_0^1(\Omega)$ by the equality (3.13).

2. Multiply j-th equality (3.13) by the number ξ_{jk}. Adding by $j = 1, ..., k$ by using the formula (3.12), we get

$$\int_\Omega \sum_{i=1}^n \left| \frac{\partial y_k}{\partial x_i} \right|^2 dx = \int_\Omega z y_k dx.$$

The term in the left-hand side of this equality is the square of the norm of the function y_k with respect to the space $H_0^1(\Omega)$ as we have the inequality

$$\left| \int_\Omega z y_k dx \right| \le \|y_k\| \|z\|_*.$$

Hence, we get the estimate

$$\|y_k\| \le \|z\|_*,$$

which is an analog of (3.12).

3. Thus, the sequence $\{y_k\}$ is bounded in the space $H_0^1(\Omega)$. Using the Banach–Alaoglu theorem, we extract a subsequence with an initial denotation such that $y_k \to y$ weakly in $H_0^1(\Omega)$. By definition of the scalar product for this space, we get

$$\int_\Omega \sum_{i=1}^n \frac{\partial y_k}{\partial x_i} \frac{\partial \mu}{\partial x_i} dx \to \int_\Omega \sum_{i=1}^n \frac{\partial y}{\partial x_i} \frac{\partial \mu}{\partial x_i} dx$$

for all functions $\mu \in H_0^1(\Omega)$. Passing to the limit in the equality (3.14) as $k \to \infty$, we have

$$\int_\Omega \sum_{i=1}^n \frac{\partial y}{\partial x_i} \frac{\partial \mu_j}{\partial x_i} dx = \int_\Omega z \mu_j dx, \ j = 1, 2, \dots.$$

An arbitrary element $\mu \in H_0^1(\Omega)$ can be interpreted as a convergent series

$$\mu = \sum_{j=1}^\infty b_j \mu_j,$$

where b_j are numbers. Multiply j-th previous equality by b_j. Adding by j by using the convergence of the series, we get

$$\int_\Omega \sum_{i=1}^n \frac{\partial y}{\partial x_i} \frac{\partial \mu}{\partial x_i} dx = \int_\Omega z \mu dx, \ \forall \mu \in H_0^1(\Omega).$$

Using Green's formula, we have the equality

$$\int_\Omega (\Delta y + f) \mu dx = 0 \ \forall \mu \in H_0^1(\Omega).$$

If the value of the arbitrary linear continuous functional at a point is equal to zero, then this point is the zero element of the space. Therefore, the element y of the space $H_0^1(\Omega)$ satisfies the equation (3.11). The boundary condition (3.10) is true by the definition of this space. Thus, for all elements $z \in H^{-1}(\Omega)$ the boundary problem (3.10), (3.11) has a solution $y \in H_0^1(\Omega)$.

102   *Optimization and Differentiation*

4. Suppose there exist two solutions y_1 and y_2 of this problem. Then its difference $y = y_1 - y_2$ is the solution of this problem for $z = 0$. Multiply the equation (3.11) for this case by this function y. After integration by using Green's formula we obtain the equality

$$\int_\Omega \sum_{i=1}^n \left| \frac{\partial y}{\partial x_i} \right|^2 dx = 0.$$

This is the square of norm for the function y of the space $H_0^1(\Omega)$. It can be zero for $y = 0$ only. Finally, we have the equality $y_1 = y_2$, i.e., the solution of the problem (3.10), (3.11) is unique. $\square$

Remark 3.27 The proof of the theorem has four steps. This is the determination of the approximate solution, obtaining of a priori estimate, the justification of passing to the limit, and the proof of the uniqueness of the solution. We shall have these steps for the analysis of other state equations of control systems.

Remark 3.28 We proved the existence of the *generalized solution* of the Dirichlet problem for the Poisson equation. Its *classical solution* is a twice continuously differentiable function that satisfies the Poisson equation for all points of the set Ω and equals to zero on the boundary of this set. However, we shall consider the extended solution only.

Remark 3.29 Theorem 3.5 gives also the justification of the estimate (3.6).

Remark 3.30 We shall consider also boundary problems for evolutional systems (see Part III). The modification of the Galerkin method for this case is the Faedo–Galerkin method. The unknown coefficients ξ_{mk} here depend upon time. Therefore, we determine it from the system of ordinary differential equations.

Now we consider an optimal control problem for the system described by the equation (3.9) with boundary condition (3.10).

3.4   Optimal control problem statement

Transform the boundary problem (3.9), (3.10) as the operator equation (3.3) of the control system. Determine the functional space $V = L_2(\Omega)$, $Y = H_0^1(\Omega)$, $Z = H^{-1}(\Omega)$. Let $A : Y \to Z$ be the operator $A = -\Delta$, and $B : V \to Z$ is the operator of embedding of the space $L_2(\Omega)$ to $H^{-1}(\Omega)$. Thus, we get the operator equation (3.3).

Remark 3.31 We know that the differentiation is the linear continuous operator from the Sobolev space $H^m(\Omega)$ to $H^{m-1}(\Omega)$. Therefore, the Laplace operator, i.e., the sum of the second derivatives, is the linear continuous operator from the space $H_0^1(\Omega)$ to $H^{-1}(\Omega)$. Thus, we have the operator $A : Y \to Z$ in reality.

Consider a non-empty convex closed subset U of the space $L_2(\Omega)$ and the functional

$$I(v) = \frac{\alpha}{2} \int_\Omega v^2 dx + \frac{1}{2} \int_\Omega \sum_{i=1}^n \left(\frac{\partial y[v]}{\partial x_i} - \frac{\partial y_d}{\partial x_i} \right)^2 dx,$$

where $y[v]$ is the solution problem (3.9), (3.10) for the control v, $\alpha > 0$, and y_d is a known function of the space $H_0^1(\Omega)$. We have the following optimal control problem.

Problem 3.2 *Find the control $u \in U$ that minimizes the functional I on the set U.*

Determine the functionals $J : V \to \mathbb{R}$ and $K : Y \to \mathbb{R}$ by the formulas

$$J(v) = \frac{\alpha}{2} \int_\Omega v^2 dx, \quad K(y) = \frac{1}{2} \int_\Omega \sum_{i=1}^n \left(\frac{\partial y}{\partial x_i} - \frac{\partial y_d}{\partial x_i} \right)^2 dx.$$

Thus, Problem 3.2 is transformed to the abstract Problem 3.1. Therefore, our previous results are applicable. Using Theorem 3.3 and Theorem 3.4, we have the following result.

Theorem 3.6 *Problem 3.2 has a unique solution.*

Proof. Check the suppositions of the considered theorems. All spaces are reflexive and even Hilbert. We know that the operator A is continuous. The operator B is continuous, because this is an embedding operator. The invertibility of the operator A follows from the unique solvability of the Dirichlet problem for the Poisson equation in the space $H_0^1(\Omega)$ for all absolute terms from the set $H^{-1}(\Omega)$. The functional J is determined by the square of the norm of Hilbert space $L_2(\Omega)$. This is the lower bounded continuous coercive functional. We proved its strict convexity before. The functional K is proportional to the square of the norm for the function $y - y_d$ with respect to the space $H_0^1(\Omega)$ where the point y_d is fixed. We can repeat here the analysis of the Dirichlet integral (see Chapter 1). Therefore, this functional is so bounded, convex, and continuous. Thus, all suppositions of Theorem 3.3 and Theorem 3.4 are true. This completes the proof of Theorem 3.6. $\square$

Theorem 3.7 *Problem 3.2 has a unique solution.*

Proof. By Theorem 2.6, it is sufficient to prove the strict uniform convexity of the given functional. This is the sum of two functionals. The functional J is determined by the square of norm of Hilbert space V. Therefore, this is the strictly uniformly convex functional (see Example 2.17). We have already proved the convexity of mapping $v \to K(y[v])$. Thus, we can apply Theorem 2.6, in reality. $\square$

Using the relation between Tihonov and Hadamard well-posedness (see Theorem 2.7), we have the following result.

104   *Optimization and Differentiation*

Theorem 3.8 *Suppose the set U is bounded. Then Problem 3.2 with parameter y_d is Hadamard well-posed.*

Proof. Tihonov well-posedness of the minimization problem for the functional I for all functions y_d is guaranteed by the previous theorem. By boundedness of the set U, the continuous dependence of the minimized functional uniform with respect to the control follows from the continuity of the norm (see also Example 2.19). Therefore, we prove Hadamard well-posedness of our problem by using Theorem 2.7. $\square$

Our next step is the justification of the necessary conditions of optimality for Problem 3.2.

3.5   Necessary conditions of optimality

Try to apply Theorem 3.2. Determine the adjoint operators for A and B, and the derivatives of the functionals J and K for Problem 3.2.

Lemma 3.4 *The following equality holds: $A^* = -\Delta$.*

Proof. We choose $-\Delta$ as the operator A that maps the space $Y = H_0^1(\Omega)$ to $Z = H^{-1}(\Omega)$. Find its adjoint operator by the equality

$$\langle p, Ay \rangle = \langle A^* p, y \rangle \ \forall y \in Y, p \in Z'.$$

Now we have

$$-\int_\Omega p \Delta y dx = \langle A^* p, y \rangle \ \forall y \in H_0^1(\Omega), p \in \left[H^{-1}(\Omega) \right]'.$$

The reflexive space $H_0^1(\Omega)$ is adjoint to $H^{-1}(\Omega)$. Therefore, we get

$$-\int_\Omega p \Delta y dx = \int_\Omega y A^* p dx \ \forall y, p \in H_0^1(\Omega). \tag{3.15}$$

Use Green's second formula

$$\int_\Omega (p \Delta y - y \Delta p) dx = \int_\Gamma \left(p \frac{\partial y}{\partial n} - y \frac{\partial p}{\partial n} \right) dx$$

with derivatives with respect to the exterior normal to the surface Γ. Using the equality to zero on the set Γ of the elements of the space $H_0^1(\Omega)$ we have

$$\int_\Omega p \Delta y dx = \int_\Omega y \Delta p dx \ \forall y, p \in H_0^1(\Omega).$$

Therefore, A^* satisfies the equality (3.15), i.e., the operator A is **self-adjoint**. □

Lemma 3.5 B^* *is the embedding operator from the space $H_0^1(\Omega)$ to $L_2(\Omega)$.*

Proof. The embedding operator of the space $L_2(\Omega)$ to $H^{-1}(\Omega)$ is the operator $B : V \to Z$. Therefore, the adjoint operator $B^* : Z' \to V'$ is determined by the equality

$$\langle q, Bv \rangle = \int_\Omega qv dx = \langle B^*, v \rangle \ \forall v \in L_2(\Omega), \ q \in \left[H^{-1}(\Omega) \right]'.$$

Then we get

$$\int_\Omega qv dx = \int_\Omega v B^* q dx \ \forall v \in L_2(\Omega), \ q \in H_0^1(\Omega).$$

This completes the proof of Lemma 3.5. □

The functional J is determined by the square of the norm of the self-adjoint Hilbert space $L_2(\Omega)$. Hence, we have the following result.

Lemma 3.6 *The following equality holds $J'(u) = \alpha u$.*

Now we prove the next lemma.

Lemma 3.7 *The following equality holds: $K'(y) = \Delta(y_d - y)$.*

Proof. Using the definition of the functional K, we have the equality

$$2K(y + \sigma h) = \|y - y_d + \sigma h\|^2 = (y - y_d + \sigma h, y - y_d + \sigma h) =$$
$$\|y - y_d\|^2 + 2\sigma(y - y_d, h) + \sigma^2 \|h\|^2$$

with scalar product of the space $H_0^1(\Omega)$. Then we get

$$\langle K'(y), h \rangle = (y - y_d, h) \ \forall h \in H_0^1(\Omega).$$

By the first Green's formula

$$(v, h) = \int_\Omega \sum_{i=1}^n \frac{\partial v}{\partial x_i} \frac{\partial h}{\partial x_i} = -\int_\Omega h \Delta v dx \ \forall v, h \in H_0^1(\Omega).$$

Therefore, we obtain

$$\langle K'(y), h \rangle = \int_\Omega h \Delta(y - y_d) dx \ \forall h \in H_0^1(\Omega).$$

This completes the proof of Lemma 3.7. □

Determine necessary conditions of optimality for Problem 3.2.

106   *Optimization and Differentiation*

Theorem 3.9 *The control u is a solution of Problem 3.2 if and only if it satisfies the variational inequality*

$$\int_\Omega (\alpha u - p)(v - u)dx \geq 0 \; \forall v \in U, \tag{3.16}$$

where p is the solution of the equation

$$-\Delta p(x) = \Delta y(x) - \Delta y_d(x), \; x \in \Omega \tag{3.17}$$

with boundary condition

$$p(x) = 0, \; x \in \Gamma. \tag{3.18}$$

Proof. By Theorem 3.2, the necessary condition of optimality for Problem 3.1 is the variational inequality

$$\langle J'(u) - B^*, v - u \rangle \geq 0 \; \forall v \in U.$$

Our problem is the partial case of the abstract one. Using Lemma 3.5 and Lemma 3.6, we transform the previous relation to (3.16). Besides, the function p satisfies the adjoint equation (3.7) that is transformed to

$$A^* p = -J'(y).$$

By Lemma 3.4 and Lemma 3.7, we get the equation (3.17). The function p is an element of the space $H_0^1(\Omega)$; therefore, the homogeneous boundary condition (3.18) is true too. $\square$

Remark 3.32 We shall give another proof of Theorem 3.9 in Section 3.7 (see also Section 3.9).

Thus, we have Dirichlet problem (3.9), (3.10) for the adjoint system (3.17), (3.18) (Dirichlet problem too) and the variational inequality (3.16) for finding three unknown functions u, y, p.

Remark 3.33 Note that we can exclude the function y if we transform the equation (3.17) to

$$\Delta p = \Delta y_d + u + f.$$

Then we have the system that consists of this equation with boundary condition (3.16) and variational inequality (3.18) with respect to two unknown functions u and p.

Remark 3.34 We shall consider an extension of Problem 3.2 in Section 3.8.

Transform necessary conditions of extremum for the special case of the set of admissible controls.

3.6 Optimal control problems with local constraints

Consider a partial case of the set of admissible control, where we have the possibility to find the exact solution of the variational inequality (3.16). Suppose we have the set

$$U = \{v \in L_2(\Omega) \,|\, a \leq v(x) \leq b,\, x \in \Omega\},$$

where a and b are given constants, as $a < b$. This constraint is called **local**, because we have the condition with respect to the value of the control at the concrete points.

Remark 3.35 There exist a lot of practical control problems with local constraints.

Determine properties of this set.

Lemma 3.8 *The set U is convex.*

Proof. Let u and v belong to the set U. Then there are the elements of the space $L_2(\Omega)$, satisfying the inequalities $a \leq u(x) \leq b$, $a \leq v(x) \leq b$ for all points of Ω. Multiplying second inequality by a number $\sigma \in (0,1)$ and first inequality by $(1 - \sigma)$. After addition we have

$$a = (1 - \sigma)a + \sigma a \leq (1 - \sigma)u(x) + \sigma v(x) \leq (1 - \sigma)b + \sigma b = b.$$

Therefore, the control $(1 - \sigma)u + \sigma v$ is an element of the set U for all $u, v \in U$ and $\sigma \in (0,1)$. $\square$

Lemma 3.9 *The set U is closed in the space $L_2(\Omega)$.*

Proof. Consider a sequence $\{v_k\}$ of elements of the set U such that $v_k \to v$ in $L_2(\Omega)$. For all convergent sequences of the space $L_2(\Omega)$ there exists a subsequence that converges almost everywhere on the set Ω. Then there exists a subsequence of $\{v_k\}$ (we save the initial denotation) such that $v_k \to v$ a.e. on Ω. Pass to the limit in the inequality $a \leq v_k(x) \leq b$. We have the condition $a \leq v(x) \leq b$ that is true a.e. on the set Ω. Therefore, we have the inclusion $v \in U$. Thus, the limit of the arbitrary sequence of the set U belongs to this set. This completes the proof of the lemma. $\square$

Lemma 3.10 *The set U is bounded in the space $L_2(\Omega)$.*

Proof. Using the definition of the set U, we have

$$\|v\|_V^2 = \int_\Omega |v|(x)|^2 dx \leq \max\{|a|, |b|\} \operatorname{mes} \Omega \ \forall v \in U,$$

where mes Ω is the measure of the set Ω. Therefore, the norm of the arbitrary

108   *Optimization and Differentiation*

element of the set U is bounded by the concrete constant. $\square$

Thus, the set U that is characterized by the local constraints satisfies the conditions of Theorem 3.1 and Theorem 3.2. Now we find the control from the optimality conditions.

Theorem 3.10 *The solution of Problem 3.2 with local constraint is*

$$u(x) = \begin{cases} a, & \text{if } p(x)/\alpha < a, \\ p(x)/\alpha, & \text{if } a \le p(x)/\alpha \le b, \\ b, & \text{if } p(x)/\alpha > b. \end{cases} \tag{3.19}$$

Proof. By Theorem 3.6 and Theorem 3.9 the unique solution of Problem 3.2 satisfies the variational inequality (3.16) that is a necessary and sufficient optimality condition. Consider the sets Ω_1 and Ω_2 of all Lebesgue points of the functions $u_1 = \alpha u - p$ and $u_2 = (\alpha u - p)u$ on Ω. Then the intersection $\Omega_* = \Omega_1 \cap \Omega_2$ is the set of Lebesgue points of both these functions. The sum of two sets with zero measure has the zero measure too. Therefore, almost all points of the set Ω belong to Ω_*.

Let $x_* \in \Omega_*$ be an arbitrary point. Consider a sequence of its neighborhoods $\{O_k\}$ such that the measure m_k of the set O_k satisfies the convergence $m_k \to 0$ as $k \to 0$. Then we have the equalities

$$\lim_{k \to \infty} \int_{O_k} \big[\alpha u(x) - p(x)\big]dx = \alpha u(x_*) - p(x_*),$$

$$\lim_{k \to \infty} \int_{O_k} \big[\alpha u(x) - p(x)\big]u(x)dx = \alpha\big[u(x_*) - p(x_*)\big]u(x_*).$$

Determine the function

$$v_k = (1 - \chi_k)u + \chi_k w,$$

where w is an arbitrary point of the interval $[a, b]$, and χ_k is the **characteristic function** of the set O_k, i.e., the function that is equal to 1 on this set, and equal to 0 outside it. The function v_k is called the **needle variation** of the control u (see Figure 3.3 for the one-dimensional case).

The function v_k belongs to the set U. Determine $v = v_k$ in the inequality (3.16); we have

$$\int_{O_k} \big[\alpha u(x) - p(x)\big]\big[w - u(x)\big]dx \ge 0.$$

Dividing by m_k and passing to the limit, we get

$$\big[\alpha u(x_*) - p(x_*)\big]\big[w - u(x_*)\big] \ge 0.$$

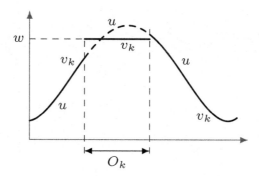

FIGURE 3.3: Needle variation.

This inequality is true for all $w \in [a, b]$ and $x_* \in \Omega_*$. Hence, we have

$$[\alpha u(x) - p(x)][w - u(x)] \geq 0 \; \forall w \in [a, b] \tag{3.20}$$

for almost all $x \in \Omega$. This is a pointwise variational inequality.

Suppose for a point x the following inequality holds: $\alpha u(x) - p(x) > 0$. By (3.20) we have $w - u(x) \geq 0$ for all $w \in [a, b]$. Hence, $u(x)$ is not greater than $w \in [a, b]$. Then $u(x)$ is not greater than a. However, $u(x)$ cannot be less than a because of the definition of the set U. Therefore, we have $u(x) = a$. Thus, if the number $p(x)$ is greater than $\alpha u(x)$ that is equal to αa then $u(x) = a$. This is true for $p(x)/\alpha < a$.

If $\alpha u(x) - p(x) < 0$, then $w - u(x) \leq 0$ for all $w \in [a, b]$ because of (3.20). Therefore, the value $u(x)$ is not less than all numbers from the interval $[a, b]$. Then this is not less than b. However, we have the inequality by definition of the set U. Hence, we have the equality $u(x) = b$. Thus, if $p(x)$ is less than $\alpha u(x)$ that is equal to αb, i.e., $p(x)/\alpha > b$, then we have the equality $u(x) = b$.

Finally, consider the equality $\alpha u(x) - p(x) = 0$. Then $u(x) = p(x)/\alpha$. However, it can be admissible if the following inequality holds: $a < p(x)/\alpha < b$. Thus, the solution of the variational inequality (3.16) is determined by the formula (3.19). □

Thus, we can find the solution of the variational inequality for the given set U. Denote it by $u = F(p)$, where the function F of the argument p is the term at the right-hand side of the equality (3.19). Put it to the equality (3.9). We obtain the nonlinear elliptic equation

$$\Delta p = y_d + F(p) + f. \tag{3.21}$$

Thus we find the solution of the equation (3.21) with boundary condition (3.18). Then we determine the optimal control by the formula (3.19) (see Figure 3.4).

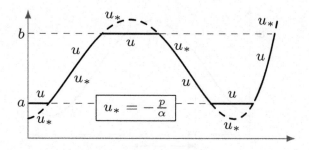

FIGURE 3.4: Optimal control for the case of local constraints.

Remark 3.36 We return to the analysis of the optimal control problem with local constraints in Section 3.8. We shall obtain another form of optimality conditions there.

Remark 3.37 We can extend all results to the systems described by general linear second order elliptic equations. Moreover, the order of equations is not important (see Part IV). We considered the easiest equation for maximal simplification of the technical transformations. The state equations can be evolutional (see Part III). Optimal control problems with general functionals will be considered in Part IV.

3.7   Lagrange multipliers method

Return to Problem 3.1. Try to prove Theorem 3.1 without its transformation to the abstract Problem 1.7. We use **Lagrange multipliers method** here.

Second proof of Theorem 3.1. Determine the map $L : V \times Y \times Z' \to \mathbb{R}$, i.e., the functional with three arguments, by the formula

$$L(v, y, p) = J(v) + K(y) + \langle p, Ay - Bv - f \rangle \quad \forall v \in V, y \in Y, p \in Z'.$$

The following equality holds:

$$L(v, y[v], p) = I(v) \ \forall v \in V, p \in Z'.$$

If the control u is a solution of Problem 3.1, then we have the inequality $I(w) - I(v) \geq 0$ for all $w \in U$. Therefore, we get

$$L(w, y[w], p) - L(u, y[u], p) \geq 0 \ \forall w \in U, p \in Z'.$$

Transform the functional L:

$$L(v, y, p) = M(v, p) + K(y) + \langle p, Ay - f \rangle \quad \forall v \in V, y \in Y, p \in Z'.$$

Linear systems

The functional M is called the **Lagrange functional**. This is determined by the formula

$$M(v, p) = J(v) - \langle B^*p, v \rangle \; \forall v \in V, p \in Z'.$$

Then we have the inequality

$$\left[M(w, p) - M(u, p) \right] + \left[K(y[w]) - K(y) \right] +$$

$$\langle A^*p, y(w) - y \rangle \geq 0 \; \forall w \in U, p \in Z',$$

where $y = y[u]$. Determine $w = v_\sigma = (1 - \sigma)u + \sigma v$, where $v \in U$, $\sigma \in (0, 1)$. We get

$$\left[M(v_\sigma, p) - M(u, p) \right] + \left[K(y_\sigma) - K(y) \right] +$$

$$\langle A^*p, y_\sigma - y \rangle \geq 0 \; \forall v \in U, \sigma \in (0, 1), p \in Z'.$$

Choose p equal to the solution of the adjoint equation (3.5). Then we obtain

$$\left[M(v_\sigma, p) - M(u, p) \right] + \left[K(y_\sigma) - K(y) - \langle K'(y), y_\sigma - y \rangle \right] \geq 0 \; \forall v \in U, \sigma \in (0, 1).$$

By invertibility of the operator A we have

$$y[v_\sigma] = A^{-1}(Bv_\sigma + f), \; y = A^{-1}(Bu + f)$$

because of the equation (3.3). Then $y[v_\sigma] = y + \sigma A^{-1}B(v - u)$. Using the definition of the functional M, we get

$$\left\{ J[u + \sigma(v - u)] - J(u) \right\} - \sigma \langle B^*p, v - u \rangle +$$

$$\left\{ K[y + \sigma A^{-1}B(v - u)] - K(y) - \sigma \langle K'(y), A^{-1}B(v - u) \rangle \right\} \geq 0.$$

Divide this inequality by σ. Passing to the limit as $\sigma \to 0$ by using the differentiability of the functionals J and K, we get

$$\langle J'(u) - B^*p, v - u \rangle \geq 0 \; \forall v \in U.$$

Thus, the variational inequality (3.6) is true. $\square$

Remark 3.38 The functional L is called a Lagrange functional too. However, we have an interest in the part of L that depends upon the control directly. This is the functional M.

Remark 3.39 The general idea of the Lagrange multipliers method is the transformation of the initial minimization problem with additional constraints (this is the state equation now) to the minimization problem for a Lagrange functional without any constraint. However, this functional depends upon additional unknown value p that is called the **Lagrange multiplier**.

Now consider Problem 3.1 with another property of the functional J. Suppose this is a convex non-differentiable functional. Therefore, we cannot prove the differentiability of the given functional I; and Theorem 1.8 is non-applicable. The second proof of Theorem 3.1 uses the inequality

$$\left[M(v_\sigma, p) - M(u, p) \right] +$$

112   *Optimization and Differentiation*

$$\left\{ K[y + \sigma A^{-1}B(v - u)] - K(y) - \sigma\langle K'(y), A^{-1}B(v - u)\rangle \right\} \geq 0.$$

Then we obtain the relation (3.6) if the functional M is differentiable with respect to the control. Of course, this is false for non-differentiable functional J. However, we can obtain another form of the condition optimality.

Theorem 3.11 *Suppose the conditions of Lemma 3.1 are true except for the differentiability of the functional J, and this functional is convex. Then the solution of Problem 3.1 satisfies the equality*

$$M(u,p) = \min_{v \in U} M(v,p). \tag{3.22}$$

Proof. The functional can be non-differentiable with respect to the control. However, this is convex as the sum of the convex functional J and a linear functional. Then we have the inequality

$$M(v_\sigma, p) \leq (1 - \sigma)M(u,p) + \sigma M(v,p) \ \forall v \in U, \sigma \in (0,1), p \in Z'.$$

Therefore, we get

$$\big[M(v_\sigma, p) - M(u,p) \big] \leq \sigma \big[M(v,p) - M(u,p) \big].$$

Determine the value

$$\eta_\sigma = \sigma^{-1}\left\{ K[y + \sigma A^{-1}B(v - u)] - K(y) - \sigma\langle K'(y), A^{-1}B(v - u)\rangle \right\}.$$

Hence, we obtain

$$\big[M(v,p) - M(u,p) \big] + \eta_\sigma \geq \sigma^{-1}\big[M(v_\sigma, p) - M(u,p) \big] + \eta_\sigma \geq 0.$$

We have the convergence $\eta_\sigma \to 0$ because of the differentiability of the functional K. Passing to the limit, we have the inequality

$$M(v,p) - M(u,p) \geq 0 \ \forall v \in U.$$

This is equivalent to (3.22). $\square$

The optimality condition (3.22) is called the **Lagrange principle**. By Theorem 3.11 we have the system that consists of the state equation (3.3) for adjoint equation (3.5) and Lagrange principle (3.22) with unknown functions u, y, p.

Remark 3.40 The Lagrange principle is a problem of finding the minimum of the Lagrange functional with respect to the first argument, i.e., the control. The function is the fixed unknown parameter here. Therefore, we could use known approximate methods of the minimization of functionals, for example, the projection gradient method. If we know the k-th iteration u_k of the control, then we find the state function $y_k = y[u_k]$ from the equation (3.3). Then we calculate the solution $_k$ of the adjoint equation (3.5) for $y = y_k$. Using the projection gradient method, we determine the next iteration of the control by the formula

$$u_{k+1} = P_U[u_k - \sigma_k M_u(u_k, p_k)]$$

if the space V is Hilbert, where P_U is the projector on the set U, $M_u(u_k, p_k)$ is the derivative of the map $v \to M(v, p_k)$ at the point u_k, and σ_k is a numerical parameter.

Linear systems

We obtained variational inequality (3.6), if the functional J is a differentiable, Lagrange principle (3.22) for the convex functional J. Determine the relation between these optimality conditions for the case of the differentiability and convexity simultaneously.

Theorem 3.12 *Under the conditions of Lemma* 3.1 *suppose the functional J is convex. Then the variational inequality* (3.6) *and Lagrange principle* (3.23) *are equivalent.*

Proof. Under the conditions of theorem the map $v \to M(v, p_k)$ is differentiable and convex simultaneously. Therefore, we have the inequality

$$\big[M(v_\sigma, p) - M(u, p)\big] \le \sigma \big[M(v, p) - M(u, p)\big].$$

Dividing by σ and passing to the limit, we get

$$\langle J'(u) - B^* p, v - u \rangle \le M(v, p) - M(u, p) \ \forall v \in U.$$

If the variational inequality (3.6) is true, then the left side of this inequality is non-negative. Then this property is true for its right side, i.e., $M(v, p) - M(u, p) \ge 0$ for all $v \in U$. This is the Lagrange principle. Suppose now the equality (3.22) holds. Choose the control $u + \sigma(v - u)$ as v in the previous inequality. After dividing by σ and passing to the limit we have the variational inequality (3.6). $\square$

Thus, if the functional is differentiable we have the variational inequality. We obtain Lagrange principle for the convex case. These optimality conditions are equivalent if we have both properties simultaneously.

3.8 Maximum principle

Apply the Lagrange principle for the analysis of the optimal control problem for the system described by the Dirichlet problem and Poisson equation. Consider again Poisson equation

$$-\Delta y(x) = v(x) + f(x), \ x \in \Omega$$

with boundary condition

$$y(x) = 0, \ x \in \Gamma.$$

Let a control v be an element of a non-empty convex closed bounded subset U of the control space $L_2(\Omega)$. For any function $v \in U$ the Dirichlet problem has a unique solution $y = y[v]$ from the set $Y = H_0^1(\Omega)$. Determine the functional

$$I(v) = \int_\Omega F(v) dx + \frac{1}{2} \int_\Omega \sum_{i=1}^n \left| \frac{\partial y[v]}{\partial x_i} - \frac{\partial y_d}{\partial x_i} \right|^2 dx,$$

114　　　　　　　　　　*Optimization and Differentiation*

where the function F is convex, lower bounded, and continuous, and y_d is a known function from the space $H_0^1(\Omega)$. Note that the function F can be non-differentiable.

Problem 3.3 *Find the control* $u \in U$ *that minimizes the functional* I *on the set* U.

Determine the functional $J : V \to R$ by the formula

$$J(v) = \int_\Omega F[v(x)]dx.$$

This is a convex, lower bounded, and continuous functional. Note that Problem 3.2 and Problem 3.3 differ in the functional J only. Determine the functional

$$M(v,p) = \int_\Omega [F(v) - vp]dx.$$

Using Theorem 3.3 and Theorem 3.9, we have the following result.

Theorem 3.13 *Problem* 3.3 *has a solution* u *that satisfies the Lagrange principle*

$$\int_\Omega [F(u) - up]dx = \min_{v \in U} \int_\Omega [F(v) - vp]dx, \qquad (3.23)$$

where the function p *satisfies the equation* (3.17) *with a homogeneous boundary condition.*

We can transform the equality (3.23) for the case of the local constraints. Consider the set

$$U = \big\{v \in L_2(\Omega)\big|\ a \le v(x) \le b,\ x \in \Omega\big\}.$$

Theorem 3.14 *The solution of Problem* 3.3 *with local constraints satisfies the equality*

$$u(x)p(x) - F[u(x)] = \max_{w \in [a,b]} \big[wp(x) - F(w)\big]. \qquad (3.24)$$

Proof. Consider the sets Ω_1 and Ω_2 of Lebesgue points for the functions $F - up$ and p on Ω. Then the intersection $\Omega_* = \Omega_1 \cap \Omega_2$ consists of Lebesgue points for both these functions. Let $\{O_k\}$ be a sequence of neighborhoods of an arbitrary point $x_* \in \Omega_*$ such that $m_k \to 0$ as $k \to \infty$, where m_k is the measure of the set O_k. By Lebesgue Theorem we have

$$\lim_{k \to \infty} \frac{1}{m_k} \int_{O_k} \big\{F[u(x)] - u(x)p(x)\big\}dx = \big\{F[u(x_*)] - u(x_*)p(x_*)\big\},$$

Linear systems

$$\lim_{k \to \infty} \frac{1}{m_k} \int_{O_k} p(x)dx = p(x_*).$$

Determine the needle variation of the control $v_k = (1 - \chi_k)u + \chi_k w$, where w is an arbitrary point of the interval $[a, b]$, and χ_k is a characteristic function of the set O_k. Determine $v = v_k$ in the equality (3.23). We get

$$\int_{O_k} [F(u) - up]dx \le \int_{O_k} [F(w) - wp]dx.$$

Dividing by m_k and passing to the limit, we have

$$\{F[u(x_*)] - u(x_*)p(x_*)\} \le [F(w) - up(x_*)],$$

where $w \in [a, b]$ is arbitrary. Therefore, the equality (3.24) is true. $\square$

Remark 3.41 The equality (3.24) is true a.e. on the set Ω. However, this is normal, because Lebesgue integrable functions are accurate within a set of zero measure.

The optimality condition (3.24) is called the **maximum principle**. This is the problem of maximization of the function $w \to [wp(x) - F(w)]$. Therefore, this is easier than minimization problem (3.23) with respect to the functional, i.e., $v \to M(v, p)$.

Remark 3.42 There exists a special case $F = 0$. Then the maximum principle (3.24) is transformed to the equality $up = \max wp$. Suppose there exists a control u such that $y(u) = y_d$. This is the optimal control, because the non-negative minimizing functional is equal to zero here. However, the right-hand side of the adjoint equation (3.17) is equal to zero for this case. Therefore, the Dirichlet problem for this equation has the zero solution. Then the maximum principle has the form $0 = 0$, and we do not have any positive information for this case. We say that the maximum principle degenerates here, and the optimal control is called **singular**.

3.9 Penalty method

Return to the analysis of Problem 3.2. This is the minimization problem of the functional

$$I(v) = \frac{\alpha}{2} \int_{\Omega} v^2 dx + \frac{1}{2} \int_{\Omega} \sum_{i=1}^{n} \left| \frac{\partial y[v]}{\partial x_i} - \frac{\partial y_d}{\partial x_i} \right|^2 dx,$$

on a non-empty convex closed subset U of the space $L_2(\Omega)$, where $y[v]$ is the solution of the equation

$$-\Delta y(x) = v(x) + f(x), \ x \in \Omega$$

with homogeneous boundary condition (3.10).

116 Optimization and Differentiation

We considered before the control v as an independent value, and the state function that depends upon control. Now we apply the **penalty method** with possessing state and control. We have the pairs (v, y) that are elements of Banach space $X = V \times Y$, where $V = L_2(\Omega)$, $Y = H_0^1(\Omega)$.

Determine the functional

$$I_\varepsilon(v, y) = \frac{\alpha}{2} \int_\Omega v^2 dx + \frac{1}{2} \int_\Omega \sum_{i=1}^n \left| \frac{\partial y}{\partial x_i} - \frac{\partial y_d}{\partial x_i} \right|^2 dx + \frac{1}{2\varepsilon} \int_\Omega (\Delta y + v + f)^2,$$

where $\varepsilon > 0$. If the function is the solution of the boundary problem (3.9), (3.10) for the control v, then the value under the third integral of the functional I_ε is equal to zero. Then we have the equality

$$I_\varepsilon(v, y[v]) = I(v) \ \forall v \in U, \varepsilon > 0.$$

Remark 3.43 If the parameter ε is small enough, then the coefficient before the third integral of the functional I_ε is large enough. Therefore, the smallness of this functional can be realized for the small values under this integral, i.e., $(\Delta y + v + f)^2$. If $\varepsilon \to 0$, we hope that this value tends to zero, i.e., the function is the solution of the problem (3.9), (3.10) for the control v, as we hope to obtain the solution of Problem 3.2 after passing to the limit because of the last equality. Certainly, this requires the justification.

Note that the last integral of the functional I_ε can be unbounded for $y \in H_0^1(\Omega)$ because we cannot guarantee that the function under the integral is an element of the space $L_2(\Omega)$. This property is true if $f \in L_2(\Omega)$ and $\Delta y \in L_2(\Omega)$. Therefore, we suppose the function f belongs to the space $L_2(\Omega)$. The functional I_ε is bounded if its second argument y is an element of the set

$$M = \{ y \in Y \, | \, \Delta y \in L_2(\Omega) \}.$$

This is a linear subspace of Y because the linear combination of the element of the set M belongs to this set.

Lemma 3.11 *If $f \in L_2(\Omega)$, then for all $v \in L_2(\Omega)$ the problem (3.9), (3.10) has a unique solution $y = y(v)$ from the set M.*

Indeed, the existence of the unique solution of this boundary problem in the space Y follows from Theorem 3.5. This is an element of the set M because the right side of the equality (3.9) belongs to the space $L_2(\Omega)$.

Remark 3.44 It is natural that a more regular solution of the boundary problem exists for a more regular value of the parameters of the system.

Determine the subset $W = U \times M$ of the space X. For any pairs (v, y) and (v', y') of the set W and arbitrary constant $\sigma \in (0, 1)$ we have the equality

$$\sigma(v, y) + (1 - \sigma)(v', y') = (\sigma v + (1 - \sigma)v', \sigma y + (1 - \sigma)y').$$

Note that W is the product of the convex set and the linear space. Therefore, the value in the right side of the last equality belongs to the set W. Thus, we have a convex set.

Consider the following extremum problem.

Linear systems   117

Problem 3.4 *Find the pair (v, y) that minimizes the functional I_ε on the set W.*

This is the problem of the minimization of the functional on the convex subset of a Hilbert space. Therefore, the methods of Chapter 1 are applicable for it. We can prove the solvability of Problem 3.4. Then it is possible to pass to the limit as $\varepsilon \to 0$ and prove that its solution tends to the solution of Problem 3.2. Necessary conditions of optimality for this problem are obtained by using Theorem 1.8. Finally, after passing to the limit we could obtain the optimality condition for Problem 3.2. However, we shall not substantiate these results, because the penalty method is far from our research.

Remark 3.45 The penalty method is used frequently if it is given a state constraint. Then it is possible to determine the functional that is an analog of I_ε and includes the special term. This is a "penalty" for the violation of this constraint.

Remark 3.46 The penalty method can be the means of the analysis of optimal control problems for the case of the absence of the information about the unique solvability of the state equation. Then we can interpret this equation as a constraint and apply the penalty method.

Remark 3.47 The penalty method is used sometimes as a means of finding the approximate solution of the problem. Particularly, if we have the convergence of the solution of Problem 3.2 as $\varepsilon \to 0$ we can determine the solution of Problem 3.4 as an approximate solution of Problem 3.2 for small enough values of the parameter ε.

3.10   Additional properties of Sobolev spaces

We considered before the homogeneous Dirichlet problem for the Poisson equation. The analysis of other problems uses additional properties of the Sobolev spaces theory. Let Ω be an open bounded set of $\mathbb{R}^n$ with a smooth enough boundary Γ.

Remark 3.48 We do not clarify this property because this does not have any relation to our general problems.

We try to determine Sobolev spaces with fractional degrees. Its easiest definition for $\Omega = \mathbb{R}^n$ is based on the Fourier transformation.

Definition 3.2 *Fourier transformation* of the function y on the set $\mathbb{R}^n$ is the function $F(y)$ that is determined by the equality

$$F(y)(x) = \int_{\mathbb{R}^n} \exp(-2\pi i x \cdot \xi) y(x) dx, \ \xi \in \mathbb{R}^n,$$

where $x \cdot \xi = x_1 \xi_1 + ... + x_n \xi_n$.

Optimization and Differentiation

Fourier transformation is the isomorphism on the space $L_2(\mathbb{R}^n)$, as

$$\|y\| = \|F(y)\|.$$

The inverse operator F^{-1} is called the **inverse Fourier transformation**. We have

$$F^{-1}(y)(x) = \int_{\mathbb{R}^n} \exp(2\pi i x \cdot \xi) y(\xi) d\xi, \ x \in \mathbb{R}^n.$$

There exists the possibility to extend the direct and the inverse Fourier transformation to the distributions. For all multi-index $\alpha = (\alpha_1, ..., \alpha_n)$ we have

$$F(D^\alpha y)(\xi) = (2\pi i)^{|\alpha|} \xi_1^{\alpha_1} ... \xi_n^{\alpha_n}, \ \forall \xi \in \mathbb{R}^n, y \in L_2(\mathbb{R})^n.$$

Consider Sobolev space

$$H^m(\mathbb{R}^n) = \left\{ y \mid \xi^\alpha F(y) \in L_2(\mathbb{R})^n, \ 0 \le |\alpha| \le m \right\},$$

where $\xi^\alpha = \xi_1^{\alpha_1} ... \xi_n^{\alpha_n}$. We can determine it also by the equality

$$H^m(\mathbb{R}^n) = \left\{ y \mid \left(1 + |\xi|^2\right)^{m/2} F(y) \in L_2(\mathbb{R})^n \right\}.$$

Then we determine the scalar product

$$(y, z) = \left(\left(1 + |\xi|^2\right)^{m/2} F(y), \left(1 + |\xi|^2\right)^{m/2} F(z) \right)_{L_2(\mathbb{R})^n} =$$

$$\int_{\mathbb{R}^n} \left(1 + |\xi|^2\right)^{m/2} F(y)(\xi) F(z)(\xi) d\xi.$$

This is equivalent to the standard definition of the scalar product for Sobolev spaces (see Chapter 1). Note that the number m can be a non-integer here. Hence, we can determine Sobolev spaces with non-integer degrees.

Definition 3.3 *For any real number s determine **Sobolev space***

$$H^s(\mathbb{R}^n) = \left\{ y \mid \left(1 + |\xi|^2\right)^{s/2} F(y) \in L_2(\mathbb{R})^n \right\}.$$

The space $H^s(\mathbb{R}^n)$ is a Hilbert with the scalar product

$$(y, z) = \left(\left(1 + |\xi|^2\right)^{s/2} F(y), \left(1 + |\xi|^2\right)^{s/2} F(z) \right)_{L_2(\mathbb{R})^n} =$$

$$\int_{\mathbb{R}^n} \left(1 + |\xi|^2\right)^{s/2} F(y)(\xi) F(z)(\xi) d\xi.$$

Its adjoint space is denoted by $H^{-s}(\mathbb{R}^n)$.

Linear systems 119

Definition 3.4 Sobolev space $H^s(\Omega)$ *is the set of all restrictions of the elements of* $H^s(\mathbb{R}^n)$ *to* Ω.

The set $H^s(\Omega)$ is a Hilbert space with the norm

$$\|y\|_{H^s(\Omega)} = \inf \|\hat{y}\|_{\mathbb{R}^n},$$

where $\hat{y} = y$ a.e. on Ω. The Sobolev space $H_0^s(\Omega)$ is the closure of $D(\Omega)$ to $H^s(\Omega)$. For $s > 0$ denote by $H^{-s}(\Omega)$ the space that is adjoint to $H_0^s(\Omega)$. For all real numbers s that are not equal to $1/2$ the differentiation $\partial/\partial x_i$ is the linear continuous operator from the space $H^s(\Omega)$ to $H^{s-1}(\Omega)$. For all real numbers s embedding of the space $H^s(\Omega)$ to $H^{s-\varepsilon}(\Omega)$ is compact, where ε is an arbitrary positive number.

We have general interest in the case $s = 1/2$ for the functions that are definite on the boundary Γ. Determine the corresponding Sobolev space. There exists a linear continuous operator $\gamma_0 : H^1(\Omega) \to L_2(\Gamma)$ that is called the **trace operator** such that the value $\gamma_0 y$ is equal to the restriction on Γ of the arbitrary function $y \in C^2(\overline{\Omega})$. The following equality holds:

$$H_0^1(\Omega) = \Big\{ y \in H^1(\Omega) \big| \ \gamma_0 y = 0 \Big\}.$$

Definition 3.5 Sobolev space $H^{1/2}(\Gamma)$ *is the image of the trace operator* $\gamma_0\big(H^1(\Omega)\big)$.

The set $H^{1/2}(\Gamma)$ is the dense subspace of $L_2(\Gamma)$ as embedding of $H^{1/2}(\Gamma)$ to $L_2(\Gamma)$ is compact. **Sobolev space** $H^{-1/2}(\Gamma)$ is the space that is adjoint with $H^{1/2}(\Gamma)$. The space $L_2(\Gamma)$ is self-adjoint. Therefore, we have continuous embedding $L_2(\Gamma) \subset H^{-1/2}(\Gamma)$.

For any function $y \in H^m(\Omega)$ determine its value and the value of the normal derivatives of the degrees that is not greater than $m - 1$. These are called the **traces**. Its properties are described by the **Trace theorem**.

Theorem 3.15 *For any function* $y \in H^m(\Omega)$ *there exist the traces*

$$y|_\Gamma, \ \frac{\partial y}{\partial n}\Big|_\Gamma, ..., \frac{\partial^{m-1} y}{\partial nm - 1}\Big|_\Gamma;$$

additionally, the **trace operator** γ_k *that is determined by the equality*

$$\gamma_k y = \frac{\partial^k y}{\partial n^k}\Big|_\Gamma$$

is the linear continuous operator from the space $H^m(\Omega)$ *to* $H^{m-k-1/2}(\Gamma)$. *This operator is the surjection.*

Remark 3.49 More exact, the map $\gamma_k : D(\Omega) \to D(\Gamma)$ can be extended to a linear continuous operator $H^m(\Omega)$ to $H^{m-k-1/?}(\Gamma)$.

120   *Optimization and Differentiation*

By Trace theorem, there exists a positive number $c = c(m, k, \Omega)$ such that

$$\left\|\frac{\partial^k y}{\partial n^k}\Big|_\Gamma\right\|_{H^{m-k-1/2}(\Gamma)} \le c\|y\|_{H^m(\Omega)}.$$

Note **Green's formula**

$$-\int_\Omega y\Delta z dx + \int_\Gamma y\frac{\partial z}{\partial n}dx = \int_\Omega \sum_{i=1}^n \frac{\partial y}{\partial x_i}\frac{\partial z}{\partial x_i}dx =$$

$$-\int_\Omega z\Delta y dx + \int_\Gamma z\frac{\partial y}{\partial n}dx \quad \forall y, z \in H^1(\Omega),$$

where the values of the considered functions on the boundary are the elements of the space $H^{1/2}(\Gamma)$, and its normal derivatives belong to $H^{-1/2}(\Gamma)$.

There exists the extension of this result. Consider the functions $a_{ij} \in C^1(\overline{\Omega})$, where $i, j = 1, ..., n$. **Conormal derivative** of the function y that is determined on the set Ω is

$$\frac{\partial y}{\partial \nu} = \sum_{i,j=1}^n a_{ij}\frac{\partial y}{\partial x_i}\cos(n, x_i),$$

where $\cos(n, x_i)$ is the i-th direction cosine of the exterior normal n to the surface Γ. Consider the following variant of **Green's formula**

$$-\int_\Omega \sum_{i,j=1}^n a_{ij}\frac{\partial y}{\partial x_j}\frac{\partial z}{\partial x_i}dx = \int_\Omega \sum_{i,j=1}^n \frac{\partial}{\partial x_i}\left(a_{ij}\frac{\partial y}{\partial x_j}\right)z dx - \int_\Gamma z\frac{\partial y}{\partial \nu}dx =$$

$$\int_\Omega \sum_{i,j=1}^n \frac{\partial}{\partial x_i}\left(a_{ji}\frac{\partial z}{\partial x_j}\right)y dx - \int_\Gamma y\frac{\partial z}{\partial \nu^*}dx \quad \forall y, z \in H^1(\Omega),$$

where

$$\frac{\partial y}{\partial \nu^*} = \sum_{i,j=1}^n a_{ji}\frac{\partial y}{\partial x_j}\cos(n, x_j).$$

Determine the linear continuous operator $L : \left[H^2(\Omega) \cap H^1_0(\Omega)\right] \to L_2(\Omega)$ by the equality

$$Ly = -\sum_{i,j=1}^n \frac{\partial}{\partial x_i}\left(a_{ij}\frac{\partial y}{\partial x_j}\right) + a_0 y.$$

Theorem 3.16 *The operator L is the isomorphism.*

Consider the equation

$$-\sum_{i,j=1}^n \frac{\partial}{\partial x_i}\left(a_{ij}\frac{\partial y}{\partial x_j}\right) + a_0 y = f, \ x \in \Omega,$$

Linear systems 121

with the homogeneous boundary condition

$$y = 0, \ x \in \Gamma.$$

By Theorem 3.16 for all functions $f \in L_2(\Omega)$ this boundary problem has a unique solution y from the space $H^2(\Omega) \cap H_0^1(\Omega)$, as

$$\|y\|_{H^2(\Omega)\cap H_0^1(\Omega)} = \|f\|_{L_2(\Omega)}.$$

Remark 3.50 By Theorem 3.5 for all $f \in H^{-1}(\Omega)$ this boundary problem is one-valued solvable in the space $H_0^1(\Omega)$. If the given function f has the stronger properties, the solution obtains the additional properties because of Theorem 3.16. This result is called the *smoothness* of the solution.

Now we have the possibility to analyze more difficult boundary problems for the linear elliptic equations.

Remark 3.51 Now we are going to consider the *nonhomogeneous Dirichlet problem*. We shall use also Sobolev spaces with fractional degrees for the analysis of the *Neumann problem* (see Chapter 4).

3.11 Nonhomogeneous Dirichlet problem for the elliptic equation

We considered the Poisson equation with the homogeneous boundary condition. Now we analyze an optimal control problem for the system described by the nonhomogeneous Dirichlet problem for the general linear elliptic equation. Let Ω be an open bounded set of $\mathbb{R}^n$ with a smooth enough boundary Γ.

Remark 3.52 We suppose that the boundary is smooth enough such that we can apply the previous results.

Consider the equation

$$-\sum_{i,j=1}^{n} \frac{\partial}{\partial x_i}\left(a_{ij}\frac{\partial y}{\partial x_j}\right) + a_0 y = f_\Omega, \ x \in \Omega \qquad (3.25)$$

with the nonhomogeneous boundary condition

$$y = f_\Gamma, \ x \in \Gamma. \qquad (3.26)$$

This is the **nonhomogeneous Dirichlet problem**. The coefficients a_{ij}, a_0 satisfy the inclusions $a_{ij} \in C^1(\overline{\Omega})$, $a_0 \in C(\overline{\Omega})$ and the inequalities

$$\sum_{i,j=1}^{n} a_{ij}(x)\xi_i\xi_j \geq \chi|\xi|^2 \ \forall \xi \in \mathbb{R}^n, \ a_0(x) \geq 0, \ x \in \Omega,$$

where $\chi > 0$.

122   *Optimization and Differentiation*

Theorem 3.17 *For all $f_\Omega \in L_2(\Omega)$, $f_\Gamma \in H^{-1/2}(\Omega)$ the problem (3.25), (3.26) has a unique solution $y \in L_2(\Omega)$.*

Proof. Determine the linear continuous operator

$$A^* : \left[H^2(\Omega) \cap H_0^1(\Omega)\right] \to L_2(\Omega)$$

by the formula

$$A^* z = -\sum_{i,j=1}^{n} \frac{\partial}{\partial x_i}\left(a_{ji} \frac{\partial z}{\partial x_j}\right) + a_0 z.$$

It has all properties of the operator L from Theorem 3.16. Then this is the isomorphism between reflexive spaces. Therefore, it has the adjoint operator that is equal to A. Hence, the operator equation

$$Ay = f \tag{3.27}$$

has a unique solution $y \in L_2(\Omega)$ for all $f \in \left[H^2(\Omega) \cap H_0^1(\Omega)\right]'$.

By Trace theorem for all function $z \in \left[H^2(\Omega) \cap H_0^1(\Omega)\right]$ we have

$$z\big|_\Gamma \in H^{3/2}(\Gamma), \ \frac{\partial z}{\partial \nu^*}\Big|_\Gamma \in H^{1/2}(\Gamma).$$

Then the equality

$$\langle f, z \rangle = \int_\Omega z f_\Omega dx - \int_\Gamma \frac{\partial z}{\partial \nu^*} f_\Gamma dx \ \forall z \in \left[H^2(\Omega) \cap H_0^1(\Omega)\right]$$

determines an element $f \in \left[H^2(\Omega) \cap H_0^1(\Omega)\right]'$.

The equation (3.27) can be transformed to the equality

$$\langle Ay, z \rangle = \int_\Omega z f_\Omega dx - \int_\Gamma \frac{\partial z}{\partial \nu^*} f_\Gamma dx \ \forall z \in \left[H^2(\Omega) \cap H_0^1(\Omega)\right].$$

Using the definition of the adjoint operator, we have

$$\langle Ay, z \rangle = \langle y, A^* z \rangle = \int_\Omega \left[-\sum_{i,j=1}^{n} \frac{\partial}{\partial x_i}\left(a_{ji} \frac{\partial z}{\partial x_j}\right) + a_0 z\right] y dx.$$

By Green's formula, we obtain

$$\int_\Omega z f_\Omega dx - \int_\Gamma \frac{\partial z}{\partial \nu^*} f_\Gamma dx = \int_\Omega \left[-\sum_{i,j=1}^{n} \frac{\partial}{\partial x_i}\left(a_{ji} \frac{\partial z}{\partial x_j}\right) + a_0 z\right] y dx =$$

$$
\int_\Omega \left[-\sum_{i,j=1}^n \frac{\partial}{\partial x_i}\left(a_{ij}\frac{\partial y}{\partial x_j}\right) + a_0 y\right] z\, dx - \int_\Gamma \frac{\partial z}{\partial \nu^*} y\, dx.
$$

Thus, there exists a unique function $y \in L_2(\Omega)$ such that

$$
\int_\Omega \left[-\sum_{i,j=1}^n \frac{\partial}{\partial x_i}\left(a_{ij}\frac{\partial y}{\partial x_j}\right) + a_0 y\right] z\, dx - \int_\Gamma \frac{\partial z}{\partial \nu^*} y\, dx =
$$

$$
\int_\Omega z f_\Omega\, dx - \int_\Gamma \frac{\partial z}{\partial \nu^*} f_\Gamma\, dx \quad \forall z \in \left[H^2(\Omega) \cap H_0^1(\Omega)\right]. \tag{3.28}
$$

Choose $\partial z/\partial \nu^* = 0$. We obtain

$$
\int_\Omega \left[-\sum_{i,j=1}^n \frac{\partial}{\partial x_i}\left(a_{ij}\frac{\partial y}{\partial x_j}\right) + a_0 y - f_\Omega\right] z\, dx = 0.
$$

The function z is arbitrary here. Then the function y satisfies the equality (3.25). Therefore, the equality (3.28) can be transformed to

$$
\int_\Gamma \frac{\partial z}{\partial \nu^*} y\, dx = \int_\Gamma \frac{\partial z}{\partial \nu^*} f_\Gamma\, dx.
$$

Using the arbitrariness of the function z we obtain the boundary condition (3.26). $\square$

Determine an optimal control problem. Consider the boundary problem

$$
-\sum_{i,j=1}^n \frac{\partial}{\partial x_i}\left(a_{ij}\frac{\partial y}{\partial x_j}\right) + a_0 y = v + f_\Omega, \ x \in \Omega, \tag{3.29}
$$

with nonhomogeneous boundary condition

$$
y = f_\Gamma, \ x \in \Gamma. \tag{3.30}
$$

The control v is an element of a convex closed subset U of the space $V = L_2(\Omega)$. By Theorem 3.17, for all $v \in V$ the boundary problem has a unique solution $y = y[v]$ from the space $Y = L_2(\Omega)$. Determine the functional

$$
I(v) = \frac{\alpha}{2}\int_\Omega v^2 dx + \frac{1}{2}\int_\Omega \left(y[v] - y_d\right)^2 dx,
$$

where $\alpha > 0$, $y_d \in L_2(\Omega)$ is a known function.

Problem 3.5 *Find a control $u \in U$ that minimizes the functional I on the set U.*

124 *Optimization and Differentiation*

Using the linearity of the system, we determine the following result that is an analog of Theorem 3.6 and Theorem 3.7.

Theorem 3.18 *Problem 3.5 has a unique solution and is Tihonov well-posed.*

Determine the differentiability of the minimizing functional.

Lemma 3.12 *The functional I for Problem 3.5 has Gâteaux derivative $I'(u) = \alpha u + p$ at the arbitrary point $u \in V$, where p is the solution of the problem*

$$- \sum_{i,j=1}^{n} \frac{\partial}{\partial x_i} \left(a_{ji} \frac{\partial p}{\partial x_j} \right) + a_0 p = y[u] - y_d, \ x \in \Omega, \tag{3.31}$$

$$p = 0, \ x \in \Gamma. \tag{3.32}$$

Proof. For all functions $h \in V$ and the number σ we have

$$\frac{I(u+\sigma h) - I(u)}{\sigma} = \frac{\alpha}{2} \int_{\Omega} (2uh + \sigma h^2) dx + \frac{1}{2} \int_{\Omega} (y[u+\sigma h] + y[u] - 2y_d) \eta_\sigma[h] dx,$$

where

$$\eta_\sigma[h] = (y[u + \sigma h] - y[u])/\sigma.$$

This function is the solution of the homogeneous Dirichlet problem for the equation

$$- \sum_{i,j=1}^{n} \frac{\partial}{\partial x_i} \left(a_{ij} \frac{\partial \eta_\sigma[h]}{\partial x_j} \right) + a_0 \eta_\sigma[h] = h. \tag{3.33}$$

Multiply this equality by an arbitrary function $\lambda \in \left[H^2(\Omega) \cap H_0^1(\Omega) \right]$. Using Green's formula we have

$$\int_{\Omega} \left[- \sum_{i,j=1}^{n} \frac{\partial}{\partial x_i} \left(a_{ji} \frac{\partial \lambda}{\partial x_j} \right) + a_0 \lambda \right] \eta_\sigma[h] dx = \int_{\Omega} \lambda h dx.$$

The term at the right-hand side of the equality (3.31) is the element of the space $L_2(\Omega)$. The problem (3.31), (3.32) is one-valued solvable in the space $H^2(\Omega) \cap H_0^1(\Omega)$. Determine $\lambda = p$ at the previous equality. We get

$$\int_{\Omega} (\alpha u + p) \eta_\sigma[h] dx = \int_{\Omega} p h dx.$$

Then we obtain the equality

$$\frac{I(u+\sigma h) - I(u)}{\sigma} = \int_{\Omega} (\alpha u + p) h dx + \frac{\alpha \sigma}{2} \int_{\Omega} h^2 dx + \frac{\sigma}{2} \int_{\Omega} \left(\eta_\sigma[h] \right)^2 dx.$$

The solution of the Dirichlet problem for the equation (3.17) is bounded in the space $L_2(\Omega)$. Pass to the limit here. We obtain

$$\langle I'(u), h \rangle = \int_{\Omega} (\alpha u + p) h dx \ \forall h \in V.$$

This completes the proof of the lemma. $\square$

Using the convexity of the given functional that is the corollary of the linearity of the system we obtain the following result.

Theorem 3.19 *The control u is the solution of Problem 3.5 if and only if it satisfies the variational inequality*

$$\int_{\Omega} (\alpha u + p)(v - u) dx \geq 0 \ \forall v \in U,$$

where p is the solution of the problem (3.31), (3.32).

Our next step is an analysis of the optimal control problems for systems described by nonlinear equations. This is the general objective of our research.

3.12 Comments

The necessary information about the functional analysis is given in the books of B. Baeuzamy [38], H. Brésis [91], H. Gajewski, K. Gröger, and K. Zacharias [200], A. Gleason [213], V. Hutson and J.S. Pym [247], K. Iosida [256], L.V. Kantorovich and G.P. Akilov [267], A.N. Kolmogorov and S.V. Fomin [276], S.G. Krein and others [283], S.S. Kutateladze [291], L.A. Lusternik and V.I. Sobolev [334], M. Reed and B. Simon [426], W. Rudin [439], L. Schwartz [447], and V.A. Trenogin [530].

The general optimal control problems are considered in the books of V.M. Alekseev, V.M. Tihomirov and S.V. Fomin [14], M. Aoki [20], A. Balakrishnan [41], A. Balakrishnan and L. Neustadt [42], R. Bellman [57], V.G. Boltyansky [78], [79]. V.V. Dikussar and A.A. Milutin [141], W.H. Fleming and R.W. Rishell [178], R. Fletcher [179], A.V. Fursikov [197], R. Gabasov and F.M. Kirillova [198], R.V. Gamkrelidze [202], P.E. Gill, W. Murrey, and M.H. Wright [212], A.D. Ioffe and V.M. Tihomirov [253], V.F. Krotov [285], V.F. Krotov, V.Z. Bukreev, and L.I. Gurman [286], V.F. Krotov and L.I. Gurman [287], G. Leitmann [307], X. Li and J. Yong [313], J.L. Lions [315], A.S. Matveev and V.A. Yakubovich [343], A.A. Milutin [359], L. Neustadt [386], L.S. Pontryagin, V.G. Boltyansky, R.V. Gamkrelidze, E.F. Mishchenko [408], B.N. Pshenichny [414], [415], L.I. Rozonoer [437], V.M. Tihomirov [522], F.P. Vasiliev [538], [539], J. Warga [553], L. Young [563], and J. Zabkzyk [565].

Optimal control problems for the distributed control systems are considered in the books of V.I. Agoshkov [6], J.-L. Armand [23], A.G. Butkovsky [101], [102], A.I. Egorov [154], [155], [447], A.V. Fursikov [197], V.I. Ivanenko and V.S. Melnik [260], I. Lasiecka and R. Triggiani [299], X. Li and J. Yong [313], J.L. Lions [314], [317], V.G. Litvinov [320], K.A. Lurie [333], A.S. Matveev and V.A. Yakubovich [343], U. Raitums [422], T.K. Sirazetdinov [503]. A. Abuladze and R. Klotzler [2], K.L. Ahmed and N.U. Teo [11], V.L. Bakke [39], M.L. Bennati

126   *Optimization and Differentiation*

[60], G. Da Prato and A. Ichikawa [135], N. Medhin [348], S.V. Morozov and V.I. Sumin [372], T. Roubiček [435], W. Schmidt [446] C. Simionescu [502], T.K. Sirazetdinov [503], and C. Trenchea [529] consider optimization problems for the integral and integro-differential equations. I. Bock and J. Lovisek [76], J.F. Bonnans and E. Casas [84], Q. Chen, D. Chu, and R.C.E. Tan [123], F. Mignot and J. P. Puel [354], F. Patrone [397], C. Saguez [444], and S.Ya. Serovajsky [496], [497] analyze optimization problems for the system described by variational inequalities.

Optimal control problems for the case of the absence of the one-value solvability of the systems are considered by A.V. Fursikov [194], [195], [196], [197], and J.L. Lions [314], [317] (see also H. Gao and N.H. Pavel [203], and S.Ya. Serovajsky [473], [475], [476], [480], [481], [485]).

Optimization problems for the infinite dimensional problems with state constraints are considered by F. Abergel and R. Temam [1], N.V. Andreev and V.S. Melnik [17], N. Arada and J.P. Raymond [21], M. Bergounioux [62], [63], J.F. Bonnans and E. Casas [84], E. Casas [110], [111], E. Casas and F. Tröltzsch [115], A. Cheng and K. Morris [124], U. Mackenroth [336], H. Maurer and H. Mittelmann [344], V.S. Melnik [350], P. Michel [353], S.Ya. Serovajsky [457], [473], M.I. Sumin [513], and others.

The theory of the linear elliptic equation is given, for example, in the book of S. Agmon [5], A.V. Bitsadze [72], V.S. Vladimirov [543], S. Farlow [168], A. Friedmann [186], O.A. Ladyzhenskaya and N. N. Uraltseva [294], J.L. Lions [315], J.L. Lions and E. Magenes [319], Ya.B. Lopatinsky [323], S.G. Mihlin [355], [356], C. Miranda [362], S. Mizohata [363], and P.K. Rashevsky [423].

The methods of optimization for the systems described by linear elliptic equations are considered by N.V. Banichuk [45], J. Baranger [47], R. Becker, H. Kapp, and R. Rannacher [56], M.F. Bidaut [69], [70], M. Bergounioux [62], [63], A. Cheng and K. Morris [124], A.I. Egorov [155], I. Ekeland [158], R.P. Fedorenko [173], N. Fujii [192], [193], I.M. Gali [201], M. Goebel [214], I.K. Gogodze [216], W.M. Hackbusch [234], M. Kostreva and A.L. Ward [278], Yu.P. Krivenkov [284], R. Li, W. Liu, H. Ma and T. Tang [312], J.L. Lions [314], [315], [318], V.G. Litvinov [321], K.A. Lurie [333], C. Meyer and A. Rösch [352], F. Murat and L. Tartar [375], A. Myslinski and J. Sokolowski [376], V.I. Plotnikov [404], C. Pucci [417], [418], U. Raitums [422], L. Tartar [515], D. Tiba [521], L. Wolfersdorf [555], [554], T. Zolezzi [571], and others. Particularly, L. Wolfersdorf [555], [554] considers optimal control problems on the complex plane; N.V. Banichuk [45], J. Baranger [47], N. Fujii [192], [193], J.L. Lions [315], [318], V.G. Litvinov [321], F. Murat and L. Tartar [375], and D. Tiba [521] analyze set control problems; J.L. Lions solves problems with impulse control and observation, the equation of four orders [315], an equation of the arbitrary order [314], and multicriterial problems [318]; A. Myslinski and J. Sokolowski [376] consider a coefficient control problem for a four order equation; I.M. Gali [201] considers infinite order equations, M.M. Kostreva and A. L. Ward [278] and I.K. Gogodze [216] consider minimax problems; M. Bergounioux [62], [63], A. Cheng and K. Morris [124] consider the well-posedness for a boundary control problem. The extension methods for the systems described by linear elliptic equations are considered by J. Baranger [47], M.F. Bidaut [69], [70], K.A. Lurie [333], U. Raitums [422], L. Tartar [515]. R. Becker, H. Kapp, and R. Rannacher [56], M.M. Kostreva, and A.L. Ward [278], R. Li, W. Liu, H. Ma, and T. Tang [312] use finite elements method for solving these problems.

J.L. Lagrange proposed the Lagrange multiplier method for the different problems of mechanics. A.D. Ioffe and V.M. Tihomirov [253] and V.M. Tihomirov [522] apply the Lagrange principle for the abstract extremum problems with constraints. C.J. Goh and X.Q. Yang [217] propose the nonlinear Lagrange function for non-convex optimization problems. Maximum principle was by proposed by L.S. Pontryagin, V.G. Boltyansky, R.V. Gamkrelidze, and E.F. Mishchenko [408] for the systems described by ordinary differential equations (see also L.D. Berkovitz [65], R. Gabasov and F.M. Kirillova [198], [199], E. Martinez [341], and L.I. Rozonoer [437]). Results of this direction were obtained earlier by L. Graves [222]. Pontryagin's maximum principle is close enough to the well-known Weierstrass condition of strong minimum of variations calculus. The singular controls theory is given by R. Gabasov and F.M. Kirillova [199] (see also Ya.M. Bershchansky [66], S.Ya. Serovajsky [482], and B.P. Yeo [561]).

Linear systems 127

Another popular optimization method is dynamic programming, developed by R. Bellman [57] (see also V.M. Alekseev, V.M. Tihomirov, and S.V. Fomin [14], V. Barbu [50], M. Bardi and S. Bottacin [54], G. Buttazzo, G. Mariano, and S. Hildebrandt [104], R. Cannarsa and O. Cârjua [107], A.K. Chaudhuri [121], A.I. Egorov [154], [155], W.H. Fleming and R.W. Rishell [178], A. Just [264], V.F. Krotov [285], J.L. Lions [315], A.J. Pritchard and M.J.E. Mayhew [413], N.N. Moiseev [364], and T.K. Sirazetdinov [503]). This is an analog of the Hamilton–Jacobi theory of the calculus of variations (see, for example, N.I. Ahiezer [7], G. Bliss [75], O. Bolza [82], G. Buttazzo, G. Mariano, and S. Hildebrandt [104], E. Elsgolts [164], I.M. Gelfand and S.V. Fomin [205], M.A. Lavrentiev and L.A. Lusternik [300], G. Leitmann [307], and L. Zlaf [570]). H.J. Pesch and R. Bulirsch [400] note that C. Caratheodory got some results in maximum principle and dynamic programming.

The *successive approximation method* is usually applied for practical solving of the system including optimality conditions (the stationary condition, the variational inequality, the maximum principle, and some others), state equation, and the adjoint equation (see, for example, P. Burgmeier and H. Jasinski [99], F.L. Chernousko and N.V. Banichuk [125], F.L. Chernousko and V.V. Kolmanovsky [126], C. Cuvelier [134], R.P. Fedorenko [173], H.J. Kelley, R.E. Kopp, and H.G. Moyer [269], I.A. Krylov and F.L. Chernousko [288], and O.V. Vasiliev [540]). Each of the above relations are solved successively one by one when found in previous iterations of the values of the considered functions. *Newton–Raphson method* is frequently used for solving necessary conditions of optimality for the systems described by the ordinary differential equations (see H.A. Antosiewicz [19], F.L. Chernousko and V.V. Kolmanovsky [126], R.P. Fedorenko [173], G. Fraser-Andrews [184], V.K. Isaev and V.V. Sonin [257], R. McGill [345], N.N. Moiseev [365], E. Polak [406], and M. Sidar [501]).

The penalty method was proposed by R. Courant [130] for the minimization problem for the multiple integral to the membrane theory. L.D. Berkovitz [65] uses it for the substantiation of the maximum principle. B.Sh. Mordukhovich [366] applies it for obtaining the optimality conditions for non-smooth problems. J.L. Lions [317] analyzes the optimal control problems for the singular systems, and N.G. Medhin [347] analyzes the minimax problems by using the penalty method. Note also its modifications. This is the adapted penalty method (see V. Barbu [49]) and approximated penalty method (see S.Ya. Serovajsky [481]). Applications of the penalty method for obtaining the conditions of optimality are considered also by A. Bensoussan [61], M. Bergounioux [62], A.E. Bryson and Ho Yu-Chi [95], T. Butler and A.V. Martin [103], J.L. Lions [314], [315], [317], and S.Ya. Serovajsky [473], [476], [481], [485], [475], [480], [496]. A. Balakrishnan [40], J. Cullum [133], N.N. Moiseev [365], E. Polak [406], B.T. Polyak [407], J. Séa [450], and M. Sibony [500] use it for practical solving of optimal control problems. W. Huyer and A. Neumaier [248] apply it for the analysis of the non-smooth extremum problem.

Fourier transform for the distribution is given in the book by L. Schwartz [448]. J.L. Lions and E. Magenes [319] describe the theory of Sobolev spaces of fractional degrees. The Trace theorem and Theorem 3.15 are considered there. The optimal control problems for the linear elliptic equations with nonhomogeneous boundary conditions are considered by J.L. Lions [315]. J.L. Lions [315], C. Meyer, P. Philip, and F. Tröltzsch [351] analyze optimization problems for the linear elliptic equations with nonlinear boundary conditions.

Chapter 4

Weakly nonlinear systems

4.1	Differentiation of operators	130
4.2	Inverse function theorem	135
4.3	Optimal control problems for weakly nonlinear systems	140
4.4	Equations with monotone operators	144
4.5	Additional results of the functional analysis	149
4.6	Nonlinear elliptic equation	152
4.7	Optimal control problems for nonlinear elliptic equations	157
4.8	Necessary conditions of optimality	159
4.9	Optimal control problems for general functionals	164
4.10	Optimal control problems for semilinear elliptic equations	168
4.11	Differentiation of the inverse operator	173
4.12	Comments	174

We considered the problem of minimization of a functional I on a convex subset U of a Banach space in Chapter 1. The necessary condition of minimum at a point u is the variational inequality

$$\langle I'(u), v - u \rangle \geq 0 \; \forall v \in U, \tag{4.1}$$

where $I'(u)$ is the Gâteaux derivative of the given functional at this point. However, the functional of the optimal control problems depends on the state function $y[v]$ of the system. This state function satisfies an equation that involves a control v as a parameter. Therefore, the determination of the Gâteaux derivative of a functional I has a relation with the analysis of the dependence $y = y[v]$.

Suppose we have the state equation $Ay = Bv + f$, where the operator A and B are linear. Then mapping $v \rightarrow y[v]$ is affine. Therefore, this is a differentiable operator. Hence, we do not have any serious difficulties in the determination of the Gâteaux derivative for our functional. We considered necessary conditions of optimality for the abstract linear control systems with application to the optimal control systems described in linear elliptic equations (see Chapter 3).

Our general subject is optimal control problems for nonlinear equations in infinite dimensional spaces. Let us have the nonlinear operator equation $Ay = Bv + f$ with invertible operator A. Then for any control v there exists its unique solution $y[v] = A^{-1}(Bv + f)$. We can obtain the differentiability of the control–state mapping if the inverse operator A^{-1} is differentiable.

129

130 *Optimization and Differentiation*

Note the important result of the nonlinear functional analysis. This is the Inverse function theorem that gives sufficient conditions of differentiability of the inverse operator. This is a means of proving differentiability for the state function with respect to the control. Using this result, we determine the derivative of the functional. Then we obtain necessary conditions of optimality by using the variational inequality (4.1). The nonlinear control systems are called weakly nonlinear if this technique is applicable. We shall consider elliptic equations with power nonlinearity with small enough value of the degree of the nonlinearity and the dimension of the set. The general case will be considered in the next chapter.

We extend our results to the general integral functionals and to the equations with general nonlinearity. We determine also the sufficient conditions of the differentiability of the inverse operator at the end of the chapter.

4.1 Differentiation of operators

We used the derivatives of the functional for obtaining the necessary conditions of optimality in Chapter 1. This is sufficient for the analysis of the optimal control problems for linear systems (see Chapter 3). Now we consider nonlinear systems. Therefore, it is necessary to analyze the dependence of the state function from the control for determination of the functional derivative. Hence, we shall consider the derivatives of the operators.

Consider an operator $L : V \to Y$, where V, Y are Banach spaces. Suppose u is the point of the space V.

Definition 4.1 *The operator L is called **Gâteaux differentiable** at a point u if there exists a linear continuous operator $L'(u)$, which is called the **Gâteaux derivative** of the operator L at this point such that $[L(u + \sigma h) - Lu]/\sigma \to L'(u)h$ in Y as $\sigma \to 0$ for all $h \in V$.*

Remark 4.1 If we have this convergence, but the limit is not a linear continuous operator with respect to h, then this is the **derivative of the operator** L at the point u **with respect to the direction** h.

For $Y = \mathbb{R}$ we have Gâteaux differentiability of the functional (see Chapter 1). Particularly, the functional I on a space V is Gâteaux differentiable at a point u if for all $h \in V$ we have

$$\lim_{\sigma \to 0} \frac{I(u + \sigma h) - I(u)}{\sigma} \to \langle I'(u), h \rangle.$$

Example 4.1 *Linear and affine operators*. If an operator L is linear, then we have

$$\frac{L(u + \sigma h) - L(u)}{\sigma} = Lh.$$

Weakly nonlinear systems 131

Thus, its derivative at the arbitrary point is equal to the operator L. Each affine operator $L : V \to Y$ is determined by the equality $Lv = L_0 v + y_0$ for all $v \in V$, where the operator L_0 is linear, and y_0 is a fixed point of the space Y. Hence, the derivative of the operator L at the arbitrary point is L_0.

Remark 4.2 We know the analogical property of the linear and affine functionals.

Example 4.2 *The set of functions with many variables.* Consider a finite set of functions $f = (f_1, ..., f_m)$ of variables $x = (x_1, ..., x_n)$. The operator f maps the Euclid space $\mathbb{R}^m$ to $\mathbb{R}^n$. Suppose the continuously differentiability of these functions. We find

$$f(x + \sigma h) = \left(f_1\big(x_1 + \sigma h_1, ..., x_n + \sigma h_n\big), ..., f_m\big(x_1 + \sigma h_1, ..., x_n + \sigma h_n\big) \right) =$$

$$\big(f_1(x_1, ..., x_n), ..., f_m(x_1, ..., x_n)\big) + \sigma \sum_{i=1}^{m} \sum_{j=1}^{n} \frac{\partial f_i(x)}{\partial x_j} h_j + \eta(\sigma),$$

where $\eta(\sigma)/\sigma \to 0$ as $\sigma \to 0$. Therefore, we have the equality

$$f'(x)h = \sum_{i=1}^{m} \sum_{j=1}^{n} \frac{\partial f_i(x)}{\partial x_j} h_j.$$

Thus, the Gâteaux derivative of the operator $f : \mathbb{R}^m \to \mathbb{R}^n$ is a matrix of the partial derivatives $(\partial f_i / \partial x_j)$. The determinant of this matrix is called the *Jacobian* of the considered transformation of Euclid spaces.

If an operator L is Gâteaux differentiable in the convex set M, then for all u, h such that $u + h \in M$ and for all λ from the adjoint space Y' we have *Lagrange formula*

$$\Big\langle \lambda, L(u + h) - Lu \Big\rangle = \Big\langle \lambda, L'\big(u + \delta h\big)h \Big\rangle,$$

where $\delta = \delta(\lambda) \in (0, 1)$.

We use also another operator derivative.

Definition 4.2 *Operator L is called **Fréchet differentiable** at a point u if there exists a linear continuous operator $L'(u) : V \to Y$, which is called a **Fréchet derivative** of the operator L at this point, such that the following equality holds:*

$$L(u + h) = Lu + L'(u)h + \eta(h),$$

where $\|\eta(h)\|_Y / \|h\|_V \to 0$ as $h \to 0$ in V.

Particularly, a functional $I : V \to \mathbb{R}$ is Fréchet differentiable at point u, if there exists a linear continuous functional $I'(u)$ on V, which is called its Fréchet derivative at this point, such that

$$I(u + h) = I(u) + \langle I'(u), h \rangle + \eta(h),$$

132 *Optimization and Differentiation*

where $|\eta(h)| = (\|h\|)$, i.e., $|\eta(h)|/\|h\| \to 0$ as $h \to 0$ in V.

The Fréchet differentiable operator at a point is continuous at this point. Passing to the limit in the equality

$$\frac{L(u + \sigma h) - Lu}{\sigma} = L'(u)h + \eta(\sigma h)/\sigma,$$

as $\sigma \to 0$ determine that Fréchet derivative $L'(u)$ is Gâteaux derivative. The Gâteaux differentiable operator can be discontinuous (see Example 1.13). Certainly, the discontinuous Gâteaux differentiable operator does not have the Fréchet derivative. Therefore, Gâteaux differentiability of operators is a weaker property than Fréchet differentiability.

If we would like to determine the Gâteaux derivative of an operator, we pass to the limit in the term $[L(u + \sigma h) - Lu]/\sigma$ as $\sigma \to 0$. If the result is a linear continuous operator with respect to h, then this operator is a Gâteaux derivative. However, the definition of the Fréchet derivative is not constructive. We can check only whether a concrete linear continuous operator or not is a Fréchet derivative. The relation between these derivatives is used usually for the practical finding of Fréchet derivatives. If there exists a Gâteaux derivative of an operator L on a neighborhood of a point u, and mapping $v \to L'(v)$ is continuous at a point u, then the operator L is Fréchet differentiable at a point u, and its Fréchet derivative is equal to the Gâteaux derivative at this point.

Remark 4.3 We used this idea in Chapter 1 for the determination of Fréchet derivatives of functionals.

Consider differential properties of an important nonlinear operator. Let Ω be an open bounded set of Euclid space $\mathbb{R}^n$. Let a function $F = F(x, y)$ be measurable with respect to $x \in \Omega$ and continuous with respect to $y \in \mathbb{R}$. Then F is called a **Caratheodory function** on a set $\Omega \times \mathbb{R}$. The operator A such that

$$Ay(x) = F[x, y(x)], \ x \in \Omega$$

is called the **Nemytsky operator**. Suppose the following inequality holds:

$$|F(x, y)| \le a(x) + b|y|^{q/r} \ \forall y \in \mathbb{R}, \text{ a.e. on } \Omega, \tag{4.2}$$

where $a \in L_r(\Omega)$, $b > 0$, $q \ge 1$, $r \ge 1$. Then the Nemytsky operator is the continuous operator from $L_q(\Omega)$ to $L_r(\Omega)$ because of **Krasnoselsky theorem**. Suppose in addition $q > r$, and there exists the continuous derivative $F_y(x, y)$ of the function F with respect to the second argument that satisfies the inequality

$$|F_y(x, y)| \le a_1(x) + b_1|y|^{(q/r)-1} \ \forall y \in \mathbb{R}, \text{ a.e. on } \Omega, \tag{4.3}$$

where $a_1 \in L_{qr/(q-r)}(\Omega)$, $b_1 > 0$. Then the operator $A : L_q(\Omega) \to L_r(\Omega)$ is Fréchet differentiable, as its derivative satisfies the equality

$$A'(y)h(x) = F_y[x, y(x)]h(x) \ \forall h \in L_q(\Omega), \ x \in \Omega.$$

Weakly nonlinear systems

Remark 4.4 Nonlinear terms of differential equations are determined frequently by Nemytsky operators.

Example 4.3 *Power operator*. Determine the operator A by the equality $Ay = |y|^\rho y$, where ρ is a positive constant. This is a partial case of a Nemytsky operator with power function $F(x,y) = |y|^\rho y$. The condition (4.2) is true here as the equality, as $a(x) = 0$, $b = 1$, $q = \rho + 2$, $q/r = \rho + 1$. Then we find $r = q' = q/(q-1)$. By Krasnoselsky theorem A is a continuous operator from $L_q(\Omega)$ to $L_{q'}(\Omega)$. By the equality $1/q + 1/q' = 1$ the space $L_{q'}(\Omega)$ is adjoint to $L_q(\Omega)$ (see Chapter 1). The derivative of the function F with respect to the second argument is equal to $(\rho + 1)|y|^\rho$. Then the condition (4.3) is true as the equality, as $a_1(x) = 0$, $b = \rho + 1$. Therefore, the power operator is Fréchet differentiable, and its derivative at point y is determined by the equality

$$A'(y)h = (\rho + 1)|y|^\rho h \ \forall h \in L_q(\Omega).$$

Remark 4.5 By the last equality, it seems that the derivative of the power is determined by the formula $f'(y) = (\rho + 1)|y|^\rho$. However, this equality is not correct. We have the operator in its left-hand side, because the derivative is an operator. But there is the function in its right-hand side. In reality, the derivative of the operator A is the multiplication of the operator argument by the value $(\rho + 1)|y|^\rho$.

Remark 4.6 We shall consider an optimal control problem for the system described by the elliptic equation. It is determined by the operator that it is the sum of the linear differential Laplace operator and the nonlinear power operator.

Suppose our functional I depends upon the state function that depends upon the control. Then it is necessary to differentiate the superposition of two operators for the determination of the functional derivative. Suppose an operator $A : X \to Y$ is Gâteaux differentiable at a point x, and an operator $B : Y \to Z$ is Fréchet differentiable at a point Ax. By the **Composite function theorem** the superposition of operators $C = BA$ that is an operator from the space X to Z is Gâteaux differentiable at the point x, as $C'(x) = B'(Ax)A'(x)$.

Example 4.4 *Fréchet example*. Suppose $X = R$, $Y = R^2$, $Z = R$, and the operators $A : X \to Y$ and $B : Y \to Z$ are determined by the equalities $Ax = (x, x^2)$, $B(y_1, y_2) = y_1$ for $y_2 = (y_1)^2$, and $B(y_1, y_2) = 0$ for $y_2 \neq (y_1)^2$. Note that $A(0) = (0,0)$, $B(A(0)) = 0$. For all $h \in R$ we have the equality $A(\sigma h) - A(0) = (\sigma h, \sigma^2 h^2)$. Therefore, the Gâteaux derivative of the operator A at the zero point is determined by the equality $A'(0)h = eh$, where e is a unit vector $(1,0)$. For all $h_1, h_2 \in R$ find the difference $B(\sigma h_1, \sigma h_2) - B(0,0) = 0$; then the Gâteaux derivative of the operator B at the point (0) is zero. Hence, we find $B'(A(0))A'(0) = 0$. But the operator $C = BA$ is an identical operator on the set of real numbers. Therefore, its derivative at the zero point is equal to 1. Thus, the Composite function theorem is not applicable for this example.

Remark 4.7 By the last example the Fréchet differentiability of the operator B is important for the Composite function theorem.

Remark 4.8 We consider an extension of the Composite function theorem at Chapter 12.

134 *Optimization and Differentiation*

There exists a stronger form of the operator differentiability.

Definition 4.3 *An operator $L : V \to Y$ is called **continuously differentiable** at a point u, if it is Gâteaux differentiable on a neighborhood of this point, as $L'(v_k)h \to L'(v)h$ in Y for all as $v_k \to v$ in V.*

If there exists a Gâteaux derivative of an operator on the neighborhood of a point, and this derivative is continuous at this point, then this operator is Fréchet differentiable there. Therefore, the continuously differentiable operator is Fréchet differentiable.

Example 4.5 *Power operator.* Consider again the power operator $A : L_q(\Omega) \to L_q(\Omega)$ that is determined by the equality $Ay = |y|^\rho y$. It has a Fréchet derivative that satisfies the equality $A'(y)h = (\rho + 1)|y|^\rho h$ for all $h \in L_q(\Omega)$. Determine the value

$$\eta = \frac{1}{\rho + 1} \left\| A'(y)h - A'(z)h \right\|_{q'}^{q'} = \int_\Omega \left| (|y|^\rho - |z|^\rho)h \right|^{q'} dx$$

for all function $y, z \in L_q(\Omega)$, where $\|v\|_p$ is the norm of the function v of the space $L_p(\Omega)$. We have the inequality

$$\int_\Omega |h|^q dx = \int_\Omega \left| |h|^{q'} \right|^{q/q'} dx < \infty.$$

Therefore, $|h|^{q'} \in L_{q/q'}(\Omega)$. The functions $|y|^\rho$, $|z|^\rho$, and its difference $\varphi = |y|^\rho - |z|^\rho$ also belong to the space $L_{q/\rho}(\Omega)$. Then $|\varphi|^{q'} \in L_{q/\rho q'}(\Omega)$. Using the equality

$$\frac{q'}{q} + \frac{\rho q'}{q} = \frac{(1 + \rho)q'}{q} = 1,$$

we determine that the space $L_{q/q'}(\Omega)$ and $L_{q/\rho q'}(\Omega)$ are mutually adjoint. Using Hölder inequality, determine

$$\eta = \int_\Omega |\varphi|^{q'} |h|^{q'} dx \leq \left\| |\varphi|^{q'} \right\|_{q/\rho q'} \left\| |h|^{q'} \right\|_{q/q'} =$$

$$\left| \int_\Omega |\varphi|^{q/\rho} dx \right|^{\rho q'/q} \left| \int_\Omega |h|^q dx \right|^{q'/q} = \|\varphi\|_{q/\rho}^{q'} \|h\|_q^{q'}.$$

Thus, we have the inequality

$$\left\| A'(y)h - A'(z)h \right\|_{q'} \leq \sqrt[q']{\rho + 1} \left\| |y|^\rho - |z|^\rho \right\|_{q/\rho} \|h\|_q.$$

The power operator B is such that $By = |y|^\rho$ is continuous. If we have $y_k \to y$ in $L_q(\Omega)$, then $|y_k|^\rho \to |y|^\rho$ is in $L_{q/\rho}(\Omega)$. Using the previous inequality, we get $A'(y)h \to A'(z)h$ in $L_{q'}(\Omega)$ for all $h \in V$ whenever $y \to z$ in $L_q(\Omega)$. Thus, the power operator is continuously differentiable.

4.2 Inverse function theorem

We consider again the control system described by the operator equation

$$Ay = Bv + f,$$

where y is the state function, v is a control, A is the state operator that maps a Banach space Y to a Banach space Z, B is a control operator that maps a Banach space V to Z, and f is a given element of Z. Suppose there exists the continuous inverse operator A^{-1}. Then for all functions $v \in V$ the state equation has a unique solution $y = y[v]$, as control-state mapping $v \to y[v]$ is continuous. This solution is determined by the formula

$$y[v] = A^{-1}(Bv + f).$$

Let the operator B be linear; but the operator A is nonlinear. Then the differentiability of the dependence $y = y[v]$ is reduced to the analogical property of the inverse operator A^{-1}.

Remark 4.9 We shall consider the optimal control problem with nonlinear control in Chapter 7.

Differentiability of the inverse operator is proved by the very important result of nonlinear functional analysis. This is the ***Inverse function theorem*** (see Figure 4.1). Consider Banach spaces Y, Z, and an operator $A : Y \to Z$ that is continuously differentiable in a neighborhood of a point $y_0 \in Y$.

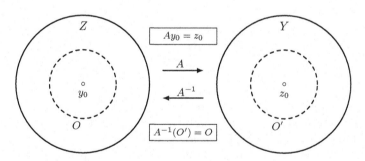

FIGURE 4.1: The Inverse function theorem.

Theorem 4.1 *Suppose there exists the continuous inverse operator A^{-1}. Then there exists an open neighborhood O of the point y_0 such that the set $O' = A(O)$ is the open neighborhood of the point $z_0 = Ay_0$, as there exists the inverse operator $A^{-1} : O \to O'$ that is continuously differentiable on the set O', and its derivative is determined by the formula*

$$(A^{-1})'(z) = \{A'[A^{-1}(z)]\} \ \forall z \in O'. \tag{4.4}$$

136 *Optimization and Differentiation*

Remark 4.10 By this theorem, $A : O \to O'$ is the continuously differentiable bijection such that its inverse operator is continuously differentiable too. An operator with this property is called **diffeomorphism**. Additionally, the sets O and O' are called **diffeomorphic**.

By the Inverse function theorem the equation $Ay = z$ has a unique solution in a neighborhood of the point $z = z_0$ that is continuously differentiable with respect to the absolute term of the equation. Additionally, we can find the derivative by the formula (4.4). The general supposition of this theorem is the invertibility of the derivative of the initial operator. This is true if the linear equation

$$A'(y_0)y = z \qquad (4.5)$$

has a unique solution $y \in Y$ for all functions $z \in Z$. This equation is called the **linearized equation**. The continuity of the inverse operator here is a corollary of the inverse operator Banach theorem.

Remark 4.11 If the operator A is linear, then its derivative at the arbitrary point is equal to initial operator. We considered the state equation with unique solution in Chapter 3. Therefore, the conditions of Inverse function theorem are true.

Consider the Inverse function theorem for the case $Y = \mathbb{R}$, $Z = \mathbb{R}$. Then A is a function of one variable $f = f(y)$. The invertibility of the operator derivative at a point y_0 is equivalent to the inequality $f'(y_0) \neq 0$. The derivative of the inverse function at a point $z = f(y)$ is determined by the formula

$$\left(f^{-1}\right)'(z) = \frac{1}{f'(y)}$$

because of (4.4).

Now consider the properties of the inverse functions for the following examples in neighborhoods of the origin of the coordinates.

Example 4.6 Consider the function $f_1(y) = y$. Its derivative at the point $y = 0$ is $f'_1(0) = 1$. This value is not equal to zero. Therefore, there exists an inverse function in the neighborhood of the point $z = f_1(0)$ (see Figure 4.2) that is differentiable because of the Inverse function theorem. In reality, the inverse function is determined by the formula $f_1^{-1} = z$. Its derivative at the point $z = 0$ is equal to $f_1^{-1}(0) = [f'_1(0)]^{-1} = 1$ by the equality (4.4).

Remark 4.12 Theorem 4.1 guarantees the existence of the differentiability of the inverse function on the neighborhood of the considered point only. However, this property is true for the arbitrary point. The set of values of this function is the set of real numbers, and its derivative is equal to zero nowhere.

Example 4.7 Consider the function $f_2(y) = (y - 1)^3 + 1$. Its derivative at the point $y = 0$ is $f'_2(0) = 3$. By the Inverse function theorem, there exists the inverse function $f_2^{-1} = 1 + \sqrt[3]{z - 1}$ in a neighborhood of a point $z = f_2(0) = 0$ (see Figure 4.3). Its derivative at a point $z = 0$ is $f_2^{-1}(0) = [f'_2(0)]^{-1} = 1/3$.

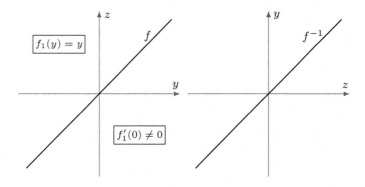

FIGURE 4.2: If $f'(0) \neq 0$, then there exists the differentiable inverse function.

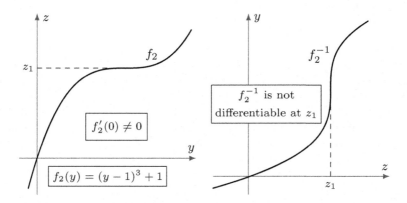

FIGURE 4.3: The non-differentiable inverse function exists everywhere.

Remark 4.13 Now the inverse function exists everywhere. But this is a non-differentiable function at the point $z_1 = 1$. Note that the derivative of the function f_2 is equal to zero at the point $y_1 = f_2^{-1}(z_1) = 1$. This is to make it agree with the Inverse function theorem.

Example 4.8 Let us have the function $f_3(y) = y^3 + y^2 - 2y$. Its derivative at the point $y = 0$ is $f_3'(0) = -2$. Then there exists the inverse function in the neighborhood of the point $z = f_3(0) = 0$ (see Figure 4.4). Its derivative at the zero point is $f_3^{-1}{}'(0) = -1/2$ by equality (4.4).

Remark 4.14 Now the inverse function is determined on the neighborhood of the zero point only. Particularly there exists a point z such that the value of the function f_3 at three points is equal to z. However, for all $z \in O'$ the value $f_3^{-1}(z) \in O$ is determined (see Figure 4.4).

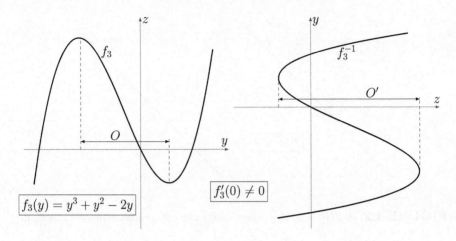

FIGURE 4.4: For the case $f'(0) = 0$ the inverse function does not exist.

Example 4.9 The derivative of the function $f_4(y) = y^2$ at the point $y = 0$ is equal to zero. Therefore, the conditions of the Inverse function theorem are false. For all small enough neighborhood O of this point and for all non-zero point z of the set $O' = f_4(O)$ we can have two cases. Maybe there exist two points y such that the value of the function f_4 at these points is equal to z; maybe we do not have any points with value z. Thus, the inverse function does not exist in the neighborhood of zero (see Figure 4.5).

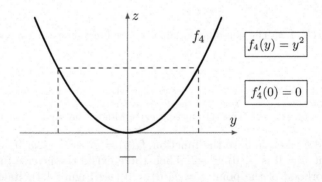

FIGURE 4.5: For the case $f'(0) = 0$ the inverse function does not exist.

Example 4.10 The derivative of the function $f_5(y) = y^3$ at the point $y = 0$ is equal to zero. Therefore, the condition of the Inverse function theorem gets

TABLE 4.1: Properties of inverse functions

functions	$f'(0)$	existence of inverse functions	differentiability of inverse functions
f_1	1	everywhere	everywhere
f_2	3	everywhere	in the neighborhood
f_3	-2	in the neighborhood	in the neighborhood
f_4	0	nowhere	—
f_5	0	everywhere	outside the neighborhood

broken. However, the inverse function is determined everywhere (see Figure 4.6) and equal to $f_5^{-1}(z) = \sqrt[3]{z}$. It can seem strange. However, its derivative at the zero point does not exist. This is conforming with the Inverse function theorem.

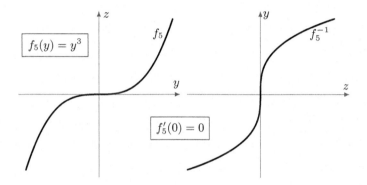

FIGURE 4.6: There exist non-differentiable inverse functions.

Remark 4.15 We see that the assertions of the Inverse function theorem can get broken in two cases (see also Table 4.1). Maybe the inverse function does not exist in neighborhoods of the given point or this is a non-differentiable function. The given function has the local extremum at the considered point for the first case. However, this is a point of inflection for the second case.

Remark 4.16 The derivative of the functions of one variable is equal to zero at the points of extremum or inflection only. However, there exist many different situations of this equality for the functions of many variables.

Remark 4.17 We shall prove that the inverse operator can be non-differentiable if the general condition of the Inverse function theorem, i.e., invertibility of the operator derivative, gets broken (see next chapter).

Remark 4.18 Theorem 4.1 in reality consists of two different results. This is the existence of the inverse operator and its differentiability. The existence of the inverse operator for our case is equivalent to the unique solvability of the state equation. We always substantiate this property by means of the theory of the concrete equations. However, the differentiability of the inverse operator is a basis of the existence of the control–state mapping derivative. We determine necessary conditions of optimality by using this result.

140 *Optimization and Differentiation*

TABLE 4.2: Control systems

chapter	control system	applicability of the Inverse function theorem	control–state mapping
3	linear	trivial	affine
4	weakly nonlinear	applicable	Gâteaux differentiable
5	strongly nonlinear	non-applicable	extendedly differentiable

We mark out the class of the nonlinear control systems such that the Inverse function theorem is applicable.

Definition 4.4 *A nonlinear control system described by the equation* (4.1) *is called* **weakly nonlinear** *if the Inverse function theorem is applicable for it. This is the* **strongly nonlinear control system** *if this theorem is not applicable.*

We consider weakly nonlinear control systems in this chapter.

Remark 4.19 Strongly nonlinear control systems are the subject of the next chapter (see Table 4.2).

Remark 4.20 We shall consider also problems with control in coefficients. Its state equations do not have the form $Ay = Bv + f$ (see Chapter 6). More difficult control systems will be considered in Chapter 7. The differentiability of the control-state mapping for this case can be proved by using the Implicit function theorem that is an extension of the Inverse function theorem.

4.3 Optimal control problems for weakly nonlinear systems

Return to the nonlinear operator equation

$$Ay = Bv + f, \tag{4.6}$$

under the supposition of the weak nonlinearity of the system. The control v is an element of the convex closed subset U of a control space V. Let us have the functional

$$I(v) = J(v) + K(y[v]),$$

where $J : V \to \mathbb{R}$, $K : Y \to \mathbb{R}$.

Problem 4.1 *Find a control $u \in U$ that minimizes the functional I on the set U.*

Weakly nonlinear systems 141

Try to determine the variational inequality (4.1). We have the need to prove differentiability of the functional I that is based on the differentiability of mapping $y[\cdot] : V \to Y$, i.e., the dependence of the solution of the equation (4.5) with respect to the control.

Lemma 4.1 *Suppose the operator A has the continuous inverse operator and is continuously differentiable in the neighborhood of the point $y = y[u]$, there exists the continuous inverse operator A^{-1}, and B is a linear continuous operator. Then mapping $y[\cdot] : V \to Y$ is Gâteaux differentiable at the point u, as the following equality holds:*

$$\langle \mu, y'[u]h \rangle = \langle B^* p_\mu[u], h \rangle \ \forall \mu \in Y', h \in V, \tag{4.7}$$

where $p_\mu[u]$ is the solution of the equation.

$$\left[A'(y) \right]^* p_\mu[u] = \mu. \tag{4.8}$$

Proof. Under our supposition for all $v \in V$ the equation (4.6) has the unique solution $y[v] = A^{-1}(Bv + f)$. By Inverse function theorem the operator A^{-1} is continuously differentiable at the point $Bu + f$, as the following equality

$$(A^{-1})'(Bu + f) = \left[A'(y) \right]^{-1}$$

is true because of (4.4), where $y = y[u]$. Using the theorem of differentiation of the composite function, we prove that mapping $y[\cdot] : V \to Y$ is Gâteaux differentiable at the point u, as its derivative is determined by the formula

$$y'[u] = \left[A'(y) \right]^{-1} B.$$

Thus, we have

$$\langle \mu, y'[u]h \rangle = \langle \mu, \left[A'(y) \right]^{-1} Bh \rangle \ \forall \mu \in Y', h \in V. \tag{4.9}$$

Using the property of the adjoint operator, we get

$$\langle \mu, \left[A'(y) \right]^{-1} Bh \rangle = \langle \{ \left[A'(y) \right]^{-1} \}^* \mu, Bh \rangle =$$

$$\langle B^* \{ \left[A'(y) \right]^{-1} \}^* \mu, h \rangle \ \forall \mu \in Y', h \in V.$$

For all $\mu \in Y'$ the equation (4.8) has the unique solution

$$p_\mu[u] = \{ \left[A'(y) \right]^* \}^{-1} \mu$$

from the Z'. Then we transform the previous equality

$$\langle \mu, \left[A'(y) \right]^{-1} Bh \rangle = \langle B^* p_\mu[u], h \rangle \ \forall \mu \in Y', h \in V.$$

Finally, using (4.9), we get the equality (4.7). $\square$

Prove the following auxiliary result.

Optimization and Differentiation

Lemma 4.2 *Suppose under the conditions of Lemma* 4.1 *the functional K is Fréchet differentiable at the point y. Then mapping $v \to K(y[v])$ has a Gâteaux derivative at the point u that is equal to $-B^*$, where p is the solution of the equation*

$$\left[A'(y)\right]^* p = K'(y). \tag{4.10}$$

Proof. The considered mapping is the superposition of the Fréchet differentiable functional and Gâteaux differentiable mapping $v \to y[v]$. Using the theorem of differentiation of the composite function, we prove the existence of the Gâteaux derivative of the considered functional that is equal to

$$(Ky)'(u) = K'(y)y'[u].$$

Then we have

$$\langle (Ky)'(u), h \rangle = \langle K'(y)y'[u], h \rangle = \langle K'(y), y'[u]h \rangle.$$

Using the equality (4.7), we find

$$\langle (Ky)'(u), h \rangle = -\langle B^* p, h \rangle \ \forall h \in V,$$

where is the solution of the equation (4.8) for $\mu = -K'(y)$, i.e., the adjoint equation (4.10). Therefore, the derivative of mapping $v \to K(y[v])$ is equal to $-B^*$. $\square$

The functional I is the sum of the functional J and mapping $v \to K(y[v])$. Hence, we get the following result.

Lemma 4.3 *If under the conditions of Lemma* 4.2 *the functional J is Gâteaux differentiable at the point u, then the functional I has a Gâteaux derivative at the point u*

$$I'(u) = J'(u) - B^* p. \tag{4.11}$$

Now we obtain the necessary conditions of optimality.

Theorem 4.2 *Under the conditions of Lemma* 4.3 *the solution u of Problem* 4.1 *satisfies the variational inequality*

$$\langle J'(u) - B^* p, v - u \rangle \geq 0 \ \forall v \in U. \tag{4.12}$$

It is sufficient to put the derivative of the functional I from the formula (4.11) in to the inequality

$$\langle I'(u), v - u \rangle \geq 0 \ \forall v \in U.$$

Thus, we have the system that involves the variational inequality (4.12), the state equation (4.6) for $v = u$, and the adjoint equation (4.10) with respect to three unknown values u, y, p. We obtained an analogical result for the linear control systems (see Theorem 3.1).

Weakly nonlinear systems 143

Remark 4.21 The unique difference between Problem 4.1 and Problem 3.1 is the nonlinearity of the operator A. Therefore, we obtain the analogical variational inequalities (3.7) and (4.12) as optimality conditions. Note the difference of the adjoint equations (3.5) and (4.10). However, if the operator A is linear, then its derivative is equal to the initial operator; and the adjoint equation (4.10) is transformed to (3.5). Thus, Theorem 3.1 is the partial of Theorem 4.1.

Remark 4.22 The serious difference between the nonlinear system from the linear one is the property of $y[\cdot] : V \to Y$. Now this is not an affine operator. Therefore, the convexity of the functionals J and K does not guarantee the convexity of the functional I. Hence, the Theorem 2.1 is not applicable. We cannot determine the sufficiency of the optimality conditions, i.e., the analog of Theorem 3.2.

Remark 4.23 We could obtain the necessary conditions of optimality if B is a nonlinear Gâteaux operator.

Remark 4.24 We could prove Theorem 4.2 by using the Lagrange multipliers method (see Chapter 3). If the operator J is convex and non-differentiable we can prove the analog of Theorem 3.9 and obtain the Lagrange principle.

Remark 4.25 We shall extend Theorem 4.2 to the case, where the operator $y[\cdot] : V \to Y$ is not Gâteaux differentiable (see next chapter).

Note that we use the supposition of the solvability of Problem 4.1 in Theorem 4.2. We could substantiate this property if we prove weak lower semicontinuity of the considered functional.

Lemma 4.4 *Suppose the operator A is invertible, the functionals J and K are convex and continuous, and mapping $y[\cdot] : V \to Y$ is weakly continuous. Then the functional I is weakly lower semicontinuous.*

Proof. Let us have the convergence $v_k \to v$ weakly in V. By weak continuity of mapping $y[\cdot] : V \to Y$ weakly in Y. Using the convexity and the continuity of the functionals J and K, we obtain its weak lower semicontinuity. Then we have

$$\inf \lim_{k \to \infty} J(v_k) \geq J(v); \quad \inf \lim_{k \to \infty} K\big(y[v_k]\big) \geq K\big(y[v]\big).$$

Therefore, we get

$$\inf \lim_{k \to \infty} I(v_k) \geq I(v).$$

Thus, the functional I is weakly lower semicontinuous. $\square$

Remark 4.26 The general supposition of Lemma 4.4 is the weak continuity of the control–state mapping. This property is the corollary of the continuity of this dependence. However, the weak continuity of the operator is an additional independent result for the nonlinear systems.

Using Theorem 2.4 and Lemma 4.4, we obtain the following result.

Theorem 4.3 *Suppose the spaces V and Y are reflexive, U is a non-empty convex closed subset of the space V, there exists the inverse operator A^{-1}, the functionals J and K are convex, bounded lower and lower semicontinuous, and mapping $y[\cdot] : V \to Y$ is weakly continuous. If also the set U is bounded or the functional J is coercive, then Problem 4.1 has a solution.*

144 *Optimization and Differentiation*

Remark 4.27 The considered spaces can be adjoined to Banach spaces.

Remark 4.28 The proof of Theorem 4.3 is not more difficult than the analogical result for linear systems (see Theorem 3.3). However, the substantiation of the weak continuity for the control–state mapping can have serious difficulties.

Remark 4.29 We cannot prove the uniqueness of the optimal control (the analog of Theorem 3.4) and well-posedness of the problem because of the nonlinearity of the state operator. The control-state mapping is not an affine operator; and we cannot obtain the strong convexity of the given functional.

We used the existence of the inverse operator A^{-1}. Consider a result that guarantees this property.

4.4 Equations with monotone operators

Determine some additional notions. Consider a separable reflexive Banach space Y and an operator $A : Y \to Y'$.

Definition 4.5 *The operator A is called **monotone**, if the following inequality holds:*
$$\langle Ay - Az, y - z \rangle \geq 0 \ \forall y, z \in Y.$$
*This operator is called **strictly monotone**, if*
$$\langle Ay - Az, y - z \rangle > 0 \ \forall y \neq z.$$

We have the classic notions of monotone (strictly increasing) and strictly monotone functions (see Figure 4.7) for the case $Y = \mathbb{R}$.

Definition 4.6 *The operator A is called **coercive** if there exists function $\alpha = \alpha(\sigma)$ that is bounded for $\sigma > 0$, and $\alpha(\sigma) \to +\infty$ as $\sigma \to +\infty$ such that the following inequality holds:*
$$\langle Ay, y \rangle \geq \alpha(\|y\|)\|y\| \ \forall y \in Y.$$

Remark 4.30 We considered in the previous chapter a linear operator A that satisfies the inequality $\langle Ay, y \rangle \geq \alpha \|y\|^2$ for all $y \in Y$, where $\alpha > 0$. This operator is coercive.

Lemma 4.5 *Suppose P is a continuous operator on Euclid space $\mathbb{R}^k$ such that $(P\xi, \xi) \geq 0$ for all vector $\xi \in \mathbb{R}^k$ with $\|\xi\| = r$, where $r > 0$. Then there exists a vector $\xi \in \mathbb{R}^k$ such that $\|\xi\| \leq r$ and $P\xi = 0$.*

Remark 4.31 This operator P is k functions of k variables. Therefore, $P\xi = 0$ is a system of k nonlinear algebraic equations. Thus, Lemma 4.5 gives the sufficient conditions of its solvability.

Weakly nonlinear systems

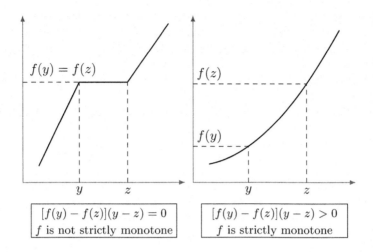

FIGURE 4.7: Monotone functions.

Remark 4.32 We shall prove the existence of the solution of the equation $Ay = z$ by the Galerkin method (see Theorem 3.5). Its approximate solution is determined by system algebraic equations. This system was linear before. The proof of its solvability is not a difficult problem. However, now we shall have the nonlinear system. We shall prove its solvability by using Lemma 4.5.

Consider a sequence of linear continuous operator $\{L_k\}$ from a Banach space Y to a linear normalized space Z. By the **Banach–Steinhaus theorem**, if the sequence $\{\|L_k y\|\}$ is bounded for all $y \in Y$, then the sequence of norms $\{\|L_k\|\}$ is bounded too.

Consider the operator equation

$$Ay = z. \tag{4.13}$$

Theorem 4.4 *If the operator A is continuous, monotone, and coercive, then the equation (4.13) has a solution $y \in Y$ for all $z \in Y'$. This solution is unique whenever A is a strictly monotone operator.*

Proof. 1. We use the Galerkin method for proving the solvability of the equation (see previous chapter). There exists a complete sequence of the linear independent elements $\{\mu_m\}$ of the space Y. Determine the sequence of elements $\{y_k\}$ of the space Y by the formula

$$y_k = \sum_{m=1}^{k} \xi_{mk} \mu_m, \ k = 1, 2, ..., \tag{4.14}$$

where the numbers $\xi_{1k}, ..., \xi_{kk}$ satisfy the equalities

$$\langle Ay_k, \mu_j \rangle = \langle z, \mu_j \rangle, \ j = 1, ..., k. \tag{4.15}$$

Optimization and Differentiation

There exists a serious difference of this system from the analogical equations, which we considered for the analysis of the Poisson equation. We had the system of linear algebraic equations before. The proof of its solvability is easy enough. However, now we have the system of nonlinear algebraic equations. The problem of its solvability is not trivial. We prove this result by using Lemma 4.5.

Determine the operator P on the space $\mathbb{R}^k$ that maps the arbitrary vector $\xi = (\xi_{1k}, ..., \xi_{kk})$ to the vector $P\xi = (\eta_1, ..., \eta_k)$, where

$$\eta_j = \langle Ay_k, \mu_j \rangle - \langle z, \mu_j \rangle, \; j = 1, ..., k.$$

Then the system nonlinear algebraic equation (4.15) is transformed to the operator equation $P\xi = 0$ (see Lemma 4.5). The operator P is continuous because of the continuity of the operator A. Find the scalar product

$$(P\xi, \xi) = \sum_{j=1}^{k} \eta_j \xi_{jk} = \langle Ay_k, y_k \rangle - \langle z, y_k \rangle.$$

Using the inequality

$$|\langle z, y_k \rangle| \leq \|z\|_{Y'} \|y_k\|_Y,$$

we have

$$(P\xi, \xi) \geq \left[\alpha(\|y_k\|_Y) - \|z\|_{Y'} \right] \|y_k\|_Y$$

because operator A is coercive.

By property of the function α, the vector ξ can have the norm r so large that the appropriate function satisfies the inequality

$$\alpha(\|y_k\|_Y) \geq \|z\|_{Y'}.$$

Then it follows from the previous inequality $(P\xi, \xi) \geq 0$. Therefore, the conditions of Lemma 4.5 are true. Hence the equation $P\xi = 0$ that is equivalent to the system of the nonlinear algebraic equation (4.15) has vector ξ as a solution. This vector determines the concrete function y_k by formula (4.14).

2. Multiply the j-th equality (4.15) by the number ξ_{jk} and sum the result by $j = 1, ..., k$. Using the formula (4.14), we get

$$\langle Ay_k, y_k \rangle = \langle z, y_k \rangle.$$

We have the inequality

$$\alpha(\|y_k\|_Y) \|y_k\|_Y \leq \langle Ay_k, y_k \rangle \leq \|z\|_{Y'} \|y_k\|_Y$$

because operator A is coercive. Thus, we obtain

$$\alpha(\|y_k\|_Y) \leq \|z\|_{Y'}.$$

Weakly nonlinear systems

If now $\|y_k\|_Y \to \infty$ as $k \to \infty$, then $\alpha(\|y_k\|_Y) \to \infty$ by the property of the function α. This contradicts the previous inequality. Therefore, there exists a constant $c > 0$ such that

$$\|y_k\|_Y \leq c. \tag{4.16}$$

Then we get the inequality

$$\langle Ay_k, y_k \rangle \leq \|z\|_{Y'}\|y_k\|_Y \leq c\|z\|_{Y'}. \tag{4.17}$$

Consider a sequence $\{\varphi_k\}$ of the space Y that tends to zero. Suppose the sequence $\{A\varphi_k\}$ is unbounded, i.e., $\|A\varphi_k\|_{Y'} \to \infty$. Determine the numbers

$$\beta_k = 1 + \|A\varphi_k\|_{Y'}\|\varphi_k\|_Y, \ k = 1, 2, \dots .$$

Using the monotony of the operator A, we have the inequality

$$(\beta_k)^{-1}\langle A\varphi_k, y \rangle \leq (\beta_k)^{-1}\big(\langle A\varphi_k, \varphi_k \rangle + \langle Ay, y - \varphi_k \rangle\big) \leq$$

$$1 + (\beta_k)^{-1}\|Ay\|_{Y'}\big(\|y\|_Y + \|\varphi_k\|_Y\big) \ \forall y \in Y.$$

The sequence of numbers $\{\beta_k\}$ does not tend to zero. Therefore, the right side of the last inequality is bounded by a constant that does not depends from k. The analogical estimate is true for the function $-y$. Thus, we get

$$\sup \lim_{k \to \infty} (\beta_k)^{-1}\langle A\varphi_k, y \rangle < \infty \ \forall y \in Y.$$

Determine the linear continuous functionals L_k on the space Y by the equality

$$L_k y = (\beta_k)^{-1}\langle A\varphi_k, y \rangle \ \forall y \in Y, \ k = 1, 2, \dots .$$

From the previous inequality there exists a constant $c' > 0$ such that

$$(\beta_k)^{-1}\|A\varphi_k\|_{Y'} \leq c'$$

by the Banach–Steinhaus theorem. Then we have the inequality

$$\|A\varphi_k\|_{Y'} \leq c'\big(1 + \|A\varphi_k\|_{Y'}\|\varphi_k\|_Y\big).$$

Choose the number k so large that the following inequality holds: $2c'\|\varphi_k\|_Y \leq 1$. Then from the previous inequality we have the estimate $\|A\varphi_k\|_{Y'} \leq 2c'$ that contradicts the convergence $\|A\varphi_k\|_{Y'} \to \infty$. Thus, from the condition $\varphi_k \to 0$ the boundedness of the sequence $\{A\varphi_k\}$ follows. Therefore, there exists a constant $c_1 > 0$ such that $\|A\varphi\|_{Y'} \leq c_1$ whenever $\|\varphi\|_Y \leq 1$.

Using the monotony of the operator A and inequalities (4.16), (4.17), we have

$$\|Ay_k\|_{Y'} = \sup_{\|\varphi\|_Y=1} \big|\langle Ay_k, \varphi \rangle\big| \leq$$

$$\sup_{\|\varphi\|_Y=1} \big(\langle Ay_k, y_k \rangle + \langle A\varphi, \varphi \rangle - \langle A\varphi, y_k \rangle\big) \leq c\|z\|_{Y'} + c_1 + cc_1$$

148   *Optimization and Differentiation*

Denoting the value in the right side by c_2 we obtain the estimate

$$\|Ay_k\|_{Y'} \leq c_2. \tag{4.18}$$

3. Using the estimates (4.16), (4.18), and the Banach–Alaoglu theorem after extracting a subsequence, we have the convergence $y_k \to y$ weakly in Y and $Ay_k \to \varphi$ weakly in Y'. Passing to the limit in the equality (4.15), we get

$$\langle \varphi, \mu_j \rangle = \langle z, \mu_j \rangle, \ j = 1, 2, \dots.$$

The arbitrary element $\mu \in Y$ can be determined by the convergent series with respect to the complete system

$$\mu = \sum_{j=1}^{\infty} b_j \mu_j,$$

where b_j is a number. Multiply the previous j-th equality by b_j and sum by j by using the convergence of the series. We obtain the equality

$$\langle \varphi, \mu \rangle = \langle z, \mu \rangle.$$

Then we have the equality $\varphi = z$ because μ is arbitrary. Therefore, we have the convergence $Ay_k \to z$ weakly in Y'.

By equality (4.15) we have

$$\lim_{k \to \infty} \langle Ay_k, y_k \rangle = \lim_{k \to \infty} \langle z, y_k \rangle = \langle z, y \rangle.$$

For all elements $\psi \in Y$ we get

$$\langle z - A\psi, y - \psi \rangle = \langle z, y \rangle - \langle z, \psi \rangle - \langle A\psi, y \rangle + \langle A\psi, \psi \rangle \geq$$

$$\sup \lim_{k \to \infty} \langle Ay_k, y_k \rangle - \sup \lim_{k \to \infty} \langle Ay_k, \psi \rangle - \sup \lim_{k \to \infty} \langle A\psi, y_k \rangle + \langle A\psi, \psi \rangle =$$

$$\sup \lim_{k \to \infty} \langle Ay_k - \psi, y_k - \psi \rangle \geq 0$$

because of the monotony of the operator A. Determine $\psi = y - \sigma h$, where $h \in Y$, $\sigma > 0$. Then we have

$$\sigma \langle z - A(y - \sigma h), h \rangle \geq 0.$$

Divide by σ and pass to the limit as $\sigma \to 0$ by using the continuity of the operator A. We obtain

$$\langle z - Ay, h \rangle \geq 0 \ \forall h \in Y.$$

Replace the arbitrary element h by $-h$. We have the equality

$$\langle z - Ay, h \rangle = 0 \ \forall h \in Y.$$

Therefore, $ay = z$. Thus, the limit y of the sequence $\{y_k\}$ determined by the formula (4.14) is a solution of the equation (4.13).

4. The operator A is strictly monotone. Suppose y_1 and y_2 are solutions of this equation. If $y_1 \neq y_2$ we have the inequality

$$\langle Ay_1 - Ay_2, y_1 - y_2 \rangle > 0.$$

However, from the equality $Ay_1 = z = Ay_2$ it follows that the left side of this inequality is equal to zero. By this contradiction, we prove the equality $y_1 = y_2$. $\square$

Remark 4.33 The proof of the solvability of the equation consists of the determination of the approximate solution, the analysis of the a priori estimates, and passing to the limit as the linear case. However, these steps are more difficult because of the nonlinearity of the state operator.

We shall apply these results for the analysis of optimal control problems for systems described by concrete nonlinear partial differential equations. Consider additional properties of the mathematical spaces theory.

4.5   Additional results of the functional analysis

Consider the set $L(Y, Z)$ of all linear continuous operators that is determined on a Banach space Y with values on a Banach space Z. This is the linear space with addition of operators A, B and multiplication of an operator A by a number a such that

$$(A + B)y = Ay + By, \quad (aA)y = aAy \ \ \forall y \in Y.$$

This space is Banach with norm

$$\|A\|_{L(Y,Z)} = \sup_{\|y\|_Y = 1} \|Ay\|_Z.$$

Particularly, the **adjoint space** Y' is the set of all linear continuous functionals on the space Y, i.e., this is the space $L(Y, R)$ with norm

$$\|\lambda\|_{Y'} = \sup_{\|y\|_Y = 1} |\langle \lambda, y \rangle|.$$

Note the inequality

$$\|Ay\|_Z \leq \|A\|_{L(Y,Z)} \|y\|_Y \ \ \forall y \in Y$$

and the equality

$$\|A\|_{L(Y,Z)} = \|A^*\|_{L(Z',Y')}.$$

150 *Optimization and Differentiation*

We shall apply also the **intersection** of functional spaces. If there exists continuous embedding of Banach spaces Y and Z to a reflexive Banach space, then its intersection $Y \cap Z$ is Banach space norm

$$\|y\|_{Y \cap Z} = \|y\|_Y + \|y\|_Z.$$

By this supposition the set

$$Y + Z = \{y + z \,|\, y \in Y, \, z \in Z\}$$

that is called the **direct sum** of these spaces is the Banach space with norm:

$$\|w\|_{Y+Z} = \inf_{y \in Y, z \in Z, y+z=w} \max\{\|y\|_Y, \|z\|_Z\}.$$

Let Y and Z be reflexive Banach spaces such that embedding of Y to Z is continuous and dense. Then embedding of the adjoint space Z' to the adjoint space Y' is continuous and dense. If there exists continuous embedding of Banach spaces Y and Z to a reflexive Banach space, and the intersection $Y \cap Z$ is dense to Y and Z, then

$$(Y \cap Z)' = Y' + Z', \ (Y + Z)' = Y' \cap Z'.$$

Therefore, we have the equalities

$$(Y \cap Z)'' = (Y' + Z')' = Y'' \cap Z'' = Y \cap Z,$$

$$(Y + Z)'' = (Y' \cap Z') = Y'' + Z'' = Y + Z.$$

Hence the intersection and the sum of reflexive spaces are reflexive. Note continuous embedding of the intersection $Y \cap Z$ to both spaces Y and Z, and continuous embedding of these spaces to the direct sum $Y + Z$.

Remark 4.34 We shall use these properties for the analysis of the boundary problem for the concrete nonlinear partial differential equation. Particularly, we shall consider the intersection $H_0^1(\Omega) \cap L_q(\Omega)$ that is the separable reflexive Banach space if $1 < q < \infty$. Its adjoint space is the direct sum $H^{-1}(\Omega) + L_{q'}(\Omega)$ that is the reflexive Banach space too, where $1/q + 1/q' = 1$.

If the sequence $\{v_k\}$ is weakly convergent on the Banach space V, then the sequence of norms $\{\|v_k\|_V\}$ is bounded.

Remark 4.35 By Banach–Alaoglu theorem, if the sequence of norms of elements of the reflexive Banach space is bounded, then there exists a subsequence of the sequence of these elements that converges weakly.

Suppose we have the convergence $u_k \to u$ strongly in V' and $v_k \to v$ weakly in V. Then

$$\left|\langle u_k, v_k \rangle\right| = \left|\langle u_k - u, v_k \rangle\right| + \left|\langle u, v_k - v \rangle\right| \le$$

$$\|u_k - u\|_{V'} + \left|\langle u, v_k - v \rangle\right|.$$

Weakly nonlinear systems

Using the boundedness of the weakly convergent sequence, we get $\langle u_k, v_k \rangle \to \langle u, v \rangle$. We obtain the analogical result if $u_k \to u$ *-weakly in V' and $v_k \to v$ strongly in V.

Consider the sequence $\{v_k(\mu)\}$ of elements of Banach space V that depends upon an element μ of a set M. We have the convergence $v_k(\mu) \to v(\mu)$ in V **uniformly** with respect to $\mu \in M$ if

$$\lim_{k \to \infty} \sup_{\mu \in M} \left\| v_k(\mu) - v(\mu) \right\|_V = 0.$$

We say that $v_k(\mu) \to v(\mu)$ in V weakly in V uniformly with respect to $\mu \in M$ if

$$\lim_{k \to \infty} \sup_{\mu \in M} \left| \langle u, v_k(\mu) - v(\mu) \rangle \right| = 0 \ \forall u \in V'.$$

Finally, let V be adjoint to a linear normalized space W. We have the convergence $v_k(\mu) \to v(\mu)$ *-weakly in V uniformly with respect to $\mu \in M$ if

$$\lim_{k \to \infty} \sup_{\mu \in M} \left| \langle v_k(\mu) - v(\mu), w \rangle \right| = 0 \ \forall w \in W.$$

Let Ω be a bounded set of R^n with a regular enough boundary, and $1 \leq p < \infty$. By the **Sobolev embedding theorem** embedding $W_p^m(\Omega) \subset W_r^s(\Omega)$ is continuous for $0 \leq s \leq r$ and $1/p - (s-k)/n \leq 1/r < 1$. If $1/p - (s-k)/n < 0$, then embedding $W_p^m(\Omega) \subset C^s(\overline{\Omega})$ is continuous. Particularly, the space $H^1(\Omega)$ (and $H_0^1(\Omega)$ too) has continuous embedding to $L_r(\Omega)$ if $r \leq 2n/(n-2)$ for $n \geq 3$, to $L_r(\Omega)$ with arbitrary value r for $n = 2$, and to $C(\overline{\Omega})$ for $n = 1$. The space $H^{-1}(\Omega)$ is adjoint to $H_0^1(\Omega)$, and $L_{r'}(\Omega)$ is adjoint to $L_r(\Omega)$, where $1/r + 1/r' = 1$. Then embedding of the space $L_{(r')}(\Omega)$ to $H^{-1}(\Omega)$ is continuous $r \leq 2n/(n-2)$ for $n \geq 3$.

Remark 4.36 Note the idea of the Sobolev theorem. It is known that the properties of functions become worse after differentiation. For example, the derivative of a continuously differentiable function is a continuous function only. Thus, the function derivative has a weaker property than the initial function. However, by definition of the space $H^1(\Omega)$, the function and its first order partial derivatives are elements of the same space of the square integrable functions. We could suppose that if the derivatives have these properties, then the initial function has stronger properties. The Sobolev embedding theorem determines correctly these properties. Note that the additional properties of the initial function depend upon the dimension of the given set that is the additional characteristic of the difficulty of the considered object.

We considered before continuous and dense embedding of spaces. Determine an additional form of embedding.

Definition 4.7 *Embedding of a Banach space Y to a Banach space Z is **compact** if $y_k \to y$ strongly in Z whenever $y_k \to y$ weakly in Y.*

Under the regular enough boundary of the set Ω, by the **Rellich–Kondrashov theorem** embedding of the space $W_p^m(\Omega)$ to $W_p^{m-1}(\Omega)$ is compact if $1 \leq p < \infty$, $m \geq 1$. The regularity of the boundary is not necessary for compact embedding of the space $\mathring{W}_p^m(\Omega)$ to $\mathring{W}_p^{m-1}(\Omega)$.

152 *Optimization and Differentiation*

Remark 4.37 We shall use often compact embedding of Sobolev spaces $H^1(\Omega)$ and $H^1_0(\Omega)$ to $L_2(\Omega)$.

Remark 4.38 We shall give additional information about compact embedding of concrete functional spaces in Part III for the analysis of evolutional partial differential equations.

Remark 4.39 We use compact embedding of functional spaces for substitution of passing to the limit in nonlinear terms.

Note the following result.

Lemma 4.6 *Suppose the set Ω is bounded, and $\{y_k\}$ is a uniformly bounded sequence of the space $L_p(\Omega)$, $1 < p < \infty$ such that $y_k \to y$ a.e. on Ω. Then we have the convergence $y_k \to y$ weakly in $L_p(\Omega)$.*

Remark 4.40 We shall use the following method of passing to the limit in nonlinear terms. At first, we prove the uniformly boundedness of the considered sequence in a narrow enough reflexive Banach space, for example, in a Sobolev space. Then we extract a subsequence that is weakly convergent. Using a compact embedding theorem, we obtain strong convergence with respect to a larger space, for example, $L_2(\Omega)$. Therefore, we determine the convergence almost everywhere. Hence we get the analogical form of convergence of nonlinear terms. If we obtain also the boundedness of these terms with respect to a space $L_p(\Omega)$ then we get the weak convergence of the sequence of nonlinear terms by Lemma 4.6.

4.6 Nonlinear elliptic equation

Let us have an open bounded set Ω of Euclid space $\mathbb{R}^n$ with boundary Γ. We consider the control system described by the ***nonlinear elliptic equation***

$$-\Delta y(x) + |y(x)|^\rho y(x) = v(x) + f(x), \ x \in \Omega \tag{4.19}$$

with the boundary condition

$$y(x) = 0, \ x \in \Gamma, \tag{4.20}$$

where v is a control, y is the state function, and f is a given function, $\rho > 0$.

Remark 4.41 The equality (4.19) has the unique difference as the Poisson equation (3.9) from the previous chapter. This is the nonlinear term in the left-hand side of the equation. This is the power operator, which was considered in Example 4.1 and Example 4.2

Transform Dirichlet problem (4.19), (4.20) to the operator equation (4.6). Consider the equation

$$-\Delta y + |y|^\rho y = f \tag{4.21}$$

with boundary condition (4.20). Multiply this equality (4.21) by the function y. After integration we have

$$\int_\Omega y \Delta y dx + \int_\Omega |y|^{\rho+2} dx = \int_\Omega yz dx.$$

Weakly nonlinear systems

Transform the first integral by using Green's formula and boundary condition (4.20). We get

$$\int_\Omega \sum_{i=1}^n \left|\frac{\partial y}{\partial x_i}\right|^2 dx + \int_\Omega |y|^{p+2} dx = \int_\Omega yz dx. \qquad (4.22)$$

The first integral in the left-hand side of this equality is a square of the norm of the function of the space $H_0^1(\Omega)$ and the second integral is the degree $q = p+2$ of the norm of the space $L_q(\Omega)$. Therefore, it will be natural to find the solution boundary problem (4.21), (4.20) in the intersection $Y = H_0^1(\Omega) \cap L_q(\Omega)$. Embedding of both spaces $H_0^1(\Omega)$ and $L_q(\Omega)$ to the Sobolev space $W_q^1(\Omega)$ that is a reflexive Banach space is continuous. Then its intersection Y is a reflexive Banach space too.

The integral on the right side of the equality (4.22) has the meaning of $y \in Y$ if the function z is an element of the adjoint space Y'. We know that the adjoint space to the intersection of Banach spaces is the sum of adjoint spaces, i.e., $Y' = [H_0^1(\Omega)]' + [L_q(\Omega)]'$. Using the equalities

$$\left[H_0^1(\Omega)\right]' = H^{-1}(\Omega), \ \left[L_q(\Omega)\right]' = L_{q'}(\Omega),$$

where $1/q + 1/q' = 1$, we determine that the absolute term z of the equation (4.21) can be an element of the space $Z = H^{-1}(\Omega) + L_{q'}(\Omega)$.

Remark 4.42 This is the preliminary analysis only for choosing the functional spaces.

Determine the operator A as a left side of the equation (4.19) by the equality

$$Ay = -\Delta y + |y|^p y \ \forall y \in Y.$$

We know that the Laplace operator maps Sobolev space $H_0^1(\Omega)$ to $H^{-1}(\Omega)$. If $y \in L_q(\Omega)$, then $|y|^p y \in L_{q'}(\Omega)$ (see Example 4.3). Thus, for all function $y \in Y$ the value Ay is the sum of two terms. The first of them belongs to the space $H^{-1}(\Omega)$ and the second term belongs to $L_{q'}(\Omega)$. Therefore, Ay is the element of the direct sum of these spaces, i.e., the space Z. Thus, we have in reality the operator $A : Y \to Z$. Choose the operator B as an embedding of the space $V = L_2(\Omega)$ to Z. Now the boundary problem (4.19), (4.20) is transformed to the operator equation (4.5).

We would like to prove the invertibility of the operator A. This is true if for all $z \in Z$ the equation $Ay = z$ has a unique solution $y \in Y$. This equation is equivalent to the boundary problem (4.20), (4.21).

Theorem 4.5 *For all $z \in Z$ the problem* (4.20), (4.21) *has a unique solution* $y \in Y$.

Proof. Prove that the operator A satisfies the suppositions of Theorem 4.4. We have the equality $Z = Y'$. Laplace operator $\Delta : H_0^1(\Omega) \to H^{-1}(\Omega)$ is continuous (see previous chapter). The continuity of mapping $y \to |y|p y$, i.e.,

154 — Optimization and Differentiation

the power operator from space $L_q(\Omega)$ to $L_{q'}(\Omega)$, follows from the Krasnoselsky theorem. Therefore, the operator $A : Y \to A'$ is continuous.

We have the equality

$$\langle Ay, y \rangle = -\int_\Omega y \Delta y \, dx + \int_\Omega |y|^{p+2} dx = \|y\|^2 + \|y\|_q^q \quad \forall y \in Y,$$

where $\|y\|$ and $\|y\|_q$ are the norms of an element y of the spaces $H_0^1(\Omega)$ and $L_q(\Omega)$. Thus, the considered operator is coercive.

Determine the value

$$\langle Ay - Az, y - z \rangle = \|y - z\|^2 + \int_\Omega \left(|y|^p y - |z|^p z \right)(y - z) dx \quad \forall y, z \in Y.$$

The first term of this equality is not negative, as this is zero for $y = z$ only. Therefore, the operator A is strictly monotone. Now the unique solvability of the considered boundary problem follows from Theorem 4.4.

Give another method of proving Theorem 4.5.

Proof. 1. We apply the Galerkin method. Consider a complete set $\{\mu_m\}$ of the space Y. Determine the sequence of the elements $\{y_k\}$ of the space Y by the equalities

$$y_k = \sum_{m=1}^{k} \xi_{mk} \mu_m, \quad k = 1, 2, ..., \tag{4.23}$$

where the numbers $\xi_{1k}, ..., \xi_{kk}$ satisfy the equalities

$$\int_\Omega \left(\sum_{i=1}^{n} \frac{\partial y_k}{\partial x_i} \frac{\partial \mu_j}{\partial x_i} + |y_k|^p y_k \mu_j \right) dx = \int_\Omega z \mu_j dx, \quad j = 1, ..., k. \tag{4.24}$$

Determine the operator P on the space R^k that maps the arbitrary vector $\xi = (\xi_{1k}, ..., \xi_{kk})$ to the vector $P\xi = (\eta_1, ..., \eta_k)$, where

$$\eta_j = \int_\Omega \left(\sum_{i=1}^{n} \frac{\partial y_k}{\partial x_i} \frac{\partial \mu_j}{\partial x_i} + |y_k|^p y_k \mu_j \right) dx - \int_\Omega z \mu_j dx, \quad j = 1, ..., k.$$

The system of the nonlinear algebraic equation (4.23) is transformed to the operator equation $P\xi = 0$ (see Lemma 4.5). We have the equality

$$(P\xi, \xi) = \sum_{j=1}^{k} \eta_j \xi_j = \int_\Omega \sum_{i=1}^{n} \left| \frac{\partial y_k}{\partial x_i} \right|^2 dx + \int_\Omega |y_k|^q dx - \int_\Omega z y_k dx.$$

Using the inequality

$$\left| \int_\Omega z y_k dx \right| \leq \|z\|_Z \|y_k\|_Y = \|z\|_Z \left(\|y_k\| + \|y_k\|_q \right),$$

Weakly nonlinear systems 155

we have
$$(P\xi,\xi) \geq \|y_k\|^2 + \|y_k\|_q^q - \|z\|_Z(\|y_k\| + \|y_k\|_q). \tag{4.25}$$

Determine vector ξ with large enough norm r such that the function y_k that is determined by the (4.23) satisfies the inequalities
$$\|y_k\| \geq \|z\|_Z, \quad \|y_k\|_q \geq \|z\|_Z^{1/(q-1)}.$$

Using (4.25), we obtain
$$(P\xi,\xi) \geq \|y_k\|\left(\|y_k\| - \|z\|_Z\right) + \|y_k\|_q\left(\|y_k\|_q^{q-1} - \|z\|_Z\right) \geq 0.$$

Therefore, Lemma 4.5 is applicable. Hence, the equation $P\xi = 0$ that is equal to the system of nonlinear algebraic equations (4.24) has a solution ξ. It determines the concrete function y_k by the formula (4.23).

2. Multiply j-th equality (4.21) by the number ξ_{jk} and sum the result by $j = 1, ..., k$. Using the formula (4.23), we get
$$\int_\Omega \sum_{i=1}^n \left|\frac{\partial y_k}{\partial x_i}\right|^2 dx + \int_\Omega |y_k|^q dx = \int_\Omega z y_k dx.$$

Thus, we have the inequality
$$\|y_k\|^2 + \|y_k\|_q^q \leq \|z\|_Z(\|y_k\| + \|y_k\|_q) \leq \frac{1}{2}\|y_k\|^2 + \frac{1}{2}\|z\|_Z^2 + \|z\|_Z\|y_k\|_q.$$

Denoting by c the norm of the given function z, we obtain
$$\frac{1}{2}\|y_k\|^2 + \|y_k\|_q^q \leq \frac{c^2}{2} + c\|y_k\|_q.$$

Then the following inequality holds:
$$\|y_k\|_q^q \leq \frac{c^2}{2} + c\|y_k\|_q.$$

Hence the sequence $\{y_k\}$ is bounded in the space $L_q(\Omega)$ because of the condition $q > 2$. From the previous inequality it follows this sequence is bounded in the space $H_0^1(\Omega)$ too.

3. Thus, the sequence $\{y_k\}$ is bounded in the reflexive Banach space Y. Using the Banach–Alaoglu theorem, extract a subsequence such that $y_k \to y$ weakly in Y. It is necessary to prove that this limit y is a solution of the considered boundary problem. We obtain this result by means of passing to the limit in the equality (4.24). We can pass to the limit in the first integral by using weak convergence in the Hilbert space $H_0^1(\Omega)$ only. We get
$$\int_\Omega \sum_{i=1}^n \frac{\partial y_k}{\partial x_i}\frac{\partial \mu}{\partial x_i} dx \to \int_\Omega \sum_{i=1}^n \frac{\partial y}{\partial x_i}\frac{\partial \mu}{\partial x_i} dx.$$

156 *Optimization and Differentiation*

However, we have some difficulties with passing to the limit in the second integral because of nonlinear terms.

By the Rellich–Kondrashov theorem, embedding of the space $H_0^1(\Omega)$ to $L_2(\Omega)$ is compact. Therefore, $y_k \to y$ strongly in $L_2(\Omega)$ whenever $y_k \to y$ weakly in $H_0^1(\Omega)$. It is known that it is possible to extract a subsequence that converges almost everywhere from the sequence that converges strongly in $L_2(\Omega)$. Then we have the convergence $y_k(x) \to y(x)$ almost for all points $x \in \Omega$, where we save the denotation of initial sequence for its subsequence. Therefore, we obtain $|y_k(x)|^\rho y_k(x) \to |y(x)|^\rho y(x)$ a.e. on Ω.

By the equality

$$\left\| |y_k|^\rho y_k \right\|_{q'}^{q'} = \int_\Omega |y_k|^{(\rho+1)q'} dx = \int_\Omega |y_k|^q dx = \|y_k\|_q^q$$

and the boundedness of the sequence $\{y_k\}$ in the space $L_q(\Omega)$ we get the boundedness of the sequence $\{|y_k|^\rho y_k\}$ in the space $L_q'(\Omega)$. Now we use Lemma 4.6. We have the convergence $|y_k|^\rho y_k \to |y|^\rho y$ weakly in $L_q'(\Omega)$. Therefore,

$$\int_\Omega |y_k|^\rho y_k \mu_j dx \to \int_\Omega |y|^\rho y \mu_j dx.$$

Passing to the limit in the equality (4.24), we get

$$\int_\Omega \left(\sum_{i=1}^n \frac{\partial y}{\partial x_i} \frac{\partial \mu_j}{\partial x_i} + |y|^\rho y \mu_j \right) dx = \int_\Omega z \mu_j dx, \ j = 1, 2, \dots .$$

The arbitrary element $\mu \in Y$ can be transformed as a convergent series

$$\mu = \sum_{j=1}^\infty b_j \mu_j,$$

where b_j are numbers. Multiply j-th previous equality by b_j. Adding by j by using the convergence of the series, we get

$$\int_\Omega \left(\sum_{i=1}^n \frac{\partial y}{\partial x_i} \frac{\partial \mu}{\partial x_i} + |y|^\rho y \mu \right) dx = \int_\Omega z \mu dx \ \forall \mu \in Y.$$

By Green's formula, we obtain the equality

$$\int_\Omega (-\Delta y + |y|^\rho y - z)) \mu dx = 0 \ \forall \mu \in Y.$$

Therefore, the function satisfies the equation (4.21). Thus, the solvability of the boundary problem (4.20), (4.21) is proved.

Weakly nonlinear systems

4. Suppose there exist two solutions y_1 and y_2 of this problem. Then we have

$$-\Delta y_1 + \Delta y_2 + |y_1|^p y_1 - |y_2|^p y_2 = 0.$$

Multiplying by $y_1 - y_2$ after integration by using Green's formula, we get

$$\int_\Omega \sum_{i=1}^n \Big| \frac{\partial(y_1 - y_2)}{\partial x_i} \Big|^2 dx + \int_\Omega \big(|y_1|^p y_1 - |y_2|^p y_2\big)(y_1 - y_2)dx = 0.$$

The value under the second integral is not negative. Thus, the sum of two non-negative values is equal to zero. It is possible only if each of them is equal to zero. The first of them is the square of the norm of the space $H_0^1(\Omega)$. Therefore, $y_1 = y_2$. Thus, the solution of the boundary problem (4.20), (4.21) is unique. $\square$

Remark 4.43 The difference between the two methods of proving Theorem 4.5 is the third step only. We substantiated passing to the limit with using the properties of the monotone operators before. Now we apply the Rellich–Kondrashov theorem about compact embedding of the functional spaces.

Remark 4.44 Suppose we have the sign plus before the Laplace operator in the equation (4.19). Then we cannot obtain the a priori estimate of the solution of the considered boundary problem. We cannot guarantee the unique solvability of this problem for arbitrary absolute terms.

Now we can analyze an optimal control problem for the considered system.

4.7 Optimal control problems for nonlinear elliptic equations

Consider the control system described by the boundary problem (4.19), (4.20). By Theorem 4.5, for all $v \in V$ it has a unique solution $y = y[v]$ from the space Y. It is given a non-empty convex closed subset U of the space $L_2(\Omega)$ and the functional

$$I(v) = \frac{\alpha}{2} \int_\Omega v^2 dx + \frac{1}{2} \int_\Omega \sum_{i=1}^n \Big(\frac{\partial y[v]}{\partial x_i} - \frac{\partial y_d}{\partial x_i} \Big)^2 dx,$$

where $\alpha > 0$, and y_d is a known function of the space $H_0^1(\Omega)$.

Problem 4.2 *Find the control $u \in U$ that minimizes the functional I on the set U.*

Remark 4.45 The unique distinction between this problem and Problem 3.2 is the nonlinear term of the state equation.

158   *Optimization and Differentiation*

Determine the functionals $J : V \to R$ and $K : Y \to R$ by the equalities

$$J(v) = \frac{\alpha}{2} \int_\Omega v^2 dx, \quad K(y) = \frac{1}{2} \int_\Omega \sum_{i=1}^n \left(\frac{\partial y}{\partial x_i} - \frac{\partial y_d}{\partial x_i} \right)^2 dx.$$

Now Problem 4.2 becomes the partial case of the abstract Problem 4.1. At first, determine the weak continuity of the dependence of the state function from the control, i.e., mapping $y[\cdot] : V \to Y$.

Lemma 4.7 *Mapping* $y[\cdot] : V \to Y$ *for the problem* (4.19), (4.20) *is weakly continuous.*

Proof. Consider a sequence $\{v_k\}$ of the space V such that $v_k \to v$ weakly in V. Multiply the equality (4.19) for $v = v_k$ by the function $y_k = y[v_k]$. After integration of the result by using Green's formula we have the equality

$$\int_\Omega \sum_{i=1}^n \left| \frac{\partial y_k}{\partial x_i} \right|^2 dx + \int_\Omega |y_k|^{p+2} dx = \int_\Omega y_k (v_k + f) dx$$

that is an analog of (4.22). Then

$$\|y_k\|^2 + \|y_k\|_q^q \le \left(\|v_k\|_Z + \|f\|_Z \right) \left(\|y_k\| + \|y_k\|_q \right).$$

By continuous embedding of the space $L_2(\Omega)$ to $L_{q'}(\Omega)$ and to Z too, embedding of the space V to Z is weakly continuous, because this is a linear continuous operator. Then the sequence $\{v_k\}$ is weakly convergent on the space Z, and this is bounded there. Repeat the transformation from the second proof of Theorem 4.5. Using the last inequality, we prove the boundedness of the sequence $\{y_k\}$ in the space Y. Extracting a subsequence, we have the convergence $y_k \to y$ weakly in Y.

Using the Rellich–Kondrashov theorem and the convergence $y_k \to y$ weakly in $H_0^1(\Omega)$, we obtain the strong convergence of this sequence in the space $L_2(\Omega)$. Extracting a subsequence, we get $y_k \to y$ a.e. on Ω. Therefore, $|y_k|_k^y \to |y|^y$ a.e. on Ω. Applying Lemma 4.6 by using the boundedness of the sequence $\{|y_k|^p y_k\}$ in the space $L_{q'}(\Omega)$ we obtain the convergence $|y_k|_k^y \to |y|^y$ weakly in $L_{q'}(\Omega)$. Multiply the equality (4.19) by the arbitrary function $\mu \in Y$. After integration we have the equality

$$\int_\Omega \left(\sum_{i=1}^n \frac{\partial y_k}{\partial x_i} \frac{\partial \mu}{\partial x_i} + |y_k|^p y_k \mu \right) dx = \int_\Omega (v_k + f) \mu dx \; \forall \mu \in Y.$$

Passing to the limit as $k \to \infty$ we have the equality

$$\int_\Omega \left(\sum_{i=1}^n \frac{\partial y}{\partial x_i} \frac{\partial \mu}{\partial x_i} + |y|^p y_k \mu \right) dx = \int_\Omega (v + f) \mu dx \; \forall \mu \in Y.$$

Thus, the function is the solution of the problem (4.19), (4.20) for the control v. This is unique by Theorem 4.5. Thus, the operator $y[\cdot] : V \to Y$ is weakly continuous. $\square$

Remark 4.46 The proof of the weak continuity of control-state mapping, particularly obtaining a priori estimates and the substantiation of passing to the limit, is the analog of the appropriate steps of the proof of the boundary problem solvability.

Now we can prove the existence of the solution of Problem 4.2. Indeed, we obtained the necessary properties of the functionals J and K in the previous chapter.

Theorem 4.6 *Problem* 4.2 *is solvable.*

Remark 4.47 We cannot prove the uniqueness of the optimal control, because we do not have any information about the convexity of the given functional. Of course, we do not prove also well-posedness of the optimal control problem.

4.8 Necessary conditions of optimality

Try to obtain the necessary optimality conditions for Problem 4.2 by Theorem 4.2. This theorem uses the following suppositions: the operator A has the continuous inverse operator and continuously differentiable in the neighborhood of the point $y = y[u]$, there exists a continuous inverse operator $A'(y)^{-1}$, B is linear continuous operator, the functional J is Gâteaux differentiable at the point u, the functional K is Fréchet differentiable at the point y. Then the necessary condition of optimality is the variational inequality (4.11)

$$\langle J'(u) - B^*p, v - u \rangle \geq 0 \ \forall v \in U,$$

where p is the solution of the adjoint equation (4.9)

$$\left[A'(y)\right]^* p = -K'(y).$$

At first, consider the boundary problem (4.19), (4.20) that is equivalent to the operator equation $Ay = z$ with a concrete operator A.

Lemma 4.8 *The operator A is continuously differentiable in the space Y, as*

$$A'(y)h = -\Delta h + (\rho + 1)|y|^\rho h \ \forall h \in Y.$$

Proof. The operator $A : Y \to Z$ is the sum of the linear Laplace operator (accurate within the sign) $(-\Delta) : H_0^1(\Omega) \to H^{-1}(\Omega)$ and the power operator $A_1 : L_q(\Omega) \to L_{q'}(\Omega)$ that is determined by the equality $A_1 y = |y|^\rho y$. The first of them is continuously differentiable, because this is the linear

160 *Optimization and Differentiation*

operator. Its derivative is equal to the initial operator. We proved the continuous differentiability of the second operator before (see Example 4.3), as $A_1'(y)h = (\rho+1)|y|^\rho h$. Hence, we determine the derivative of the operator A. $\square$

Now we would like to prove the differentiability of the inverse operator by use of the Inverse function theorem. It is necessary to determine the invertibility of the derivative $A'(y_0)$ of the operator A at the point $y_0 = y[v_0]$. This property is equivalent to the unique solvability of the linearized equation $A'(y_0)y = z$ in the space Y for all $z \in Z$. By Lemma 4.8 the linearized equation (4.5) is transformed to the linear elliptic equation

$$-\Delta y + (\rho + 1)|y_0|^\rho y = z \tag{4.26}$$

with the homogeneous boundary condition (Dirichlet problem). Multiply this equality by the function and integrate formally the result. Using Green's formula, we have

$$\int_\Omega \left| \frac{\partial y}{\partial x_i} \right|^2 dx + (\rho + 1) \int_\Omega |y_0|^\rho y^2 dx = \int_\Omega yz dx. \tag{4.27}$$

The first term of this equality is the square of the norm of the function with respect to the space $H_0^1(\Omega)$. Therefore, we have the possibility to obtain a priori estimate of the solution of the linearized equation in this space. However, we cannot use the second integral for obtaining a priori estimate of the solution in the space $L_q(\Omega)$. Suppose we prove the solvability of the linearized boundary problem in the space $H_0^1(\Omega)$ only. Then the integral on the right-hand side of the equality (4.27) makes sense if the function z is an element of the space $H^{-1}(\Omega)$ that is narrower than Z.

Thus, we could prove the unique solvability of the linearized equation (4.26) not in space Y, and in a larger set $H_0^1(\Omega)$ and for the absolute term of the equation not from a space Z, and of the more narrow set $H^{-1}(\Omega)$ (see Figure 4.8 and Lemma 4.9). This is not sufficient for the invertibility of the operator $A'(y_0)$ and using the Inverse function theorem too. However, we can obtain the result under additional supposition.

Suppose the condition

$$n = 2 \text{ or } \rho \le 4/(n-2) \text{ for } n = 3. \tag{4.28}$$

Therefore, we have continuous embedding $H_0^1(\Omega) \subset L_q(\Omega)$ by the Sobolev embedding theorem. Then we have embedding of adjoint spaces $L_{q'}(\Omega) \subset H^{-1}(\Omega)$. Hence, we obtain the equalities $Y = H_0^1(\Omega)$, $Z = H^{-1}(\Omega)$. Now we obtain a priori estimate of the solution of the linearized equation in the space Y for all elements z from the space Z because of the equality (4.27).

Lemma 4.9 *Under the condition* (4.28) *the operator $A'(y_0)$ for Problem 4.2 is invertible.*

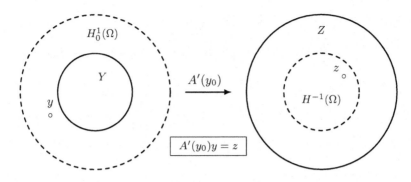

FIGURE 4.8: The solvability of the linearized equation.

Proof. It is sufficient to prove that for all functions $z \in H^{-1}(\Omega)$ the equation (4.26) has a unique solution $y \in H_0^1(\Omega)$. Consider the sequence $\{y_k\}$ that is determined by the equalities

$$y_k = \sum_{m=1}^{k} \xi_{mk} \mu_m, \quad k = 1, 2, ...,$$

where the numbers $\xi_{1k}, ..., \xi_{kk}$ satisfy the equalities

$$\int_\Omega \sum_{i=1}^{n} \frac{\partial y_k}{\partial x_i} \frac{\partial \mu_j}{\partial x_i} dx + (\rho+1) \int_\Omega |y_0|^\rho y_k \mu_j dx = \int_\Omega z \mu_j dx, \quad j = 1, ..., k. \quad (4.29)$$

This system of linear algebraic equations has a unique solution.

Multiply j-th equality (4.29) by ξ_j and sum the result by $j = 1, ..., k$. We obtain the equality

$$\int_\Omega \sum_{i=1}^{n} \left|\frac{\partial y_k}{\partial x_i}\right|^2 dx + (\rho+1) \int_\Omega |y_0|^\rho |y_k|^2 dx = \int_\Omega y_k z dx$$

that is an analog of (4.27). Then we have the estimate $\|y_k\| \leq \|z\|_*$, where $\|z\|_*$ is the norm of element z of the space $H^{-1}(\Omega)$. By boundedness of the sequence $\{y_k\}$ in the space $H_0^1(\Omega)$, there exists its subsequence such that $y_k \to y$ weakly in $H_0^1(\Omega)$. Using continuous embedding $H_0^1(\Omega) \subset L_q(\Omega)$, we have the convergence $y_k \to y$ weakly in $L_q(\Omega)$. The functions y_0 and μ_j belong to the space $H_0^1(\Omega)$ and to $L_q(\Omega)$ too. Then we have the inclusion $|y_0|^\rho \in L_{q/\rho}(\Omega)$. Therefore, $|y_0|^\rho \mu_j \in L_s(\Omega)$, where $1/s = \rho/n + 1/q = 1/q'$. Thus, we get $|y_0|^\rho \mu_j \in L_{q'}(\Omega)$. Passing to the limit in the equality (4.29) as $k \to \infty$, we obtain

$$\int_\Omega \sum_{i=1}^{n} \frac{\partial y}{\partial x_i} \frac{\partial \mu_j}{\partial x_i} dx + (\rho+1) \int_\Omega |y_0|^\rho y \mu_j dx = \int_\Omega z \mu_j dx, \quad j = 1, 2, ...$$

Therefore, the following equality holds:

$$\int_\Omega \sum_{i=1}^n \frac{\partial y}{\partial x_i} \frac{\partial \mu}{\partial x_i} dx + (\rho+1) \int_\Omega |y_0|^\rho y\mu dx = \int_\Omega z\mu dx, \ \forall \mu \in H_0^1(\Omega).$$

Thus, y is a solution of the homogeneous Dirichlet problem of the equation (4.26).

Suppose there exist two solutions y_1 and y_2 of this problem. Then its difference $y = y_1 - y_2$ is the solution of this problem for $z = 0$. Multiply the equation (4.26) for this case by the function y. Integrate the result by using Green's formula. We obtain the equality

$$\int_\Omega \sum_{i=1}^n \left|\frac{\partial y}{\partial x_i}\right|^2 dx + (\rho+1) \int_\Omega |y_0|^\rho y^2 dx = 0.$$

Therefore, $y_1 = y_2$. Thus, this equation has the unique solution in the space $H_0^1(\Omega)$ and the operator $A'(y_0)$ is invertible. $\square$

Remark 4.48 This control system is weakly nonlinear if the Inverse function theorem is applicable. Under the supposition (4.28), which guarantees the applicability of this result, the nonlinear term has no effect on the domain and codomain of the state operator. The solution of the state equation without a nonlinear term (see previous chapter) is a point of a larger functional space for the general case, as this is true for the absolute term of the narrower set. However, the linear and weakly nonlinear state operator have the same domain and codomain.

Remark 4.49 In reality, for all functions z from a larger set the equation (4.26) has a more regular solution (see next chapter). It is important that this results even without the condition (4.28). This will be the basis for obtaining necessary conditions of optimality for Problem 4.2 without supposition (4.28).

Using the last result we can prove the differentiability of mapping $y[\cdot] : V \to Y$ by Lemma 4.1. Its derivative satisfies the equality (4.7) with function p_μ that is the solution of the adjoint equation (4.8). Now we determine an operator that is adjoint to $A'(y_0)$.

Lemma 4.10 *The operator $[A'(y_0)]^*$ is determined by the formula*

$$\big[A'(y_0)\big]^* p = -\Delta p + (\rho+1)|y_0|^\rho p \ \forall p \in Z'.$$

Proof. By the definition of the operator $[A'(y_0)]^*$ we have

$$\langle p, A'(y_0)h \rangle = \int_\Omega p\big[-\Delta h + (\rho+1)|y_0|^\rho h\big] dx \ \ \forall h \in Y, p \in Z'.$$

Using Green's formula and the homogeneous boundary conditions, we obtain the equality

$$\big\langle \big[A'(y_0)\big]^* p, h \big\rangle = \int_\Omega h\big[-\Delta p + (\rho+1)|y_0|^\rho p\big] dx \ \forall h \in Y, p \in Z'.$$

Then we get the formula of the adjoint operator. $\square$

Weakly nonlinear systems

Remark 4.50 By Lemma 4.10, the operator $A'(y_0)$ is self-adjoint.

Lemma 4.11 *Under the condition* (4.28) *the operator* $y[\cdot] : V \to Y$ *has a Gâteaux derivative at the arbitrary point* $u \in V$ *that satisfies the equality*

$$\int_\Omega \mu y'[u]h dx = \int_\Omega h p_\mu[u] dx \quad \forall \mu \in Y', h \in V, \tag{4.30}$$

where $p_\mu[u]$ *is the solution of the homogeneous Dirichlet problem for the equation*

$$-\Delta p_\mu[u] + (\rho + 1)\big|y[u]\big|^\rho p_\mu[u] = \mu. \tag{4.31}$$

Indeed, the operator B^* is embedding of the space $Z' = Y$ to $V' = V$. Then the equality (4.6) that determines the derivative of the control-state mapping can be transformed to the equality (4.30) with adjoint equation (4.31).

By the equality $y[v] = A^{-1}(Bv + F)$ we have

$$y'[u] = (A^{-1})'(Bu + f).$$

Thus, the difference between the derivatives of mapping $y[\cdot] : V \to Y$ at the point u and inverse operator $A^{-1} : Z \to Y$ at the point $z = Bu + f$ is the operator B that is embedding of the space $L_2(\Omega)$ to $H^{-1}(\Omega)$. Therefore, we have the following result.

Lemma 4.12 *Under the condition* (4.28) *the operator* $A^{-1} : Z \to Y$ *has a Gâteaux derivative at the point* $z = u + f$ *with arbitrary* $u \in V$ *such that*

$$\int_\Omega \mu(A^{-1})'(z) dx = \int_\Omega h p_\mu[u] dx \quad \forall \mu \in Y', h \in Z, \tag{4.32}$$

where $p_\mu[u]$ *is the solution of the homogeneous Dirichlet problem for the equation* (4.31).

Find the derivative of the given functional.

Lemma 4.13 *Under the condition* (4.28) *the functional* I *has Gâteaux derivative* $I'(u) = \alpha u - p$ *at the point* u, *where is the solution of the equation*

$$-\Delta p(x) + (\rho + 1)|y(x)|^\rho p(x) = \Delta y(x) - \Delta y_d(x), \quad x \in \Omega \tag{4.33}$$

with $y = y[u]$ *and the boundary condition*

$$p(x) = 0, \quad x \in \Gamma. \tag{4.34}$$

Proof. By the formula (4.10) we have the equality $I'(u) = J'(u) - B^*p$. The derivative of the functional J is $J'(u) = \alpha \alpha u$ (see Lemma 3.6). Then we find $I'(u) = \alpha u - p$. The function p satisfies the equation (4.31) with absolute term $\mu = -K'(y)$. By Lemma 3.7 we find the derivative $K'(y) = \Delta y - \Delta y_d$. Therefore, the function p is the solution of the boundary problem (4.33), (4.34). ⊔

164 *Optimization and Differentiation*

Remark 4.51 In reality, our functional is Gâteaux differentiable even without the condition (4.28). However, this result can be obtained if we replace the Gâteaux derivative of the inverse operator by its weaker analog.

Now we can use Theorem 4.2 and obtain necessary conditions of optimality for Problem 4.2.

Theorem 4.7 *Under the condition* (4.28) *the solution u of Problem 4.2 satisfies the variational inequality*

$$\int_\Omega (\alpha u - p)(v - u)dx \geq 0 \ \forall v \in U. \tag{4.35}$$

Indeed, put the derivative of the functional (see Lemma 4.2) to the variational inequality (4.11).

Thus, we have the system that consists of the initial boundary problem (4.19), (4.20) for $v = u$, the adjoint boundary problem (4.33), (4.34), and the variational inequality (4.25) with respect to the unknown functions u, y, p.

Remark 4.52 Note the equality of the variational inequalities (3.15) and (4.35). Indeed, we have the same availability of the control in the state equation and the cost functional for both problems. If we exclude the nonlinear term from the state equation (4.19), then it is transformed to (3.9). Then the adjoint equation (4.33) takes on form (3.16). Therefore, we have the same optimality conditions.

Remark 4.53 We have the serious difference between Theorem 3.7 and Theorem 4.7 because we do not have sufficiency of the optimality conditions for the nonlinear case.

Remark 4.54 If we have the set of admissible controls that is determined by local constraints, then we can transform the variational inequality (4.35) to the formula (3.19) (see Theorem 3.8). We can prove also the analog of Theorem 3.12 and obtain the maximum principle as the necessary condition of optimality.

4.9 Optimal control problems for general functionals

Extend our results to the general integral functionals. Consider again the control system described by the boundary problem (4.19), (4.20) with the same set of admissible controls. Consider the functional

$$I(v) = \int_\Omega F\big(x; y[v](x), \nabla y[v](x), v(x)\big)dx,$$

where F is a given function.

Problem 4.3 *Find a control $u \in U$ that minimizes the functional I on the set U.*

Weakly nonlinear systems

We use the following result for its analysis.

Theorem 4.8 *Let Ω be an open n-dimensional set, and F be a Caratheodory function $F : \Omega \times \mathbb{R}^{l+m} \to \mathbb{R}$ such that $F(x; \varphi, \cdot)$ is convex, $F(x; \varphi, \psi) \geq \Phi(|\psi|)$ a.e. on Ω, for all $\varphi \in \mathbb{R}^l$, $\psi \in \mathbb{R}^m$, where $\Phi : \mathbb{R}_+ \to \mathbb{R}_+$ is an increasing convex lower semicontinuous function such that $\Phi(\sigma)/\sigma \to \infty$ as $\sigma \to \infty$. If $\{y_k\}$ is a sequence of measurable functions such that $y_k \to y$ a.e. on Ω, and $z_k \to z$ weakly in $\left[L_1(\Omega)\right]^m$, then*

$$\int_\Omega F\big(x; y(x), z(x)\big)\, dx \leq \inf \lim_{k \to \infty} \int_\Omega F\big(x; y_k(x), z_k(x)\big)\, dx.$$

Prove the solvability of Problem 4.3.

Theorem 4.9 *Suppose $F : \Omega \times \mathbb{R}^{n+2} \to \mathbb{R}$ is a Caratheodory function, as $F(x; \varphi, \cdot)$ is convex, $F(x; \varphi, \psi) \geq \Phi(|\psi|)$ a.e. on Ω, for all $\varphi \in \mathbb{R}$, $\psi \in \mathbb{R}^n$, where $\Phi : \mathbb{R}_+ \to \mathbb{R}_+$ is an increasing convex lower semicontinuous function such that $\Phi(\sigma)/\sigma \to \infty$ as $\sigma \to \infty$. Then Problem 4.3 is solvable.*

Proof. Let $\{v_k\}$ be a minimizing sequence. Using the boundedness of the set of admissible controls, after extracting a subsequence we obtain $v_k \to v$ weakly in V. Then $v \in U$ because of the convexity and the closeness of the set U. Using Lemma 4.7, determine the convergence $y[v_k] \to y[v]$ weakly in Y, and $\nabla y[v_k] \to \nabla y[v]$ weakly in $\left[L_2(\Omega)\right]^n$. By the Rellich–Kondrashov theorem, $y[v_k] \to y[v]$ strongly in $L_2(\Omega)$ and a.e. on Ω. Embedding of the space $L_2(\Omega)$ to $L_1(\Omega)$ is continuous. Therefore, $z_k \to z$ weakly in $\left[L_1(\Omega)\right]^{n+1}$, where

$$z_k = \big(\nabla y[v_k], v_k\big), \quad z = \big(\nabla y[v], v\big).$$

Using Theorem 4.8, we get

$$I(v) \leq \inf \lim_{k \to \infty} I(v_k).$$

This completes the proof of Theorem 4.9. $\square$

Prove the differentiability of the given functional.

Lemma 4.14 *Under the assumptions of Theorem 4.9 and (4.28) suppose there exist the continuous partial derivatives $F_0'(x; \zeta)$, $F_1'(x; \zeta),..., F_{n+1}'(x; \zeta)$ of the function $F(x; \cdot)$ at the arbitrary point $\zeta \in \mathbb{R}^{n+2}$ for almost all $x \in \Omega$, as for all $\varphi \in \mathbb{R}$, $\omega \in \mathbb{R}^{n+1}$ the following inequalities hold:*

$$\big|F(x; \varphi, \omega)\big| \leq a(x) + b\Big(|\varphi|^r + \sum_{i=1}^{n+1} |\omega_i|^2\Big),$$

$$\big|F_0'(x; \varphi, \omega)\big| < a_0(x) + b_0\Big(|\varphi|^{r/r'} + \sum_{i=1}^{n+1} |\omega_i|^{2/r'}\Big),$$

Optimization and Differentiation

$$\left|F_j'(x;\varphi,\omega)\right| \le a_j(x) + b_j\left(|\varphi|^{r/2} + \sum_{i=1}^{n+1}|\omega_i|\right), \ j = 1,...,n+1,$$

where $a \in L_1(\Omega)$, $a_0 \in L_r(\Omega)$, $a_j \in L_2(\Omega)$, $b > 0$, $b_0 > 0$, $b_j > 0$, $j = 1,...,n+1$, r is arbitrary if $n = 2$, and $r = 2n/(n-2)$ for $n > 2$. Then the functional I has Gâteaux derivative

$$I'(u) = p + F_{n+1}'(x;y,\nabla y,u) \tag{4.36}$$

at the arbitrary point $u \in V$, where $y = y[u]$, p is the solution of the homogeneous Dirichlet problem for the equation

$$-\Delta p + (\rho+1)|y|^p p = F_0'(x;y,\nabla y,u) - div\, F_{\nabla y}'(x;y,\nabla y,u), \tag{4.37}$$

and

$$div\, F_{\nabla y}'(x;y,\nabla y,u) = \sum_{i=1}^{n}\frac{\partial F_i'(x;y,\nabla y,u)}{\partial x_i}.$$

Proof. Determine the Nemytsky operator

$$A : L_r(\Omega) \times \left[L_2(\Omega)\right]^{n+1} \to L_1(\Omega)$$

by the equality

$$(Az)(x) = F\big(x;z(x)\big) \ \forall z \in L_r(\Omega) \times \left[L_2(\Omega)\right]^{n+1}, \ x \in \Omega.$$

By the Krasnoselsky theorem, this is Fréchet differentiable, as its derivative $A'(z)$ at the arbitrary point $z \in L_r(\Omega) \times \left[L_2(\Omega)\right]^{n+1}$ is determined by the equality

$$\left[A'(z)h\right](x) = \sum_{i=0}^{n+1} F_i'\big(x;z(x)\big)h_i(x). \tag{4.38}$$

Using Lemma 4.11, we have the convergence

$$\sigma^{-1}\big(y[u+\sigma h] - y[u]\big) \to y'[u]h \ \text{ in } H_0^1(\Omega)$$

for all $h \in V$ as $\sigma \to 0$, where the derivative $y'[u]h$ satisfies the equality (4.30). Then

$$\frac{1}{\sigma}\left(\frac{\partial y[u+\sigma h]}{\partial x_i} - \frac{\partial y[u]}{\partial x_i}\right) \to \frac{\partial y'[u]h}{\partial x_i} \ \text{ in } L_2(\Omega), \ i = 1,...,n.$$

Using the Sobolev theorem, we have continuous embedding $H_0^1(\Omega) \subset L_r(\Omega)$. Therefore,

$$\sigma^{-1}\big(y[u+\sigma h] - y[u]\big) \to y'[u]h \ \text{ in } L_r(\Omega).$$

Determine the operator

$$B : L_2(\Omega) \to L_r(\Omega) \times \left[L_2(\Omega)\right]^{n+1}$$

by the equality
$$Bv = (y[v], \nabla y[v], v).$$

It has a Gâteaux derivative at the arbitrary point $u \in V$ such that
$$B'(u)h = (y'[u]h, \nabla y'[u]h, h) \ \forall h \in V. \tag{4.39}$$

Determine the operator $C : V \to L_1(\Omega)$ by the equality $C + AB$. By the Composite function theorem it has a Gâteaux derivative, as
$$C'(u)h = A'(y, \nabla y, v)B'(u)h \ \forall h \in V.$$

Using (4.38), (4.39), we get
$$C'(u)h = F_0'(x; y, \nabla y, u)y'[u]h+$$
$$\sum_{i=0}^{n} F_i'(x; y, \nabla y, u)\frac{\partial y'[u]h}{\partial x_i} + F_{n+1}'(x; y, \nabla y, u)h \ \forall h \in V.$$

The functional I is the composition of the integration operator and the operator C. It has a Gâteaux derivative at the arbitrary point $u \in V$ such that
$$\langle I'(u), h \rangle = \int_{\Omega} [C'(u)h](x)dx \ \forall h \in V.$$

Using Green's formula, we find
$$\langle I'(u), h \rangle = \int_{\Omega} \left[F_0'(x; y, \nabla y, u) - \operatorname{div} F_{\nabla y}'(x; y, \nabla y, u) \right] y'[u]hdx+$$
$$\int_{\Omega} F_{n+1}'(x; y, \nabla y, u)hdx \ \forall h \in V.$$

The equality (4.31) for
$$\mu = F_0'(x; y, \nabla y, u) - \operatorname{div} F_{\nabla y}'(x; y, \nabla y, u)$$

can be transformed to (4.37). Using (4.30), we have
$$\langle I'(u), h \rangle = \int_{\Omega} [p + F_{n+1}'(x; y, \nabla y, u)] hdx.$$

Therefore, the formula (4.36) is true. $\square$

Now we can obtain the necessary conditions of optimality.

Theorem 4.10 *Under the assumptions of Lemma 4.14 the optimal control for Problem 4.3 satisfies the variational inequality*
$$\int_{\Omega} [p + F_{n+1}'(x; y, \nabla y, u)](v - u)dx \geq 0 \ \forall v \in U.$$

4.10 Optimal control problems for semilinear elliptic equations

We consider the equation with power nonlinearity and try to extend our results to the equation with general nonlinearity. Let Ω be an open bounded set of Euclid space $\mathbb{R}^n$ with the boundary Γ. Consider the system described by nonlinear elliptic equation

$$-\Delta y(x) + a(x, y(x)) = f(x), \ x \in \Omega \tag{4.40}$$

with boundary condition

$$y(x) = 0, \ x \in \Gamma, \tag{4.41}$$

where a and f are given functions.

Theorem 4.11 *Suppose $a = a(x)$ is a Caratheodory function that satisfies the inequalities*

$$a(x, \varphi)\varphi \geq \chi|\varphi|^q, \tag{4.42}$$

$$[a(x, \varphi) - a(x, \psi)](\varphi - \psi) \geq 0, \tag{4.43}$$

$$|a(x, \varphi)| \leq a_0(x) + b|\varphi|^{q-1} \tag{4.44}$$

for all $\varphi \in \mathbb{R}$, $\psi \in \mathbb{R}$ and almost for all $x \in \Omega$, where $q > 2$, $a_0 \in L_{q'}(\Omega)$, $b > 0$, $\chi > 0$. Then the problem (4.40), (4.41) has a unique solution y from the space $Y = H_0^1(\Omega) \cap L_q(\Omega)$ for all f from the space $Z = Y' = H^{-1}(\Omega) + L_{q'}(\Omega)$, where $1/q + 1/q' = 1$.

Proof. Determine the operator A on the space Y by the equality $A = -\Delta + L$, where L is the Nemytsky operator that is determined by the equality

$$(Ly)(x) = a(x, y(x)), \ x \in \Omega.$$

The Laplace operator is the continuous operator from the space $H_0^1(\Omega)$ to $H^{-1}(\Omega)$. Using the Krasnoselsky theorem and the inequality (4.44), we determine the continuity of the map $L : L_q(\Omega) \to L_{q'}(\Omega)$. Hence, the operator $A : Y \to Y'$ is continuous. Using (4.42), we obtain the inequality

$$\langle Ay, y \rangle = \int_\Omega \left[-\Delta y + a(x, y) \right] y dx =$$

$$\int_\Omega \sum_{i=1}^n \left(\frac{\partial y}{\partial x_i} \right)^2 dx + \int_\Omega a(x, y) y dx \geq \|y\|^2 + \chi\|y\|_q^q \ \forall y \in Y.$$

Then the operator A is coercive. Finally, we have

$$\langle Ay - Az, y - z \rangle = \int_\Omega \left[\Delta z - \Delta y + a(x, y) - a(x, z) \right](y - z) dx =$$

Weakly nonlinear systems

$$\int_\Omega \sum_{i=1}^n \left[\frac{\partial(y-z)}{\partial x_i}\right]^2 dx + \int_\Omega \left[a(x,y) - a(x,z)\right](y-z)dx.$$

This value is non-negative because of the condition (4.43). Besides, we can have the equality here for $y = z$ only. Hence, the operator A is strictly monotone. Thus, all assumptions of Theorem 4.4 are true. This completes the proof of Theorem 4.11.

Remark 4.55 The power function $a(x,y) = |y|^{q-2}y$ satisfies the suppositions of Theorem 4.11. Therefore, the equation (4.40) is the extension of the elliptic equation (4.19).

Consider the control system described by the boundary problem

$$-\Delta y + a(x,y) = f + v \ \text{ in } \Omega \qquad (4.45)$$

with boundary condition

$$y = 0 \ \text{ in } \Gamma, \qquad (4.46)$$

where the control v belongs to the non-empty convex closed subset U of the space $V = L_2(\Omega)$. Under the conditions of Theorem 4.11 for all $v \in V$ it has a unique solution $y = y[v]$ from the space Y. Determine the functional

$$I(v) \ = \ \frac{\alpha}{2} \int_\Omega v^2 dx \ + \ \frac{1}{2} \int_\Omega \sum_{i=1}^n \left(\frac{\partial y[v]}{\partial x_i} - \frac{\partial y_d}{\partial x_i}\right)^2 dx,$$

where $\alpha > 0$, and y_d is a known function of the space $H_0^1(\Omega)$.

Problem 4.4 *Find the control $u \in U$ that minimizes the functional I on the set U.*

Prove the weak continuity of control–state mapping.

Lemma 4.15 *Mapping $y[\cdot] : V \to Y$ for the problem (4.45), (4.46) is weakly continuous.*

Proof. Consider a sequence $\{v_k\}$ of the space V such that $v_k \to v$ weakly in V. Multiply the equality (4.45) for $v = v_k$ by the function $y_k = y[v_k]$. After integration of the result by using Green's formula we have the equality

$$\int_\Omega \sum_{i=1}^n \left|\frac{\partial y_k}{\partial x_i}\right|^2 dx + \int_\Omega a_k y_k dx = \int_\Omega y_k(v_k + f)dx,$$

where $a_k(x) = a(x, y_k(x))$. Then we obtain

$$\|y_k\|^2 + \chi\|y_k\|_q^q \le \|v_k + f\|_Z (\|y_k\| + \|y_k\|_q) \le$$

$$\frac{1}{2}\|y_k\|^2 + \frac{1}{2}\|v_k + f\|_Z^2 + \|v_k + f\|_Z \|y_k\|_q.$$

170 *Optimization and Differentiation*

By continuous embedding of the space $L_2(\Omega)$ to $L_{q'}(\Omega)$ and to Z too, embedding of the space V to Z is weakly continuous. Then the sequence $\{v_k\}$ is weakly convergent on the space Z, and this is bounded in this space. By previous inequality, the sequence $\{y_k\}$ is bounded in Y. After extracting a subsequence, we have the convergence $y_k \to y$ weakly in Y.

By the Rellich–Kondrashov theorem, from the convergence $y_k \to y$ weakly in $H_0^1(\Omega)$ the strong convergence in $L_2(\Omega)$ and a.e. on Ω follows. Then $a_k \to a$ a.e. on Ω, where $a = a(x, y(x))$. From the inequality (4.44) it follows the boundedness of the sequence $\{a_k\}$ in $L_{q'}(\Omega)$. Using Lemma 4.6, we have $a_k \to a$ weakly in $L_{q'}(\Omega)$. Multiply the equality (4.45) by an arbitrary function $\mu \in Y$. After integration, we have

$$\int_{\Omega} \Big(\sum_{i=1}^{n} \frac{\partial y_k}{\partial x_i} \frac{\partial \mu}{\partial x_i} + a_k \mu \Big) dx = \int_{\Omega} (v_k + f) \mu dx \ \forall \mu \in Y.$$

Passing to the limit as $k \to \infty$ we have the equality

$$\int_{\Omega} \Big(\sum_{i=1}^{n} \frac{\partial y}{\partial x_i} \frac{\partial \mu}{\partial x_i} + a\mu \Big) dx = \int_{\Omega} (v + f) \mu dx \ \forall \mu \in Y.$$

Therefore, the function y is the solution of the problem (4.45), (4.46) for the control v. This is unique because of Theorem 4.11. Thus, the operator $y[\cdot] : V \to Y$ is weakly continuous. $\square$

Remark 4.56 This result is the extension of Lemma 4.7.

Using Lemma 4.15, we determine the weak lower semicontinuity of the minimizing functional that guarantees the solvability of the considered problem.

Theorem 4.12 *Problem* 4.4 *is solvable.*

It is necessary to differentiate the functional for obtaining the optimality conditions. This is true, if the dependence of the state function with respect to the control is differentiable. Suppose $A : Y \to Z$ is a differentiable operator. By the Inverse function theorem, the solution of the equation $Ay = v$ is differentiable at point $v_0 \in Z$, whenever the linearized equation $A'(y_0)y = z$ has a unique solution $y \in Y$ for all $z \in Z$, where $Ay_0 = v_0$.

Suppose there exists the continuous partial derivative $a_y(x, y)$ of the function $a(x, \cdot)$ at an arbitrary point $y \in \mathbb{R}$ for almost all $x \in \Omega$, as

$$\big| a_y(x, y) \big| \leq a_1(x) + b_1 |y|^{q-2}, \tag{4.47}$$

where $a_1 \in L_{q/(q-2)}(\Omega)$. By the Krasnoselsky theorem, the Nemytsky operator that is determined by the function a is Fréchet differentiable. Then the operator $A = -\Delta + L$ has the derivative that satisfies the equality

$$A'(y_0)y(x) = -\Delta y(x) + a_y(x, y_0(x)y(x).$$

Weakly nonlinear systems

Thus, the linearized equation is the homogeneous Dirichlet problem for the equation

$$-\Delta y + a_y(x, y_0)y = z \tag{4.48}$$

that is the extension of (4.26).

We use again the condition (4.28)

$$n = 2 \text{ or } \rho \leq 4/(n-2) \text{ for } n > 2.$$

Lemma 4.16 *Under the assumptions of Theorem 4.11 suppose the conditions (4.28) and (4.47) are true. Then the operator $y[\cdot] : V \to Y$ for the problem (4.45), (4.46) has a Gâteaux derivative at the arbitrary point $u \in V$ such that*

$$\int_\Omega \mu y'[u]h dx = \int_\Omega p_\mu[u]h dx \ \forall \mu \in Y', h \in V, \tag{4.49}$$

where $p_\mu[u]$ is the solution of the homogeneous Dirichlet problem for the equation

$$-\Delta p_\mu[u] + a_y(x, y_0)p_\mu[u] = \mu. \tag{4.50}$$

Proof. Prove that all conditions of the Inverse function theorem are true. Multiply the equality for $v_0 = u$ by the function $y = y[u]$. After integration we have

$$\int_\Omega \sum_{i=1}^n \left|\frac{\partial y}{\partial x_i}\right|^2 dx + \int_\Omega a_y(x, u)y^2 dx = \int_\Omega yz dx. \tag{4.51}$$

Consider the monotony condition (4.43). Determine $y = z + h$ there, where h is a positive number. Divide by h and pass to the limit as $h \to 0$. We obtain the inequality $a_y(x, z) \geq 0$ for all z. Therefore, the second integral at the equality (4.50) is non-negative. Under the condition (4.28) we have continuous embedding $H_0^1(\Omega) \subset L_q(\Omega)$ because of the Sobolev theorem. Hence, $Y = H_0^1(\Omega)$. We have also embedding of adjoint spaces $L_{q'}(\Omega) \subset H^{-1}(\Omega)$ and the equality $Z = H^{-1}(\Omega)$. Hence, from the equality (4.51) it follows that

$$\|y\|_Y^2 = \|y\|^2 \leq \left|\int_\Omega yz dx\right| \leq \|y\|\|z\|_* = \|y\|_Y\|z\|_Z.$$

Thus, for all $z \in Z$ we have the a priori estimate of the solution of the boundary problem for the equation (4.48) in the space Y. Using the standard theory of the linear elliptic equations (see Chapter 3), prove the one-value solvability of the considered boundary in this space for all $z \in Z$. Thus, the applicability of Inverse function theorem is proved. Therefore, the dependence of the solution of the boundary problem (4.45), (4.46) with respect to the control is differentiable.

Find the derivative from the equality

$$(A^{-1})'(u)h = y'[u]h = [A'(y)]^{-1}h \ \forall h \in Z.$$

172   *Optimization and Differentiation*

For all $\mu \in Y'$, $h \in V$ we have

$$\langle \mu, y'[u]h \rangle = \left\langle \mu, \left[A'(y)\right]^{-1}h \right\rangle = \left\langle \left\{ \left[A'(y)\right]^{-1} \right\}^{*} \mu, h \right\rangle = \left\langle \left\{ \left[A'(y)\right]^{*} \right\}^{-1} \mu, h \right\rangle.$$

The operator $A'(y)$ is self-adjoint. Then $A'(y)^{*}p_{\mu}[u]$ is equal to the left-hand side of the equality (4.50). This is the analog of the linearized equation (4.48). Therefore, we obtain

$$\langle \mu, y'[u]h \rangle = \langle p_{\mu}[u], h \rangle \ \forall \mu \in Y', h \in V.$$

That is the equality (4.49). $\square$

Find the derivative of the functional.

Lemma 4.17 *Under the suppositions of Lemma* 4.16 *the functional I for Problem* 4.4 *has Gâteaux derivative* $I'(u) = \alpha u - p$ *at the point u, where p is the solution of the homogeneous boundary problem for the equation*

$$-\Delta p + a_{y}(x, y_0)p = \Delta y[u] - \Delta y_d. \tag{4.52}$$

Proof. Using the Composite function theorem, we have

$$\langle I'(u), h \rangle = \int\limits_{\Omega} \alpha u h dx + \int\limits_{\Omega} \sum_{i=1}^{n} \left(\frac{\partial y[u]}{\partial x_i} - \frac{\partial y_d}{\partial x_i} \right) \frac{\partial y'[u]h}{\partial x_i} dx =$$

$$\int\limits_{\Omega} \alpha u h dx + \int\limits_{\Omega} (\Delta y[u] - \Delta y_d) y'[u]h dx \ \forall h \in V.$$

This completes the proof of Lemma 4.17. $\square$

Put the value of the functional derivative in to the standard variational inequality. We obtain the necessary condition of optimality.

Theorem 4.13 *Under the suppositions of Lemma* 4.16 *the solution u of Problem* 4.4 *satisfies the variational inequality*

$$\int\limits_{\Omega} (\alpha u - p)(v - u)dx \geq 0 \ \forall v \in U.$$

Remark 4.57 Theorem 4.13 and Lemmas 4.16, 4.17 are the extension of Theorem 4.7 and Lemmas 4.11, 4.13 for the control systems with semilinear equations.

Remark 4.58 We could replace the Laplace operator by the general linear elliptic operator (see Section 3.11).

Weakly nonlinear systems   173

4.11 Differentiation of the inverse operator

Consider an operator $A : Y \rightarrow Z$, where Y and Z are Banach spaces. Choose a point $y_0 \in Y$, and $z_0 = Ay_0$. Determine the differentiability of the inverse operator at the point z_0 without using the Inverse function theorem. Consider the following supposition.

Assumption 4.1 *The operator A is invertible at a neighborhood O of the point z_0.*

Denote the inverse operator A^{-1} by L. Consider the equalities

$$ALz_\sigma = z_0 + \sigma h, \quad ALz_0 = z_0$$

for all functions $h \in Z$ and the number σ so small that $z_\sigma \in O$, where $z_\sigma = z_0 + \sigma h$. Then

$$ALz_\sigma - ALz_0 = \sigma h.$$

Assumption 4.2 *The operator A is Gâteaux differentiable.*

Using Lagrange formula, we have

$$\langle \lambda, Ay - Ay_0 \rangle = \langle \lambda, A'[y_0 + \delta(\lambda)(y - y_0)](y - y_0) \rangle \ \forall \lambda \in Z',$$

where $\delta(\lambda) \in [0, 1]$. Determine the linear continuous operator $G(z, \lambda)$ by the formula

$$G(z, \lambda) = A'[y_0 + \delta(\lambda)(Lz - Lz_0)],$$

where $z \in Z$. Transform the previous equality. We get

$$\langle \lambda, G(z_\sigma, \lambda)(Lz_\sigma - Lz_0) \rangle = \sigma \langle \lambda, h \rangle \ \forall \lambda \in Z'.$$

Then

$$\langle G(z_\sigma, \lambda)^* \lambda, (Lz_\sigma - Lz_0)/\sigma \rangle = \langle \lambda, h \rangle \ \forall \lambda \in Z'. \tag{4.53}$$

The adjoint operator is a linear continuous operator from the space Z' to Y' here.

Consider the operator equation

$$G(z_\sigma, p_\mu[z])^* p_\mu[z] = \mu, \tag{4.54}$$

where $\mu \in Y'$. This is the linear equation

$$A'(y_0)^* p_\mu[z_0] = \mu \tag{4.55}$$

for $\sigma = 0$.

174 *Optimization and Differentiation*

Assumption 4.3 *For any $h \in Z$ the equation* (4.54) *has a unique solution* $p_\mu[z] \in Z'$.

Choose $\lambda = p_\mu[z]$ for small enough σ at the equality (4.53). We have

$$\langle \mu, [L(z_0 + \sigma h) - Lz_0]/\sigma \rangle = \sigma \langle p_\mu[z_\sigma], h \rangle \ \forall \mu \in Y', h \in Z. \tag{4.56}$$

Assumption 4.4 *For any $h \in Z$ we have the convergence $p_\mu[z_\sigma] \to p_\mu[z_0]$ $*$-weakly in Z' uniformly with respect to μ from the subset M of the space Y' with unit norm as $\sigma \to 0$.*

Theorem 4.14 *Under Assumptions* 4.1 − 4.4 *the operator A^{-1} has Gâteaux derivative D at the point z_0 such that*

$$\langle \mu, Dh \rangle = \sigma \langle p_\mu[z_0], h \rangle \ \forall \mu \in Y', h \in Z. \tag{4.57}$$

Proof. The operator D from the equality (4.57) is a linear continuous operator from Z to Y. Using the properties of the norm, from (4.56) and (4.57) we have

$$\left\| \frac{L(z_0 + \sigma h) - Lz_0}{\sigma} - Dh \right\|_Z = \sup_{\|\mu\|_{Y'}=1} \left| \left\langle \mu, \frac{L(z_0 + \sigma h) - Lz_0}{\sigma} - Dh \right\rangle \right| =$$

$$\sup_{\|\mu\|_{Y'}=1} \left| \langle p_\mu[z_\sigma] - p_\mu[z_0], h \rangle \right|.$$

By Assumption 4.4, $p_\mu[z_\sigma] \to p_\mu[z_0]$ $*$-weakly in Z' uniformly with respect to $\mu \in M$ for all $h \in Z$. Pass to the limit as $\sigma \to 0$ for all $h \in Z$. We obtain the convergence $[L(z_0 + \sigma h) - Lz_0]/\sigma \to Dh$ in Z. Thus, the operator D is in reality a Gâteaux derivative of the operator A^{-1} at the point z_0. $\square$

We shall use the technique of proving Theorem 4.8 at the next chapter for the analysis of the more difficult systems.

Our next step is the analysis of Problem 4.2 without the condition (4.28). We shall obtain the necessary conditions of optimality for this case on the basis of weaker operator derivatives.

4.12 Comments

The differentiation of the operators in the normalized spaces is considered, for example, by J.P. Aubin and I. Ekeland [31], Yu.G. Borisovich, V.G. Zvyagin, and Yu.I. Sapronov [86], J. Dieudonné [140], N.V. Ivanov [261], L.V. Kantorovich and G.P. Akilov [267], A.N. Kolmogorov and S.V. Fomin [276], M.A. Krasnoselsky and P.P. Zabreiko [282], S.G. Krein and others [283]. The analogical results for the non-normalized spaces are given by V.I. Averbuh and O.G. Smolyanov [36], [37], A. Frölicher and W. Bucher [190], J. Gil de Lamadrid [209],

Weakly nonlinear systems 175

H.H. Keller [268], J. Sebastiãoe e Silva [451], A. Shapiro [498], and C. Ursescu [534]. The mean value theorem is given in V.I. Averbuh and O.G. Smolyanov [36]; and the Krasnoselsky theorem is considered in M.A. Krasnoselsky and P.P. Zabreiko [282], and S.G. Krein and others [283]. The proof of the Composite function theorem is given, for example, in L.V. Kantorovich and G.P. Akilov [267]; and Example 4.4 is considered by V.I. Averbuh and O.G. Smolyanov [37].

Theorem 4.1 is proved by J.P. Aubin and I. Ekeland [31] and J. Dieudonné [140]. The Inverse function theorem is one of the important results of the nonlinear functional analysis. Its finite dimensional version was installed at the stage of the emergence of the classic differential calculus. One of the first from its infinite dimensional analog is the theorem of L. Graves [223]. The different variants of the inverse function theorem are proved by A.V. Arutunov [26], [27], J.P. Aubin and I. Ekeland [31], J.P. Aubin and H. Frankowska [32], E.V. Avakov and A.V. Arutunov [35], V.I. Averbuh and O.G. Smolyanov [36], [37], R.G. Bartle [55], M. Cristea [132], V.F. Demyanov and A.M. Rubinov [137], A. Domokos [143], L. Doyen [145], E. Dubuc and J.Zilber [148], J. Eells [153], M. Fečkan [175], H. Frankowska [182], [183], X. Guo and J. Ma [228], M. Kong, G. Wu, and Z. Shen [277], S. Lang [296], K. Lorenz [324], J. Milnor [358], R. Radulescu and M. Radulescu [420], and S.Ya. Serovajsky [466], [467].

The Inverse function theorem is, in reality, the linearization result, because we obtain the properties of the nonlinear object (the nonlinear equation) on the basis of the properties of its linear approximation (the linearized equation). The meaning of the differentiation is the local linearization of the considered object. Particularly, the smooth curve can be local approximated by its tangent. By J. Dieudonné, *the general idea of the differential calculus is the local approximation of the function by the linear function* [140]. The analogical assertion was given earlier by M. Fréchet.

The Inverse function theorem is applied frequently for the substantiation of the linearization method that is denoted by the *general method of the analysis* (see D. Henry [240]). The linearized technique is used also in calculus mathematics (Newton–Kantorovich method and other linearization algorithms; see L.V. Kantorovich and G.P. Akilov [267]), dynamic system theory (Lyapunov's theorem about the stability with respect to the linear approximation; see A.M. Lyapunov [335]), differential geometry (Riemann geometry is the theory of the space with local structure of the Euclid space; see E.G. Poznyak and E.V. Shikin [411]), differential topology (the smooth variety has the local structure of the linear space; see S. Lang [296]), Lie group theory (the Lie group is associated with Lie algebra that is the tangent space at the unit; see N. Bourbaki [87]), and others.

The classification of the abstract nonlinear operators is given, for example, in H. Gajewski, K. Gröger, and K. Zacharias [200]. R.I. Kachurovsky [266], G. Minti [361], and others give the basis of the monotone operator theory. These results are given also in H. Gajewski, K. Gröger, and K. Zacharias [200], J.L. Lions [316], R.E. Showalter [499], and M.M. Vainberg [535]. The proof of Lemma 4.5 and Theorem 4.5 is given in H. Gajewski, K. Gröger, and K. Zacharias [200] and J.L. Lions [316]. The different embedding theorems of the functional spaces are considered by R. Adams [4], O.V. Besov, V.P. Ilin, and S.M. Nikolsky [67], J.L. Lions and E. Magenes [319], J. Peetre [398], and S.L. Sobolev [508]. The theory of the nonlinear elliptic equations are described in the books of H. Gajewski, K. Gröger, and K. Zacharias [200], D. Gilbarg N. S. Trudinger [210], O.A. Ladyzhenskaya and N.N. Uraltseva [294], J.L. Lions [316], and others.

The different optimal control problems for the systems described by the nonlinear elliptic equations are considered, for example, by F. Abergel and R. Temam [1], J.J. Alibert and J.P. Raymond [15], N.V. Andreev and V.S. Melnik [17], N. Arada and J.P. Raymond [21], M. Bardi and S. Bottacin [54], M. Bergounioux [62], [63], J.F. Bonnans [83], J.F. Bonnans and E. Casas [84], A. Cañada, J.L. Gámez, and J.A. Montero [105], E. Casas [110], E. Casas and L. A. Fernández [112], E. Casas and M. Mateos [113], E. Casas and F. Tröltzsch [115], A.I. Egorov [155], N. Fujii [192], [193], A.V. Fursikov [197], M. Goebel and U. Raitums [215], L. Gong and P. Fiu [219], J.P. Gossez [220], V.I. Ivanenko and V.S. Melnik [260], U. Ledzewicz and A. Nowakowsky [302], J.L. Lions [315], H.V. Lou [327], [328], [329], F.V. Lubyshev and M.E. Fairuzov [330], V.S. Melnik [350], P. Michel [353], U. Raitums [421], [422], A. Rösch and F. Tröltzsch [433], S.Ya. Serovajsky [455], [460], [461], [466],

176 *Optimization and Differentiation*

[472], [473], [478], [479], [480], [485], T. Slawing [504], M.I. Sumin [513], and M.D. Voisie [544].

Particularly, N.V. Andreev and V.S. Melnik [260] and M.D. Voisie [544] consider extremum problems for the abstract systems with monotone operators. A.V. Fursikov [197], H. Gao and N.H. Pavel [203], J.L. Lions [314], [317], and S.Ya. Serovajsky [461], [473], [475], [476], [480], [481], [485] analyze optimal control problems for singular elliptic equations. H.W. Lou [329] considers degenerate elliptic equations. L. Gong and P. Fiu [219] and S.Ya. Serovajsky [478], [480], [481], [485] solve extremum problems for the equations with non-smooth nonlinearity. U. Ledzewicz and A. Nowakowsky [302] consider an optimization problem for the nonlinear elliptic equation on the complex plane. F. Abergel and R. Temam [1], N.V. Andreev and V.S. Melnik [17], N. Arada and J.P. Raymond [21], M. Bergounioux [62], [63], J.F. Bonnans and E. Casas [84], E. Casas [110], E. Casas and L. A. Fernández [112], E. Casas and F. Tröltzsch [115], U. Mackenroth [336], V.S. Melnik [350], P. Michel [353], S.Ya. Serovajsky [457], [473], and M.I. Sumin [513] consider nonlinear elliptic equations with state constraints. J.F. Bonnans [83], E. Casas and M. Mateos [113], and E. Casas and F. Tröltzsch [115] obtain second conditions of optimality. J.F. Bonnans and E. Casas [84] determine optimality conditions by using the Ekeland principle. The extension methods for the optimization problems for nonlinear elliptic equations are considered by N. Arada and J.P. Raymond [21], U. Raitums [422], S.Ya. Serovajsky [479], and M.I. Sumin [513]. Optimization problems for the stationary Navier–Stokes equations are analyzed by A.V. Fursikov [197], J.F. Gossez [220], M. Gunzburger, L. Hou, and T. Svobodny [226], [225], K. Ito and S.S. Ravindran [259], J.L. Lions [317], and G. Wang [547]. Optimization problems for the system, described by the nonlinear elliptic variational inequalities, are considered by I. Bock and J. Lovisek [76], Q. Chen, D. Chu, and R. C. E. Tan [123], F. Patrone [397], and S.Ya. Serovajsky [496], [497].

The methods of the transformation of the nonlinear control systems to the linear control systems are considered by S.Ya. Serovajsky [457].

Chapter 5

Strongly nonlinear systems

5.1   Introduction 178
5.2   Absence of the differentiability of the inverse operator 180
5.3   Extended differentiation of operators 187
5.4   Necessary conditions of optimality for the strongly nonlinear
      system 195
5.5   Boundary control for Neumann problem 199
5.6   Extended differentiability of the inverse operator 206
5.7   Comments 209

The basic result of the extremum theory is the necessary condition of minimum for the differentiable functions. By the stationary condition, the derivative of the function at the point of minimum is equal to zero. We can extend this result to the problem of minimization for the smooth functionals. Particularly, the Gâteaux derivative of the functional at the point of its minimum is equal to zero. Then the necessary condition of minimum for the differentiable functional on a convex set is the variational inequality (see Chapter 1).

Note that the minimized functional depends frequently upon the unknown function, i.e., control nondirectly. It depends upon the state function; besides, the state function satisfies an equation that includes the control as a parameter. Therefore, we need to analyze the dependence of the state function from the control, i.e., control–state mapping for the determination of the functional derivative.

Suppose we have the state equation $Ay = Bv + f$, where the operators A and B are linear. Then the dependence $y = y(v)$ is affine. Finding of the Gâteaux derivative of a minimized functional is not difficult for this case. We obtained necessary conditions of optimality for abstract linear control systems (see Chapter 3). An optimization control problem for the system described by a linear elliptic equation was considered as an example.

We consider nonlinear operator equation $Ay = Bv + f$ in Chapter 4. If the operator A is invertible, then the solution of the state equation is $y = A^{-1}(Bv + f)$. By the Composite function theorem, the state function is differentiable with respect to the control if the inverse operator A^{-1} is differentiable. We can prove this property by using the Inverse function theorem. This result is applicable for the weakly nonlinear systems. The elliptic equation with power nonlinearity was considered as an application of this theory (see Figure 5.1). If the dimension of the given set and the degree of the

177

nonlinearity are small enough, then the considered control system is weakly nonlinear because of the Sobolev embedding theorem. Then we can determine necessary conditions of optimality.

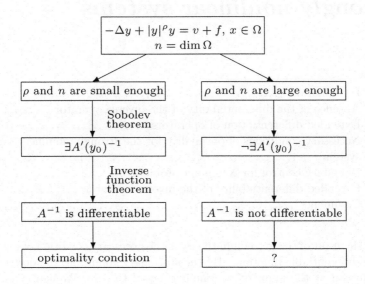

FIGURE 5.1: Unique stationary point is the point of minimum.

Now we consider strongly nonlinear systems. Suppositions of the Inverse function theorem get broken for this case. We prove that the dependence of the state function for the considered nonlinear elliptic equation is not Gâteaux differentiable with respect to the control. We determine the criterion of Gâteaux differentiability of control–state mapping for this system and the abstract case too.

Then we propose the extended operator derivative. This is an extension of the Gâteaux derivative. The dependence of the solution of the considered equation is extendedly differentiable with respect to the control without any suppositions. This property is sufficient for obtaining necessary conditions of optimality. We obtain this result for the Dirichlet problem with the distributed control and for the Neumann problem with the boundary control.

5.1 Introduction

Return to the analysis of the considered nonlinear elliptic equation. Let Ω be an open bounded set of Euclid space $\mathbb{R}^n$ with a boundary Γ. Consider the control system, described by the equation (4.19).

Strongly nonlinear systems 179

$$-\Delta y(x) + |y(x)|^\rho y(x) = v(x) + f(x), \; x \in \Omega \tag{5.1}$$

with the boundary condition

$$y(x) = 0, \; x \in \Gamma. \tag{5.2}$$

The parameter ρ is positive here. The control v is a point of a non-empty convex closed subset U of the space $L_2(\Omega)$. By Theorem 4.4 for all functions $v \in V$ and f from the space $Z = H^{-1}(\Omega) + L_{q'}(\Omega)$, the boundary problem (5.1), (5.2) has a unique solution $y = y[v]$ from the space $Y = H_0^1(\Omega) \cap L_q(\Omega)$, where $q = \rho + 2$, $1/q + 1/q' = 1$.

Determine the operator $A : Y \to Z$ such that Ay is the term at the left-hand side of the equality (5.1). Then the boundary problem (5.1), (5.2) can be interpreted as the operator equation $Ay = Bv + f$, where B is the embedding operator of the space V to Z. Its unique solvability is equivalent to the existence of the inverse operator A^{-1}. Then the solution of this equation is equal to $y[v] = A^{-1}(Bv + f)$. We considered before the minimization problem for the functional

$$I(v) = \frac{\alpha}{2} \int_\Omega v^2 dx + \frac{1}{2} \int_\Omega \sum_{i=1}^n \left(\frac{\partial y[v]}{\partial x_i} - \frac{\partial y_d}{\partial x_i}\right)^2 dx$$

on the set U, where $\alpha > 0$, and y_d is a known function of the space $H_0^1(\Omega)$ (see Problem 4.2). We used the differentiability of the mapping $v \to y[v]$ for obtaining necessary conditions of optimality. The operator B is linear. Therefore, the wishful property is true if the inverse operator A^{-1} is differentiable.

Using the Inverse function theorem, we proved Gâteaux differentiability of the inverse operator and the map $y[\cdot] : V \to Y$ under the supposition

$$n = 2 \; \text{or} \; \rho \le 4/(n-2) \; \text{for} \; n \ge 3. \tag{5.3}$$

Thus, we obtain necessary conditions of optimality for the considered problem for small enough value of the dimension n of the set Ω and the degree ρ of the nonlinear term. The control system is easy enough for this assumption. Note that the condition (5.3) guarantees continuous embedding

$$H_0^1(\Omega) \subset L_q(\Omega). \tag{5.4}$$

This is a corollary of the Sobolev theorem. Under this condition, we have the equality $Y = H_0^1(\Omega)$ and $Z = H^{-1}(\Omega)$. Note that the domain and codomain of the state operator are determined by the linear terms of the equation (5.1) only for this case. Therefore, we denote this system as weakly nonlinear.

However, what are the properties of the strongly nonlinear systems? Let the condition (5.4) be false. By Theorem 4.6, the considered optimization problem is solvable for this case too. Hence, the analysis of the system (5.1), (5.2) for the general case is an important enough problem.

5.2 Absence of the differentiability of the inverse operator

We proved the differentiability of the inverse operator under the supposition of the applicability of the Inverse function theorem. We use the hypothesis (5.4) because we cannot prove the necessary properties of the linearized equation $A'(y_0)y = z$ that guarantees the applicability of this theorem. However, maybe these properties in reality are true, but we cannot substantiate it by using known methods.

The linearized equation for our case is the homogeneous Dirichlet problem for the linear elliptic equation

$$-\Delta y + |y|^p y = z. \tag{5.5}$$

The invertibility of the derivative $A'(y_0)$ is equivalent to the one-valued solvability of the equation (5.5) in the space Y for all $z \in Z$. However, we have the following result.

Lemma 5.1 *If the supposition* (5.4) *gets broken, then there exist points* $y_0 \in Y$, $z \in Z$ *such that the equation* (5.5) *does not have the solution* $y \in Y$.

Proof. For any $y \in Y$ we have $\Delta y \in H^{-1}(\Omega)$. If the function $y_0 \in Y$ is continuous, then the term $|y_0|^p y_0$ belongs to the space $L_q(\Omega)$ and to $L_2(\Omega)$ too. Thus, for all $y \in Y$ the term at the left-hand side of the equality (5.5) belongs to $Z_* = H^{-1}(\Omega)$. This space is not equal to Z if the inclusion (5.4) is false. Therefore, the equation (5.5) cannot have a solution $y \in Y$ if $z \in (Z \setminus Z_*)$. $\square$

Thus, there exists a function $y_0 \in Y$ such that the derivative $A'(y_0)$ is not invertible. Hence, the difficulties of the equation (5.5) are principled because the wishful properties of the linearized equation can be broken if the inclusion (5.4) is false. Note that the optimal state y_0 is unknown. Therefore, we cannot exclude the possibility of its continuity as the considered situation can be realized not only for the case of the continuity of the function y_0.

Remark 5.1 The heightened regularity of the function y_0 is important, in reality. Then the term $|y_0|^p y_0$ is an element from a space that is narrower than $L_{q'}(\Omega)$. Therefore, the codomain of the derivative of the operator A is narrower than Z. Thus, this derivative is not surjection because there exists a function $z \in Z$ such that the equation $A'(y_0)y = z$ is not solvable in the space Y. The suppositions of the Inverse function theorem get broken for this case. We can obtain this result for the function $y_0 \in L_r(\Omega)$ if $r > q$ too. Thus, the Inverse function theorem is not applicable for non-typical points, i.e., the objects with heightened regularity (see Figure 5.2 and Theorem 5.2). We cannot find any possibilities to exclude this case. Therefore, we humble the principal non-applicability of the Inverse function theorem if the inclusion (5.4) gets broken.

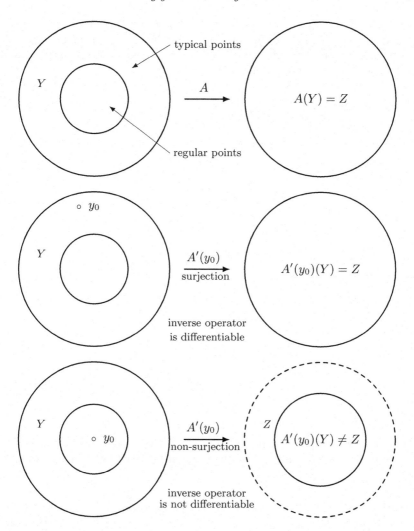

FIGURE 5.2: The absence of the differentiability of the inverse operator at points with heightened regularity.

Remark 5.2 The equality of the derivative codomain to the codomain of the given operator is called sometimes the ***Lusternik condition***, and the point, where this condition is true, is called the ***normal point*** of the operator. By Lemma 5.1, the point with heightened regularity is not the normal point of the operator.

Thus, the difficulty of the application of the Inverse function theorem for the considered example is serious enough. However, our problem is not the Inverse function theorem. We have an interest in the differentiability of the operator A^{-1}. Inverse function theorem gives a sufficient condition of this

182 *Optimization and Differentiation*

result. The necessity of this supposition is not obvious. We could suppose that the differentiability of the solution of the boundary problem (5.1), (5.2) with respect to the absolute term can be true without the condition (5.3). However, we have the following result.

Lemma 5.2 *If the condition* (5.4) *is false, then the operator* A^{-1} *for the problem* (5.1), (5.2) *is not Gâteaux differentiable.*

Proof. Suppose the operator $A^{-1} : Z \to Y$ is Gâteaux differentiable at the arbitrary point $z_0 = A y_0$. Then for all $z \in Z$ we have the convergence

$$\frac{A^{-1}(z_0 + \sigma h) - A^{-1} z_0}{\sigma} \to (A^{-1})'(z_0) z \ \text{ in } Y$$

as $\sigma \to 0$. Consider the equality

$$-\Delta \Big[A^{-1}(z_0 + \sigma h) - A^{-1} z_0 \Big] +$$

$$\Big[\big| A^{-1}(z_0 + \sigma h) \big|^p A^{-1}(z_0 + \sigma h) - \big| A^{-1} z_0 \big|^p A^{-1} z_0 \Big] = \sigma z.$$

Dividing it by σ and passing to the limit, we get

$$-\Delta (A^{-1})'(z_0) z + (\rho + 1) |y_0|^p (A^{-1})'(z_0) z = z.$$

Compare this equality with (5.5). Then for all $z \in Z$ the linearized equation has the solution

$$y = (A^{-1})'(z_0) z$$

from Y. However, this is impossible for $z \in (Z \setminus Z_*)$ because of Lemma 5.1. Therefore, our supposition about the differentiability of the inverse operator is false. $\square$

Thus, the difficulties of the given optimization problem without the condition (5.4) are principled because control–state mapping is not differentiable for the general case.

Remark 5.3 The considered operator A is invertible and continuously differentiable. However, its inverse operator is not differentiable. Thus, this operator is not a diffeomorphism for the strongly nonlinear case. Note that we can prove that the solution of the boundary problem (5.1), (5.2) is continuous with respect to the absolute term. Therefore, the operator A^{-1} is continuous. Thus, the operator is the continuous bijection such that its inverse operator is continuous too. An operator with the following properties is called the *homeomorphism*. Hence, we have the differentiable homeomorphism that is not the diffeomorphism.

Remark 5.4 If a continuously differentiable operator does not have the differentiable inverse operator, then its inverse operator is not differentiable and has the continuously differentiable inverse operator.

Using Lemma 4.9 and Lemma 5.2, we have

Strongly nonlinear systems 183

Theorem 5.1 *The inverse operator for the problem* (5.1), (5.2) *is Gâteaux differentiable at the arbitrary point if and only if the supposition* (5.4) *is true.*

Theorem 5.1 gives the criterion of the differentiability of the inverse operator A^{-1} at the arbitrary point of the space Z. However, we can have the differentiability of the inverse operator at a point and the absence of the differentiability at the other point. Consider the criterion of the differentiability of the inverse operator at the concrete point.

Theorem 5.2 *The operator* A^{-1} *of the problem* (5.1), (5.2) *is Gâteaux differentiable at the point* $z_0 \in Z$ *if and only if the derivative of the operator* A *at the point* $y_0 = A^{-1}z_0$ *is the surjection.*

Proof. Suppose the derivative $A'(y_0)$ is the surjection. If this is not the bijection, then there exists a point $z \in Z$ such that the linearized equation $A'(y_0)y = z$ has two solutions at least. Denote by y the difference between these solutions. Then we have the equality

$$-\Delta y + (\rho + 1)|y_0|^\rho y = 0.$$

After multiplication by y and integration we have

$$\int_\Omega \sum_{i=1}^n \left(\frac{\partial y}{\partial x_i}\right)^2 dx + (\rho + 1) \int_\Omega |y_0|^\rho y^2 dx = 0.$$

Thus, the function y is equal to zero. Therefore, the equation $A'(y_0)y = z$ has a unique solution. Using Banach inverse operator theorem, we determine the existence of the continuous inverse operator $A'(y_0)^{-1}$. Using the Inverse function theorem, we prove the differentiability of the inverse operator at the given point. If the derivative $A'(y_0)$ is not surjection, we can repeat the proof of Lemma 5.2. Then the inverse operator is not Gâteaux differentiable at the point z_0. $\square$

Thus, the absence of the differentiability of the inverse operator is realized at the points, where the considered derivative is not surjection. Extend this result to the abstract case.

Lemma 5.3 *Suppose an operator* $A : Y \to Z$ *is Fréchet differentiable at a point* $y_0 \in Y$ *and invertible in a neighborhood of this point. If its derivative at the point* y_0 *is not surjection, then the inverse operator* A^{-1} *is not Gâteaux differentiable at the point* $z_0 = A_0$.

Proof. Suppose there exists a Gâteaux derivative D of the inverse operator at the point z_0. We have the equality

$$A\left[A^{-1}(z_0 + \sigma z)\right] - AA^{-1}z_0 - \sigma z$$

184 *Optimization and Differentiation*

for arbitrary point $z \in Z$ and small enough number σ. After dividing by σ and passing to the limit as $\sigma \to 0$ by using the theorem of differentiation of composite function, we obtain

$$A'(y_0)Dz = z.$$

Thus, for all $z \in Z$ there exists the element $y = Dz$ of the space Y such that $A'(y_0)y = z$. Therefore, the derivative of the operator A at the point y_0 is the surjection that contradicts the condition of the lemma. Hence, the inverse operator is not differentiable. $\square$

Remark 5.5 Lemma 5.2 is the direct corollary of Lemma 5.3.

By the Inverse function theorem, if the derivative of the operator is invertible, then the inverse operator is differentiable. By Lemma 5.3, the inverse operator is not differentiable if the operator derivative is not invertible. We can interpret this result as an inverse proposition of the Inverse function theorem. From Lemma 5.3 and the Inverse function theorem, it follows the criterion of the differentiability of the inverse operator. Let an operator be continuously differentiable at a neighborhood of the point.

Theorem 5.3 *Suppose there exists an open neighborhood O of the point y_0 such that the set $O' = A(O)$ is the open neighborhood of the point $z_0 = Ay_0$ as there exists the inverse operator $A^{-1} : O' \to O$. Suppose also there exists a positive constant c such that the following inequality holds:*

$$\left\|A'(y_0)y\right\|_Z \geq c\|y\|_Y \; \forall y \in Y. \tag{5.6}$$

Then the inverse operator is Gâteaux differentiable at the point z_0 if and only if the derivative $A'(y_0)$ is the surjection.

Proof. If the derivative $A'(y_0)$ is the surjection, then from inequality (5.6) and the Banach inverse operator theorem it follows that $A'(y_0)$ has a continuous inverse operator. Then Gâteaux differentiability of the operator A^{-1} at the point z_0 is the corollary of the Inverse function theorem. Now suppose the operator A^{-1} has a Gâteaux derivative at this point. If the derivative $A'(y_0)$ is not the surjection, then the inverse operator is not Gâteaux differentiable at the point z_0 because of Lemma 5.3. This contradiction proves that the derivative $A'(y_0)$ is the surjection. $\square$

Thus, the absence of Gâteaux differentiability of control–state mapping is the widespread phenomenon.

Remark 5.6 The obtained results give the method of finding examples of equations with non-differentiable dependence of the solution with respect to a parameter. It is necessary to determine the operator derivative and choose the point of the differentiation so regular that the codomain of the derivative is heightened regular (see Figure 5.2). If this is possible, then the codomain of the derivative is a narrower set than the codomain of the initial operator. Then this derivative is not surjection, and the inverse operator is not Gâteaux differentiable.

Strongly nonlinear systems

Remark 5.7 We suppose that the majority of points is not heightened regular. However, we do not know which point is typical or regular. Therefore, the derivative of the state operator at the optimal control can be non-surjection, and control–state mapping can be non-differentiable.

Remark 5.8 We shall determine the analogical results for more general systems.

Example 5.1 *Power operator.* Consider the easier power operator $L : L_q(\Omega) \to L_{q'}(\Omega)$ that is determined by the equality $Ly = |y|^\rho y$ for all $y \in L_q(\Omega)$, where $q = \rho + 2$. It has the inverse operator $L^{-1}z = |z|^{-\rho/(\rho+1)}z$ for all $z \in L_{q'}(\Omega)$. The derivative of the operator L at a point y_0 satisfies the equality $L'(y_0)y = (\rho + 1)|y_0|^\rho y$ for all $y \in L_q(\Omega)$. The point y_0 is a typical point of the space $L_q(\Omega)$ if there does not exist a number $\alpha > 0$ such that $y_0 \in L_{q+\alpha}(\Omega)$. If this point is not typical, then the codomain of the derivative of the operator L at this point is a proper subspace of $L_q(\Omega)$, i.e., the derivative $A'(y_0)$ is not the surjection. Then the inverse operator is not differentiable at the point $z_0 = Ly_0$ because of Lemma 5.3. Thus, the operator L^{-1} is not differentiable for all non-typical points.

Remark 5.9 It will be interesting to determine the complete set of non-differentiability of the inverse operator.

Theorem 5.3 uses the inequality (5.6). Prove that this is true for the operator of the system (5.1), (5.2) under the condition (5.3). We use the denotation $\| \cdot \|$, $\| \cdot \|_*$, and $\| \cdot \|_p$ for the norms of the spaces $H_0^1(\Omega)$, $H^{-1}(\Omega)$, and $L_p(\Omega)$ as before.

Lemma 5.4 *Under the condition* (5.3) *the operator A of the problem* (5.1), (5.2) *satisfies the inequality* (5.6).

Proof. The norm of the space $Z = H^{-1}(\Omega) + L_{q'}(\Omega)$ is determined by the equality (see Section 4.5)

$$\|z\|_Z = \inf_{z_1 \in H^{-1}(\Omega), \, z_2 \in L_{q'}(\Omega), \, z_1 + z_2 = z} \max \left\{ \|z_1\|_*, \|z_2\|_{q'} \right\}.$$

For any function $y \in Y$ we have the inclusions $\Delta y \in H^{-1}(\Omega)$, $|y_0|^\rho y \in L_{q'}(\Omega)$. Then we obtain the equality

$$\left\| A'(y_0)y \right\|_Z = \max \left\{ \|\Delta y\|_*, (\rho + 1)\||y_0|^\rho y\|_{q'} \right\} \geq \|\Delta y\|_* \quad \forall y \in Y. \quad (5.7)$$

Using the Sobolev embedding theorem and the condition (5.6), we have the equalities $Y = H_0^1(\Omega)$, $Z = H^{-1}(\Omega)$. By Theorem 3.5 for all functions $z \in H^{-1}(\Omega)$ Poisson equation $\Delta y = z$ has a unique solution $y \in H_0^1(\Omega)$, as $\|y\| \leq \|z\|_*$. Then for all functions $y \in Y$ we have the inequality $\|y\| \leq \|\Delta y\|_*$ because of the Poisson equation. Using (5.7), we get the inequality (5.6) with the constant $c = 1$. $\square$

TABLE 5.1: Relations between the properties of the operator derivative and the differentiability of the inverse operator

$A'(y_0)$ is the injection	$A'(y_0)$ is the surjection	differentiablility of A^{-1} at the point $z = Ay$
Yes	Yes	Yes
Yes	No	No
No	No	No
No	Yes	?

Remark 5.10 If the derivative has a continuous inverse operator, then the condition (5.6) is true. Therefore, Lemma 5.4 is applicable. Thus, we can have three different situations (see Figure 5.3 and Table 5.1) because of Theorem 5.3. If the derivative has a continuous inverse operator, then the considered inverse operator is Gâteaux differentiable by the Inverse function theorem. If this derivative is not the surjection, then the inverse operator is not differentiable because of Lemma 5.3. Finally, if the derivative is the surjection, but the condition (5.6) is false, then we cannot use the Inverse function theorem and Lemma 5.3 also. Therefore, we do not have any information about differential properties of the inverse operator. Note that this case is impossible for the considered example because of Theorem 5.2.

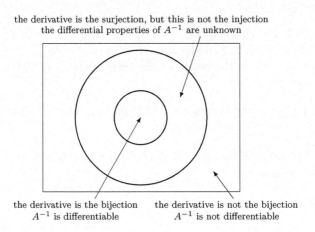

FIGURE 5.3: The relations between the properties of the operator derivative and the differentiability of the inverse operator.

Thus Gâteaux differentiability of the solution of the given boundary problem with respect to the absolute term is not realized for the general case, i.e., for the arbitrary values of the dimension of the set and the degree of the nonlinearity. However, the considered optimization problem is solvable for the general case. Therefore, it is necessary to find the method of the analysis for optimization control problems that uses weaker notions of the derivative.

5.3 Extended differentiation of operators

We return to consider the linearized equation (5.5). Inverse function theorem is not applicable if the condition (5.5) is false. However, this equation can have some positive properties for the general case. Multiply the equality (5.5) by the function y. After the formal integration by using Green's formula we get

$$\int_\Omega \sum_{i=1}^n \left(\frac{\partial y}{\partial x_i}\right)^2 dx + (\rho+1)\int_\Omega |y_0|^\rho y^2 dx = \int_\Omega yz dx. \qquad (5.8)$$

The first term in the left-hand side here is the square of norm of the space $H_0^1(\Omega)$. However, the second term has some relation with a norm. Determine the set

$$Y_0 = H_0^1(\Omega) \cap \left\{y\mid |y_0|^{\rho/2}y \in L_2(\Omega)\right\}.$$

This is a Hilbert space with the scalar product

$$(y,z)_{Y_0} = \int_\Omega \sum_{i=1}^n \frac{\partial y}{\partial x_i}\frac{\partial z}{\partial x_i}dx + (\rho+1)\int_\Omega |y_0|^\rho yz dx$$

and the square of norm

$$\|y\|_{Y_0}^2 = \int_\Omega \sum_{i=1}^n \left(\frac{\partial y}{\partial x_i}\right)^2 dx + (\rho+1)\int_\Omega |y_0|^\rho y^2 dx.$$

Then the value at the left-hand side of the equality (5.8) is the square of norm of the space Y_0. The integral at the right-hand side of this equality makes sense if the function z belongs to the space Z_0 that is adjoint to Y_0.

The set Y_0 is the intersection of two spaces. Therefore, the space Z_0 is the direct sum of its adjoint spaces. The first of them is the space $H^{-1}(\Omega)$. The second space is the set of the functions z such that the following inequality holds:

$$\int_\Omega yz dx < \infty$$

under the condition $|y_0|^{\rho/2}y \in L_2(\Omega)$. This property is true if $z = |y_0|^{\rho/2}\varphi$, where $\varphi \in L_2(\Omega)$. Indeed, we have

$$\int_\Omega yz dx = \int_\Omega y|y_0|^{\rho/2}\varphi dx \le \left(\int_\Omega |y_0|^\rho y^2 dx\right)^{1/2}\left(\int_\Omega \varphi^2 dx\right)^{1/2} < \infty.$$

Thus, we obtain

$$Z_0 = H^{-1}(\Omega) + \left\{z\mid z = |y_0|^{\rho/2}\varphi,\ \varphi \in L_2(\Omega)\right\}.$$

188   *Optimization and Differentiation*

Therefore, each element z of the space Z_0 is a sum $\psi + |y_0|^{\rho/2}\varphi$, where $\varphi \in L_2(\Omega)$.

Lemma 5.5 *For all $z \in Z_0$ the equation (5.5) has a unique solution $y \in Y_0$.*

Proof. We use the Galerkin method. Consider a complete set $\{\mu_m\}$ of the space Y. Determine an approximate solution y_k by the formula

$$y_k = \sum_{m=1}^{k} \xi_m \mu_m,$$

where the numbers $\xi_1, ..., \xi_m$ satisfy the system of linear algebraic equations

$$\int_\Omega \sum_{i=1}^{n} \frac{\partial y_k}{\partial x_i} \frac{\partial \mu_j}{\partial x_i} dx + (\rho+1)\int_\Omega |y_0|^\rho y_k \mu_j dx = \int_\Omega z\mu_j dx, \ j = 1, ..., k. \quad (5.9)$$

Hence, we can find the function $y_k \in Y$. After multiplication of the j-th equality (5.9) by ξ_j and addition by j we have the equality

$$\int_\Omega \sum_{i=1}^{n} \left(\frac{\partial y_k}{\partial x_i}\right)^2 + (\rho+1)\int_\Omega |y_0|^\rho y_k^2 dx = \int_\Omega y_k z dx.$$

The term at the left-hand side here is the square of the norm of the function y_k of the space Y_0 and the right-hand side is the value of the linear continuous functional z, i.e., the element of the space Z_0 at the point y_k. Then we have the inequality

$$\|y_k\|_{Y_0}^2 \le \|y_k\|_{Y_0} \|z\|_{Z_0}.$$

Therefore, we get

$$\|y_k\|_{Y_0} \le \|z\|_{Z_0}.$$

By the definition of the norm of the space Y_0 the sequences $\{y_k\}$ and $\{|y_0|^{\rho/2}y_k\}$ are bounded in the spaces $H_0^1(\Omega)$ and $L_2(\Omega)$ correspondingly.

Using the Banach–Alaoglu theorem, we extract a subsequence such that we have the convergence $y_k \to y$ weakly in $H_0^1(\Omega)$ and $|y_0|^{\rho/2}y_k \to \varphi$ weakly in $L_2(\Omega)$. Applying the Rellich–Kondrashov theorem, we have the convergence $y_k \to y$ strongly in $L_2(\Omega)$ and a.e. on Ω. Then we have the convergence $|y_0|^{\rho/2}y_k \to |y_0|^{\rho/2}y$ a.e. on Ω and the equality $\varphi = |y_0|^{\rho/2}y$. The functions y_0 and μ_j are the elements of the space $L_q(\Omega)$. Therefore, we get $|y_0|^{\rho/2} \in L_{2q/\rho}(\Omega)$ and $|y_0|^{\rho/2}\mu_j \in L_2(\Omega)$. Hence, we have the convergence

$$\int_\Omega |y_0|^\rho y_k \mu_j dx \to \int_\Omega |y_0|^\rho y \mu_j dx.$$

Passing to the limit at the equality (5.9), we get

$$\int_\Omega \sum_{i=1}^{n} \frac{\partial y}{\partial x_i} \frac{\partial \mu_j}{\partial x_i} dx + (\rho+1)\int_\Omega |y_0|^\rho y \mu_j dx = \int_\Omega z\mu_j dx, \ j = 1, 2, ...$$

Then we obtain (see Theorem 3.5, Theorem 4.4, and Lemma 4.9)

$$\int_\Omega \sum_{i=1}^n \frac{\partial y}{\partial x_i} \frac{\partial \mu}{\partial x_i} dx + (\rho+1) \int_\Omega |y_0|^\rho y\mu dx = \int_\Omega z\mu dx \ \forall \mu \in Y.$$

After the transformation of the first integral by using Green's formula, we prove that the function $y \in Y_0$ is the solution of the equation (5.5) in reality. The proof of the uniqueness of this solution is standard because of the linearity of the equation (see Theorem 4.4). □

Remark 5.11 We proved before the following weaker property (see Lemma 4.9). The equation (5.5) has a unique solution $y \in H_0^1(\Omega)$ for all $z \in H^{-1}(\Omega)$.

Remark 5.12 Note that the approximate solution y_k of the considered problem is an element of the space Y, but the exact solution belongs to the larger space $Y_* = H_0^1(\Omega)$. This situation is typical enough for the approximation theory and numerical methods. Objects that are more regular approximate the unknown solution. However, the convergence is proved with respect to the weaker topology only.

Thus, we can prove the solvability of the linearized equation in the space Y_0 that is larger than Y for all absolute terms from the space Z_0 that is narrower than Z (see Lemma 5.5). Note that the considered spaces depend upon the function y_0 that is the image of the point of differentiation z_0 of the inverse operator. Therefore, the functional space that characterizes the differential dependence of control–state mapping will depend upon the point v_0, where we differentiate the map $v \to y(v)$. These spaces change, but there exists a boundary of its change. Indeed, determine the spaces $Y_* = H_0^1(\Omega)$, $Z_* = H^{-1}(\Omega)$. We obtain the following continuous embedding (see Figure 5.4): $Y \subset Y_0 \subset Y_*$, $Z_* \subset Z_0 \subset Z$. If the condition (5.4) is true, we have the equalities $Y = Y_0 = Y_*$, $Z_* = Z_0 = Z$.

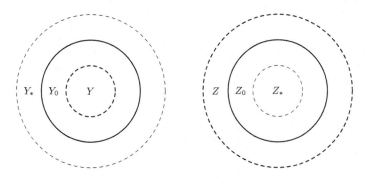

FIGURE 5.4: The spaces that characterize the linearized equation and the extended derivative.

Lemma 5.5 gives the stronger property than the results of the previous chapter. This is not sufficient for using the Inverse function theorem that is

Optimization and Differentiation

not applicable if the condition (5.4) gets broken. However, we can interpret the obtained result as a weaker form of suppositions of the Inverse function theorem because we prove the solvability of the linearized equation in the weaker sense and for a narrower class of the absolute terms. Therefore, we suppose that the dependence of the solution of the equation from the control can have some positive property that is weaker than Gâteaux differentiability.

Consider again the operator $A : Y \to Z$ that is determined by the formula $Ay = -\Delta y + |y|^p y$. Denote its inverse operator by L. For all functions $z_0, h \in Z$ and numbers σ we have the equality

$$\left(-\Delta L z_\sigma + |L z_\sigma|^p L z_\sigma\right) - \left(-\Delta L z_0 + |L z_0|^p L z_0\right) = \sigma h, \qquad (5.10)$$

where $z_\sigma = z_0 + \sigma h$. Using the Lagrange formula, we get

$$-\Delta\left(L z_\sigma - L z_0\right) + g(z_\sigma)^2\left(L z_\sigma - L z_0\right) = \sigma h,$$

where $g(z)^2 = (\rho+1)|y_0 + \varepsilon(Lz - L z_0)|^p$, $y_0 = L z_0$, $\varepsilon = \varepsilon(x) \in [0,1]$. Multiply the previous equality by a smooth enough function λ. After integration we get

$$-\int_\Omega \Delta\left(L z_\sigma - L z_0\right)\lambda dx + \int_\Omega g(z_\sigma)^2\left(L z_\sigma - L z_0\right)\lambda dx = \sigma \int_\Omega h\lambda dx. \quad (5.11)$$

Consider the equation

$$-\Delta p + g(z)^2 p = \mu \qquad (5.12)$$

and Hilbert spaces

$$Y(z) = H_0^1(\Omega) \cap \left\{y\mid g(z)y \in L_2(\Omega)\right\}, \quad Z(z) = Y(z)'$$

that are equal to Y_0 and Z_0 for $\sigma = 0$. Determine the scalar product

$$(y, z)_{Y(z)} = \int_\Omega \sum_{i=1}^n \frac{\partial y}{\partial x_i}\frac{\partial z}{\partial x_i} dx + \int_\Omega g(z)^2 yz dx$$

and the norm

$$\|y\|_{Y(z)}^2 = \int_\Omega \sum_{i=1}^n \left(\frac{\partial y}{\partial x_i}\right)^2 dx + \int_\Omega g(z)^2 y^2 dx$$

of the space $Y(z)$. Note embedding $Y \subset Y(z) \subset Y_*$, $Z_* \subset Z(z) \subset Z$. Multiply the equality (5.12) by the function p. After integration, we have the inequality

$$\|p\|_{Z(z)'} \le \|\mu\|_{Y(z)'}.$$

Using this estimate, we prove the following analog of Lemma 5.5.

Lemma 5.6 *For all functions* $z \in Z$, $\mu \in Y(z)'$ *the equation* (5.12) *has a unique solution* $p = p_\mu[z]$ *from the space* $Z(z)'$.

Consider the integrals of the equality (5.11). Using the inclusion $(Lz_\sigma - Lz_0) \in H^{-1}(\Omega)$ we determine that the first integral exists if $\lambda \in H_0^1(\Omega)$. The solution of the problem (5.1), (5.2) is an element of the space $L_q(\Omega)$. Therefore, we have the inclusion $g(z_0 + \sigma h) \in L_{2q/\rho}(\Omega)$ for all functions $h \in Z$. Then the function $g(z_\sigma)(Lz_\sigma - Lz_0)$ belongs to the space $L_2(\Omega)$. Thus, the second integral of the equality (5.11) exists if $g(z_\sigma)\lambda \in L_2(\Omega)$. Hence, the value at the left-hand side of the equality (5.11) has a sense for all functions λ from the set $Y(z_\sigma) = Z(z_\sigma)'$. The integral at the right-hand side of the equality (5.11) exists for all λ from the space Z' because $h \in Z$. We choose the function h from the narrower set Z_*. Then the right-hand side of the equality (5.11) has a sense for all $\lambda \in Z_*'$ even more so for than $\lambda \in Z(z_\sigma)'$. Therefore, we can choose λ equal to the solution $p_\mu^{\sigma h} = p_\mu[z_0 + \sigma h]$ of the equation (5.12). Then we get the equality

$$\int_\Omega \mu \frac{L(z_0 + \sigma h - Lz_0)}{\sigma} dx = \int_\Omega h p_\mu^{\sigma h} dz \ \forall h \in Z_*, \ \mu \in Y(z_\sigma)'. \tag{5.13}$$

Consider the value under the integral at the left-hand side of this equality. If we could pass to the limit in the second multiplier here, we obtain the Gâteaux derivative. However, this is impossible because our inverse operator $A^{-1} = L$ is not differentiable by Lemma 5.2.

Remark 5.13 We could determine the Gâteaux derivative from the equality (5.13) if However, these functions have weaker properties.

Try nevertheless to pass to the limit at the equality (5.13).

Lemma 5.7 *If* $\sigma \to 0$, *then we have the convergence* $p_\mu^{\sigma h} \to p_\mu$ *weakly in* Z_* *for all* $h \in Z_*$, $\mu \in Y_*'$, *where* p_μ *is a solution of the homogeneous Dirichlet problem for the equation*

$$-\Delta p_\mu + (\rho + 1)|y_0|^\rho p_\mu = \mu. \tag{5.14}$$

We shall prove Lemma 5.7 by using the following result.

Lemma 5.8 $Lz_\sigma \to Lz_0$ *in* Y_* *for all* $h \in Z_*$ *as* $\sigma \to 0$.

Proof. Consider again the equality (5.10) for $h \in Z_*$. Multiply it by $Lz_\sigma - Lz_0$ and integrate. We get

$$\int_\Omega \sum_{i=1}^n \left[\frac{\partial(Lz_\sigma - Lz_0)}{\partial x_i} \right]^2 dx +$$

$$\int_\Omega \left[|Lz_\sigma|^\rho Lz_\sigma - |Lz_0|^\rho Lz_0 \right](Lz_\sigma - Lz_0)dx = \int_\Omega (Lz_\sigma - Lz_0)hdx.$$

The second integral is non-negative because of the monotony of the function $f(y) = |y|^p y$. We obtain the inequality

$$\|Lz_\sigma - Lz_0\|^2 \leq \sigma \left| \int_\Omega (Lz_\sigma - Lz_0)hdx \right| \leq \sigma \|Lz_\sigma - Lz_0\| \|h\|_*.$$

Then the following estimate holds:

$$\|Lz_\sigma - Lz_0\| \leq \sigma \|h\|_*.$$

Therefore, we have the convergence $Lz_\sigma \to Lz_0$ in Y_*. $\square$

Remark 5.14 It is possible to prove the continuity of the inverse operator without any suppositions. Therefore, operator A is the homeomorphism. However, we do not use this property.

Proof of Lemma 5.7. We have the convergence $Lz_\sigma \to Lz_0$ in Y_* as $\sigma \to 0$ for all $h \in Z_*$ because of Lemma 5.8. Multiply the equality

$$-\Delta Lz_\sigma + |Lz_\sigma|^p Lz_\sigma = z_\sigma$$

by the function Lz_σ. After the integration we obtain

$$\|Lz_\sigma\|^2 + \|Lz_\sigma\|_q^q = \int_\Omega (z_0 + \sigma h)Lz_\sigma dx.$$

The point z_0 belongs to the set space Z. Therefore, this is a sum $\varphi + \psi$, where $\varphi \in H^{-1}(\Omega)$, $\psi \in L_{q'}(\Omega)$. Then we get the inequality

$$\left| \int_\Omega (z_0 + \sigma h)Lz_\sigma dx \right| \leq \|z_0 + \sigma h\|_* \|Lz_\sigma\| + \|\psi\|_{q'} \|Lz_\sigma\|_q.$$

Hence, the set $\{Lz_\sigma\}$ is bounded in the space $L_q(\Omega)$. Then $\{g(z_\sigma)\}$ is bounded in $L_{2q/\rho}(\Omega)$.

Multiplying the equality (5.12) for $z = z_0 + \sigma h$ by the function $p_\mu^{\sigma h}$ and integrating, we obtain the inequality

$$\|p_\mu^{\sigma h}\|_{Z'_*}^2 + \|g(z_\sigma)p_\mu^{\sigma h}\|_2^2 \leq \|p_\mu^{\sigma h}\|_{Z'_*} \|\mu\|_{Y'_*}.$$

Then the following estimates hold:

$$\|p_\mu^{\sigma h}\|_{Z'_*} \leq \|\mu\|_{Y'_*}, \quad \|g(z_\sigma)p_\mu^{\sigma h}\|_2 \leq \|\mu\|_{Y'_*} \ \forall h \in Z_*.$$

Using the inequality

$$\|g(z_\sigma)^2 p_\mu^{\sigma h}\|_{q'} \leq \|g(z_\sigma)\|_{2q/\rho} \|g(z_\sigma)p_\mu^{\sigma h}\|_2,$$

we prove the boundedness of the set $\{g(z_\sigma)^2 p_\mu^{\sigma h}\}$ in the space $L_{q'}(\Omega)$.

Extracting the subsequence, we have the convergence $p_\mu^{\sigma h} \to p$ weakly in Z'_* as $\sigma \to 0$ for all $\mu \in Y'_*$, $h \in Z_*$. Using the Rellich–Kondrashov theorem, we have $p_\mu^{\sigma h} \to p$ strongly in $L_2(\Omega)$ and a.e. on Ω. Then we have the convergence $g(z_\sigma)^2 p_\mu^{\sigma h} \to (\rho+1)|y_0|^\rho p$ a.e. on Ω. By the lemma of the weakly convergence in the spaces of integrable functions (see Lemma 4.6) by using the obtained estimate, we have $g(z_\sigma)^2 p_\mu^{\sigma h} \to (\rho+1)|y_0|^\rho p$ weakly in $L_{q'}(\Omega)$. By the equality (5.12) after passing to the limit, we prove that the function p is in reality the solution of the homogeneous Dirichlet problem for the equation (5.14).

Determine the linear continuous operator $D : Z_0 \to Y_0$ by the equality

$$\int_\Omega \mu D h dx = \int_\Omega h p_\mu dx \quad \forall h \in Z_0, \, \mu \in Y'_0. \tag{5.15}$$

Lemma 5.9 *If $\sigma \to 0$, then for all $h \in Z_*$ we have the convergence*

$$\frac{L(z_0 + \sigma h) - L z_0}{\sigma} \to D h \text{ weakly in } Y_*.$$

Proof. The equalities (5.13) and (5.15) make sense for $\mu \in Y'_*$, $h \in Z_*$. Therefore, the following equality holds:

$$\int_\Omega \mu \left[\frac{L(z_0 + \sigma h) - L z_0}{\sigma} - D h \right] dx = \int_\Omega h \big(p_\mu^{\sigma h} - p_\mu\big) dx \quad \forall \mu \in Y'_*, \, h \in Z_*.$$

Using Lemma 5.7, we prove the convergence $p_\mu^{\sigma h} \to p_\mu$ weakly in Z_* for all $h \in Z_*$, $\mu \in Y'_*$. Passing to the limit in the last equality, we finish the proof of the lemma. $\square$

Thus, the considered inverse operator, i.e., the map $v \to y(v)$, has a property that is a weaker analog of Gâteaux differentiability. This is the general notion of this chapter.

Definition 5.1 *An operator $L : Z \to Y$ is called $(Z_0, Y_0; Z_*, Y_*)$-**extendedly Gâteaux differentiable** or more simply **extendedly differentiable** at a point $z_0 \in Z$ if there exist linear topological spaces $Z_0, Y_0; Z_*, Y_*$ that satisfy the following continuous embedding (see Figure 5.4): $Z_* \subset Z_0 \subset Z$, $Y \subset Y_0 \subset Y_*$ and a linear continuous operator $D : Z_0 \to Y_0$ that is called the **extended derivative** such that*

$$\frac{L(z_0 + \sigma h - L z_0)}{\sigma} \to D h \text{ in } Y_*$$

as $\sigma \to 0$ for all $h \in Z_*$. If there exists a linear continuous operator $D : Z_0 \to Y_0$ such that

$$L(z_0 + h) = L z_0 + D h + \eta(h),$$

*where $\|\eta(h)\|_{Y_*}/\|h\|_{Z_*} \to 0$ as $h \to 0$ in Z_*, then D is called $(Z_0, Y_0; Z_*, Y_*)$-**extended Fréchet derivative** of the operator L at the point z_0.*

194   *Optimization and Differentiation*

Note that the $(Z, Y; Z, Y)$-extended derivative of the operator is equal to the usual Gâteaux derivative.

Remark 5.15 It is obvious that the $(Z, Y; Z_*, Y)$-extended Gâteaux derivative is the derivative with respect to the subspace Z_* (see Chapter 1).

Remark 5.16 It is easy to verify that the relation between extended Gâteaux and Fréchet derivatives are analogical as the relation between these classical derivatives.

Using Lemma 5.9, we obtain the following result.

Theorem 5.4 *The operator* $A^{-1} : Z \to Y$ *is* $(Z_0, Y_0; Z_*, Y_*^w)$*-extendedly Gâteaux differentiable at the arbitrary point* $z_0 \in Z$*, where* Y_*^w *is the space* Y_* *with the weak topology. Its extended derivative* D *is determined by the equality* (5.15).

Remark 5.17 Note that the spaces from the definition of the extended derivative depend upon the point of differentiation.

Thus, the inverse operator is extendedly differentiable at the arbitrary point for the general case, although it does not have the Gâteaux derivative if the condition (5.3) is false. The solvability of the linearized equation with respect to the weaker sense is combined with positive properties of the inverse operator.

The application of the extended differentiability gives exacter properties of the dependence of the equation solution from the absolute term. Indeed, the solution y of the equation

$$-\Delta y + |y|^\rho y = z$$

is Gâteaux differentiable with respect to the absolute term z for small enough values of the dimension n of the set and the degree of nonlinearity ρ if the condition (5.3) is true by Lemma 4.12. However, this is non-differentiable dependence for large enough values of these parameters if this condition is false because of Lemma 5.2. Therefore, the property of this dependence changes with a jump with respect to the parameters of the equation. Theorem 5.4 gives the appreciably more exact result. The extended differentiability is always true. However, the spaces of the definition of the extended derivative depend upon the parameters ρ and n (and upon the point of differentiation too). The extended derivative is equal to the classic one for small enough values of these parameters. However, the spaces of the definition of the extended derivative differ from the given spaces as increasing of the parameters ρ and n that determine the degree of the difficulty of the equation.

The condition (5.3) gets broken for large enough values of these parameters. Therefore, the inverse operator does not have a Gâteaux derivative. However, this is an extendedly differentiable operator (see Figure 5.5) as the difference between standard and extended derivatives increases with increasing of the parameters ρ and n. Therefore, the differential properties of the

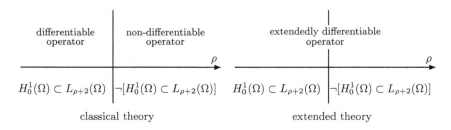

FIGURE 5.5: Classical and extendedly differentiable operators.

dependence of the equation solution from the absolute term gradually improve if the difficulties of the equation increase. We do not have any jumps. This result seems more exact.

We suppose that the more exact comprehension of the phenomenon can give the results that are not accessible by the grosser technique of the classical derivatives.

Theorem 5.4 gives weaker properties of the inverse operator for the boundary problem (5.1), (5.2) without any suppositions. This is positive information about control-state mapping. Try to use this result for obtaining necessary conditions of optimality.

5.4 Necessary conditions of optimality for the strongly nonlinear system

We return to the analysis of the optimization control problem for the system described by the nonlinear elliptic equation. We have the equation (5.1)
$$-\Delta y + |y|^p y = v + f$$
in the open bounded set Ω of the Euclid space $\mathbb{R}^n$ with homogeneous first order boundary condition (5.2). The known function f belongs to the space $Z = H^{-1}(\Omega) + L_{q'}(\Omega)$, where $1/q + 1/q' = 1$, $q = p + 2$. The control v is a point of a non-empty convex closed subset U of the space $V = L_2(\Omega)$. By Theorem 4.5, for all functions $v \in V$, $f \in Z$ the problem (5.1), (5.2) has a unique solution $y = y[v]$ from the space $Y = H_0^1(\Omega) \cap L_q(\Omega)$. Determine the functional
$$I(v) = \frac{\alpha}{2} \int_\Omega v^2 dx + \frac{1}{2} \int_\Omega \sum_{i=1}^n \left(\frac{\partial y[v]}{\partial x_i} - \frac{\partial y_d}{\partial x_i}\right)^2 dx,$$
where $\alpha > 0$, y_d is a given function from the space $H_0^1(\Omega)$.

196 *Optimization and Differentiation*

Problem 5.1 *Find the control that minimizes the functional I on the set U.*

We have in reality Problem 4.2 that is solvable by Theorem 4.5. Under condition (5.4), the Inverse function theorem is applicable, and the operator $y[\cdot] : V \to Y$ is Gâteaux differentiable (see Lemma 4.11). Then the minimized functional is differentiable (see Lemma 4.13). By Theorem 4.7, the optimal control satisfies the variational inequality

$$\int_\Omega (\alpha u - p)(v - u)dx \geq 0 \ \forall v \in U, \tag{5.16}$$

where p is the solution of the equation

$$-\Delta p + (\rho + 1)|y[u]|^\rho p = \Delta y[u] - \Delta y_d \tag{5.17}$$

with homogeneous first order boundary conditions.

If the condition (5.4) is false, then the inverse operator is not differentiable by Lemma 5.2. Therefore, the optimization methods from the previous chapter are not applicable. However, the operator $y[\cdot] : V \to Y$ is extendedly differentiable because of Theorem 5.4. We suppose that this property will be sufficient for the analysis of Problem 5.1.

Using Theorem 5.4, we prove the following result.

Lemma 5.10 *The operator $y[\cdot] : V \to Y$ has $\left(V, Y_0; V, Y_*^w\right)$-extended Gâteaux derivative at the arbitrary point $u \in V$ that satisfies the equality*

$$\int_\Omega \mu y'[u]hdx = \int_\Omega hp_\mu[u]dx \ \forall \mu \in Y_0', \ h \in V, \tag{5.18}$$

where $p_\mu[u]$ is a solution of the homogeneous Dirichlet problem for the equation (5.14) with $y_0 = y[u]$ and the spaces Y_0 and Y_*^w are determined in Theorem 5.4.

Proof. The function $p_\mu[u]$ belongs to the space Z_0' and the larger set V for all $\mu \in_0' Y$. Then the integral at the right-hand side of the equality (5.18) makes sense. Therefore, this equality determines a linear continuous operator $y'[u] : V \to Y_0$. Prove that this is the Gâteaux extended derivative of the dependence $y = y[v]$.

The solution of the boundary problem (5.1), (5.2) is determined by the formula $y[v] = A^{-1}(Bv + f)$, where Ay is the left-hand side of the equality (5.1), and B is the embedding operator of the space V to Z. For all functions $u, h \in V$ we have

$$\frac{y[u + \sigma h] - y[u]}{\sigma} = \frac{A^{-1}(Bu + f + \sigma h) - A^{-1}(Bu + f)}{\sigma}.$$

By Theorem 5.4, the operator A^{-1} has the extended derivative D at the point

Bu + f that satisfies the equality (5.15). Under the conditions of Lemma 5.10, we have the same function $p_\mu[u]$ at the equalities (5.15) and (5.18). These formulas are equal for $D = y'[u]$ if $h \in V$. Therefore, the operator $y'[u] : V \to Y_0$ from the equality (5.18) is in reality the extended derivative of the operator $y[\cdot] : V \to Y$.

Remark 5.18 We could determine a larger domain of the operator $y'[u]$ as determining the extended derivative, we could pass to the limit for a larger class of the directions h. However, this is not necessary, and the concrete control space V is given. But we could consider Problem 5.1 with larger control space.

We obtained the necessary optimality conditions by using Gâteaux differentiability of the minimized functional. We shall prove that this property is true if control-state mapping is extendedly differentiable. Prove at first the following result.

Lemma 5.11 *The map $y[\cdot] : V \to Y_*$ is continuous.*

Proof. For all $u, v \in V$ we have

$$\left(-\Delta y[u] + |y[u]|^p y[u] \right) - \left(-\Delta y[v] + |y[v]|^p y[v] \right) = u - v.$$

After the multiplication of this equality by $y[u] - y[v]$ and the integration by using Green's formula and the homogeneous boundary condition, we get

$$\int_\Omega \sum_{i=1}^n \left(\frac{\partial y[u]}{\partial x_i} - \frac{\partial y[v]}{\partial x_i} \right)^2 dx + \int_\Omega \left(|y[u]|^p y[u] - |y[v]|^p y[v] \right) dx =$$

$$\int_\Omega \left(y[u] - y[v] \right) (u - v) dx.$$

Using the non-negativity of the value under the integral of the second term at the left-hand side of this equality, we have

$$\left\| y[u] - y[v] \right\|^2 \leq \left| \int_\Omega \left(y[u] - y[v] \right) (u - v) dx \right| \leq \left\| y[u] - y[v] \right\| \| u - v \|_*.$$

Therefore, the operator $y[\cdot] : V \to Y_*$ is continuous. $\square$

Lemma 5.12 *The functional I of Problem 5.1 has Gâteaux derivative $I'(u) = \alpha u - p$, where p is the solution of the homogeneous Dirichlet problem for the equation (5.17).*

Proof. We have the equality

$$I(u + \sigma h) - I(u) = \frac{\alpha}{2} \int_\Omega \left[(u + \sigma h)^2 - u^2 \right] dx +$$

$$\frac{1}{2}\int_\Omega \sum_{i=1}^n \left\{ \left(\frac{\partial y[u+\sigma h]}{\partial x_i} - \frac{\partial y_d}{\partial x_i} \right)^2 + \left(\frac{\partial y}{\partial x_i} - \frac{\partial y_d}{\partial x_i} \right)^2 \right\} dx,$$

where $y = y[u]$. Transform the second integral

$$\int_\Omega \sum_{i=1}^n \left\{ \left(\frac{\partial y[u+\sigma h]}{\partial x_i} - \frac{\partial y_d}{\partial x_i} \right)^2 + \left(\frac{\partial y}{\partial x_i} - \frac{\partial y_d}{\partial x_i} \right)^2 \right\} dx =$$

$$\int_\Omega \left(\frac{\partial y[u+\sigma h]}{\partial x_i} + \frac{\partial y}{\partial x_i} - 2\frac{\partial y_d}{\partial x_i} \right) \left(\frac{\partial y[u+\sigma h]}{\partial x_i} - \frac{\partial y}{\partial x_i} \right) dx.$$

Using the definition of the scalar product for the space $H_0^1(\Omega)$, we get

$$\frac{I(u+\sigma h) - I(u)}{\sigma} = \frac{\alpha}{2} \int_\Omega (2uh + \sigma h^2) dx +$$

$$\frac{1}{2} \left(\frac{y[u+\sigma h] - y[u]}{\sigma}, y[u+\sigma h] + y[u] - 2y_d \right).$$

By Lemma 5.10, we have the convergence

$$\frac{y[u+\sigma h] - y[u]}{\sigma} \to y'[u]h \text{ weakly in } Y_*$$

as $\sigma \to 0$ for all $h \in V$. Using Lemma 5.11, we obtain $y[u+\sigma h] \to y[u]$ strongly in Y_*.

It is known (see Section 2.3) that if $\varphi_k \to \varphi$ weakly in Hilbert space H and $\psi_k \to \psi$ strongly in H, then we have the convergence of the scalar products $(\varphi_k, \psi_k) \to (\varphi, \psi)$. Therefore, passing to the limit at the last equality, we get

$$\lim_{\sigma \to 0} \frac{I(u+\sigma h) - I(u)}{\sigma} = \int_\Omega \alpha uh dx + (y'[u]h, y[u] - y_d) =$$

$$\int_\Omega \alpha uh dx - \int_\Omega (\Delta y[u] - \Delta y_d) y'[u]h dx.$$

Using the formula (5.18) for $\mu = \Delta y[u] - \Delta y_d$, we obtain the equality

$$\int_\Omega \mu y'[u]h dx = \int_\Omega hp dx.$$

Therefore, the following equality holds:

$$\langle I'(u), h \rangle = \int_\Omega (\alpha u - p)h dx.$$

Strongly nonlinear systems 199

This completes the proof of Lemma 5.12. □

Thus, we have the minimization problem of a Gâteaux differentiable functional on a convex subset of a Banach space. By Theorem 1.8 its solution satisfies the variational inequality

$$\langle I'(u), v - u \rangle \geq 0 \ \ \forall v \in U.$$

Using the value of the functional derivative, we obtain the formula (5.16). We have the following result.

Theorem 5.5 *The solution of Problem* 5.1 *satisfies the variational inequality* (5.16).

Thus, the necessary conditions of optimality for Problem 5.1 are realized for the general case without any supposition about the dimension of the set and the degree of the nonlinearity. The extended derivative of the control-state mapping is sufficient for obtaining of the necessary.

Remark 5.19 Of course, the necessary conditions of optimality (5.16) for the strongly nonlinear case and (4.35) for the weakly nonlinear case are equal among themselves. Therefore, Theorem 5.5 is the extension of Theorem 4.7.

5.5 Boundary control for Neumann problem

We considered before the first boundary problems with the distributed control. Now we consider the second boundary problem. Let Ω be an open bounded set of $\mathbb{R}^n$ with smooth enough boundary Γ. We have the equation on the set Ω

$$-\sum_{i,j=1}^{n} \frac{\partial}{\partial x_i}\left(a_{ij}\frac{\partial y}{\partial x_j}\right) + a_0 y + |y|^\rho y = f_\Omega \tag{5.19}$$

with the second order boundary condition on Γ

$$\frac{\partial y}{\partial \nu} = f_\Gamma, \tag{5.20}$$

where ν is the conormal. The problem (5.19), (5.20) is called the **Neumann problem**. The functions a_{ij}, a_0, f_Ω, and f_Γ satisfy the conditions

$$a_{ij} \in C^1(\overline{\Omega}), \ a_0 \in C(\overline{\Omega}), \ f_\Omega \in L_2(\Omega), \ f_\Gamma \in H^{-1/2}(\Omega)$$

and the inequalities

$$\sum_{i,j=1}^{n} a_{ij}(x)\xi_i\xi_j \geq \chi|\xi|^2 \ \forall \xi \in \mathbb{R}^n, \ a_0(x) \geq \xi, \ x \in \Omega,$$

200   *Optimization and Differentiation*

where $\chi > 0$. The parameter ρ is positive. Determine the space $Y_1 = H^1(\Omega)$ and the space $Y = Y_1(\Omega) \cap L_q(\Omega)$ with the norm

$$\|y\|_Y = \|y\|_{Y_1} + \|y\|_q,$$

where $q = \rho + 2$.

Theorem 5.6 *The problem* (5.19), (5.20) *has the unique solution from the space* Y.

    Proof. 1. Let $\{\mu_l\}$ be a complete set of the space Y. Determine the sequence $\{y_k\}$ by the equalities

$$y_k = \sum_{l=1}^{k} \xi_{lk}\mu_l, \ k = 1, 2, ..., \tag{5.21}$$

where the numbers $\xi_{1k}, ..., \xi_{kk}$ satisfy the conditions

$$\int_\Omega \left(\sum_{i,j=1}^{n} a_{ij} \frac{\partial y_k}{\partial x_j} \frac{\partial \mu_l}{\partial x_i} + a_0 y_k \mu_l + |y_k|^\rho y_k \mu_l \right) dx = \int_\Omega f_\Omega \mu_l dx + \int_\Gamma f_\Gamma \mu_l dx, \tag{5.22}$$

$l = 1, ..., k$.

    Determine the operator P on the space $\mathbb{R}^k$ that maps the arbitrary vector $\xi = (\xi_{1k}, ..., \xi_{kk})$ to the vector $P\xi = (\eta_1, ..., \eta_k)$, where

$$\eta_l = \int_\Omega \left(\sum_{i,j=1}^{n} a_{ij} \frac{\partial y_k}{\partial x_j} \frac{\partial \mu_l}{\partial x_i} + a_0 y_k \mu_l + |y_k|^\rho y_k \mu_l \right) dx - \int_\Omega f_\Omega \mu_l dx - \int_\Gamma f_\Gamma \mu_l dx,$$

$l = 1, ..., k$. The system of nonlinear algebraic equations (5.22) is equivalent to the operator equation $P\xi = 0$. Determine the scalar product

$$(P\xi, \xi) = \sum_{l=1}^{k} \eta_l \xi_l = \int_\Omega \left(\sum_{i,j=1}^{n} a_{ij} \frac{\partial y_k}{\partial x_j} \frac{\partial y_k}{\partial x_i} + a_0 |y_k|^2 + |y_k|^q \right) dx -$$

$$\int_\Omega f_\Omega y_k dx - \int_\Gamma f_\Gamma y_k dx \geq \chi \|y_k\|_{Y_1}^2 + \|y_k\|_q^q - \|f_\Omega\|_2 \|y_k\|_2 - \|f_\Gamma\|_{\Gamma*} \|y_k\|_\Gamma,$$

where $\| \cdot \|_\Gamma$ and $\| \cdot \|_{\Gamma*}$ are the norms of the spaces $H^{1/2}(\Gamma)$ and $H^{-1/2}(\Gamma)$.

    We have the inequality

$$\|y_k\|_2 \leq \|y_k\|_{Y_1}.$$

By the Trace theorem, there exists a constant $c_1 > 0$ such that

$$\|y_k\|_\Gamma \leq c_1 \|y_k\|_{Y_1}.$$

Then we obtain

$$(P\xi,\xi) \geq \chi\|y_k\|_{Y_1}^2 + \|y_k\|_q^q - c_2\|y_k\|_{Y_1} = \|y_k\|_{Y_1}\left(\chi\|y_k\|_{Y_1} - c_2\right) + \|y_k\|_q^q,$$

where $c_2 = \|f_\Omega\|_2 + c_1\|f_\Gamma\|_{\Gamma*}$.

Determine a vector ξ with a large enough norm such that the function y_k of the formula (5.21) satisfies the inequality $\chi\|y_k\|_{Y_1} \geq c_2$. Then we obtain $(P\xi,\xi) \geq 0$. Using Lemma 4.5, we prove the solvability of the equation $P\xi = 0$. Therefore, the problem (5.22) has a solution ξ_{lk} that determines the function y_k by the formula (5.21).

2. Multiply l-th equality (5.22) by the number ξ_{lk}. After summing we get

$$\int_\Omega \left(\sum_{i,j=1}^n a_{ij}\frac{\partial y_k}{\partial x_j}\frac{\partial y_k}{\partial x_i} + a_0|y_k|^2 + |y_k|^q\right)dx = \int_\Omega f_\Omega y_k dx + \int_\Gamma f_\Gamma y_k dx.$$

Then we have the inequality

$$\chi\|y_k\|_{Y_1}^2 + \|y_k\|_q^q \leq \|f_\Omega\|_2\|y_k\|_2 + \|f_\Gamma\|_{\Gamma*}\|y_k\|_\Gamma \leq [\|f_\Omega\|_2 + c_1\|f_\Gamma\|_{\Gamma*}]\|y_k\|_{Y_1}.$$

Therefore, the sequence $\{y_k\}$ is bounded in the space Y.

3. Using the Banach–Alaoglu theorem, extract a subsequence of $\{y_k\}$ such that $y_k \to y$ weakly in Y. Then

$$\int_\Omega \left(\sum_{i,j=1}^n a_{ij}\frac{\partial y_k}{\partial x_j}\frac{\partial \mu_l}{\partial x_i} + a_0 y_k\mu_l\right)dx \to \int_\Omega \left(\sum_{i,j=1}^n a_{ij}\frac{\partial y}{\partial x_j}\frac{\partial \mu_l}{\partial x_i} + a_0 y\mu_l\right)dx.$$

By Rellich–Kondrashov theorem, embedding of the space $H^1(\Omega)$ to $L_2(\Omega)$ is compact. From the convergence $y_k \to y$ weakly in $H^1(\Omega)$ it follows that $y_k \to y$ strongly in $L_2(\Omega)$ and a.e. on Ω. Then $|y_k(x)|^p y_k(x) \to |y(x)|^p y(x)$ a.e. on Ω. By the boundedness of the sequence $\{y_k\}$ in $L_q(\Omega)$, the sequence $\{|y_k|^p y_k\}$ is bounded in $L_q'(\Omega)$. Using Lemma 4.6, we have $|y_k(x)|^p y_k(x) \to |y(x)|^p y(x)$ weakly in $L_q'(\Omega)$. Therefore,

$$\int_\Omega |y_k|^p y_k\mu_l dx \to \int_\Omega |y_k|^p y\mu_l dx.$$

Passing to the limit in (5.22), we have

$$\int_\Omega \left(\sum_{i,j=1}^n a_{ij}\frac{\partial y}{\partial x_j}\frac{\partial \mu_l}{\partial x_i} + a_0 y\mu_l + |y|^p y\mu_l\right)dx = \int_\Omega f_\Omega\mu_l dx + \int_\Gamma f_\Gamma\mu_l dx, \, l = 1, 2, \dots.$$

Then the following equality holds:

$$\int_\Omega \left(\sum_{i,j=1}^n a_{ij}\frac{\partial y}{\partial x_j}\frac{\partial \mu}{\partial x_i} + a_0 y\mu_l + |y|^p y\mu\right)dx = \int_\Omega f_\Omega\mu dx + \int_\Gamma f_\Gamma\mu dx \, \forall\mu \in Y.$$

202   *Optimization and Differentiation*

Choose $\mu = 0$ in the boundary Γ. We obtain

$$\int_\Omega \left[-\sum_{i,j=1}^n \frac{\partial}{\partial x_i} \left(a_{ij} \frac{\partial y}{\partial x_j} \right) + a_0 y + |y|^p y - f_\Omega \right] \mu dx = 0.$$

Hence, the limit function y satisfies the equality (5.19). From the previous equality and Green's formula, it follows that

$$\int_\Gamma \left(\frac{\partial y}{\partial \nu} - f_\Gamma \right) \mu dx = 0.$$

Then the equality (5.20) is true.

4. Suppose there exist two solutions y_1 and y_2 of the considered problem. Then its difference y satisfies the equalities

$$-\sum_{i,j=1}^n \frac{\partial}{\partial x_i} \left(a_{ij} \frac{\partial y}{\partial x_j} \right) + a_0 y + |y_1|^p y_1 - |y_2|^p y_2 = 0 \ \text{in} \ \Omega,$$

$$\frac{\partial y}{\partial \nu} = 0 \ \text{in} \ \Gamma.$$

Multiply the first equality by y. After integration by using Green's formula and the boundary condition we get

$$\int_\Omega \left(\sum_{i,j=1}^n a_{ij} \frac{\partial y}{\partial x_j} \frac{\partial y}{\partial x_i} + a_0 |y|^2 \right) dx + \int_\Omega \left(|y_1|^p y_1 - |y_2|^p y_2 \right) (y_1 - y_2) dx = 0.$$

The value under the second integral is non-negative. Therefore, this equality can be true for $y_1 = y_2$ only. Then the solution of our problem is unique. $\square$

Remark 5.20 The space of the infinite differentiable functions $D(\Omega)$ is not dense in $H^1(\Omega)$. Then we cannot identify the adjoint space of $H^1(\Omega)$ with a subspace of the distribution space $D'(\Omega)$ Therefore, we choose $f_\Omega \in L_2(\Omega)$. However, it is possible to consider a larger class of the absolute terms.

Remark 5.21 We can prove Theorem 5.6 by using monotone operators theory.

Consider an optimization control problem for the considered system. We analyzed before the distributed controls only, because the controls were the absolute terms of the state equations. Now we consider the boundary control. We have the equation (5.19) with the boundary condition

$$\frac{\partial y}{\partial \nu} = v + f_\Gamma, \tag{5.23}$$

where the control v belongs to the non-empty convex closed subset U of the space $V = L_2(\Gamma)$. By Theorem 5.6, for all $v \in V$ there exists a unique solution $y = y[v]$ of the system (5.19), (5.23) from the space Y.

$$
\text{Strongly nonlinear systems} \qquad 203
$$

Consider the functional

$$
I(v) = \frac{\alpha}{2} \int_\Omega v^2 dx + \frac{1}{2} \int_\Omega \big(y[v] - y_d \big)^2 dx,
$$

where $\alpha > 0$, $y_d \in L_2(\Omega)$ is a known function.

Problem 5.2 *Find a control $u \in U$ that minimizes the functional I on the set U.*

Prove the weak continuity of control–state mapping.

Lemma 5.13 *The map $y[\cdot] : V \to Y$ for the problem $(5.19), (5.23)$ is weakly continuous.*

Proof. Suppose $v_k \to v$ weakly in V. Multiply the equality (5.19) for $v = v_k$ by the function $y_k = y[v_k]$. After the integration by using (5.23) we obtain

$$
\int_\Omega \Big(\sum_{i,j=1}^n a_{ij} \frac{\partial y_k}{\partial x_j} \frac{\partial y_k}{\partial x_i} + a_0 |y_k|^2 + |y_k|^q \Big) dx = \int_\Omega f_\Omega y_k dx + \int_\Gamma \big(v_k + f_\Gamma \big) y_k dx.
$$

Then we have

$$
\chi \|y_k\|_{Y_1} + \|y_k\|_q^q \le \|f_\Omega\|_2 \|y_k\|_2 + \big[\|f_\Omega\|_{\Gamma*} + \|v_k\|_{\Gamma*} \big] \|y_k\|_\Gamma.
$$

Using continuous embedding $L_2(\Gamma)$ and Trace theorem, we prove the existence of the constant $c > 0$ such that

$$
\chi \|y_k\|_{Y_1} + \|y_k\|_q^q \le c \|y_k\|_{Y_1}.
$$

Then the sequence $\{y_k\}$ is bounded in Y. Extracting a subsequence, we have $y_k \to y$ weakly in Y. Using the technique of proving Theorem 5.6, we obtain the equality $y = y[v]$. $\square$

By Lemma 5.13, the minimizing functional is weakly lower semicontinuous. Then we obtain the following result.

Theorem 5.7 *Problem 5.2 is solvable.*

Try to prove that the map $y[\cdot] : V \to Y$ is extended differentiably at an arbitrary point $u \in V$. For any $h \in V$ and the number σ the function $\eta_\sigma[h] = (y[u + \sigma h] - y[u])/\sigma$ is the solution of the boundary problem

$$
-\sum_{i,j=1}^n \frac{\partial}{\partial x_i} \Big(a_{ij} \frac{\partial \eta_\sigma[h]}{\partial x_j} \Big) + a_0 \eta_\sigma[h] + \big(g_\sigma[h] \big)^2 \eta_\sigma[h] = 0 \text{ in } \Omega, \qquad (5.24)
$$

$$\frac{\partial \eta_\sigma[h]}{\partial \nu} = h \ \text{in} \ \Gamma, \tag{5.25}$$

where

$$\left(g_\sigma[h]\right)^2 = (\rho + 1)\big|y[u + \sigma h]\big| + \varepsilon\big(y[u + \sigma h] - y[u]\big)\big|^\rho, \ \varepsilon \in [0, 1].$$

Determine the spaces

$$Y_\sigma = \{y \,|\, y \in Y_1, \ g_\sigma[h]y \in L_2(\Omega)\}, \quad Y_\sigma' = \{\lambda + g_\sigma[h]\xi \,|\, \lambda \in Y_1', \ \xi \in L_2(\Omega)\}.$$

For all $\eta \in Y_\sigma$ the value at the left-hand side of the equality (5.24) for $\eta = \eta_\sigma[h]$ belongs to the space Y_σ'. Multiply this equality by the arbitrary function $\lambda \in Y_\sigma$. After integration by using Green's formula and the condition (5.25), we have

$$\int_\Omega \left[-\sum_{i,j=1}^n \frac{\partial}{\partial x_i}\left(a_{ji}\frac{\partial \lambda}{\partial x_j}\right) + a_0\lambda + \left(g_\sigma[h]\right)^2\lambda\right]\eta_\sigma[h]dx - \int_\Gamma \frac{\partial \lambda}{\partial \nu^*}\eta_\sigma[h]dx = \int_\Gamma \lambda h dx. \tag{5.26}$$

Consider the boundary problem

$$-\sum_{i,j=1}^n \frac{\partial}{\partial x_i}\left(a_{ji}\frac{\partial p}{\partial x_j}\right) + a_0 p + \left(g_\sigma[h]\right)^2 p = \mu_\Omega \ \text{in} \ \Omega, \tag{5.27}$$

$$\frac{\partial p}{\partial \nu^*} = \mu_\Gamma \ \text{in} \ \Gamma. \tag{5.28}$$

For $\sigma = 0$ we have

$$-\sum_{i,j=1}^n \frac{\partial}{\partial x_i}\left(a_{ji}\frac{\partial p_\mu[u]}{\partial x_j}\right) + a_0 p_\mu[u] + (\rho + 1)\big|y[u]\big|^\rho p_\mu[u] = \mu_\Omega \ \text{in} \ \Omega, \tag{5.29}$$

$$\frac{\partial p_\mu[u]}{\partial \nu^*} = \mu_\Gamma \ \text{in} \ \Gamma, \tag{5.30}$$

where $\mu = (\mu_\Omega, \mu_\Gamma)$.

Lemma 5.14 *For all numbers* σ *and the functions* $\mu_\Omega \in L_2(\Omega)$, $\mu_\Gamma \in H^{-1/2}(\Gamma)$ *the problem* (5.27), (5.28) *has a unique solution* $p = p_\mu^\sigma[h]$ *from the space* Y_σ, *as* $p_\mu^\sigma[h] \to p_\mu[u]$ *weakly in* Y_1 *as* $\sigma \to 0$.

Proof. Multiply the equality (5.27) by the function p. After the integration by using Green's formula and the equality (5.28), we get

$$\int_\Omega \left[\sum_{i,j=1}^n a_{ij}\frac{\partial p}{\partial x_i}\frac{\partial p}{\partial x_j} + a_0 p^2\right]dx + \int_\Omega \left(g_\sigma[h]p\right)^2 dx = \int_\Omega \mu_\Omega p dx - \int_\Gamma \mu_\Gamma p dx.$$

Then

$$\chi\|p\|_{Y_1}^2 + \big\|g_\sigma[h]p\big\|_2^2 \leq \|\mu_\Omega\|_2\|p\|_2 + \|\mu_\Gamma\|_{\Gamma*}\|p\|_\Gamma. \tag{5.31}$$

Therefore, we obtain a priori estimate of the solution of the problem (5.27), (5.28) in Y_σ. Using the linear elliptic equations theory (see Chapter 3), we prove that for all numbers σ and the functions $h \in V$, $\mu_\Omega \in L_2(\Omega)$, $\mu_\Gamma \in H^{-1/2}(\Gamma)$ this problem has a unique solution $p = p_\mu^\sigma[h]$ from Y_σ.

Suppose the convergence $\sigma \to 0$. Using the weak continuity of the operator $y[\cdot] : V \to Y$, we have $y[u + \sigma h] \to y[u]$ weakly in Y. Then the set $\{y[u + \sigma h]\}$ is bounded in $L_q(\Omega)$, and the set $\{g_\sigma[h]\}$ is bounded in $L_{2q/\rho}(\Omega)$. By the inequality (5.31), the set $\{p_\mu^\sigma[h]\}$ is bounded in Y_1, and the set $\{g_\sigma[h]p_\mu^\sigma[h]\}$ is bounded in $L_2(\Omega)$. By Hölder inequality, we get

$$\left\| \left(g_\sigma[h]\right)^2 p_\mu^\sigma[h] \right\|_{q'} \leq \left\| g_\sigma[h] \right\|_{2q/\rho} \left\| g_\sigma[h]p_\mu^\sigma[h] \right\|_2.$$

Then the set $\left\{ \left(g_\sigma[h]\right)^2 p_\mu^\sigma[h] \right\}$ is bounded in $L_{q'}(\Omega)$. After extracting subsequences we have $y[u + \sigma h] \to y[u]$ and $p_\mu^\sigma[h] \to p_\mu$ weakly in Y_1. Using the Rellich–Kondrashov theorem, we obtain these convergences strongly in $L_2(\Omega)$ and a.e. on Ω. Then $\left(g_\sigma[h]\right)^2 p_\mu^\sigma[h] \to (\rho + 1)|y[u]|^\rho p_\mu$ a.e. on Ω. Therefore, $\left(g_\sigma[h]\right)^2 p_\mu^\sigma[h] \to (\rho + 1)|y[u]|^\rho p_\mu$ weakly in $L_{q'}(\Omega)$. Pass to the limit in the equality (5.27) for $p = p_\mu^\sigma[h]$ as $\sigma \to 0$, we obtain the equality (5.29). Then $p_\mu = p_\mu[u]$. This completes the proof of Lemma 5.14. $\square$

Lemma 5.15 *The operator $y[\cdot] : V \to Y$ for the problem (5.19), (5.23) has $(V, V; V, V^w)$-extended derivative at the arbitrary point $u \in V$ such that*

$$\int_\Omega \mu_\Omega y'[u]h dx = \int_\Gamma p_\mu[u]h dx \ \forall h, \mu_\Omega \in V, \qquad (5.32)$$

where V^w is the space V with the weak topology.

Proof. The function $p_\mu[u]$ belongs to Y_1. By the Trace theorem, we have $\gamma p_\mu[u] \in H^{1/2}(\Gamma)$. Then the equality (5.32) determines in reality a linear continuous operator on the set V. Choose $\lambda = p_\mu^\sigma[h]$ in the equality (5.26). We obtain

$$\int_\Omega \mu_\Omega \frac{y[u + \sigma h] - y[u]}{\sigma} dx = \int_\Gamma p_\mu^\sigma[h]h dx.$$

From the equality (5.32) and the last formula, it follows that

$$\int_\Omega \mu_\Omega \left(\frac{y[u + \sigma h] - y[u]}{\sigma} - y'[u]h \right) dx = \int_\Gamma \left(p_\mu^\sigma[h] - p_\mu[u] \right) h dx.$$

Using Lemma 5.8 and the Trace theorem we have $\gamma p_\mu^\sigma[h] \to \gamma p_\mu[u]$ weakly in $H^{1/2}(\Gamma)$. After passing to the limit in the previous equality, we prove that $y'[u]$ is in reality the extended derivative of the considered operator. $\square$

Remark 5.22 We can prove as before that the operator $y[\cdot] : V \to Y$ is not Gâteaux differentiable, if the inclusion $H^1(\Omega) \subset L_q(\Omega)$ is false.

206   *Optimization and Differentiation*

From Lemma 5.15 it follows that the minimizing functional is differentiable.

Lemma 5.16 *The functional I for Problem 5.2 has Gâteaux derivative $I'(u) = \alpha u - \gamma p$ at the arbitrary point $u \in V$, where p is the solution of the equation*

$$-\sum_{i,j=1}^{n} \frac{\partial}{\partial x_i}\left(a_{ji}\frac{\partial p}{\partial x_j}\right) + a_0 p + (\rho + 1)\big|y[u]\big|^{\rho} p = y_d - y[u] \qquad (5.33)$$

in the set Ω with the boundary condition

$$\frac{\partial p}{\partial \nu^*} = 0. \qquad (5.34)$$

Now we determine the necessary conditions of optimality.

Theorem 5.8 *The solution u of Problem 5.2 satisfies the inequality*

$$\int_{\Omega} (\alpha u - p)(v - u)dx \geq 0 \ \forall v \in U. \qquad (5.35)$$

Thus, we have the state system (5.19), (5.23) for $v = u$, the adjoint system (5.33), (5.34), and the variational inequality (5.35) for finding the optimal control.

Remark 5.23 We extend our results to the general integral functionals and to the equations with general nonlinearity (see Chapter 3).

5.6 Extended differentiability of the inverse operator

Under the condition (5.4), the inverse operator is Gâteaux differentiable. If this condition gets broken, this is the extendedly differentiable operator. Try to determine extended differentiability of the inverse operator such that Theorem 5.4 will be a partial case of this result. We apply here the technique of proving the theorem of the differentiability of the inverse operator (see Theorem 4.14).

Consider Banach spaces Y, Z, an operator $A : Y \to Z$, and points $y_0 \in Y$, $z_0 = Ay_0$. Let us have a Banach space Z_* that is a subspace of Z and its neighborhood O_* of zero. Then the set $O = z_0 + O_*$ is the neighborhood of the point z_0. Consider the following assumption.

Assumption 5.1 *The operator A is invertible on the set O.*

Strongly nonlinear systems

207

Denote the inverse operator by L. Consider the equalities

$$ALz_\sigma = z_0 + \sigma h, \quad ALz_0 = z_0$$

for an arbitrary function $h \in Z_*$ and a number σ that is so small that the following inclusion holds: $z_\sigma \in O$, where $z_\sigma = z_0 + \sigma h$. Then we have

$$ALz_\sigma - ALz_0 = \sigma h. \tag{5.36}$$

If the operator A is Gâteaux differentiable, then we obtain the equality

$$\langle \lambda, Ay - Ay_0 \rangle = \langle \lambda, A'[y_0 + \delta(\lambda)(y - y_0)] \rangle \quad \forall y \in Y, \lambda \in Z',$$

because of the Lagrange formula (see Section 4.1), where $\delta(\lambda) \in (0,1)$.

For all $z \in O$, $\lambda \in Z'$ determine the linear continuous operator $G(z, \lambda) : Y \to Z$ by the formula

$$G(z, \lambda)y = A'[y_0 + \delta(\lambda)(Lz - Lz_0)]y \,\forall y \in Y.$$

Note the equality $G(z_0, \lambda) = A'(y_0)$ for all $\lambda \in Z'$. From (5.36) we have

$$\langle \lambda, G(z_\sigma, \lambda)(Lz_\sigma - Lz_0) \rangle = \sigma \langle \lambda, h \rangle \quad \forall \lambda \in Z', h \in Z_*. \tag{5.37}$$

Consider an operator $G : Y \to Z$ and a set $\overline{Y}$ that contains Y.

Definition 5.2 *The operator $\overline{G}$ with domain $\overline{Y}$ is called the* **continuation** *of the operator G on the set $\overline{Y}$ if $\overline{G}y = Gy$ for all $y \in Y$.*

For all $z \in O$ consider Banach spaces $Z(z)$ and $Y(z)$ such that embedding of Y, Z_*, and $Z(z)$ to $Y(z)$, $Z(z)$, and Z are continuous and dense correspondingly (see Figure 5.6).

Assumption 5.2 *The operator A is Gâteaux differentiable, as for any $z \in O$, there exists a continuous continuation $\overline{G}(z, \lambda)$ of the operator $G(z, \lambda)$, on the set $Y(z)$ such that its codomain is a subset $Z(z)$ (see Figure 5.6).*

We have the conditions $Lz \in Lz_o + Y(z)$ and $Z' \subset Z(z)'$ for all $z \in O$. From equality (5.37) it follows that

$$\left\langle \overline{G}(z_\sigma, \lambda)^* \lambda, (Lz_\sigma - Lz_0)/\sigma \right\rangle = \langle \lambda, h \rangle \,\forall \lambda \in Z(z_\sigma)', h \in Z_*. \tag{5.38}$$

Determine the equation

$$\left[\overline{G}(z_\sigma, p_\mu[z])\right]^* p_\mu[z] = \mu. \tag{5.39}$$

This is the linear equation

$$\left[\overline{A}'(y_0)\right]^* p_\mu = \mu \tag{5.40}$$

for $z = z_0$, where $\overline{A}'(y_0)$ is a continuation of the operator $A'(y_0)$ on the set $Y(z_0)$.

Consider a Banach space Y_* that is continuously and dense includes to the space $Y'(z)$ for all $z \in O$ (see Figure 5.6). Suppose the following assumption.

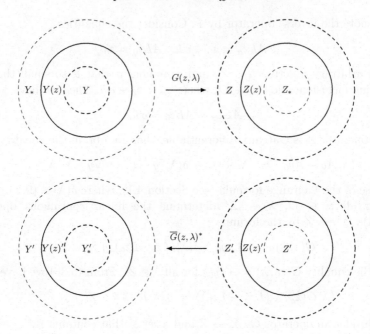

FIGURE 5.6: The operators $G(z,\lambda)$, $\overline{G}(z,\lambda)$, and $\overline{G}(z,\lambda)^*$.

Assumption 5.3 *For all $z \in O$, $\mu \in Y(z)'$ the equation (5.39) has a unique solution $p_\mu[z] \in Z(z)'$.*

Return to the equality (5.38). Determine $\lambda = p_\mu[z_\sigma]$ for small enough σ. We get

$$\left\langle \mu, [L(z_0 + \sigma h) - Lz_0]/\sigma \right\rangle = \langle p_\mu[z_\sigma], h \rangle \; \forall \mu \in Y(z_\sigma)', h \in Z_*. \quad (5.41)$$

Assumption 5.4 *We have the convergence $p_\mu[z_0 + \sigma h] \to p_\mu$ *-weakly in Z'_* as $\sigma \to 0$ for all $\mu \in Y'_*$, $h \in Z_*$, where p_μ is a solution of the equation (5.40).*

Now prove the following assertion.

Theorem 5.9 *Under the assumptions 5.1 – 5.4, the operator A^{-1} has $(Z(z_0), Y(z_0); Z_*, Y_*^w)$-extended Gâteaux derivative D at the point z_0 that satisfies the equality*

$$\langle \mu, Dh \rangle = \langle p_\mu, h \rangle \; \forall \mu \in Y(z_0)', \, h \in Z_*, \quad (5.42)$$

where Y_*^w is the space Y_* with weak topology.

Proof. The equality (5.42) determines in reality a linear continuous operator $D : Z(z_0) \to Y(z_0)$. From the equalities (5.41), (5.42) it follows that

$$\left\langle \mu, [L(z_0 + \sigma h) - Lz_0]/\sigma - Dh \right\rangle = \langle p_\mu[z_\sigma] - p_\mu, h \rangle \; \forall \mu \in Y'_*, h \in Z_*.$$

By Assumption 5.4 we have the convergence $p_\mu[z_\sigma] \to p_\mu$ *-weakly in Z'_* as $\sigma \to 0$. After the passing to the limit we obtain the convergence $[L(z_0 + \sigma h) - Lz_0]/\sigma \to Dh$ weakly in Y_* for all $h \in Z_*$. Therefore, D is in reality the extended derivative of the considered inverse operator. $\square$

Prove the applicability of Theorem 5.6 for the equation (5.1).

Lemma 5.17 *Operator A of the equation* (5.1) *satisfies the assumptions of Theorem 5.6.*

Proof. Assumption 5.1 is the solvability of the equation (5.1) in the neighborhood of the given point that is true. We have also Gâteaux differentiability of the operator A. Determine the operator $G(z)$ by the formula $G(z)y = -\Delta y + (g(z)^2 y$ for all $\in Y$, where $g(z)^2 = (\rho+1)|y_0 + \varepsilon(Lz - Lz_0)|^\rho$. Let the spaces $Y_*, Z_*, Y(z), Z(z)$ be determined as before. Determine the operator $\overline{G}(z, \lambda) : Y(z) \to Z(z)$ by the equality $\overline{G}(z)y = -\Delta y + g(z)^2 y$ for all $\in Y(z)$. Therefore Assumption 5.2 is true as Assumption 5.3 and Assumption 5.4 are corollaries of Lemma 5.6 and Lemma 5.7 correspondingly. $\square$

Remark 5.24 We shall return to the analysis of the extended differentiability of the inverse in the final chapter.

The obtained results can be extended to other nonlinear control systems (see next chapter).

5.7 Comments

The example of the absence of the solution of nonlinear elliptic equations with respect to the absolute term was given by S.Ya. Serovajsky [467], [484], [470], [468], [486], [488]. He proposes also the extended derivative [470], [486], [484], [467]. V.I. Averbuh and O.G. Smolyanov determine the following requirements for the operator differentiation [37]. The derivative is the linear continuous map. The derivative of the linear operator is equal to the initial one. If the domain of the operator is equal to the set of real numbers, then its derivative is equal to the Gâteaux derivative of the corresponding abstract function. The theorem of differentiation of composite function is true. The operation of differentiation is unique. All these properties are true for the extended derivative.

J. Baranger [47] proves the solvability of the optimization control problems for the system described by elliptic equations with nonlinearity in the general part of the operator. The necessary conditions of optimality for the nonlinear elliptic equations on the basis of the extended derivatives were obtained by S.Ya. Serovajsky [460], [472], [468], [486]. This is also the analogical technique for the analysis of extremum problems for nonlinear parabolic equations [469], [458], [488] and hyperbolic too [488], [464]. The extended differentiation is used also for the substantiation of Pareto optimality [462], the analysis of the optimization problems for non-smooth strong nonlinear systems [478], the convergence of the gradient method for the optimization problems for nonlinear elliptic equations [472], [484], and the analysis of the extremum problems on smooth subvarieties of Banach spaces.

Optimization and Differentiation

The extended differentiability of the inverse operator in Banach spaces was proved by S.Ya. Serovajsky [467], []. The analogical result for the case of non-normalized spaces was obtained in [466]. The **extended differentiable subvariety** was proposed in [471], [474]. This is an analog of the standard differential subvariety of differential topology with change of the classic differential by the extended one.

The extended differentiation is applied also for the proof of the convergence of the analog of the Newton–Kantorovich method [463], and for the analysis of the stability of the infinite dimensional systems with respect to the linear approximation [465].

Chapter 6

Coefficients optimization control problems

6.1	Coefficients control problem	212
6.2	Derivative with respect to a convex set	214
6.3	Optimization control problem for bilinear systems	216
6.4	Analysis of the coefficient optimization problem	220
6.5	Nonlinear system with the control at the coefficients	224
6.6	Differentiability with respect to a convex set for abstract systems	230
6.7	Strongly nonlinear systems with the control at the coefficients .	234
6.8	Comments	240

We considered before optimization problems for the control systems $Ay = Bv + f$. The control and the state function belong to the different terms of the given equation here. However, we can have other forms of the control systems. This is true, for example, for the systems with the control in the coefficients of the state operator. The additional difficulty of these problems is the solvability of the state equation on the set of the admissible controls only. This makes difficult the determination of the derivative of the state function with respect to the control.

The general notion of this chapter is the derivative with respect to a convex set. This chapter has the analogical structure as the previous chapters (see Figure 6.1). At first, we consider systems that are linear with respect to the state functions. We determine the derivative of the functional with respect to a convex set (see Section 6.3 and Section 6.4). Then we analyze nonlinear systems with the control in the coefficients. We determine optimality conditions by using the extended derivative with respect to the convex set of the control–state mapping (see Section 6.5). Then we extend these results to the abstract operator equations (see Section 6.6). Finally, we consider the nonlinear systems with extended differentiability with respect to the convex set of the control–state mapping for the case, where its usual differentiability with respect to the convex set is absent.

211

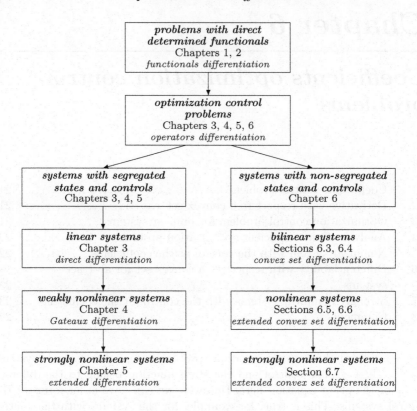

FIGURE 6.1: The structure of Chapter 6.

6.1 Coefficients control problem

Let us have an open bounded set Ω of Euclid space $\mathbb{R}^n$ with a boundary Γ. Consider the equation

$$-\Delta y(x) + v(x)y(x) = f(x), \ x \in \Omega \qquad (6.1)$$

with the boundary condition

$$y(x) = 0, \ x \in \Gamma, \qquad (6.2)$$

where v is the control, y is the state function of the system, and f is the known function of the space $H^{-1}(\Omega)$.

Consider the optimization control problem for the equation (6.1) with the boundary condition (6.2). Determine the control space $V = L_2(\Omega)$ and the set of the admissible controls

$$U = \left\{ v \in V \,\middle|\, a \leq v(x) \leq b, \ x \in \Omega \right\},$$

where a and b are constants, as $b > a > 0$.

Multiply the equality (6.1) by the function y. After the formal integration by using Green's formula we have

$$\int_{\Omega} \sum_{i=1}^{n} \left(\frac{\partial y}{\partial x_i} \right)^2 dx + \int_{\Omega} v y^2 dx = \int_{\Omega} y f dx. \tag{6.3}$$

By the definition of the set U, we get the inequality

$$\int_{\Omega} \sum_{i=1}^{n} \left(\frac{\partial y}{\partial x_i} \right)^2 dx + a \int_{\Omega} y^2 dx \leq \left| \int_{\Omega} y f dx \right|.$$

Then we obtain

$$\|y\| \leq \|f\|_*. \tag{6.4}$$

Using this a priori estimate and the linearity of the equation (6.1), we can repeat the proof of Theorem 3.5 and prove the solvability of the boundary problem (6.1), (6.2).

Theorem 6.1 *For all functions $v \in U$ the problem (6.1), (2.2) has a unique solution $y = y[v]$ from the space $H_0^1(\Omega)$.*

Note the important property. If the control v has negative values, then the second integral at the left-hand side of the equality (6.3) can be negative. This prohibits obtaining the inequality (6.4). Therefore, we cannot prove the one-valued solvability of the considered boundary problem. Thus, we can guarantee this result for the set of admissible controls only. Prove the importance of this assumption.

Example 6.1 *Boundary problems without the one-valued solvability.*
Consider the one-dimensional case, i.e., $n = 1$. Determine $\Omega = (0, \pi)$, $v(x) = -\lambda^2$, $f(x) = 1$, where λ is a positive constant. We have the following boundary problem:

$$y''(x) + \lambda^2 y(x) = 1, \ 0 < x < \pi, \tag{6.5}$$

$$y(0) = 0, \ y(\pi) = 0. \tag{6.6}$$

The general solution of the equation (6.5) is

$$y(x) = c_1 \sin \lambda x + c_2 \cos \lambda x + 1/\lambda^2,$$

where c_1, c_2 are arbitrary constants.

214 *Optimization and Differentiation*

Determine, for example, $\lambda = 1$. From the equality (6.6) it follows that

$$c_2 + 1 = 0, \quad -c_2 + 1 = 0.$$

These equalities cannot be true simultaneously. Therefore, the problem (6.5), (6.6) is insolvable for $\lambda = 1$.

Now suppose $\lambda = 2$. Then both boundary conditions (6.6) have the form

$$c_2 + 1/4 = 0.$$

Hence the function

$$y(x) = c_1 \sin 2x + (\cos 2x - 1)/4$$

is the solution of the boundary problem (6.5), (6.6) for all constants c_1. Thus, this problem has the infinite set of solutions for $\lambda = 2$.

Remark 6.1 We see that the absence of the a priori estimates of the boundary problem is a troubling. We can have negative properties of the system for this case.

Thus, the one-valued solvability of the boundary problem (6.1), (6.2) can get broken if the control does not belong to the set U. Now let us consider a minimization problem of a functional I for this system. Try to determine its Gâteaux derivative at a point u. Consider the difference $I(u + \sigma h) - I(u)$ for arbitrary functions h. We need to determine the state $y[u+\sigma h]$. Unfortunately, the function $u+\sigma h$ is not an admissible control for some value h, because it can be negative. Therefore, we cannot guarantee the existence and the uniqueness of the solution of the boundary problem. This hinders the calculation of the Gâteaux derivative, and we cannot determine the optimality conditions. The extended derivatives are not applicable also for this situation. Therefore, we use another form of the differentiation.

Remark 6.2 We could not use the Gâteaux derivative in the previous chapter because of the difficulty of passing to the limit. Now we cannot even determine the term for passing to the limit, i.e., we have another form of difficulty.

Remark 6.3 The absence of the one-valued solvability of the state equation for each admissible control is not the obstacle of the consideration of the optimization control problem. We could minimize the cost functional on the set of admissible pairs, i.e., the pairs "control–state" that satisfy the state equation.

Determine a new form of operator derivative.

6.2 Derivative with respect to a convex set

Consider an operator A that is determined on a convex subset U of a Banach space V with values from a Banach space Y. Suppose we would like

Coefficients optimization control problems

to find its Gâteaux derivative at a point $u \in U$. Determine the difference $A(u + \sigma h) - Au$ for an arbitrary direction $h \in V$. We cannot guarantee that the value $u + \sigma h$ with arbitrary h belongs to the set U. Therefore, $A(u + \sigma h)$ does not make sense because the operator A is determined on the set U only. Hence, we cannot calculate the Gâteaux derivative. Define a weaker form of the operator derivative. Let a point $u \in U$ be arbitrary. Determine the set of the difference

$$\delta(u) = -u + U = \{v - u | v \in U\}.$$

Definition 6.1 *The operator A is called **differentiable with respect to the convex set** U at the point $u \in U$ if there exists an affine continuous operator $A'_U(u) : \delta(u) \to Y$ such that*

$$\frac{A[u + \sigma(v - u)] - Au}{\sigma} \to A'_U(u)(v - u)$$

in Y for all $v \in U$ as $\sigma \to 0$.

The derivative with respect to the set U at the point u is determined on the set of differences $v - u$ only. If $U = V$, then we can choose v equal to $u + h$, where h is an arbitrary element of the space V. The derivative with respect to the convex set V is the usual Gâteaux derivative.

Remark 6.4 If the set U is an affine variety $u + W$, where W is a subspace V, the operator derivative at the point u with respect to the convex set is called the *derivative with respect to the subspace W*. This is true for the functional derivative with respect to the subspace from Chapter 1.

Consider the minimization problem of a functional on a convex set as an application.

Problem 6.1 *Find a point $u \in U$ that minimizes a functional I on a convex subset U of Banach space V if this functional is determined on a set U only.*

Problem 6.1 is analogical to Problem 1.7 of minimization of the functional I on the convex subset U of the space V. By Theorem 1.8, the necessary condition of minimum of the point u here is the variational inequality

$$\langle I'(u), v - u \rangle \geq 0 \ \forall v \in U \tag{6.7}$$

if the functional I is Gâteaux differentiable at the point u. However, we cannot obtain this result because the value of this functional at the point $u + \sigma h$ does not make sense for the arbitrary element $h \in V$. Therefore, we use the derivative with respect to the convex set.

Theorem 6.2 *If the functional I has the derivative $I'_U(u)$ with respect to the convex set U at the point u that is a solution of Problem 6.1, then the following inequality holds:*

$$\langle I'_U(u), v - u \rangle \geq 0 \ \forall v \in U. \tag{6.8}$$

216   *Optimization and Differentiation*

Proof. Let u by a point of minimum of the functional I on the set U. Then we have the inequality

$$I\big(u + \sigma(v - u)\big) - I(u) \geq 0 \; \forall v \in U.$$

Divide this inequality by σ. After passing to the limit as $\sigma \to 0$ by using the definition of the derivative with respect to the convex set, we obtain the inequality (6.8). $\square$

If the functional is Gâteaux differentiable, then its Gâteaux derivative is equal to the derivative with respect to the convex set V, i.e., the whole space. Therefore, the variational inequality (6.8) is transformed to (6.7). Thus, Theorem 6.2 is an extension of Theorem 1.8 to the functionals that are differentiable with respect to the convex set only.

Remark 6.5 The supposition of Gâteaux differentiability of the functional is excessively strong for obtaining the considered form of the necessary conditions of optimality. It is sufficient to apply the differentiability with respect to the convex set here.

Remark 6.6 We used the derivative with respect to the subspace for necessary conditions of minimum of the functional on an affine variety (see Chapter 1).

Remark 6.7 The derivative with respect to the convex set $U = u + W$ is equal to the derivative with respect to the subspace W. Therefore, Theorem 6.1 is the extension of Theorem 1.7 that gives the necessary condition of extremum for a functional that is differentiable with respect to the subspace.

Remark 6.8 The variational inequality (6.8) is the necessary and sufficient condition of minimum for the convex functionals (see Theorem 2.1).

We can use Theorem 6.2 for the analysis of the coefficient optimization problem for bilinear systems. The boundary problem (6.1), (6.2) is the example of these systems.

6.3   Optimization control problem for bilinear systems

Consider Banach spaces V, Y, Z and a non-empty convex closed subset of admissible controls U of the space V. For any control $v \in U$ determine a linear state operator $A(v) : Y \to Z$ that depends upon control.

Remark 6.9 Particularly, we have $A(v)y = -\Delta y + vy$ for the equation (6.1).

Consider the system described by the equation

$$A(v)y = f, \tag{6.9}$$

where $f \in Z$ is known. Suppose the map $v \to A(v)y$ is affine and continuous for all $y \in Y$. Then the control system is called a ***bilinear system***.

Coefficients optimization control problems

The state operator depends upon the control for the bilinear systems. Therefore, these systems have special peculiarities. This is a combination of linear and nonlinear properties. We have a linear state equation for the fixed control. Therefore, the proof of the solvability of this system is not a serious problem. However, the dependence of the state function upon the control is not linear. Therefore, we can have some difficulties proving the solvability of the optimization problem and the optimality conditions.

Let the operator $A(v)$ be invertible for all control $v \in U$. Therefore, there exists a unique point $y = y[v]$ of the space Y that satisfies the equality (6.9).

Remark 6.10 We do not suppose the solvability of the state equation outside the set U. Therefore, we apply the derivative with respect to the convex set here.

Determine the functional

$$I(v) = J(v) + K(y[v]),$$

where $J : V \to \mathbb{R}$, $K : Y \to \mathbb{R}$. We have the following optimization control problem.

Problem 6.2 *Find a control $u \in U$ that minimizes the functional I on the set U of the bilinear system described by the equation* (6.9).

Problem 6.2 is a partial case of Problem 6.1. Therefore, we can try to use Theorem 6.2 for its analysis. However, we have the need to substantiate the differentiability with respect to the convex set of control–state mapping for the equation (6.9). Prove, at first, the following preliminary result.

Lemma 6.1 *Suppose the operator $A(v)$ is invertible for all control $v \in U$, as there exists such a constant $c > 0$ that $\left\|A(v)^{-1}\right\|_{L(Y,Z)} \leq c$. Then the map $y[\cdot] : U \to Y$ is continuous.*

Proof. We have the equality

$$A(v)y[v] = A(u)y[u] \ \forall u, v \in U.$$

Then we have

$$A(v)\big(y[v] - y[u]\big) = \big[A(u) - A(v)\big]y[u].$$

Using the assumption of the lemma, we get the inequality

$$\left\|y[v] - y[u]\right\|_Y \leq c\left\|\big[A(u) - A(v)\big]y[u]\right\|_Z.$$

The map $v \to A(v)y[v]$ is continuous here. Therefore, the operator $y[\cdot] : U \to Y$ is continuous at the arbitrary point $u \in U$. $\square$

Prove the differentiability of the state function with respect to the control.

218 *Optimization and Differentiation*

Lemma 6.2 *Under the assumptions of Lemma* 6.1 *the operator* $y[\cdot] : U \to Y$ *has the derivative with respect to the convex set* U *at the arbitrary point* $u \in U$ *that satisfies the equality*

$$\big\langle \mu, y'_U[u](v-u) \big\rangle = \big\langle p_\mu[u], [A(u)-A(v)]y[u] \big\rangle \ \forall v \in U, \mu \in Y', \quad (6.10)$$

where $p_\mu[u]$ *is the solution of the equation*

$$A(u)^* p_\mu[u] = \mu. \quad (6.11)$$

Proof. The operator $A(u) : Y \to Z$ is invertible. Then adjoint operator $A(u)^* : Z' \to Y'$ is invertible too. Therefore, for all $\mu \in Y'$ the equation (6.11) has a unique solution $p_\mu[u]$ from the space Z'. Then the term at the right-hand side of the equality (6.10) has the sense for all $v \in U$, $\mu \in Y'$. We have an operator $y'_U[u]$ there that is determined for all $v - u$ for $v \in U$. Hence its domain is the set of difference $\delta(u) = -u + U$. This operator has the values from the space Y. This is an affine operator also by the properties of the map $v \to A(v)$.

Consider sequence $\{v_k\}$ of elements of the set U such that $v_k \to v$ in V. Using the closeness of the set U, we have $v \in U$. The norm of Banach space Y satisfies the formula

$$\|y\|_Y = \sup_{\|\mu\|_{Y'}=1} \big|\langle \mu, y \rangle\big|.$$

From the equality (6.11) it follows that

$$\Big\|y'_U[u](v_k-u) - y'_U[u](v-u)\Big\|_Y = \sup_{\|\mu\|_{Y'}=1} \Big|\Big\langle \mu, y'_U[u](v_k-u) - y'_U[u](v-u)\Big\rangle\Big| =$$

$$\sup_{\|\mu\|_{Y'}=1} \Big|\Big\langle p_\mu[u], [A(v_k)-A(v)]y[u] \Big\rangle\Big| \le \sup_{\|\mu\|_{Y'}=1} \big\|p_\mu[u]\big\|_{Z'} \big\|[A(v_k)-A(v)]y[u]\big\|_Z.$$

Pass to the limit as $k \to \infty$ by using the continuity of the map $v \to A(v)y[u]$. We have the convergence $y'_U[u](v_k - u) \to y'_U[u](v - u)$ in Y. Thus, the affine operator $y'_U[u] : W \to Y$ is continuous.

Consider the equalities

$$A(v_\sigma)y[v_\sigma] = f, \ A(u)y[u] = f$$

for arbitrary value $v \in U$, where the point $v_\sigma = u + \sigma(v - u)$ is the element of the set U. Using the linearity of the operator $A(v_\sigma)$ we have the equality

$$A(u)\big(y[v_\sigma] - y[u]\big) + A(v_\sigma)y[v_\sigma] - A(u)y[v_\sigma] = 0.$$

The map $v \to A(v)y$ is affine; therefore, we get

$$A\big[(1-\sigma)u + \sigma v\big]y = (1-\sigma)A(u)y + \sigma A(v)y \ \forall y \in Y.$$

Then we transform the previous equality

$$A(u)\big(y[v_\sigma] - y[u]\big) + \sigma\big\{A(v)y[v_\sigma] - A(u)y[v_\sigma]\big\} = 0.$$

We obtain

$$\Big\langle \lambda, A(u)\frac{y[v_\sigma] - y[u]}{\sigma}\Big\rangle = \big\langle \lambda, [A(u) - A(v)]y[v_\sigma]\big\rangle \ \forall v \in U, \lambda \in Z'.$$

Determine $\lambda = p_\mu[u]$. We have

$$\Big\langle \mu, \frac{y[v_\sigma] - y[u]}{\sigma}\Big\rangle = \big\langle p_\mu[u], [A(u) - A(v)]y[v_\sigma]\big\rangle \ \forall v \in U, \mu \in Y'.$$

Using the equality (6.10), we get

$$\Big\|\frac{y[v_\sigma] - y[u]}{\sigma} - y_U'[u](v-u)\Big\|_Y = \sup_{\|\mu\|_{Y'}=1}\Big|\Big\langle \mu, \frac{y[v_\sigma] - y[u]}{\sigma}\Big\rangle - y_U'[u](v-u)\Big\rangle\Big| =$$

$$\sup_{\|\mu\|_{Y'}=1}\Big|\Big\langle p_\mu[u], [A(u) - A(v)]\big(y[v_\sigma] - y[u]\big)\Big\rangle\Big| \ \forall v \in U, \mu \in Y'.$$

From the convergence $v_\sigma \to u$ in V as $\sigma \to 0$ and Lemma 6.1 it follows that $y[v_\sigma] \to y[u]$ in Y. Using the continuity of the linear operator $A(v)$ for all $v \in U$ we have the convergence $[A(u) - A(v)]\big(y[v_\sigma] - y[u]\big) \to 0$ in Z. Pass to the limit at the previous equality. We get

$$\frac{y\big[u + \sigma(v - u)\big] - y[u]}{\sigma} \to y_U'[u](v - u)$$

in Y. Thus, the operator $y_U'[u]$ is in reality the derivative of the map $v \to y[v]$ with respect to the set U at the point u. $\square$

Now determine the differentiability of the minimized functional for Problem 6.2.

Lemma 6.3 *Under the assumptions of Lemma 6.1 suppose the functional J is Gâteaux differentiable at the point $u \in U$ and the functional K is Fréchet differentiable at the point $y = y[u]$. Then the functional I is differentiable with respect to the set U at the point u, as*

$$\langle I_U'[u], v - u\rangle = \langle J'(u), v - u\rangle + \big\langle p, [A(v) - A(u)]y\big\rangle \ \forall v \in U, \qquad (6.12)$$

where p is the solution of the adjoint equation

$$A(u)^* p = -K'(y). \qquad (6.13)$$

Proof. Find the difference

$$I(v_\sigma) - I(u) = \big[J(v_\sigma) - J(u)\big] + \big[K(y_\sigma) - K(y)\big],$$

where $v_\sigma = u + \sigma(v - u)$, $y_\sigma = y[v_\sigma]$, σ is a constant, and v is an arbitrary element of the set U. From Fréchet differentiability of the functional the equality

$$K(y_\sigma) - K(y) = \big\langle K'(y), y_\sigma - y \big\rangle + \eta(y_\sigma - y)$$

follows, where $\eta(\varphi)/\|\varphi\|_Y \to 0$ as $\|\varphi\|_Y \to 0$. Divide the last equality by σ and pass to the limit by using Lemma 6.2 and continuity of the state function with respect to the control. We have

$$\lim_{\sigma \to 0} \frac{K(y_\sigma) - K(y)}{\sigma} = \big\langle K'(y), y'_U[u](v - u) \big\rangle = \big\langle p, \big[A(v) - A(u)\big]y[u] \big\rangle$$

for all $v \in U$ where p is the solution of the equation (6.7). Now we get

$$\langle I'_U(u), v - u \rangle = \lim_{\sigma \to 0} \frac{I(v_\sigma) - I(u)}{\sigma} =$$

$$\langle J'(u), v - u \rangle + \big\langle p, \big[A(v) - A(u)\big]y[u] \big\rangle \ \forall v \in U.$$

This completes the proof of Lemma 6.3. $\square$

Remark 6.11 We do not find the derivative of control–state mapping directly. However, the equality (6.10) is sufficient for the determination of the derivative of the minimized.

Now we apply Theorem 6.2 for obtaining the necessary conditions of optimality for Problem 6.2.

Theorem 6.3 *Under the assumptions of Lemma 6.2, if the point is a solution of Problem 6.2, then the following variational inequality holds:*

$$\langle J'(u), v - u \rangle + \big\langle p, \big[A(v) - A(u)\big]y[u] \big\rangle \ge 0 \ \forall v \in U. \tag{6.14}$$

Indeed, put the value of the functional derivative from the formula (6.12) in to the inequality (6.8). Then the variational inequality is true.

Apply Theorem 6.3 for the analysis of the coefficient optimization problem for the elliptic equation.

6.4 Analysis of the coefficient optimization problem

Return to the analysis of boundary problem (6.1), (6.2). Transform it to the operator equation (6.9). Determine the spaces $Y = H_0^1(\Omega)$, $Z = H^{-1}(\Omega)$. For all control $v \in U$ determine the operator $A(v)$ on the set Y such that

Coefficients optimization control problems

$A(v)y$ is equal to the left-hand side of the equality (6.1). By definition of the set of admissible controls, the operator $A(v)$ has the values from the space Z. Thus, the boundary problem (6.1), (6.2) is transformed to the operator equation (6.9). The bilinearity of the considered system is obvious.

Consider the functional

$$I(v) = \frac{\alpha}{2} \int_\Omega v^2 dx + \frac{1}{2} \int_\Omega \sum_{i=1}^n \left(\frac{\partial y[v]}{\partial x_i} - \frac{\partial y_d}{\partial x_i} \right)^2 dx,$$

that coincides with functionals of the previous optimization control problems, where $y[v]$ is the solution of the boundary problem (6.1), (6.2) for the control v, $\alpha > 0$, and y_d is a given function from the space $H_0^1(\Omega)$.

Problem 6.3 *Find the control that minimizes the functional I on the set U.*

Remark 6.12 The unique distinction between this problem and optimization control problems of the previous chapters is the position of the control. We had the control in the right-hand side of the state equation before. Now the control is a coefficient of the state operator.

Remark 6.13 Coefficient optimization problems make serious practical sense. Let us have a mathematical model of a phenomenon. Using lows of physics, chemistry, etc., we can determine state equations. However, there exist parameters of the considered object that are included in the state equation. Note that we have the possibility to obtain the results of experiments for state functions (temperature, velocity, concentration, etc.). Unfortunately, coefficients of the state operator (coefficients of the heat conductivity, diffusion, viscosity, and others) cannot be determined by the direct experiment. We determine it as a rule by *inverse problems*. There are the problems of finding the unknown parameters such that the difference between the solution of state equation for these parameters and the results of the experiment will be soon small. There are very optimization control problems from the point of view of mathematics although we do not control this system. We solve the *identification problem* of the system.

Lemma 6.4 *The operator $y[\cdot] : U \to Y$ for the problem (6.1), (6.2) is weakly continuous.*

Proof. Consider the sequence $\{v_k\}$ of the set U such that $v_k \to v$ weakly in V. Using the convexity and the closeness of the set U, we have $v \in U$. Then we obtain the estimate

$$\|y_k\| \leq \|f\|_*$$

that is an analog of the inequality (6.4), where $y_k = y[v_k]$. After extracting a subsequence we have $y_k \to y$ weakly in Y. Using the Rellich–Kondrashov theorem, we obtain $y_k \to y$ strongly in $L_2(\Omega)$. For any function $\lambda \in D(\Omega)$ we have the inequality

$$\left| \int_\Omega v_k y_k \lambda dx - \int_\Omega vy\lambda dx \right| \leq \left| \int_\Omega (v_k - v)y\lambda dx \right| + \left| \int_\Omega v_k(y_k - y)\lambda dx \right|.$$

Using the inclusion $y\lambda \in L_2(\Omega)$, boundedness of the sequence $\{v_k\lambda\}$ in the

222 *Optimization and Differentiation*

space $L_2(\Omega)$, and the convergence of the sequences $\{v_k\}$ and $\{y_k\}$ weakly and strongly accordingly in this space, we have

$$\int_\Omega v_k y_k \lambda dx \to \int_\Omega v y \lambda dx.$$

Multiply the equality (6.1) for $v = v_k$ by the arbitrary function $\lambda \in D(\Omega)$. Integrating the result and passing to the limit as $k \to \infty$, we get $y = y[u]$. $\square$

Remark 6.14 The equation (6.1) is linear. Therefore, the proof of the solvability of the boundary problem for the fixed control is easy enough. However, we have the necessity to pass to the limit at the term $v_k y_k$ for the substantiation of the weak continuity of the solution of the system with respect to the control. We use the technique of nonlinear systems here. Thus, bilinear systems have some nonlinear peculiarities.

Theorem 6.4 *Problem* 6.3 *is solvable.*

Indeed, the minimized functional for Problem 6.3 is equal to the functionals of previous optimization problems. We proved the weak continuity of control–state mapping. Therefore, we can prove the weak semicontinuity of the functional and apply Theorem 2.4 for proving the solvability of Problem 6.3.

Remark 6.15 The presence of the control at the coefficient of the state operator does not permit to prove the convexity of the minimized functional that is used for the proof of the uniqueness of the optimal control. This is additional evidence of the closeness of the coefficient optimization problems to the nonlinear systems.

Apply Theorem 6.3 for obtaining necessary conditions of optimality. Now we would like to prove the differentiability of the solution of the boundary problem with respect to the control by using Lemma 6.2. At first, verify the truth of the assumptions of Lemma 6.1.

Lemma 6.5 *The assumptions of Lemma* 6.1 *are true for the problem* (6.1), (6.2).

Proof. Determine boundedness of the inverse operator $A(v)^{-1}$ for the considered boundary problem uniformly with respect to $v \in U$. By Theorem 6.1, this problem has a unique solution $y[v] = A(v)^{-1}f$ as the estimate (6.4) is true, i.e.,

$$\left\|A(v)^{-1}f\right\|_Y \le \|f\|_Z \ \forall f \in Z$$

for all $v \in U$. $\square$

By Lemma 6.2 and Lemma 6.5, we obtain the following result.

Lemma 6.6 *The operator* $y[\cdot] : U \to Y$ *for the problem* (6.1), (6.2) *has the derivative with respect to the set* U *at the arbitrary point* $u \in U$ *such that*

$$\int_\Omega \mu y'_U[u](v-u)dx = \int_\Omega (u-v)p_\mu[u]y[u]dx \ \forall \mu \in H^{-1}(\Omega), \ v \in U,$$

Coefficients optimization control problems 223

where $p_\mu[u]$ is a solution of the homogeneous Dirichlet problem for the equation

$$-\Delta p_\mu[u] + u p_\mu[u] = \mu.$$

Using Lemma 6.3 and Lemma 6.6, find the derivative with respect to the convex set for the minimized functional of Problem 6.3.

Lemma 6.7 *The minimized functional for Problem 6.3 has the derivative with respect to the set U at the arbitrary point $u \in U$ such that*

$$\langle I'_U(u), v - u \rangle = \int_\Omega (\alpha u + py)(v - u)dx \quad \forall v \in U, \tag{6.15}$$

where $y = y[u]$ and is the solution of the equation

$$-\Delta p(x) + u(x)p(x) = \Delta y(x) - \Delta y_d(x), \quad x \in \Omega \tag{6.16}$$

with the boundary condition

$$p(x) = 0, \quad x \in \Gamma. \tag{6.17}$$

From Theorem 6.2 it follows the theorem.

Theorem 6.5 *The solution to Problem 6.3 is*

$$u(x) = \begin{cases} a, & \text{if } -y(x)p(x) < \alpha a, \\ -y(x)p(x)/\alpha, & \text{if } \alpha a \le -y(x)p(x) \le \alpha b, \\ b, & \text{if } -y(x)p(x) > \alpha b. \end{cases} \tag{6.18}$$

Indeed, by Theorem 6.2 and Lemma 6.7, we have the variational inequality

$$\int_\Omega (\alpha u + yp)(v - u)dx \ge 0 \quad \forall v \in U$$

that is the necessary condition of optimality for Problem 6.3. The transformation to the formula (6.18) is standard (see previous chapters).

Remark 6.16 This assertion gives only the necessary condition of optimality. We cannot prove it sufficiently because of the absence of the information about the convexity of the minimized functional. This is a quadratic functional, and our state equation (6.1) is linear for the fixed control. However, control–state mapping of this equation is not affine because of the presence of the bilinear term vy.

Thus, we can find the solution of Problem 6.3 from the system that includes the equalities (6.1), (6.2) for $v = u$ and the formulas (6.16)–(6.18). Denote by $F(y, p)$ the term at the right-hand side of the equality (6.18). We obtain the system of two nonlinear elliptic equations

$$-\Delta y + F(y, p)y = f,$$

224   *Optimization and Differentiation*

$$-\Delta p + F(y,p)p = \Delta y - \Delta y_d$$

with homogeneous boundary conditions. Transform the first of them by using the second equality. We get the system

$$-\Delta y + F(y,p)y = f,$$

$$-\Delta p + F(y,p)(p-y) = -(f + \Delta y_d).$$

Find the solution of this system; then we determine the control by the formula (6.18).

6.5 Nonlinear system with the control at the coefficients

We considered before the nonlinear state equation with the control at the absolute term and the linear equation with control as a coefficient of the state operator. Now we analyze the control system with both these peculiarities. Let Ω be an open bounded set of Euclid space $\mathbb{R}^n$ with the boundary Γ. Consider the equation

$$-\Delta y(x) + |y(x)|^\rho y(x) + v(x)y(x) = f(x), \ x \in \Omega \qquad (6.19)$$

with the boundary condition

$$y(x) = 0, \ x \in \Gamma, \qquad (6.20)$$

where v is a control, y is the state function, and f is a given function, $\rho > 0$.

Remark 6.17 The equation (6.19) differs from (6.1) by the presence of the nonlinear term. This differs from the nonlinear equation (4.19) by the presence of the control as a coefficient of the state operator.

Multiply the equality (6.19) by the function y. After the formal integration by using Green's formula and boundary condition (6.20) we get

$$\int_\Omega \sum_{i=1}^n \left|\frac{\partial y}{\partial x_i}\right|^2 dx + \int_\Omega |y|^{\rho+2} dx + \int_\Omega vy^2 dx = \int_\Omega fy dx. \qquad (6.21)$$

The first term at the left-hand side of the equality (6.21) is the square of the norm of the space $H_0^1(\Omega)$, and the second term is the degree of the norm of the space $L_q(\Omega)$, where $q = \rho + 2$. Determine the control space $V = L_2(\Omega)$ and the set of admissible controls

$$U = \{v \in V \mid a \le v(x) \le b, \ x \in \Omega\},$$

Coefficients optimization control problems

where a and b are constant, as $b > a > 0$. Then the third term at the left-hand side of the equality (6.21) is not negative. Determine the spaces $Y = H_0^1(\Omega) \cap L_q(\Omega)$, $Z = H^{-1}(\Omega) + L_{q'}(\Omega)$, where $1/q + 1/q' = 1$. Let the function f be an element of the set Z. Determine the operator $A(v) : Y \to Z$ such that $A(v)y$ is equal to the term at the left-hand side of the equality (6.19).

Determine the inequality

$$\|y\|^2 + \|y\|_q^q \le \|f\|_Z \left([\|y\| + \|y\|_q) \le \frac{1}{2}\|y\|^2 + \frac{1}{2}\|f\|_Z^2 + \|f\|_Z\|y\|_q.$$

Then we have the estimate of the solution of the problem (6.19), (6.20) in the space Y for all $f \in Z$.

Theorem 6.6 *For all $v \in U$ the problem* (6.19), (6.20) *has a unique solution* $y = y[v]$ *from the space Y.*

Remark 6.18 If $v \in V$, $y \in Y$, then the product vy is an element from a narrower space than $L_{q'}(\Omega)$. The assertions of Theorem 6.3 can be true for weaker functional properties of the control such that $vy \in L_{q'}(\Omega)$ for all $v \in V$, $y \in Y$ (see final section of this chapter).

The general difference of Theorem 6.6 from Theorem 4.5 is the one-valued solvability of the boundary problem on the set of admissible controls only. Therefore, we use the derivative with respect to the convex set instead of the Gâteaux derivative.

Consider the functional

$$I(v) = \frac{\alpha}{2} \int_\Omega v^2 dx + \frac{1}{2} \int_\Omega \sum_{i=1}^n \left(\frac{\partial y[v]}{\partial x_i} - \frac{\partial y_d}{\partial x_i} \right)^2 dx,$$

where $\alpha > 0$, and y_d is a known function of the space $H_0^1(\Omega)$.

Problem 6.4 *Find the control $u \in U$ that minimizes the functional I on the set U.*

Now determine the following result.

Lemma 6.8 *Mapping $y[\cdot] : V \to Y$ for the problem* (6.19), (6.20) *is weakly continuous.*

Proof. Consider a sequence $\{v_k\}$ of the set U such that $v_k \to v$ weakly in V. We have $v \in U$. Using the standard technique, we prove the boundedness of the sequence $\{y_k\}$ with respect to the space Y, where $y_k = y[v_k]$. Extracting a subsequence, we have the convergence $y_k \to y$ weakly in Y. Then we obtain (see the proof of Lemma 4.7) $|y_k|^p y_k \to |y|^p y$ weakly in $L_{q'}(\Omega)$. Repeat the transformations of the proof of Lemma 6.4; for all functions $\lambda \in D(\Omega)$ we have the convergence

$$\int_\Omega v_k y_k \lambda dx \to \int_\Omega vy\lambda dx.$$

Multiply the equality (6.19) for $v = v_k$ by the arbitrary function $\lambda \in D(\Omega)$. After integration and passing to the limit as $k \to \infty$ we get $y_k = y[v_k]$. ⊔

Optimization and Differentiation

Remark 6.19 We have two difficulties with the proof of the weak continuity of the solution of the problem (6.19), (6.20) with respect to the control. This is passing to the limit at the second and the third terms of the equation (6.19) for $v = v_k$. We pass to the limit at the nonlinear term by using the technique from the proof of Lemma 4.7, where we analyze the nonlinear equation without the control as a coefficient of the state operator. We pass to the limit at the bilinear term by means of the method from the proof of Lemma 6.4, where we consider the linear equation with control at the coefficient.

Using Lemma 6.8, we prove the existence of the optimal control as an analog of Theorem 6.4.

Theorem 6.7 *Problem* 6.4 *is solvable.*

Now apply Theorem 6.2 for obtaining the necessary conditions of optimality. We would like to have the differentiability of the state function with respect to the control for it.

Consider the equality

$$-\Delta y_\sigma + |y_\sigma|^p y_\sigma + v_\sigma y_\sigma + \Delta y - |y|^p y - vy = 0,$$

where $y = y[u]$, $y_\sigma = y[v_\sigma]$. Then we have

$$-\Delta(y_\sigma - y) + g(v_\sigma)^2(y_\sigma - y) + u(y_\sigma - y) = \sigma(u - v)y_\sigma,$$

where

$$g(v)^2 = (\rho + 1)|y + \varepsilon(y[v] - y)|^p, \ \varepsilon \in [0, 1].$$

Multiply the previous equality by the smooth enough function λ that is equal to zero on the boundary Γ. After integration by using Green's formula we get

$$\int_\Omega \left[-\Delta\lambda + g(v_\sigma)^2\lambda + u\lambda\right]dx = \sigma \int_\Omega (u - v)y_\sigma \lambda dx. \tag{6.22}$$

Consider the equation

$$-\Delta p_\mu[v_\sigma] + g(v_\sigma)^2 p_\mu[v_\sigma] + up_\mu[v_\sigma] = \mu. \tag{6.23}$$

If $\sigma = 0$, then for all $v \in U$ it can be transformed to

$$-\Delta p_\mu[u] + (\rho + 1)|y[u]|^p p_\mu[u] + up_\mu[u] = \mu. \tag{6.24}$$

Now we determine the following result.

Lemma 6.9 *We have the convergence* $y[u + \sigma(v - u)] \to y[u]$ *in* $H_0^1(\Omega)$ *as* $\sigma \to 0$ *for all* $v \in U$.

Proof. Choose $\lambda = y_\sigma - y$ in the equality (6.22). Then we have

$$\|y_\sigma - y\|^2 + \int_\Omega \left[g(v_\sigma)(y_\sigma - y)\right]^2 dx + \int_\Omega u(y_\sigma - y)^2 dx = \sigma \int_\Omega (u - v)y_\sigma(y_\sigma - y)dx.$$

We obtain the inequality

$$\|y_\sigma - y\| \le \sigma \|(u-v)y_\sigma\|_* ,\qquad (6.25)$$

If $\sigma \to 0$, then the set $\{y_\sigma\}$ is bounded in $H_0^1(\Omega)$ and in $L_2(\Omega)$ too. Hence, we have the convergence $y_\sigma \to y$ a.e. on Ω (see Lemma 6.8). Then $\sigma(u-v)y_\sigma \to 0$ a.e. on Ω. We have the inequality

$$\int_\Omega \left[(u-v)y_\sigma\right]^2 dx \le b^2 \int_\Omega (y_\sigma)^2 dx$$

by the definition of the set U. Then the set $\{(u-v)y_\sigma\}$ is bounded in $L_2(\Omega)$. We have the convergence $\sigma(u-v)y_\sigma \to 0$ weakly in $L_2(\Omega)$ and strongly in the space $H^{-1}(\Omega)$. Using the estimate (6.25), we get $y_\sigma \to y$ strongly in $H_0^1(\Omega)$. This completes the proof of the lemma. $\square$

Determine the space (see the previous chapter)

$$Y(v) = H_0^1(\Omega) \cap \{\varphi|\, g(v)\varphi \in L_2(\Omega)\}.$$

This is a Hilbert with the scalar product

$$(\varphi,\psi)_{Y(v)} = (\varphi,\psi)_{H_0^1(\Omega)} + \int_\Omega g(v)\varphi\psi dx.$$

It has the adjoint Hilbert in the space $Y(v)'$ of functions $\varphi + g(v)\psi$, where $\varphi \in H^{-1}(\Omega)$, $\psi \in L_2(\Omega)$.

Lemma 6.10 *For all $\sigma \ge 0$, $v \in U$, $\mu \in Y(v_\sigma)'$ the homogeneous Dirichlet problem for the equation (6.23) has a unique solution $p_\mu[v_\sigma]$ from the space $Y(v_\sigma)$, as for all $v \in U$, $\mu \in H^{-1}(\Omega)$ we have the convergence $p_\mu[v_\sigma] \to p_\mu[u]$ weakly in $H_0^1(\Omega)$ as $\sigma \to 0$.*

Proof. Multiply the equality (6.23) by the function $p_\mu[v_\sigma]$. After integration we have

$$\left\|p_\mu[v_\sigma]\right\|^2 + \left\|g(v_\sigma)p_\mu[v_\sigma]\right\|_2^2 + \int_\Omega u\big(p_\mu[v_\sigma]\big)^2 dx = \int_\Omega \mu p_\mu[v_\sigma]dx. \qquad (6.26)$$

The function u is non-negative. Using the definition of the norm space $Y(v_\sigma)$, we get

$$\left\|p_\mu[v_\sigma]\right\|_{Y(v_\sigma)}^2 \le \left\|\mu\right\|_{Y(v_\sigma)'}\left\|p_\mu[v_\sigma]\right\|_{Y(v_\sigma)}.$$

We obtain the a priori estimate of the solution of the equation (6.23) in the space $Y(v_\sigma)$. From linearity of this equation, one-valued solvability follows in this space.

228   *Optimization and Differentiation*

Suppose the convergence $\sigma \to 0$. From the equality (6.25) it follows that

$$\left\|p_\mu[v_\sigma]\right\|^2 + \left\|g(v_\sigma)p_\mu[v_\sigma]\right\|_2^2 \le \|\mu\|_* \|p_\mu[v_\sigma]\|.$$

Thus, we have the boundedness of the sequences $\{p_\mu[v_\sigma]\}$ and $\{g(v_\sigma)p_\mu[v_\sigma]\}$ in the spaces $H_0^1(\Omega)$ and $L_2(\Omega)$. After extraction of subsequence we obtain the convergence $p_\mu[v_\sigma] \to \varphi_\mu$ weakly in $H_0^1(\Omega)$. Note that $y[v_\sigma] \to y[u]$ strongly in $H_0^1(\Omega)$. Using the Rellich–Kondrashov theorem, we get $p_\mu[v_\sigma] \to \varphi_\mu$ and $y[v_\sigma] \to y[u]$ a.e. on Ω. Therefore, we have $g(v_\sigma)^2 p_\mu[v_\sigma]\} \to (\rho+1)|y[u]|^\rho \varphi_\mu$ a.e. on Ω.

Consider the equality

$$\left\|g(v_\sigma)\right\|_{2q/\rho}^{2q/\rho} = \sqrt{\rho+1} \int_\Omega |y[u] + \varepsilon(y[v_\sigma] - y[u])|^q dx.$$

Using boundedness of $\{y[v_\sigma]\}$ in the space $L_q(\Omega)$, we obtain the boundedness of $\{g(v_\sigma)\}$ in the space $L_{2q/\rho}(\Omega)$. By Hölder inequality, we have

$$\left\|g(v_\sigma)^2 p_\mu[v_\sigma]\right\|_{q'} \le \|g(v_\sigma)\|_{2q/\rho} \|g(v_\sigma)p_\mu[v_\sigma]\|_2$$

because of the equality $\frac{1}{q'} = \frac{1}{2} + \frac{\rho}{2q}$. Thus, the set $\{g(v_\sigma)^2 p_\mu[v_\sigma]\}$ is bounded in the space $L_{q'}(\Omega)$. Then $g(v_\sigma)^2 p_\mu[v_\sigma] \to (\rho+1)|y[u]|^\rho \varphi_\mu$ weakly in $L_{q'}(\Omega)$. Multiply the equality (6.23) by a smooth enough function λ. After integration and passing to the limit we determine φ_μ. This completes the proof of Lemma 6.10. $\square$

Remark 6.20   We could prove additional properties for the equation (6.23) (see Lemma 6.16).

Consider an additional form of the operator differentiation.

Definition 6.2 *The operator $L : V \to Y$ is called $(V_0, Y_0; V_*, Y_*)$-**extended differentiable** (or more easily **extended differentiable**) **with respect to the convex set** U at a point $u \in U$ if there exist linear topological spaces V_0, Y_0, V_*, Y_* that satisfy the continuous embedding $V_* \subset V_0 \subset V$, $Y \subset Y_0 \subset Y_*$ and an affine continuous operator D on the set $\delta(U) = \{v - u | v \in U\} \cap V_0$ with codomain Y_0 such that $\{L[u + \sigma(v - u)] - Lu\}/\sigma \to D(v - u)$ in Y_* as $\sigma \to 0$ for all $v \in U$ that satisfy the inclusion $(v - u) \in V_*$.*

If $V_* = V_0 = V$, $Y = Y_0 = Y_*$, then we have the differentiability with respect to the convex set. We obtain extended differentiability for $U = V$.

Determine the space

$$Y(u) = H_0^1(\Omega) \cap \{\varphi | \; |y[u]|^{\rho/2}\varphi \in L_2(\Omega)\}.$$

Denote by Y_*^w the space $Y_* = H_0^1(\Omega)$ with weak topology.

Coefficients optimization control problems

Lemma 6.11 *The operator $y[\cdot] : V \to Y$ for the problem* $(6.19), (6.20)$ *has $(V, Y(u); V, Y_*^w)$-extended Gâteaux derivative $y_U'[u]$ with respect to the set U at the arbitrary point $u \in U$ that satisfies the equality*

$$\int_\Omega \mu y_U'[u](v - u)dx = \int_\Omega (u - v)p_\mu[u]y[u]dx \ \ \forall v \in U, \ \mu \in H^{-1}(\Omega), \quad (6.27)$$

where $p_\mu[u]$ is the solution of the homogeneous Dirichlet problem for the equation (6.24).

Proof. The equality (6.27) determines in reality an affine continuous operator $y_U'[u] : \delta(u) \to Y(u)$. Return to the equality (6.22). Determine $\lambda = p_\mu[v_\sigma]$. Then we get

$$\int_\Omega \mu\{y[u + \sigma(v - u)] - y[u]\}dx = \sigma \int_\Omega (u - v)y[u + \sigma(v - u)]p_\mu[v_\sigma]dx.$$

Hence, using (6.27), we obtain

$$\int_\Omega \mu\left\{\frac{y[u + \sigma(v - u)] - y[u]}{\sigma} - y_U'[u](v - u)\right\}dx =$$

$$\sigma \int_\Omega (u - v)\left\{y[u + \sigma(v - u)]p_\mu[v_\sigma] - y[u]p_\mu[u]\right\}dx \ \ \forall v \in U, \mu \in Y_*'. \quad (6.28)$$

From Lemma 6.9 it follows that the convergence $y[u + \sigma(v - u)] \to y[u]$ in $H_0^1(\Omega)$ as $\sigma \to 0$. By Lemma 6.10, we have $p_\mu[v_\sigma] \to p_\mu[u]$ weakly in $H_0^1(\Omega)$. After extracting a subsequence we obtain this convergence almost everywhere on Ω. Therefore, $(u - v)p_\mu[v_\sigma] \to (u - v)p_\mu[u]$ a.e. on Ω. Using the definition of the set U, we have the inequality

$$\int_\Omega \left\{(u - v)p_\mu[v_\sigma]\right\}^2 dx \le 2b \int_\Omega \left(p_\mu[v_\sigma]\right)^2 dx.$$

Hence, the set $\{(u - v)p_\mu[v_\sigma]\}$ is bounded in the space $L_2(\Omega)$. Then we have the convergence $(u - v)p_\mu[v_\sigma] \to (u - v)p_\mu[u]$ weakly in $L_2(\Omega)$. Pass to the limit in the equality (6.28). We get $\{y[u + \sigma(v - u)] - y[u]\}/\sigma \to y_U'[u](v - u)$ weakly in Y_*. This complete the proof of Lemma 6.11. $\square$

Determine the following result as an analog of Lemma 6.7.

Lemma 6.12 *The functional I for Problem 6.4 is differentiable with respect to the set U at the arbitrary point $u \in U$ as its derivative satisfies the equality*

$$I_U'[u](v - u) = \int_\Omega (\alpha u = py)(v - u)dx \ \ \forall v \in U,$$

230   *Optimization and Differentiation*

where $y = y[u]$ and p is the solution of the homogeneous Dirichlet problem for the equation

$$-\Delta p + (\rho + 1)|y|^\rho p + up = \Delta y - \Delta y_d. \tag{6.29}$$

Using Theorem 6.2 and Lemma 6.12 determine the necessary condition of optimality as an analog of Theorem 6.5.

Theorem 6.8 *The solution to Problem 6.4 is*

$$u(x) = \begin{cases} a, & \text{if } -y(x)p(x) < \alpha a, \\ -y(x)p(x)/\alpha, & \text{if } \alpha a \le -y(x)p(x) \le \alpha b, \\ b, & \text{if } -y(x)p(x) > \alpha b. \end{cases} \tag{6.30}$$

Remark 6.21 The equalities (6.29), (6.30) are the analogs of the formulas (6.16), (6.18) for the equation (6.19).

Thus, the solution of Problem 6.4 satisfies the system that involves the equation (6.19) for $v = u$ and (6.29) with homogeneous boundary conditions, and the equality (6.30). Denote by $F(y, p)$ the term at the right-hand side of the equality (6.30). We obtain the system of two nonlinear elliptic equations

$$-\Delta y + |y|^\rho y + F(y, p)y = f,$$

$$-\Delta p + (\rho + 1)|y|^\rho + F(y, p)p = \Delta y - \Delta y_d$$

with homogeneous boundary conditions. If we find its solutions y and p, then we can find the solution of Problem 6.4 by the formula (6.30). We obtained optimality conditions for the control system described by nonlinear elliptic equations by using extended differentiability of the control–state mapping in the previous chapter. Then we obtained this property for the abstract system. Now we obtain the necessary conditions of optimality for the nonlinear elliptic equations with the control at the coefficient of the state operator by using extended differentiability with respect to the convex set. Try to extend this property to the abstract system.

6.6   Differentiability with respect to a convex set for abstract systems

Consider an abstract nonlinear system with the control in coefficients. Let us have Banach spaces V, Y, Z and a non-empty convex closed subset of admissible controls U of the space V. For all control $v \in U$ it is given a nonlinear state operator $A(v) : Y \to Z$ that depends upon the control. Consider the system described by the equation

$$A(v)y = f, \tag{6.31}$$

where $f \in Z$ is known. Let the map $v \to A(v)y$ be affine for all $y \in Y$.

Coefficients optimization control problems

Consider an arbitrary point $u \in U$. Suppose the following condition is true.

Assumption 6.1 *For all v from a convex neighborhood O of the point u the operator $A(v)$ is invertible.*

Thus, for all $v \in O$ there exists a unique solution $y[v] = A(v)^{-1}f$ of the equation (6.31) from the space Y. Consider the equality

$$A(v_\sigma)y[v_\sigma] - A(u)y = 0,$$

where $v_\sigma = u + \sigma(v - u)$, $y = y[u]$ for the arbitrary function $v \in U$ and the positive number σ that is so small that we have the inclusion $v_\sigma \in O$. Then we have

$$A(v_\sigma)y[v_\sigma] - A(v_\sigma)y = A(u)y - A(v_\sigma)y. \tag{6.32}$$

Suppose the operator $A(v)$ is Gâteaux differentiable. Using Lagrange formula, we have the equality

$$\left\langle \lambda, A(v_\sigma)y[v_\sigma] - A(v_\sigma)y \right\rangle = \left\langle \lambda, A_y\Big\{v_\sigma, y + \varepsilon(v_\sigma, \lambda)\big(y[v_\sigma] - y\big)\Big\}\big(y[v_\sigma] - y\big) \right\rangle$$

for all $v \in O$, $\lambda \in Z'$, where $A_y(v, z)$ is the derivative of the operator $A(v)$ at the point z, $\varepsilon(v_\sigma, \lambda) \in (0, 1)$. For all $v \in O$, $\lambda \in Z'$ determine the linear continuous operator $G(v, \lambda)$ by the formula

$$G(v, \lambda)\varphi = A_y\Big\{v, y + \varepsilon(v, \lambda)\big(y[v] - y\big)\Big\}\varphi.$$

The dependence of the operator $A(v)$ from the control is affine. Therefore, we can transform the equality (6.32). We obtain

$$\left\langle \lambda, G(v_\sigma, \lambda)\big(y[v_\sigma] - y\big) \right\rangle = \sigma\langle \lambda, A(u)y - A(v)y \rangle \ \forall \lambda \in Z' \tag{6.33}$$

that is an analog of (5.20).

For all $v \in O$ consider Banach spaces Z_*, $Z(v)$, and $Y(v)$ such that embedding of Y, Z_*, and $Z(v)$ to $Y(v)$, $Z(v)$, and Z are continuous and dense correspondingly. Suppose the following property:

Assumption 6.2 *The operator $A(v)$ is Gâteaux differentiable, as for any $v \in O$, $\lambda \in Z'$ there exists a continuous continuation $\overline{G}(v, \lambda)$ of the operator $G(v, \lambda)$ on the set $Y(v)$ such that its codomain is a subset $Z(v)$.*

From the equality (6.33) the analog of (5.21) follows.

$$\left\langle \overline{G}^*(v_\sigma, \lambda)\lambda, \big(y[v_\sigma] - y\big)/\sigma \right\rangle = \langle \lambda, A(u)y - A(v)y \rangle \ \forall \lambda \in Z(v_\sigma)'. \tag{6.34}$$

Using the known technique (see Theorem 5.6) consider the operator equation

$$\overline{G}\big(v, p_\mu[v]\big)^* p_\mu[v] = \mu \tag{6.35}$$

232 *Optimization and Differentiation*

that is an analog of (5.22), where $\mu \in Y(v_\sigma)'$. For $v = u$ this is the linear equation

$$\overline{A}_y(u, y)^* p_\mu[u] = \mu \tag{6.36}$$

that is an analog of (5.23), where $\overline{A}_y(u, y)$ is a continuation of the operator $A_y(u, y)$ on the set Y(u).

Consider a Banach space Y_* that is continuously and dense includes to the space $Y(v)$ for all $v \in O$. Suppose the following assumption.

Assumption 6.3 *For all $\mu \in Y(v)'$ the equation (6.35) has a unique solution $p_\mu[v] \in Z(v)'$.*

Return to the equality (6.34). Determine $\lambda = p_\mu[v_\sigma]$ for small enough σ. We get

$$\Big\langle \mu, (y[v_\sigma] - y)/\sigma \Big\rangle = \Big\langle p_\mu[v_\sigma], A(u)y - A(v)y \Big\rangle \ \forall \mu \in Y(v_\sigma)', v \in U \tag{6.37}$$

that is an analog of the equality (5.24). If we have the possibility of passing to the limit as $\sigma \to 0$, then we obtain the derivative of the dependence $y = y[v]$ with respect to the convex set U at the left-hand side of this equality. However, we have the necessity to pass to the limit at the right-hand side of this equality.

Assumption 6.4 *We have the convergence $p_\mu[v_\sigma] \to p_\mu[u]$ ∗-weakly in Z_*' as $\sigma \to 0$ for all $v \in U$.*

Denote by Y_*^w the space Y_* with weak topology. Prove the following result.

Theorem 6.9 *Under the assumptions $6.1 - 6.4$ the operator $y[\cdot] : U \to Y$ has $(V, Y(u); V, Y_*^w)$-extended Gâteaux derivative $y'_U[u]$ with respect to the set U at the arbitrary point $u \in U$ that satisfies the equality*

$$\Big\langle \mu, y'_U[u](v - u) \Big\rangle = \Big\langle p_\mu[u], A(u)y[u] - A(v)y[u] \Big\rangle \ \forall \mu \in Y_*', v \in U. \tag{6.38}$$

Proof. Note that the operator $y'_U[u]$ is affine here. Its domain is the set

$$\delta(u) = -u + U = \big\{ v - u \mid v \in U \big\}$$

of the space V, and the codomain is Y. Using the equalities (6.37), (6.38), we have

$$\Big\langle \mu, (y[v_\sigma] - y)/\sigma - y'_U[u](v - u) \Big\rangle =$$

$$\Big\langle p_\mu[v_\sigma] - p_\mu[u], A(u)y[u] - A(v)y[u] \Big\rangle \ \forall \mu \in Y_*', v \in U.$$

By Assumption 6.4 we have the convergence $p_\mu[v_\sigma] \to p_\mu[u]$ ∗-weakly in Z_*' for all $v \in U$. Pass to the limit at the previous equality. We have the convergence $(y[v_\sigma] - y)/\sigma \to y'_U[u](v - u)$ weakly in Y for all $v \in U$. Thus, the operator $y'_U[u](v - u)$ is in reality the derivative of the map $y[\cdot] : U \to Y$ at the point u with respect to the set U. $\square$

Remark 6.22 The last result is the natural generalization of Theorem 5.6 to the systems with the control in the coefficients and the derivatives with respect to the convex set.

Remark 6.23 Using Theorem 6.9, we could determine the necessary condition of optimality abstract control systems.

Make sure that the assumptions of Theorem 6.9 are true for the considered nonlinear elliptic equation.

Lemma 6.13 *The assumptions of Theorem 6.9 are true for the boundary problem* (6.19), (6.20).

Proof. Determine the state operator by the equality

$$A(v)y = -\Delta y + |y|^p y + vy.$$

Consider the spaces V, Y, Z from the boundary problem (6.19), (6.20). The operator $A(v)$ is invertible by Theorem 6.6, i.e., Assumption 6.1 is true. For all admissible control v its derivative at the arbitrary point y is determined by the formula

$$A_y(v,y)\varphi = -\Delta \varphi + (p+1)|y|^p \varphi + v\varphi \ \forall \varphi \in Y.$$

For all functions $v \in U$, $\lambda \in Z'$ determine the operator $G(v, \lambda)$ by the equality

$$G(v,\lambda)\varphi = -\Delta \varphi + g(v)^2 \varphi + v\varphi \ \forall \varphi \in Y.$$

Determine the spaces

$$Y(v) = H_0^1(\Omega) \cap \{\varphi| \ g(v)\varphi \in L_2(\Omega)\}, \quad Z(v) = Y(v)'.$$

Determine the linear continuous operator $\overline{G}(v, \lambda)$ by the formula

$$\overline{G}(v,\lambda)\varphi = -\Delta \varphi + g(v)^2 \varphi + v\varphi \ \forall \varphi \in Y(v).$$

Hence, Assumption 6.2 is true. Finally, Assumption 6.3 and Assumption 6.4 follow from Lemma 6.10. This completes the proof of Lemma 6.11. $\square$

Thus, Lemma 6.11 that is the basis of obtaining the necessary conditions of optimality for Problem 6.4 is the corollary of Theorem 6.9.

We had the optimality conditions for the strongly nonlinear systems without the differentiability of the control–state mapping in the previous chapter. Try to obtain the analogical result for the nonlinear equation with the control at the coefficient of the state operator.

6.7 Strongly nonlinear systems with the control at the coefficients

Consider again the equation (6.19)

$$-\Delta y + |y|^\rho y + vy = f, \ x \in \Omega$$

with homogeneous boundary condition (6.20). Let the spaces Y, Z and the operator $A(v)$ be defined in the same way as for Problem 6.4. By the Rellich–Kondrashov theorem, embedding of the space $H_0^1(\Omega)$ to $L_r(\Omega)$ is compact, where r, $2n/(n-2)$ for $n > 2$ and r is arbitrary for $n = 2$. Determine the control space $V = L_\beta(\Omega)$, where $\beta = \max\{r', s\}$, $s = 1 + 2/\rho$, $1/r + 1/r' = 1$. For the large enough values of the parameter of nonlinearity ρ and the dimension of the set n the space V is larger than $L_r(\Omega)$. Using Hölder inequality, determine that for all functions $v \in L_r(\Omega)$, $y \in L_q(\Omega)$ we have the inclusion $vy \in L_{q'}(\Omega)$, where $q = \rho + 2$, $1/q + 1/q' = 1$. Then for all the operators $A(v)$ maps the space Y to Z. Determine the set of admissible controls by the formula

$$U = \{v \in V \mid 0 < a \le v(x), \ x \in \Omega\},$$

where a is known constant. For all $v \in U$ the problem (6.19), (6.20) has a unique solution $y = y[v]$ from the space Y.

Remark 6.24 We use the space $L_{r'}(\Omega)$ for the proof of the solvability of the optimization problem (see Lemma 6.13).

Remark 6.25 We change the set of admissible controls for the proof of the absence of the differentiability of the state function with respect to the control for the strongly nonlinear case.

Determine the functional

$$I(v) = \frac{\alpha}{\beta} \int_\Omega |v|^\beta dx + \frac{1}{2} \int_\Omega \sum_{i=1}^n \left(\frac{\partial y[v]}{\partial x_i} - \frac{\partial y_d}{\partial x_i} \right)^2 dx,$$

$\alpha > 0$, and y_d is a known function of the space $H_0^1(\Omega)$. We have the following optimization control problem.

Problem 6.5 *Find the control $u \in U$ that minimizes the functional I on the set U.*

Remark 6.26 The control is an element of the larger set here. Therefore, we change the first term of the minimized functional.

Remark 6.27 In principle, the situation, where the functional is determined on the part of the set of admissible control only, makes sense. We could minimize it on this part. This idea is the basis for solving optimization control problems without one-valued solvability of the state equation on the whole set of admissible controls (see Comments).

Coefficients optimization control problems 235

Remark 6.28 The additional difference between Problem 6.4 and Problem 6.5 is unlimitedness of the set of admissible controls. Determine the weak continuity of the control–state mapping that is used to prove the solvability of Problem 6.5.

Determine the weak continuity of the control–state mapping that is used to prove the solvability of Problem 6.5.

Lemma 6.14 *The map* $y[\cdot] : U \to Y$ *for Problem* 6.5 *is weakly continuous.*

Proof. Consider sequence $\{v_k\}$ of the set U such that $v_k \to v$ weakly in V and in $L_{r'}(\Omega)$ too. Then (see Lemma 6.8) we have the convergence $y_k \to y$ weakly in Y, and strongly in $L_r(\Omega)$ because of the Rellich–Kondrashov theorem, where $y_k = y[v_k]$. Therefore (see the proof of Lemma 4.7), $y_k|y_k|^p \to |y|^p y$ weakly in $L_{q'}(\Omega)$. For all functions $\lambda \in D(\Omega)$ we have the inequality

$$\left| \int_\Omega v_k y_k \lambda dx - \int_\Omega vy \lambda dx \right| \leq \left| \int_\Omega (v_k - v) y \lambda dx \right| + \left| \int_\Omega v_k (y_k - y) \lambda dx \right|.$$

Using the inclusion $y\lambda \in L_r(\Omega)$, the boundedness of the sequence $\{v_k \lambda\}$ in the space $L_{r'}(\Omega)$, weak convergence of the sequence $\{v_k\}$ in the space $L_{r'}(\Omega)$, and strong convergence of $\{y_k\}$ in $L_r(\Omega)$ we get

$$\int_\Omega v_k y_k \lambda dx \to \int_\Omega vy \lambda dx.$$

Multiply the equality (6.19) for $v = V_k$ by the arbitrary function $\lambda \in D(\Omega)$. After the integration and passing to the limit as $k \to \infty$ we have $y = y[v]$. $\square$

The last lemma guarantees the following result.

Theorem 6.10 *Problem* 6.5 *is solvable.*

Remark 6.29 The absence of the boundedness of the set of admissible controls is not the obstacle for the proof of the solvability of the problem because of the coercivity of the minimized functional (see Theorem 2.4).

We would like to obtain necessary conditions of optimality for this problem. Therefore, we have the necessity to prove the differentiability of the state function system with respect to the control. We obtained this result by using the Inverse function theorem (see Chapter 4). This theorem uses the properties of the linearized equation. The linearized equation for our system is

$$-\Delta y + (\rho + 1)|y_0|^p y + v_0 y = z \tag{6.39}$$

that is an analog of the equations (4.26) and (5.5), where it is the solution of the boundary problem (6.19), (6.20) for the control. Determine the following result, which is an analog of Lemma 5.1.

236 *Optimization and Differentiation*

Lemma 6.15 *If embedding $H_0^1(\Omega) \subset L_q(\Omega)$ gets broken, then there exists functions $v_0 \in U$, $z \in Z$ such that the homogeneous Dirichlet problem for the equation (6.39) does not have any solution.*

Proof. Suppose the condition $H_0^1(\Omega) \subset L_q(\Omega)$ is false. Consider arbitrary functions $v_0 \in L_\delta(\Omega) \cap U$, $y_0 \in H_0^1(\Omega) \cap L_\gamma(\Omega)$, where the parameters γ, δ satisfy the equalities

$$\frac{1}{\gamma} = \frac{1}{q} - \varepsilon_1, \quad \frac{1}{\delta} = \frac{1}{\beta} - \varepsilon_1 - \varepsilon_2,$$

and ε_1, ε_2 are small enough positive constants. Then we have the inequalities

$$\gamma > q, \ \delta > \beta \geq s, \tag{6.40}$$

$$\frac{1}{\delta} - \frac{1}{\gamma} < \frac{1}{\beta} - \frac{1}{q} \leq \frac{s}{1} - \frac{1}{q} = \frac{q-3}{q}. \tag{6.41}$$

Determine the function $f_0 = \Delta y_0 + (\rho + 1)|y_0|^\rho y_0 + v_0 y_0$. Hence, y_0 is the solution of the boundary problem (6.19), (6.20) for the control v_0 if $f = f_0$. Then for all $y \in Y$ we have

$$\Delta y \in H^{-1}(\Omega), \ |y_0|^\rho y \in L_\eta(\Omega), \ v_0 y \in L_\pi(\Omega),$$

where

$$\rho/\gamma + 1/q = 1/\eta, \ 1/\delta + 1/q = 1/\pi.$$

From the inequalities (6.40) it follows that

$$1/\eta < (q-2)/q + 1/q = 1/q'.$$

Then we have $\eta > q'$. Using the inequalities (6.40), (6.41), we get

$$1/\pi < (q-3)/q + 1/\gamma + 1/q < (q-1)/q.$$

Therefore, $\pi > q'$. Thus, for all $y \in Y$ the term at the left-hand side of the equality (6.39) belongs to the space $Z_1 = H^{-1}(\Omega) + L_\omega(\Omega)$, where $\omega = \min\{\eta, \pi\} > q'$. This set is narrower than Z. Thus, for $z \in Z \setminus Z_1$, specifically for $z \in L_{q'}(\Omega) \setminus L_\omega(\Omega)$, the equation (6.39) does not have the solution from the space Y. $\square$

Consider the analog of Lemma 5.2.

Lemma 6.16 *If embedding $H^{-1}(\Omega) \subset L_q(\Omega)$ gets broken, and $1/\rho \leq 2(r-1)$, then the map $y[\cdot] : U \to Y$ for Problem 6.5 is not differentiable with respect to the set U.*

Proof. If $\rho \leq 2(r-1)$, then $s \geq q'$. Therefore, $\beta = s$. Let the functions f_0 and v_0 be chosen in Lemma 6.15. Suppose the inclusion $H^{-1}(\Omega) \subset L_q(\Omega)$ is false, but the operator $y[\cdot] : U \to Y$ is differentiable with respect to the set U at the point v_0. Hence, we have the convergence $\{y[v_0+\sigma(v-v_0)]-y[v_0]\}/\sigma \to \varphi$ in Y as $\sigma \to 0$ for all $v \in U$, where $\varphi = y'_U[v_0](v-v_0)$. From the equality (6.19) for the controls $v = v_0 + \sigma(v - v_0)$ and $v = v_0$ after dividing by σ and passing to the limit as $\sigma \to 0$ it follows that

$$-\Delta\varphi + (\rho+1)|y_0|^\rho\varphi + v_0\varphi = -(v - v_0)y_0. \tag{6.42}$$

We have the conditions $y_0 \in L_\gamma(\Omega)$, $v - v_0 \in L_s(\Omega)$ here. Then the value at the right-hand side of the equality (6.42) belongs to the space $L_\chi(\Omega)$ where $1/s + 1/\gamma = 1/\chi$. From the inequalities (6.40) it follows that

$$1/\chi = (q - 2)/q + 1/\gamma < (q - 1)/q,$$

i.e., $\chi > q'$.

We have

$$\frac{1}{\eta} - \frac{1}{\chi} = \frac{\rho}{\gamma} + \frac{1}{q} - \frac{1}{\gamma} - \frac{1}{s} = \frac{q-2}{\gamma} + \frac{1}{q} - \frac{1}{\gamma} - \frac{q-2}{q} = (q-3)\left(\frac{1}{\gamma} - \frac{1}{q}\right) < 0$$

because of the first condition (6.40) and inequality $q > 3$, which is the corollary of the suppositions of the lemma. Therefore, $\chi < \eta$. We obtain also

$$\frac{1}{\pi} - \frac{1}{\chi} = \frac{1}{\delta} + \frac{1}{q} - \frac{1}{\gamma} - \frac{1}{s} = \left(\frac{1}{\gamma} - \frac{1}{q}\right) - \frac{q-3}{q} < 0$$

because of the inequality (6.41). Hence, $\chi < \pi$. Thus, we get $\chi < \omega$. Then $L_\chi(\Omega) \setminus L_\omega(\Omega) \neq \emptyset$. Therefore, there exists the point $z \in Z \setminus Z_1$ that is $z = -(v - v_0)y_0$ for a value v such that the equation (6.39) has a solution $y = y'_U(v_0)(v - v_0)$ from the set Y. However, this contradicts Lemma 6.14. Thus, the supposition about the differentiability of the solution of the problem (6.19), (6.20) with respect to the control is false. $\square$

Remark 6.30 The inequality $1 < \rho \leq (2r-1)$ does not contradict to embedding $H^{-1}(\Omega) \subset L_q(\Omega)$. Therefore, we can have in reality the absence of the differentiability of control–state mapping for Problem 6.5.

Remark 6.31 We change the control space for obtaining the last result only.

Prove the extended differentiability of the solution of the boundary problem (6.19), (6.20) with respect to the control at a point $u \in U$. By the Sobolev theorem, we have continuous embedding of the space $H^{-1}(\Omega)$ to $L_\theta(\Omega)$, where θ is arbitrary for $n = 2$ and equal to $2n/(n-2)$ if $n > 2$. Determine the parameter ψ from the equality $1/\psi = 1/q + 1/\theta$. Consider the space

$$P(u) = \left\{p \in H_0^1(\Omega)\,\middle|\, |y[u]|^{\rho/2}p \in L_2(\Omega),\ \sqrt{u}p \in L_2(\Omega)\right\}.$$

This is a Hilbert space with scalar product

$$(p, \varphi)_{P(u)} = \int_\Omega \sum_{i=1}^n \frac{\partial p}{\partial x_i}\frac{\partial \varphi}{\partial x_i}dx + (\rho+1)\int_\Omega |y[u]|^\rho p\varphi dx + \int_\Omega up\varphi dx.$$

Denote by Y_*^w the space Y_* with weak topology.

238   *Optimization and Differentiation*

Lemma 6.17 *The operator* $y[\cdot] : U \to Y$ *is* $(L_{\psi'}(\Omega), P(u); L_{\psi'}(\Omega), Y_*^w)$-
extended differentiable with respect to the set U *at the arbitrary point* $u \in U$,
as its derivative satisfies the equality

$$\int_\Omega \mu y_U'[u](v-u)dx = \int_\Omega (u-v)p_\mu[u]y[u]dx \ \forall \mu \in P(u)', \ v \in \delta(u), \quad (6.43)$$

where $\delta(u) = \{v \in U| \ (v-u) \in L_{\psi'}(\Omega)\}$ *and* $p_\mu[u]$ *is the solution of the
homogeneous Dirichlet problem for the equation*

$$-\Delta p_\mu[u] + (\rho+1)|y[u]|^\rho p_\mu[u] + up_\mu[u] = \mu. \quad (6.44)$$

Proof. For all $v \in U$, $\sigma \in [0,1]$ we have the equality

$$-\Delta(y_\sigma - y) + g(v_\sigma)^2(y_\sigma - y) + v_\sigma(y_\sigma - y) = \sigma(u-v)y,$$

where

$$v_\sigma = u + \sigma(v-u), \ y_\sigma = y[v_\sigma], \ y = y[u], \ g(v)^2 = (\rho+1)|y + \varepsilon(y[v] - y)|^\rho,$$

$\varepsilon \in [0,1]$. Multiply the previous equality by the smooth enough function λ.
After integration we have

$$\int_\Omega \left[-\Delta\lambda + g(v_\sigma)^2\lambda + v_\sigma\lambda\right](y_\sigma - y)dx = \int_\Omega \sigma(u-v)ydx \quad (6.45)$$

that is an analog of (6.22).

Determine the equation

$$-\Delta p_\mu[v] + g(v)^2 p_\mu[v] + vp_\mu[v] = \mu, \quad (6.46)$$

which is an analog of (6.23). Consider the homogeneous Dirichlet problem for
it. If $v = u$, then the equation (6.46) is transformed to (6.44), which is equal
to (6.24). After multiplication of the equality (6.46) by the function $p_\mu[v]$ and
integration we get

$$\int_\Omega \sum_{i=1}^n \left(\frac{\partial p_\mu[v]}{\partial x_i}\right)^2 dx + \int_\Omega \{g(v)p_\mu[v]\}^2 dx + \int_\Omega v(p_\mu[v])^2 dx = \int_\Omega \mu p_\mu[v]dx.$$

Determine Hilbert space

$$P(v) = \left\{p \in H_0^1(\Omega)| \ g(v)p \in L_2(\Omega), \ \sqrt{v}p \in L_2(\Omega)\right\}$$

with norm

$$\|p\|_{P(v)}^2 = \int_\Omega \sum_{i=1}^n \left(\frac{\partial p}{\partial x_i}\right)^2 dx + \int_\Omega [g(v)p]^2 dx + \int_\Omega vp^2 dx.$$

Coefficients optimization control problems

This space is equal to $P(u)$ for $v = u$. Then we have

$$\|p\|_{P(v)}^2 = \langle \mu, p \rangle.$$

Therefore, we have the estimate

$$\|p\|_{P(v)} \leq \|\mu\|_{P(v)'}.$$

Thus, for all $\mu \in P(v)'$ the elliptic equation (6.46) has a unique solution $p_\mu[v] \in P(v)$. Return to the equality (6.45). Determine $\lambda = p_\mu[v]$ here. We get

$$\int_\Omega \mu(y_\sigma - y)/\sigma dx = \int_\Omega (u - v)p_\mu[v_\sigma]y[u]dx. \tag{6.47}$$

Consider the equality (6.43) that is an analog of (6.27). The function $p_\mu[u]$ belongs to the space $P(u)$ for all $\mu \in P(v)'$ here. Then we have $p_\mu[u] \in L_\theta(\Omega)$. Using Hölder inequality, we have $p_\mu[u]y[u] \in L_\psi(\Omega)$. Therefore, the equality (6.43) makes sense for $(v - u) \in L_{\psi'}(\Omega)$. Thus, the equality (6.43) is true for all $\mu \in P(u)'$ and $v \in U$ if $(v - u) \in L_{\psi'}(\Omega)$. Determine an affine continuous operator $y_U'[u]$ on the set $\delta_0(u) = \{v - u \mid v \in U\} \cap V_0$ with codomain $Y_0 = P(u)$, where $V_0 = L_{\psi'}(\Omega)$.

Consider the convergence $\sigma \to 0$. Then $v_\sigma \to u$ in V, and $y_\sigma \to y$ weakly in Y. We have the inequalities

$$\left\| p_\mu[v_\sigma] \right\| \leq \|\mu\|_*, \quad \left\| g(v_\sigma)p_\mu[v_\sigma] \right\|_2 \leq \|\mu\|_*, \quad \left\| \sqrt{v_\sigma} p_\mu[v_\sigma] \right\|_2 \leq \|\mu\|_*.$$

Hence, we obtain the convergence $p_\mu[v_\sigma] \to p_\mu$ weakly in Y_*. Using the boundedness of the set $\{y_\sigma\}$ in the space $L_q(\Omega)$, we determine the boundedness of $\{g(v_\sigma)^2 p_\mu(v_\sigma)\}$ in $L_{q'}(\Omega)$. Using the Rellich–Kondrashov theorem, we have $y_\sigma \to y$, $p_\mu[v_\sigma] \to p_\mu$ strongly in $L_r(\Omega)$ and a.e. on Ω. Therefore, $g(v_\sigma)^2 p_\mu(v_\sigma) \to (\rho + 1)|y|^\rho p_\mu$ a.e. on Ω. Hence, we get $g(v_\sigma)^2 p_\mu(v_\sigma) \to (\rho + 1)|y|^\rho p_\mu$ weakly in $L_{q'}(\Omega)$. Using the convergence $v_\sigma \to u$ strongly in $L_{r'}(\Omega)$ for all functions $\lambda \in D(\Omega)$ we have

$$\int_\Omega v_\sigma p_\mu(v_\sigma)\lambda dx \to \int_\Omega u p_m u\lambda dx.$$

After multiplication of the equality (6.46) by an arbitrary function $\lambda \in D(\Omega)$, integration, and passing to the limit we prove that $p_\mu = p_\mu[u]$.

From the equalities (6.43), (6.47) it follows that

$$\int_\Omega \mu\{(y_\sigma - y)/\sigma - y_U'[u](v - u)\}dx = \int_\Omega (p_\mu[v_\sigma] - p_\mu[u])(u - v)y[u]dx.$$

For all $\mu \in Y_*'$ we have the convergence $p_\mu(v_\sigma) \to p_\mu[u]$ weakly in $L_\theta(\Omega)$. If $(v - u) \in L_{\psi'}(\Omega)$, then we have the inclusion $y[u](v - u) \in L_\theta(\Omega)$ because of the equality $1/\psi = 1/q + 1/\theta$. Then we have $(y_\sigma - y)/\sigma \to y_U'[u](v - u)$

240   *Optimization and Differentiation*

weakly in Y_*. Thus, $y'_U[u]$ is $\big(L_{\psi'}(\Omega), P(u); L_{\psi'}(\Omega), Y_*^w\big)$-extended derivative of the operator $y[\cdot] : V \to Y$ with respect to the set U at the point u. $\square$

Using Lemma 6.17, we obtain the following result.

Lemma 6.18 *The functional I for Problem 6.5 is $\big(L_{\psi'}(\Omega), \mathbb{R}; L_{\psi'}(\Omega), \mathbb{R}\big)$-extended differentiable with respect to the set U at the arbitrary point $u \in U$, as its derivative satisfies the equality*

$$I'_U(u)(v - u) = \int_\Omega \big(\alpha|u|^{\beta-2}u + py\big)(v - u)dx \ \forall v \in \delta_0(u),$$

where p is the solution of the homogeneous Dirichlet problem for the equation (6.29).

Remark 6.32 Lemma 6.16 and Lemma 6.18 are analog of Lemma 6.11 and Lemma 6.12.

Using Lemma 6.18, we obtain the following result that is an extension of Theorem 6.8.

Theorem 6.11 *The solution u of Problem 6.5 satisfies the variational inequality*

$$\int_\Omega \big(\alpha|u|^{\beta-2}u + py\big)(v - u)dx \ \forall v \in U.$$

Thus, we can obtain necessary conditions of optimality by using extended derivatives with respect to the convex set if control–state mapping is not differentiable with respect to the convex set.

6.8   Comments

The coefficients optimization problems are important for the coefficient inverse problems (see, for example, A.L. Bukhgeim [98], S.I. Kabanihin[265], A. Lorenzi [325]) and set control problems (see N. Fujii [192], [193], V.G. Litvinov [320], [321], W.B. Liu, P. Neittaanmäki and D. Tiba [322], F. Murat and L. Tartar [375], A. Myslinski and J. Sokolowski [376], T. Slawing [504], and D. Tiba [521]). Application of the coefficients optimization problems is considered also by J.-L. Armand [23], N.V. Banichuk [45], V.G. Litvinov [321], and K.A. Lurie [333].

The coefficients optimization problems can be non-solvable (see J.L. Lions [315], K.A. Lurie [333], F. Murat [373], [374], and L. Tartar [515]). The sequential extension of these problems is proposed by S.Ya. Serovajsky [477], [479]. Optimization control problems described by the linear elliptic equations with coefficient control are considered by R. Acar [3], J. P. Aubin and I. Ekeland [31], J. Baranger [47], A.S. Bratus [89], I. Ekeland and R. Temam [162], M. Goebel [214], K. Kunisch and L.W. White [290], J.L. Lions [315], V.G. Litvinov [320], [321], K.A. Lurie [333], F. Murat and L. Tartar [375], C. Pucci [417], [418], U. Raitums [421], [422], S.Ya. Serovajsky [486], T. Tartar [515], C. Trenchea [529], E.M. Vaisbord [536], C. Yu and J. Yu [564], T. Zolezzi [571], and others.

Coefficients optimization control problems 241

The analogical problems for the nonlinear systems are analyzed by N. Fujii [193], V.I. Ivanenko and V.S. Melnik [260], F.V. Lubyshev and M.E. Fairuzov [330], U. Raitums [422], S.Ya. Serovajsky [466], [483], [486], and T. Slawing [504]. A. Myslinski and J. Sokolowski [376] consider a coefficients optimization problem for a four order elliptic equation. I.G. Gogodze [216] considers coefficients optimization problems with integral constraints. K. Kunisch and L.W. White [290] use the Galerkin method for the approximate solving of coefficients optimization problems for the linear elliptic equations.

The derivative with respect to the convex set and the extended derivative with respect to the convex set were proposed by S.Ya. Serovajsky [483].

Chapter 7

Systems with nonlinear control

7.1	Implicit function theorem	243
7.2	Optimization control problems for abstract systems	246
7.3	Weakly nonlinear control systems	248
7.4	Extended differentiability of the implicit function	254
7.5	Strongly nonlinear control systems	256
7.6	Comments	260

We considered before optimization control problems for systems described by linear and nonlinear equations. However, these equations were linear with respect to the control. Now we analyze systems such that the state operator is nonlinear with respect to the control. Therefore, we shall use an additional technique. This is the Implicit function theorem that is an extension of the Inverse function theorem. This result can guarantee the differentiability of the implicit operator that is control–state mapping. However, this theorem uses the solvability of the linearized equation in the natural spaces that is an analog of the general assumption of the Inverse function theorem. If this property is false, then the dependence of state function from the control can be non-differentiable. However, we can determine its extended derivatives. We consider optimization control problems for the systems described by a nonlinear elliptic equation with a nonlinear control as an application.

7.1 Implicit function theorem

Let us have Banach spaces V, Y, Z, non-empty convex subset U, space V, and the operator $A : V \times Y \to Z$. For all control $v \in U$ the state of the system $y[v] \in Y$ satisfies the equation

$$A\big(v, y[v]\big) = 0. \tag{7.1}$$

This is the most general form of the state equation of the control systems. It determines the map $v \to y[v]$ that is the dependence of the state function with respect to the control. This is called the *implicit operator* or the *implicit function* of the equation (7.1).

243

244 *Optimization and Differentiation*

We have the need to determine the differentiability of this operator, i.e., the map $y[\cdot] : V \to Y$. This result uses the partial derivative of operators. Consider a nonlinear operator A on a set W of the space $V \times Y$ with the space Z as the codomain. Let the point $v \in V$ be fixed. Consider the operator $B(v) = A(v, \cdot)$ that maps the set W^v of points $y \in Y$ such that $(v, y) \in W$ to the space Z, i.e., $B(v)y = A(v, y)$ for all $y \in Y$. Now fix the point $y \in Y$. Consider the operator $C(y) = A(\cdot, y)$ that maps the set W_y of points $v \in V$ such that $(v, y) \in W$ to the space Z, i.e., $C(y)v = A(v, y)$ for all $v \in V$. The derivative of the operator $B(v)$ at an interior point y of the set W^v is called the **partial derivative** with respect to y of the operator A at the point (v, y). The derivative of the operator $C(y)$ at an interior point v of the set W_y is called the partial derivative with respect to v of the operator A. We denote these partial derivatives by $A_y(v, y)$ and $A_v(v, y)$.

We shall use an important result of the nonlinear functional analysis, the **Implicit function theorem**, for proving the differentiability of the map $y[\cdot] : V \to Y$, i.e., implicit operator. Consider an operator A that is determined on a neighborhood O of the point (v_0, y_0) with the domain $V \times Y$ and the codomain Z. Suppose the equality $A(v_0, y_0) = 0$ and the existence of the partial derivatives of the operator A on the set O that are continuous at the point (v_0, y_0).

Theorem 7.1 *If the derivative $A_y(v_0, y_0)$ has a continuous inverse operator, then there exists a neighborhood O' of the point v_0 such that there exists an operator $y[\cdot] : O' \to Z$ (implicit operator) such that $A(v, y[v]) = 0$ for all $v \in O'$ that is Fréchet differentiable at the point v_0 besides*

$$y'(v_0) = -\left[A_y(v_0, y_0)\right]^{-1} A_v(v_0, y_0). \tag{7.2}$$

The general supposition of this theorem is the invertibility of the operator $A_y(v_0, y_0)$. It is necessary to prove that for all $z \in Z$ the equation

$$A_y(v_0, y_0)y = z \tag{7.3}$$

has a unique solution $y \in Y$. This is the **linearized equation**.

Consider the easiest applications of the Implicit function theorem.

Example 7.1 Implicit function. Determine $V = \mathbb{R}$, $Y = \mathbb{R}$, $Z = \mathbb{R}$. We have the function of two variables $f = f(v, y)$. Suppose $f(v_0, y_0) = 0$; the function f is continuously differentiable at the point (v_0, y_0), as $f_y(v_0, y_0) \neq 0$. Then there exists a neighborhood O' of the point v_0 that is the domain if the implicit function $y = y(v)$ such that $f(v, y(v)) = 0$ for all $v \in O'$. This implicit function is differentiable at the point v_0, as $y'(v_0) = -f_v(v_0, y_0)/f_y(v_0, y_0)$. Consider now the concrete example. Determine the function $f(v, y) = v^2 + y^2 - 1$ and the point $M_0 = (v_0, y_0) = (0, 1)$. The equation (7.1) is $v^2 + y^2 = 1$. This is the circle (see Figure 7.1). We have the equalities $f_v(0, 1) = 2v_0 = 0$, $f_y(0, 1) = 2y_0 = 2$. By Theorem 7.1 there exists an implicit function $y = y(v)$

on the neighborhood of the point v_0 (see Figure 7.1) that is differentiable at this point. Its derivative is equal to $y'(0) = -f_v(0,1)/f_y(0,1) = 0$. In reality, the implicit function is $y(v) = \sqrt{1-v^2}$. Its derivative at zero is equal to zero. Now consider the point $M_1 = (v_1, y_1) = (1, 0)$ that satisfies the equality $f(v_1, y_1) = 0$. The Implicit function theorem is not applicable now because of the equality $f_y(1,0) = 0$. Indeed, the implicit function is not determined at the neighborhood of the point v_1 (see Figure 7.1). Particularly, for all values v that are close enough to v_1 and less than v_1 there exist two points on the circle, i.e., two solutions of the equation $f(v, y) = 0$, as for all v that are close enough to v_1 and greater than v_1 this equation is not solvable.

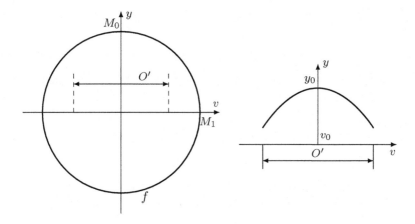

FIGURE 7.1: The implicit function exists at the neighborhood of the point M_0; this function does not exist at the neighborhood of the point M_1.

Remark 7.1 The implicit function is determined on the neighborhood O' of the point v_0 (see Figure 7.1). Thus, the Implicit function theorem gives the local result only. Note also that for all points $v \in O'$ there exist two values y such that $f(v,y) = 0$. However, one of them only includes the small neighborhood O of the point y_0. But for all v from each small neighborhood of the point M_1 we can have two situations. Maybe the equation has two solutions, but maybe this is an insolvable equation.

Remark 7.2 Note that it is not necessary to know the implicit function to determine its derivative at the concrete point. We had the analogical result for the Inverse function theorem.

Example 7.2 Inverse operator. Suppose the equality $Z = V$ and the operator A is determined by the formula $A(v, y) = By - v$, where B is an operator with domain Y and codomain V. Its partial derivative with respect to v is equal to E, where E is a unit operator. Consider a point $v_0 \in V$ and $y_0 = Bv_0$. Suppose the operator B is continuously differentiable at the point y_0, as its derivative $B'(y_0)$ is invertible. Using the equality

246 *Optimization and Differentiation*

$A_y(v_0, y_0) = B'(y_0)$, we prove the applicability of the Implicit function theorem. Then we can determine the implicit operator $y[\cdot] : V \to y$ on the neighborhood of the point v_0 that is differentiable at this point. We have the equality $A(v, y[v]) = By[v] - v = 0$. Therefore, $By[v] = v$, i.e., $y[\cdot]$ the inverse operator to B. By the formula (7.2) we have the equality $y'[v_0] = \left[B'(y_0)\right]^{-1}$ that is the corollary of the Inverse function theorem. Hence, this result is the partial case of the Implicit function theorem.

Remark 7.3 Thus, the previous results can be deduced from the results of this chapter.

7.2 Optimization control problems for abstract systems

Consider the system described by the equation (7.1). Suppose for all $v \in V$ this equation has a unique solution $y[v]$ from the space Y, i.e., there exists an implicit operator for the system (7.1) that is determined on the control space. Determine the state functional

$$I(v) = J(v) + K(y[v])$$

with functionals $J : V \to \mathbb{R}$ and $K : Y \to \mathbb{R}$. Let a convex subset U of the space V be the set of admissible controls. We have the following optimization control problem.

Problem 7.1 *Find a control $u \in U$ that minimizes the functional I on the set U for the system described by the equation (7.1).*

Remark 7.4 We could consider the general functional $I = I(v, y)$. However, this is not fundamental. By the way, we shall consider general functionals in Part IV.

Suppose Problem 7.1 has a solution u. Denote by y the solution of the equation (7.1) for the control u, i.e., $y = y[u]$. The necessary condition of the minimum of the differentiable functional I at the point u is the variational inequality

$$\langle I'(u), v - u \rangle \geq 0 \ \forall v \in U. \tag{7.4}$$

We could use this result for the given problem if the functionals J, K, and the dependence of the state function from the control, i.e., the map $y[\cdot] : V \to Y$, are differentiable.

Using the Implicit function theorem, we prove the following result.

Lemma 7.1 *Suppose the operator A has the partial derivatives at the point (u, y) such that $A(u, y) = 0$ are continuous at this point, as the derivative*

Systems with nonlinear control

$A_y(u, y)$ *has a continuous inverse operator. Then the map* $y[\cdot] : V \to Y$ *is Fréchet differentiable at the point* u, *as the following equality holds:*

$$\langle \mu, y'[u]h \rangle = -\langle [A_v(u, y)]^* p_\mu[u], h \rangle \quad \forall h \in V, \ \mu \in Z', \tag{7.5}$$

where $p_\mu[u]$ *is the solution of the equation*

$$[A_y(u, y)]^* p_\mu[u] = \mu. \tag{7.6}$$

Proof. By the Implicit function theorem, the derivative satisfies the formula

$$y'[u] = -[A_y(u, y)]^{-1} A_v(u, y).$$

Then we have

$$\langle \mu, y'[u]h \rangle = -\langle \mu, [A_y(u, y)]^{-1} A_v(u, y)h \rangle \quad \forall h \in V, \ \mu \in Z'.$$

Using the definition of the adjoint operator, we get

$$\langle \mu, [A_y(u, y)]^{-1} A_v(u, y)h \rangle = \langle \{ [A_y(u, y)]^{-1} \}^* \mu, A_v(u, y)h \rangle =$$

$$\langle \{ [A_y(u, y)]^* \}^{-1} \mu, A_v(u, y)h \rangle = \langle [A_v(u, y)]^* \{ [A_y(u, y)]^* \}^{-1} \mu, h \rangle.$$

Therefore, we obtain the equality

$$\langle \mu, y'[u]h \rangle = -\langle [A_v(u, y)]^* \{ [A_y(u, y)]^* \}^{-1} \mu, h \rangle.$$

The adjoint operator $[A_y(u, y)]^*$ is invertible because of the invertibility of the operator $A_y(u, y)$. Then for all $\mu \in Y'$ the equation (7.6) has the unique solution

$$p_\mu[u] = \{ [A_y(u, y)]^* \}^{-1} \mu$$

from the space Z'. Finally, the formula (7.5) follows from the previous equality. $\square$

Now find the derivative of the minimized functional.

Lemma 7.2 *Under the assumptions of Lemma 7.1 suppose the functional J is Gâteaux differentiable at the point* u, *and the functional K is Fréchet differentiable at the point* y. *Then the Gâteaux derivative of the functional I at the point* u *is determined by the formula*

$$I'(u) = J'(u) + [A_v(u, y)]^* p, \tag{7.7}$$

where p *is the solution of the equation*

$$[A_y(u, y)]^* p = -K'(y). \tag{7.8}$$

248 *Optimization and Differentiation*

Proof. Find the Gâteaux derivative of the minimized functional from the equality

$$\langle I'(u), h \rangle = \langle J'(u), h \rangle + \langle K'(y), y'[u]h \rangle \quad \forall h \in V.$$

Choose $\mu = -K'(y)$ at the equality (7.6). We have the equation (7.8). Then transform the previous formula

$$\langle I'(u), h \rangle = \langle J'(u), h \rangle + \langle [A_v(u, y)]^* p, h \rangle \quad \forall h \in V.$$

Therefore, the formula (7.7) is true. $\square$

Remark 7.5 Lemma 7.1 and Lemma 7.2 are the extensions of Lemma 4.1 and Lemma 4.3 for the state equation (7.1). Indeed, the value $A(v, y)$ is equal to $Ay - Bv - f$ in Chapter 4. Hence, it has the partial derivatives $A_y(u, y) = A'(y)$, $A_v(u, y) = -B$. Therefore, the equalities (7.5)–(7.8) can be transformed to (4.7), (4.8), (4.10), and (4.11).

Put the functional derivative in to the inequality (7.4). We obtain the following result.

Theorem 7.2 *Under the assumptions of Lemma 7.2 the solution to Problem 7.1 satisfies the variational inequality*

$$\langle J'(u) + [A_v(u, y)]^* p, v - u \rangle \geq 0 \quad \forall v \in U. \tag{7.9}$$

Remark 7.6 Theorem 7.2 is the extension of Theorem 4.2 to the state equation (7.1). Particularly, the formula (4.12) follows from the variational inequality (7.9).

Apply Theorem 7.2 for the analysis of an optimization control problem with nonlinear control.

7.3   Weakly nonlinear control systems

Let Ω be an open bounded set of the Euclid space $\mathbb{R}^n$ with the boundary Γ. Consider the equation

$$-\Delta y(x) + |y(x)|^\rho y(x) + b(x, v(x))y(x) = f(x, v(x)), \ x \in \Omega \tag{7.10}$$

with the boundary condition

$$y(x) = 0, \ x \in \Gamma, \tag{7.11}$$

where v is a control, and y is the state function, $\rho > 0$. The functions b and f are known here. We have the system with nonlinear control.

Systems with nonlinear control

Remark 7.7 We try to consider easy enough non-trivial problems. Therefore, we could analyze a system with the unique term that depends upon the control. Suppose the function f only depends upon the control. Then we determine a new control by the formula $w(x) = f(x, v(x))$. Thus, we obtain the system with the control w as the absolute term of the state equation. It will be the subject of Chapter 4 and Chapter 5. If the function b only depends upon the control, then we determine a new control by the formula $w(x) = b(x, v(x))$. Hence, we get the system with the control at the coefficient of the state equation. This is the subject of Chapter 6. Therefore, we consider now the case where both functions b and f depend upon control.

It is necessary to have some suppositions about the functions b and f. We denote by $f(x, \cdot)$ the map $v \to f(x, v)$ with the fixed value x and by $f(\cdot, v)$ the map $x \to f(x, v)$ with the fixed value v. The map $f : \Omega \times \mathbb{R} \to \mathbb{R}$ is called the Caratheodory function, if the function $f(x, \cdot)$ is continuous for a.e. value $x \in \Omega$, and the function $f(\cdot, v)$ is measurable for all $v \in \mathbb{R}$ on the set Ω.

Consider again the functional spaces $V = L_2(\Omega)$, $Y = H_0^1(\Omega) \cap L_q(\Omega)$, $Z = H^{-1}(\Omega) + L_{q'}(\Omega)$, where $q = \rho + 2$, $1/q + 1/q' = 1$. By Sobolev theorem we have continuous embedding $H_0^1(\Omega) \subset L_r(\Omega)$ and $L_{r'}(\Omega) \subset H^{-1}(\Omega)$, where $r \leq 2n/(n - 2)$ for $n \geq 3$ and r is arbitrary for $n = 2$.

Determine Nemytsky operators B and F on the control space V by the equalities

$$Bv(x) = b(x, v(x)), \ Fv(x) = f(x, v(x)).$$

Determine the operator A on the product $V \times Y$ by the formula

$$A(v, y) = -\Delta y + |y|^\rho y + Bvy - Fv \ \forall v \in V, y \in Y.$$

Lemma 7.3 *Suppose b and f are Caratheodory functions, as for all $v \in \mathbb{R}$ and a.e. on Ω the following inequalities hold:*

$$0 < c_1 \leq b(x, v) \leq b_0(x) + c_2|v|^{2/s}, \tag{7.12}$$

$$|f(x, v)| \leq f_0(x) + c_3|v|^{2/r'}, \tag{7.13}$$

where $c_2 > 0$, $c_3 > 0$, $b_0 \in L_s(\Omega)$, $f_0 \in L_{r'}(\Omega)$ and the parameter s satisfies the equality $1/q + 1/s = 1/r'$. Then the operator $A : V \times Y \to Z$ is continuous.

Proof. We considered before the first and the second terms of the operator A. Particularly, the Laplace operator is the continuous map. Therefore, the operator $\Delta : Y \to Z$ is continuous too. The power operator of the equation (7.10) is the continuous map from the space $L_q(\Omega)$ to $L_{q'}(\Omega)$ and from Y to Z too. The condition (7.13) and Krasnoselsky theorem guarantee the continuity of the operator $F : L_2(\Omega) \to L_{r'}(\Omega)$. Hence, the operator $F : V \to Z$ is continuous. Then the operator $B : L_2(\Omega) \to L_s(\Omega)$ is continuous because of the inequality (7.12). Therefore, $(v, y) \to Bvy$ is the continuous map from the space $V \times Y$ to $L_{r'}(\Omega)$ and to Z too by Hölder inequality. Then for all $v \in V$, $y \in Y$ each term of the equality (7.10) is an element of the space Z, as the operator $A : V \times Y \to Z$ is continuous. $\square$

250 *Optimization and Differentiation*

Theorem 7.3 *Under the assumptions of Lemma 7.3 for all function $v \in V$ the problem* (7.10), (7.11) *has a unique solution $y = y[v]$ from the space Y.*

Indeed, for the fixed control the equation (7.10) can be transformed to (6.19). Therefore, Theorem 7.3 is the corollary of Theorem 6.6. The suppositions of this assertion are true because of the given assumptions.

Remark 7.8 By the inequality (7.12) the function b does not have any negative values. Therefore, the solvability of the boundary problem is true on the whole control space. Hence, we can use Gâteaux derivatives and extended derivatives in place of derivatives with respect to convex sets.

Consider a non-empty convex closed subset U of the space V and the functional

$$I(v) = \frac{\alpha}{2} \int_\Omega v^2 dx + \frac{1}{2} \int_\Omega \sum_{i=1}^n \left(\frac{\partial y[v]}{\partial x_i} - \frac{\partial y_d}{\partial x_i} \right)^2 dx,$$

where $y[v]$ is the solution of the problem (7.10), (7.11) for the control v, $\alpha > 0$, and y_d is a known function of the space $H_0^1(\Omega)$.

Problem 7.2 *Find the control $u \in U$ that minimizes the functional I on the set U.*

Suppose there exists a solution u of Problem 7.2 and $y = y[u]$.

Remark 7.9 We have some difficulties in proving the solvability of Problem 7.2 because of the substantiation of the weak continuity of control–state mapping. Suppose we have the convergence $v_k \to v$ weakly V. Using the technique of Lemma 4.7 and Lemma 6.8, we can determine the convergence $y[v_k] \to y$ weakly in Y. However, we have the need to prove the convergence $Bv_k y_k \to Bvy$ and $Fv_k \to Fv$ for obtaining the equality $y = y[v]$. We cannot obtain these properties. Therefore, we use the supposition of the existence of the optimal control.

Remark 7.10 We could prove the solvability of the optimization problem if we choose the control from the Sobolev space. Then we use the Rellich–Kondrashov theorem. Using the weak convergence of the controls in this space, we get its strong convergence in the space of the integrable functions. Then we use the strong continuity of the Nemytsky operator.

Find the partial derivatives of the operator A by using the Implicit function theorem.

Lemma 7.4 *Under the assumptions of Lemma 7.3 suppose the functions $b(x, \cdot)$ and $f(x, \cdot)$ a.e. on Ω have the derivatives that are Caratheodory functions and satisfy the inequalities*

$$|b_v(x, v)| \le b_1(x) + d_1 |v|^{2/(s-1)}, \quad |f_v(x, v)| \le f_1(x) + d_2 |v|^{2/(r'-1)} \quad (7.14)$$

for all $v \in \mathbb{R}$ and a.e. on Ω, where $d_1 > 0$, $d_2 > 0$, $b_1 \in L_{2s/(2-s)}(\Omega)$, $f_1 \in L_{2r'/(2-r')}(\Omega)$, $r = 1 + 2/\rho$, $s = 2q/\rho$. Then the operator A has the partial derivatives at the arbitrary point (v_0, y_0) such that

$$A_y(v_0, y_0)y = -\Delta y + (\rho + 1)|y_0|^\rho + b(\cdot, v_0)y \quad \forall y \in Y,$$

$$A_v(v_0, y_0)v = b_v(\cdot, v_0)y_0 v - f_v(\cdot, v_0)v \quad \forall v \in V.$$

Proof. Consider Nemytsky operators B and F on the space V that are determined by the equalities

$$Bv(x) = b(x, v(x)), \ Fv(x) = f(x, v(x)) \ \forall v \in V.$$

By Krasnoselsky theorem these operators are Fréchet differentiable with domain $L_2(\Omega)$ codomains and $L_s(\Omega)$ and $L_r(\Omega)$, correspondingly. Its derivatives at a point v_0 can be determined by the formulas

$$B'(v_0)v(x) = b_v(x, v_0(x))v(x), \ F'(v_0)v(x) = f_v(x, v_0(x))v(x) \ \forall v \in V.$$

By Hölder inequality from $Bv_0 \in L_r(\Omega)$ that is a corollary of (7.12) and $y \in L_q(\Omega)$ it follows that its product belongs to the space $L_p(\Omega)$, where $1/p = 1/r + 1/q$. Using the values of r and q, we have $1/p = \rho/(\rho+2) + 1/(\rho+2)$, then $p = q'$. Therefore, we get $Bv_0 y \in L_{q'}(\Omega)$. Hence, the partial derivative $A_y(v_0, y_0)$ of the operator A at the arbitrary point is in reality an element of the space Z.

Analogically, for all $v \in L_2(\Omega)$ we have $F'(v_0)v \in L_{q'}(\Omega)$ because of (7.13). Finally, from the first condition (7.12) it follows that $F'(v_0)v \in L_r(\Omega)$. By Hölder inequality we have the inclusion $B'(v_0)vy_0 \in L_{q'}(\Omega)$. Therefore the codomain of the derivative $A_v(v_0, y_0)$ is the space Z. $\square$

The general assumption of the Implicit function theorem is the property of the linearized equation (7.3). Now this is the homogeneous Dirichlet problem for the equation

$$-\Delta y + (\rho + 1)|y_0|^\rho y + b(x, v_0)y = z. \tag{7.15}$$

After multiplication by y and integration by using Green's formula we have the equality

$$\int\limits_\Omega \sum\limits_{i=1}^n \left(\frac{\partial y}{\partial x_i}\right)^2 dx + (\rho + 1) \int\limits_\Omega |y_0|^\rho y^2 dx + \int\limits_\Omega b(x, v_0)y^2 dx = \int\limits_\Omega yz dx$$

that is the analog of (4.26). We cannot obtain here the a priori estimate of the solution of the equation (7.15) in the space $L_{q'}(\Omega)$. Therefore, we have the known difficulty (see Chapter 4).

Remark 7.11 It is clear because the equation (4.19) is the partial case of (7.10).

Consider the following supposition

$$n = 2 \ \text{ or } \ \rho \le 4/(n-2) \ \text{for } n = 3 \tag{7.16}$$

that is equal to (4.21). By the Sobolev embedding theorem, we have $H_0^1(\Omega) \subset L_q(\Omega)$; therefore, $Y = H_0^1(\Omega)$. Then we have embedding of the adjoint spaces $L_{q'}(\Omega) \subset H^{-1}(\Omega)$ and the equality $Z = H^{-1}(\Omega)$. We prove the following result that is an analog of Lemma 4.9.

252 *Optimization and Differentiation*

Lemma 7.5 *Under the assumptions of Lemma 7.4 and (7.16) the operator $A_y(v_0, y_0)$ for problem (7.10), (7.11) is invertible.*

Remark 7.12 It is necessary to prove that for all $z \in Z$ the equation (7.15) has a unique solution $y \in Y$. We can obtain this result by using a priori estimates of the solution of this equation. Under the condition (7.16), the previous integral equality guarantees this property.

Thus, the Implicit function theorem is applicable for our system if the supposition (7.16) is true.

Lemma 7.6 *Under the assumptions of Lemma 7.5 the map $y[\cdot] : V \to Y$ for the boundary problem (7.10), (7.11) is Fréchet differentiable at the arbitrary point $u \in V$, as*

$$\int_\Omega \mu y'[u] h dx = \int_\Omega \{f_v(x, u) - b_v(x, u)y[u]\}p_\mu[u] dx \ \forall \mu \in Y', h \in V, \quad (7.17)$$

where $p_\mu[u]$ is the solution of the homogeneous Dirichlet problem for the equation

$$-\Delta p_\mu[u] + (\rho + 1)|y[u]|^\rho p_\mu[u] + b(x, u)p_\mu[u] = \mu. \quad (7.18)$$

Proof. By Lemma 7.5 the considered map is differentiable because of the Implicit function theorem. as its derivative satisfies the equality (7.5). Prove that the equalities (7.5), (7.6) can be transformed to (7.17), (7.18). Find the adjoint operator of the partial derivatives of the operator A. We have

$$\left\langle \left[A_y(u, y)\right]^* p, h \right\rangle = \left\langle p, A_y(u, y)h \right\rangle = \int_\Omega \left[-\Delta h + (\rho+1)|y|^\rho h + b(x, u)h\right] p dx =$$

$$\int_\Omega \left[-\Delta p + (\rho + 1)|y|^\rho p + b(x, u)p\right] h dx \ \forall h \in Y, \ p \in Z^*.$$

Therefore, we have the equality

$$\left[A_y(u, y)\right]^* p = -\Delta p + (\rho + 1)|y|^\rho p + b(x, u) \ \forall p \in Z^*.$$

Then the equation (7.6) is transformed to (7.18).

Analogically, we get

$$\langle p, A_v(u, y)v \rangle = \left\langle \left[A_v(u, y)\right]^* p, v \right\rangle =$$

$$\int_\Omega \{b_v(x, u)y[u] - f_v(x, u)\}vp dx \ \forall v \in V, \ p \in Z^*.$$

Then we find

$$\left[A_v(u, y)\right]^* p = \{b_v(x, u)y[u] - f_v(x, u)\}p \ \forall p \in Z^*.$$

Now the equality (7.5) is transformed (7.17). $\square$

Using Lemma 7.2, we find the derivative of the minimized functional.

Systems with nonlinear control

Lemma 7.7 *Under the assumption of Lemma 7.5 the derivative of the functional I for Problem 7.2 at the point u is*

$$I'(u) = \alpha u + \{b_v(x, u)y[u] - f_v(x, u)\}p, \tag{7.19}$$

where $y = y[u]$ and p is the solution of the equation

$$-\Delta p(x) + (\rho + 1)|y(x)|^\rho p(x) + b(x, u(x))p = \Delta y(x) - \Delta y_d(x), \ x \in \Omega \tag{7.20}$$

with the boundary condition

$$p(x) = 0, \ x \in \Gamma. \tag{7.21}$$

Indeed, put the adjoint operators in to the partial derivatives of the operator A in to the equality (7.7) and (7.7), and use Lemma 7.2.

Theorem 7.4 *Under the assumption of Lemma 7.5 suppose the existence of the solution of Problem 7.2. Then the following variational inequality holds:*

$$\int_\Omega \Big\{\alpha u + [b_v(x, u)y[u] - f_v(x, u)]p\Big\}(v - u)dx \geq 0 \ \forall v \in U. \tag{7.22}$$

Indeed, it is sufficient to apply Theorem 7.3, where the inequality (7.9) is transformed to (7.22).

Thus, we obtain the system that includes the boundary problem (7.10), (7.11) for $v = u$, the adjoint system (7.20), (7.21), and the variational inequality (7.22) as the necessary condition of optimality.

Remark 7.13 Of course, we cannot find directly the control for the condition (7.22) for the general case of the functions b and f. By the way, the easiest problem of minimization for the function of one variable does not have an analytic solution for the general case too.

Remark 7.14 If $b(x, v) = 0$, $f(x, v) = v$, then the variational inequality (7.22) can be transformed to (4.35). If $b(x, v) = v$, $f(x, v) = f(x)$, then we have the optimization problem of Chapter 6. Therefore, Problem 7.2 is the extension of all examples, which were considered before.

Remark 7.15 Suppose the function b is positive on the set of admissible control only. Then we change the Gâteaux derivative by the derivative with respect to the convex set (see Chapter 6).

Now we would like to consider Problem 7.2 for the strongly nonlinear case, where the condition (7.16) can be false. The state equation (7.10) is the extension of (5.1) and (6.28). By Lemma 5.2 and Lemma 6.16, the solution of the considered boundary problem can be non-differentiable with respect to the control. Therefore, we try to apply an extended derivative in place of the Gâteaux derivative. We obtain at first the result for abstract systems.

7.4 Extended differentiability of the implicit function

Determine sufficient conditions of the extended differentiability of the implicit operator for the equation (7.1). This is an extension of Theorem 5.6 about the extended differentiability of the inverse operator.

Consider Banach space V, Y, Z and an operator $A : V \times Y \to Z$. Let points $v_0 \in V$, $y_0 \in Y$ satisfy the equality $A(v_0, y_0) = 0$. Consider Banach space V_* that is a subspace V with a neighborhood O_* of zero. Then the set $O = v_0 + O_*$ is the neighborhood of the point v_0. Consider the following supposition.

Assumption 7.1 *For all $v \in O$ there exists a unique value $y[v] \in Y$ such that $A\big(v, y[v]\big) = 0$.*

Remark 7.16 Assumption 7.1 guarantees the existence of the implicit operator on the neighborhood of the given point. This is an analog of Assumption 5.1 of Theorem 5.6.

Denote the map $v \to y[v]$, i.e., implicit operator, by L. From Assumption 7.1 it follows that the equality

$$A(v_\sigma, Lv_\sigma) = 0, \ A(v_0, Lv_0) = 0,$$

where $v_\sigma = v_0 + \sigma h$, $h \in V_*$ and the number σ is small enough such that $v_\sigma \in O$. Then we have the equality

$$A(v_\sigma, Lv_\sigma) - A(v_\sigma, Lv_0) = A(v_0, y_0) - A(v_\sigma, y_0). \qquad (7.23)$$

If for all $v \in O$ the operator $A(v, \cdot)$ is Gâteaux differentiable, then we obtain the equality

$$\big\langle \lambda, A(v_\sigma, y) - A(v_\sigma, y_0) \big\rangle = \big\langle \lambda, A_y \big[v_\sigma, y_0 + \delta(\lambda)(y - y_0)\big](y - y_0) \big\rangle \ \forall y \in Y, \lambda \in Z',$$

because of the Lagrange formula, where $\delta(\lambda) \in (0, 1)$.

For all $v \in O$, $\lambda \in Z'$ determine the linear continuous operator $G(v, \lambda) : Y \to Z$ by the formula

$$G(v, \lambda)y = A_y \big[v_\sigma, y_0 + \delta(\lambda)(Lv - Lv_0)\big]y \ \forall y \in Y.$$

Note the equality $G(v_0, \lambda) = A_y(v_0, y_0)$ for all $\lambda \in Z'$. From (7.23) we have the equality

$$\big\langle \lambda, G(v_\sigma, \lambda)(Lv_\sigma - Lv_0) \big\rangle = \big\langle \lambda, A(v_0, y_0) - A(v_\sigma, y_0) \big\rangle \ \forall \lambda \in Z', h \in V_* \quad (7.24)$$

that is an analog of (5.20).

For all $v \in O$ consider Banach spaces $Z(v)$, $Y(v)$, V_0, Y_*, and Z_* such that the following embeddings,

$$V_* \subset V_0 \subset V, \ Y \subset Y(v) \subset Y_*, \ Z_* \subset Z(v) \subset Z,$$

are continuous and dense.

Systems with nonlinear control

Assumption 7.2 *The operator $A(v, \cdot)$ is Gâteaux differentiable, as for any $\lambda \in Z'$ there exists a continuous continuation $\overline{G}(v, \lambda)$ of the operator $G(v, \lambda)$ on the set $Y(v)$ such that its codomain is a subset $Z(v)$.*

Assumption 7.3 $A_v(v, y_0)$ *belongs to* $A_v(v_0, y_0) + Z_*$ *for all* $v \in O$.

We have the conditions $Lv \in Lv_0 + Y(v)$ and $Z' \subset Z(v)'$ for all $v \in O$. From equality (7.24) it follows that

$$\left\langle [\overline{G}(v_\sigma, \lambda)]^* \lambda, Lv_\sigma - Lv_0 \right\rangle = \left\langle \lambda, A(v_0, y_0) - A(v_\sigma, y_0) \right\rangle \forall \lambda \in Z(v)', h \in V_*. \tag{7.25}$$

This is an analog of (5.21).

Determine the equation

$$\left[\overline{G}(v, p_\mu[v]) \right]^* p_\mu[v] = \mu. \tag{7.26}$$

This is the linear equation

$$\left[\overline{A}_y(v_0, y_0) \right]^* p_\mu = \mu \tag{7.27}$$

for $v = v_0$, where $\overline{A}_y(v_0, y_0)$ is a prolongation of the operator $A_y(v_0, y_0)$ on the set $Y(v_0)$. The equations (7.26) and (7.27) are extensions of the equations (5.22]) and (5.23).

Assumption 7.4 *For all* $v \in O$, $\mu \in Y(v)'$ *the equation (7.26) has a unique solution* $p_\mu[v] \in Z(v)'$.

Return to the equality (7.25). Determine $\lambda = p_\mu[v_\sigma]$ for small enough σ. We get

$$\left\langle \mu, [L(v_0 + \sigma h) - Lv_0]/\sigma \right\rangle =$$

$$\left\langle p_\mu[v_\sigma], A(v_0, y_0) - A(v_\sigma, y_0) \right\rangle \forall \mu \in Y(v_\sigma)', h \in V_*. \tag{7.28}$$

This is an analog of the equality (5.24).

Assumption 7.5 *Operator* $A(\cdot, y_0)$ *is* $(V_0, Z(v_0); V_*, Z_*)$*-extended differentiable at the point* v_0.

Assumption 7.6 *We have the convergence* $p_\mu[v_0 + \sigma h] \to p_\mu$ *-weakly in* Z'_* *as* $\sigma \to 0$ *for all* $\mu \in Y'_*$, $h \in V_*$, *where* p_μ *is a solution of the equation (7.27).*

Let Y^w_* be the space Y_* with weak topology, and $A_v(v_0, y_0)$ be the extended derivative of the operator $A(\cdot, y_0)$ at the point v_0. Now prove the following assertion.

Theorem 7.5 *Under the assertions 7.1–7.6, the operator* $y[\cdot] : V \to Y$ *has* $(V_0, Y(v_0); V_*, Y^w_*)$*-extended Gâteaux derivative D at the point v_0 that satisfies the equality*

$$\langle \mu, Dh \rangle = \langle p_\mu, A_v(v_0, y_0)h \rangle \ \forall \mu \in Y(v_0)', h \in V_*. \tag{7.29}$$

256   *Optimization and Differentiation*

Proof. From the equalities (7.28), (7.29) it follows that

$$\left\langle \mu, \left[L(v_0 + \sigma h) - Lv_0\right]/\sigma - Dh \right\rangle = \left\langle p_\mu[v_\sigma], \left[A(v_0, y_0) - A(v_\sigma, y_0)\right]/\sigma \right\rangle -$$

$$\left\langle p_\mu, A_v(v_0, y_0)h \right\rangle \quad \forall \mu \in Y_*, h \in V_*. \tag{7.30}$$

By Assumption 7.6 we have the convergence $p_\mu[v_\sigma] \to p_\mu$ *-weakly in Z'_* as $\sigma \to 0$. Therefore, we get

$$\left\langle p_\mu[v_\sigma] - p_\mu, A_v(v_0, y_0)h \right\rangle \to 0.$$

By Assumption 7.5 we obtain

$$\left[A(v_0, y_0) - A(v_\sigma, y_0)\right]/\sigma \to A_v(v_0, y_0)h$$

in Z_*. Pass to the limit at the equality (7.30). We have $\left[L(v_0 + \sigma h) - Lv_0\right]/\sigma \to Dh$ weakly in Y_* for all $h \in V_*$. Therefore, D is in reality the extended derivative of the operator $y[\cdot] : V \to Y$ at the point v_0. $\square$

Prove that Theorem 5.6 about the extended differentiability of the inverse operator follows from this result.

Lemma 7.8 *Under the suppositions of Theorem 7.5 the assumptions of Theorem 5.6 are true.*

Proof. For Theorem 7.5 we have the equality $V = Z$, and the value $A(v, y)$ is equal to $Ay - v$. Assumption 7.1 is equal to Assumption 5.1 here. Then Assumption 7.2 is equal to Assumption 5.2. The equation (7.26) is transformed to (5.22). Therefore, Assumption 7.4 and Assumption 7.6 can be transformed to Assumption 5.3 and Assumption 5.4. $\square$

Consider an application of Theorem 7.5.

7.5   Strongly nonlinear control systems

Consider Problem 7.1 for the case of the extended differentiability of only the implicit operator. We would like to find a control u from the convex subset U of the space V that minimizes there the functional I that is determined by the equality

$$I(v) = J(v) + K\big(y[v]\big),$$

where the functional $J : V \to \mathbb{R}$ and $K : Y \to \mathbb{R}$ are Gâteaux and Fréchet differentiable, correspondingly, and the function $y[v]$ is the solution of the equation (7.1)

$$A\big(v, y[v]\big) = 0.$$

Suppose this problem has a solution u, and $y = y[v]$.

The following assertion is obvious.

Lemma 7.9 *Under the assumptions of Theorem 7.5 the functional I has $(V_0, \mathbb{R}; V_*, \mathbb{R})$-extended derivative at the point u*

$$I'(u) = J'(u) - [A_v(u,y)]^* p,$$

where p is the solution equation

$$[A_y(u,y)]^* p = -K'(y). \tag{7.31}$$

Then we obtain the following result.

Theorem 7.6 *Under the assumptions of Theorem 7.5 the solution u of Problem 7.1 satisfies the variational inequality*

$$\langle J'(u) - [A_v(u,y)]^* p, v - u \rangle \ \forall v \in U. \tag{7.32}$$

Use this result for the analysis of Problem 7.2 without the condition (7.16) that guarantees the applicability of the Implicit function theorem for the equation (7.10). Determine the spaces $Y_* = H_0^1(\Omega)$, $Z_* = H^{-1}(\Omega)$.

Lemma 7.10 *For all $h \in V$ we have the convergence $y[v_0 + \sigma n] \to y[v_0]$ in Y_* for the solution of the equation (7.10) as $\sigma \to 0$.*

Proof. Denote $v_\sigma = v_0 + \sigma h$, $y_0 = y[v_0]$, $h_\sigma = y[v_\sigma] - y_0$. We have the equality

$$-\Delta h_\sigma + g(v_\sigma)^2 h_\sigma + B v_\sigma h_\sigma = (F v_\sigma - F v_0) - (B v_\sigma - B v_0) y_0,$$

where

$$g(v)^2 = (\rho + 1)|y + \varepsilon(y[v] - y)|^\rho, \ \varepsilon \in [0, 1].$$

After multiplication by the function h_σ and integration we get

$$\int_\Omega \sum_{i=1}^n \left(\frac{\partial h_\sigma}{\partial x_i} \right)^2 + \int_\Omega [g(v_\sigma) h_\sigma]^2 dx +$$

$$\int_\Omega B v_\sigma (h_\sigma)^2 dx = \int_\Omega [(F v_\sigma - F v_0) - (B v_\sigma - B v_0) y_0] h_\sigma dx.$$

By Lemma 7.3, the operator F maps the space $V = L_2(\Omega)$ to $L_{r'}(\Omega)$ and to the space Z_* that is adjoint to Y_*. This operator is continuous as the codomain of the map $(v, y) \to B v y$ is the space $L_{r'}(\Omega)$. The map $v \to B v y_0$ is the continuous operator from V to Z_*. Thus, we have the convergence $F v_\sigma \to F v_0$ in Z_* and $B v_\sigma y_0 \to B v_0 y_0$ in Z_*. From the previous integral equality the estimate follows:

$$\|h_\sigma\| \leq \|F v_\sigma - F v_0\|_* + \|(B v_\sigma - B v_0) y_0\|_*.$$

This completes the proof of Lemma 7.10. $\square$

258 *Optimization and Differentiation*

Remark 7.17 This result is an extension of Lemma 5.11.

Lemma 7.11 *Under the assumptions of Lemma 7.4 the suppositions of Theorem 7.5 are true for the equation (7.10).*

Proof. 1. Consider the operator A as before. The Assumption 7.1, i.e., the existence of the implicit function, follows from Theorem 7.2.

2. The operator $A(v, \cdot)$ has a Gâteaux derivative that satisfies the equality

$$A_y(v, y)h = -\Delta h + (\rho + 1)|y|^\rho h + Bvh \quad \forall h \in V.$$

For all $v \in O$, $\lambda \in Z'$ the operator $G(v, \lambda) \in L(Y, Z)$ can be determined by the formula

$$G(v, \lambda)h = -\Delta h + g(v)^2 h + Bvh \quad \forall h \in V.$$

Determine the spaces

$$Y(v) = H_0^1(\Omega) \cap \{\varphi \mid g(v)\varphi \in L_2(\Omega), \sqrt{Bv}\varphi \in L_2(\Omega)\}, \ Z(v) = Y(v)'.$$

The continuation of the operator $G(v, \lambda)$ on the set $Y(v)$ satisfies the equality

$$\overline{G}(v, \lambda)h = -\Delta h + g(v)^2 h + Bvh \quad \forall h \in Y(v).$$

Therefore, Assumption 7.2 is true.

3. Consider the equality

$$A(v, y_0) - A(v_0, y_0) = (Bv_0 - Bv)y_0 - (Fv_0 - Fv).$$

Using the properties of the operators F and B (see the previous lemma), we determine the inclusion $A(v, y_0) \in A(v_0, y_0) + Z_*$ for all control $v \in V$ that guarantees Assumption 7.3.

4. The operator $A(\cdot, y_0)$ has a Gâteaux derivative that satisfies the equality

$$A_v(v_0, y_0)h = B'(v_0)y_0 h - F'(v_0)h \quad \forall h \in V.$$

Then Assumption 7.5 is true.

5. We have the equation (7.26)

$$-\Delta p_\mu[v] + g(v)^2 p_\mu[v] + Bv p_\mu[v] = \mu. \tag{7.33}$$

Particularly, for $v = v_0$ we get the equation (7.27)

$$-\Delta p_\mu[v_0] + (\rho + 1)|y_0|^\rho p_\mu[v_0] + Bv_0 p_\mu[v_0] = \mu. \tag{7.34}$$

Multiply the equality (7.33) by the function $p_\mu[v]$. After integration we obtain the equality

$$\int_\Omega \sum_{i=1}^n \left(\frac{\partial p_\mu[v]}{\partial x_i}\right)^2 dx + (\rho) \int_\Omega \left[g(v)p_\mu[v]\right]^2 dx +$$

$$\int_\Omega Bv \big(p_\mu[v]\big)^2 dx = \int_\Omega \mu p_\mu[v] dx. \qquad (7.35)$$

Then we have the inequality

$$\|p_\mu[v]\|_{Z(v)'} \leq \|\mu\|_{Y(v)'}.$$

Using the standard theory of the linear elliptic equations, we obtain that for all functions $\mu \in Y(v)'$ the equation (7.33) has the unique solution $p_\mu[v] \in Z(v)'$. Therefore, Assumption 7.4 is true.

6. Determine the space $V_* = V$. Suppose $\sigma \to 0$. By Lemma 7.11 for all $h \in V$ we have the convergence $y[v_\sigma] \to y[v_0]$ in Y_*, where $v_\sigma = v + \sigma h$. Using (7.35), we obtain the inequality

$$\big\|p_\mu[v_\sigma]\big\|_{Z'_*}^2 + \big\|g(v_\sigma)p_\mu[v_\sigma]\big\|_2^2 + \big\|\sqrt{Bv_\sigma}p_\mu[v_\sigma]\big\|_2^2 \leq \big\|p_\mu[v_\sigma]\big\|_{Z'_*} \|\mu\|_{Y'_*}.$$

Then we have

$$\big\|p_\mu[v_\sigma]\big\|_{Z'_*} \leq \|\mu\|_{Y'_*}, \ \big\|g(v_\sigma)p_\mu[v_\sigma]\big\|_2 \leq \|\mu\|_{Y'_*}, \ \big\|\sqrt{Bv_\sigma}p_\mu[v_\sigma]\big\|_2 \leq \|\mu\|_{Y'_*}.$$

From Hölder inequality it follows that

$$\big\|g(v_\sigma)^2 p_\mu[v_\sigma]\big\|_2 \leq \big\|g(v_\sigma)\big\|_{2q/\rho} \big\|g(v_\sigma)p_\mu[v_\sigma]\big\|_2.$$

By the boundedness of the set $\{y[v_\sigma]\}$ in the space $L_q(\Omega)$ we obtain the boundedness of $\{g(v_\sigma)\}$ in $L_{2q/\rho}(\Omega)$. Then the set $\{g(v_\sigma)^2 p_\mu[v_\sigma]\}$ is bounded in the space $L_{q'}(\Omega)$. We have also the inequality

$$\big\|Bv_\sigma p_\mu[v_\sigma]\big\|_{r_1} \leq \big\|\sqrt{Bv_\sigma}\big\|_{2s} \big\|\sqrt{Bv_\sigma}p_\mu[v_\sigma]\big\|_2,$$

where $1/r_1 = 1/2s + 1/2$. Using the boundedness of $\{Bv_\sigma\}$ in the space $L_s(\Omega)$ we determine the boundedness of $\{\sqrt{Bv_\sigma}\}$ is $L_{2s}(\Omega)$. Using the last inequalities, we have the boundedness of $\{Bv_\sigma p_\mu[v_\sigma]\}$ in the space $L_{r_1}(\Omega)$.

After extracting of the subsequences we obtain the convergence $p_\mu[v_\sigma] \to p_\mu$ weakly in $H_0^1(\Omega)$, $g(v_\sigma)^2 p_\mu[v_\sigma] \to q_\mu$ weakly in $L_{q'}(\Omega)$, and $Bv_\sigma p_\mu[v_\sigma] \to r_\mu$ weakly in $L_{r_1}(\Omega)$. Using the Rellich–Kondrashov theorem, we get $p_\mu[v_\sigma] \to p_\mu$ strongly in $L_2(\Omega)$ and a.e. on Ω. By Lemma 7.11, we have strongly $y[v_\sigma] \to y[v_0]$ in $H_0^1(\Omega)$ and a.e. on Ω. Then $g(v_\sigma)^2 \to (\rho+1)|y_0|^\rho$ a.e. on Ω. The function $b(x,\cdot)$ is continuous. Therefore, $Bv_\sigma \to Bv_0$ a.e. on Ω. Now we have the convergence $g(v_\sigma)^2 p_\mu[v_\sigma] \to (\rho+1)|y_0|^\rho p_\mu$ and $Bv_\sigma p_\mu[v_\sigma] \to Bv_0 p_\mu$ a.e. on Ω. Using Lemma 4.5, we have $g(v_\sigma)^2 p_\mu[v_\sigma] \to (\rho+1)|y_0|^\rho p_\mu$ weakly in $L_{q'}(\Omega)$ and $Bv_\sigma p_\mu[v_\sigma] \to Bv_0 p_\mu$ weakly in $L_{r_1}(\Omega)$. Hence, we obtain the equalities $q_\mu = (\rho+1)|y_0|^\rho p_\mu$, $r_\mu = Bv_0 p_\mu$. Multiply the equality (7.33) by the arbitrary function $\lambda \in Y$. After the integration and passing to the limit we prove that the function p_μ satisfies the equation (7.34), i.e., $p_\mu = p_\mu[v_0]$. Thus, Assumption 7.6 is true. This complete the proof of Lemma 7.11. $\square$

Using Theorem 7.5 and Lemma 7.11, we obtain the following result.

260 *Optimization and Differentiation*

Lemma 7.12 *Under the assumptions of Lemma 7.4 the map $y[\cdot] : V \to Y$ for the equation (7.10) has the extended derivative $y'[u]$ at the point u that satisfies the equality*

$$\int_\Omega \mu y'[u]h dx = \int_\Omega \left[f_v(x,u) - b_v(x,u)y \right] p_\mu h dx \ \forall h \in V, \ \mu \in \mu \in Y(u)', \quad (7.36)$$

where p_μ is the solution of the homogeneous Dirichlet problem for the equation (7.34) with $v_0 = u$.

Indeed, the equality (7.29) can be transformed now to (7.36), and the adjoint equation (7.27) is equal to (7.34). Now we use Theorem 7.6 for obtaining necessary conditions of optimality for Problem 7.2.

Theorem 7.7 *Under the suppositions of Lemma 7.4 and existence of the optimal control the solution of Problem 7.2 satisfies the variational inequality (7.22).*

This is the extension of Theorem 7.3 for the strongly nonlinear case of equation (7.10).

Remark 7.18 We could extend our results to systems with general nonlinearity and general functionals, boundary control, and Neumann problems.

Our next step is optimization control problems for non-stationary systems.

7.6   Comments

The Implicit function theorem has some relations with first results of the differential calculus. Particularly, P. Fermat analyzes the tangent to the function that is determined nondirectly. I. Newton considers the implicit function as a series. G.W. Leibniz differentiates the implicit function. Note the theorem of J. Nash that is an infinite dimensional analog of the Implicit function theorem [380]. The proof of Theorem 7.1 is given, for example, in the book of L.V. Kantorovich and G.P. Akilov [267]. The different variants of the Implicit function theorem are considered by A.V. Arutunov [26], [27], V.F. Demyanov, and A.M. Rubinov [137], A. Domokos [143], [144], B.D. Gelman [207], K. Lorenz [324], S. Hiltunen [242], A.D. Ioffe and V.M. Tihomirov [253], J. Nash [380], L. Nirenberg [387], B.N. Pshenichny [416], and S.M. Robinson [431]. S. Hiltunen [242] proves an Implicit function theorem for the non-normalized spaces. The extended differentiability of the implicit operator is proved by S.Ya. Serovajsky [459].

Part III

Evolutional systems

Chapter 8

Linear first-order evolution systems

8.1	Abstract functions	264
8.2	Ordinary differential equations	270
8.3	Linear first order evolutional equations	271
8.4	Optimization control problems for linear evolutional equations .	278
8.5	Optimization control problems for the heat equation	283
8.6	Optimization control problems with the functional that is not dependent from the control	287
8.7	Non-conditional optimization control problems	290
8.8	Non-conditional optimization control problems for the heat equation	294
8.9	Hamilton–Jacobi equation	297
8.10	Bellman method for an optimization problem for the heat equation	301
8.11	Comments	307

We continue the analysis of the extremum problems by means of the differentiation theory. The necessary condition of minimum at a point u of Gâteaux differentiable functional I on a convex subset U of a Banach space V is the variational inequality

$$\langle I'(u), v - u \rangle \geq 0 \ \forall v \in U,$$

where $I'(u)$ is a Gâteaux derivative of the given functional at this point, and $\langle \lambda, v \rangle$ is the value of the linear continuous functional λ at the point v. Suppose we have the functional $I = I(v, y[v])$, where $y[v]$ is the state function of the system for the control v. Then we differentiate the dependence $y = y[v]$, i.e., control–state mapping. Therefore, we use properties of the state equations.

We considered before the stationary systems, described by boundary problems for elliptic equations. Now we consider evolutional systems. Its state depends upon the time. Therefore, we shall use an additional mathematical technique, in general, abstract functions theory.

We consider first order linear evolutional equations, for example, the classical heat equation. We give general properties of the abstract linear evolutional equations. The dependence $y = y[v]$ is affine for this case. Hence, we have the differentiable operator. The analysis of the optimization problem is easy enough for these systems. If the state functional does not depend upon

263

264 *Optimization and Differentiation*

the control, then the optimal control is singular. We can use the Tihonov regularization method for these problems.

If we have a non-conditional optimization problem, then the necessary condition of extremum is the stationary condition

$$I'(u) = 0.$$

The system of the optimality conditions is linear for this case. Then we can determine the optimal control not only as a function of the independent variables; we can solve also the synthesis problem with determination of the dependence of the optimal control as a function of the state system as we consider the Hamilton–Jacobi equation. The Bellman equation is its analog for optimization problems with constraints.

8.1 Abstract functions

The nonstationary phenomenon is described by functions $y = y(x,t)$, where independent variable x is the space coordinate, and t is the time. The point x belongs to a space set Ω, and t is a point of an interval S. We consider as a rule the time interval $(0,T)$. Sometimes, we consider the map $x \to y(x,t)$ with fixed value t. We denote it by $y(t)$. This is a function of the space variable x. Therefore, this is an element of a functional space W.

Definition 8.1 *The dependence $t \to y(t)$ that maps a number $t \in S$ to an element $y(t)$ of the set W is called the **abstract function** on the interval S with the values in the space W.*

Remark 8.1 If W is a subset of the n-dimensional Euclid space, then the abstract function is a standard function of n variables.

Remark 8.2 We consider abstract evolutional equations. Its solutions are abstract functions.

We shall consider abstract functions with special properties. Particularly, we shall differentiate and integrate the abstract functions.

Definition 8.2 *An abstract function y is called **differentiable** at a point t, if there exists an element $y'(t)$ that is called its **derivative** at this point such that we have the convergence*

$$\frac{y(t+\tau) - y(t)}{\tau} \to y'(t) \ in \ W$$

as $\tau \to 0$.

$$\text{Linear first-order evolution systems} \qquad 265$$

If the derivative $y'(t)$ of an abstract function y is determined at the arbitrary $t \in (0, T)$, then this abstract function is called **differentiable**. Besides, an element y' is an abstract function. Then we can determine its derivative at a point t that is called the **second derivative** of the abstract function y and is denoted by $y''(t)$. We could determine also high order derivatives of the abstract functions.

Remark 8.3 We shall use second derivatives of abstract functions for the analysis of the second order evolutional differential equations (see Chapter 10).

Determine sets of the abstract functions.

Definition 8.3 *The **space** $C^m(S; W)$ is the set of the abstract functions of the closed interval S with values in a space W that has the continuous m order derivative.*

If S is the interval $[0, T]$, then we use the short denotation $C^m(0, T; W)$ in place of $C^m([0, T]; W)$. If $m = 0$, then we have the space of the continuous abstract functions that is denoted by $C(S; W)$ or $C(0, T; W)$. If W is a Banach space, then the space $C(0, T; W)$ is Banach with the norm

$$\|y\|_{C(0,T;W)} = \max_{t \in [0,T]} \|y(t)\|_W.$$

Then the space $C^m(0, T; W)$ is Banach too with the norm

$$\|y\|_{C^m(0,T;W)} = \sum_{j=0}^{m} \max_{t \in [0,T]} \|y^{(j)}(t)\|_W,$$

where $y^{(j)}$ is j order derivative of the abstract function y at the point t.

We shall consider integrable abstract functions. An abstract function y on an interval S with values from a Banach space W is called **simple**, if there exists a finite quantity of non-intersecting Lebesgue measurable sets $S_1, ..., S_k$ with finite Lebesgue measure such that $y(t) = y_i$ for all $t \in S_i$, where $y_i \in W$, as y is equal to zero outside of these sets. Determine the **Bochner integral** of the simple function by the formula

$$\int_S y(t)dt = \sum_{i=1}^{k} \mu(S_i)y_i,$$

where $\mu(S_i)$ is the measure of the set S_i. An abstract function y is **Bochner measurable**, if there exists a sequence of the simple functions $\{y_k\}$ such that $y_k(t) \to y(t)$ in W almost for all $t \in S$.

Definition 8.4 *A Bochner measurable abstract function y is called a **Bochner integrable** on the interval S, if the following equality holds:*

$$\lim_{k \to \infty} \int_S \|y_k(t) - y(t)\|_W dt = 0.$$

The value

$$\int_S y(t)dt = \lim_{k \to \infty} \int_S y_k(t)dt$$

is called the **Bochner integral** *of the function y.*

If S is the interval $(0,T)$, then we denote the Bochner integral by

$$\int_0^T y(t)dt.$$

Remark 8.4 If W is a closed subset of the Euclid space, then the Bochner integral is the usual Lebesgue integral.

The Bochner integral of the abstract function with values in the space W is an element of the set W. A Bochner measurable abstract function y is Bochner integrable if and only if the map $t \to \|y(t)\|_W$ is Lebesgue integrable, as

$$\left\| \int_S y(t)dt \right\|_W \le \int_S \|y(t)\|_W dt.$$

If an abstract function y is Bochner integrable on the interval $(0,T)$, then the abstract function that is determined by the equality

$$z(t) = \int_0^t y(\tau)d\tau$$

is differentiable a.e. on $(0,T)$, as $z'(t) = y(t)$ almost for all $t \in (0,T)$.

Definition 8.5 *The* **space** $L_p(S;W)$ *is the set of all Bochner measurable abstract functions on the interval S with values in the space W such that*

$$\int_S \|y(t)\|_W^p dt < \infty.$$

The space $L_p(S;W)$ with the norm

$$\|y\|_{L_p(S;W)} = \left[\int_S \|y(t)\|_W^p dt \right]^{1/p}$$

is the Banach space. If the space W is reflexive and separable, then $L_p(S;W)$ is a reflexive Banach space if $1 < p < \infty$, as its adjoint space is $L_{p'}(S;W')$, where $1/p + 1/p' = 1$. If W is a Hilbert space, then $L_2(S;W)$ is a Hilbert space too with scalar product

$$(y,z)_{L_2(S;W)} = \int_S (y(t), z(t))_W dt.$$

Linear first-order evolution systems

A Bochner measurable abstract function y is called **essentially bounded**, if there exists a positive constant c such that $\|y(t)\|_W \le c$ almost for all $t \in S$. The lower bound of the set of these constants c is denoted by $\operatorname{vrai\,max}_{t \in S} \|y(t)\|_W$.

Definition 8.6 *The **space** $L_\infty(S; W)$ is the set of all essentially bounded abstract functions on the interval S with the values in the space W.*

The set $L_\infty(S; W)$ is a Banach space with the norm

$$\|y\|_{L_\infty(S;W)} = \operatorname{vrai\,max}_{t \in S} \|y(t)\|_W.$$

If W is reflexive and separable, then $L_\infty(S; W)$ is an adjoint space to $L_1(S; W')$.

If $1 \le p < \infty$, then Hölder inequality holds

$$|\langle \lambda, y \rangle| = \left| \int_S \langle \lambda(t), y(t) \rangle_W dt \right| \le \|\lambda\| \|y\| \ \forall y \in L_p(S; W), \lambda \in L_{p'}(S; W').$$

We denote $L_p(0, T; W)$ the space $L_p(S; W)$ for $S = (0, T)$.

Let S be an open finite interval, and $D(S)$ is the set of all infinite differentiable functions on S with closed supports.

Definition 8.7 *The linear continuous operator from $D(S)$ to the space W with the weak topology is called the **distribution** on the interval S with values in Banach space W.*

The set $D'(S; W)$ of the distributions on S with values in W involves the linear operators $\lambda : D(S) \to W$ such that $\langle \lambda, y_k \rangle \to \langle \lambda, y \rangle$ weakly in W whenever $y_k \to y$ in $D(S)$.

For all $\lambda \in D'(S; W)$ we can determine the generalized derivative $\lambda' \in D'(S; W)$ by the equality

$$\langle \lambda', y \rangle = -\langle \lambda, y' \rangle \ \forall y \in D(S). \tag{8.1}$$

We have

$$\langle \lambda', ay + bz \rangle = -\langle \lambda, ay' + by' \rangle = -a\langle \lambda, y' \rangle - b\langle \lambda, z' \rangle = a\langle \lambda', y \rangle + b\langle \lambda', z \rangle$$

for all $a, b \in \mathbb{R}$, $y, z \in D(S)$. Thus, $y \to \langle \lambda', y \rangle$ is a linear continuous operator from $D(S)$ to W. If $y_k \to y$ in $D(S)$, then we have the convergence $y'_k \to y'$ in $D(S)$. Therefore, $\langle \lambda, y'_k \rangle \to \langle \lambda, y' \rangle$ weakly in W for $\lambda \in D'(S; W)$. Using the equality (8.1), we get $\langle \lambda', y_k \rangle \to \langle \lambda', y \rangle$ weakly in W. Hence, the object λ' that is determined by the equality (8.1) is in reality the linear continuous operator from the set $D(S)$ to the space W with weak topology. Therefore, this is the distribution. Thus, the map $\lambda \to \lambda'$ is the continuous operator on the space $D'(S; W)$.

Remark 8.5 The distributions of Chapter 1 have analogical properties.

Determine special spaces of distributions. Consider a separable reflexive Banach space W and Hilbert space H such that embedding $W \subset H$ is continuous and dense. Then we have continuous and dense embedding for the adjoint spaces $H' \subset W'$. By the Riesz Theorem we can identify the space H with its adjoint space. Therefore, we have continuous and dense embedding $W \subset H \subset W'$.

Let p and q be real numbers such that $1 < p \le q < \infty$. Consider the intersection

$$X = L_p(S; W) \cap L_q(S; H).$$

We know (see Chapter 4) that the intersection of Banach spaces is a Banach space, as its norm is determined by the equality

$$\|y\|_X = \|y\|_{L_p(S;W)} + \|y\|_{L_q(S;H)}.$$

By the definition of the adjoint space of the intersection of Banach spaces, X has adjoint reflexive Banach space

$$X' = L_{p'}(S; W') + L_{q'}(S; H)$$

with the norm

$$\|\lambda\|_X = \inf_{\mu,\nu} \left\{ \|\mu\|_{L_{p'}(S;W')}, \|\nu\|_{L_{q'}(S;H)} \right\},$$

where we choose $\mu \in L_{p'}(S; W')$, $\nu \in L_{q'}(S; H)$, as $\mu + \nu = \lambda$. The value of a linear continuous functional λ at a point $y \in X$ is

$$\langle \lambda, y \rangle = \int_S \langle \mu(t), y(t) \rangle_W dt + \int_S \left(\nu(t), y(t) \right)_H dt.$$

We consider also the space

$$Y = \left\{ y \,\middle|\, y \in X,\, y' \in X' \right\}$$

with the norm

$$\|y\|_Y = \|y\|_X + \|y'\|_{X'}.$$

Theorem 8.1 *The set Y is a Banach space.*

Suppose $p = q$ here. Determine $X = L_p(0, T; W)$ and the space

$$Y = \left\{ y \,\middle|\, y \in L_p(0, T; W),\, y' \in L_{p'}(0, T; W') \right\}.$$

Theorem 8.2 *If W is a Hilbert space, then*

$$Y = \left\{ y \,\middle|\, y \in L_2(0, T; W),\, y' \in L_2(0, T; W') \right\}$$

is a Hilbert space too.

Linear first-order evolution systems 269

Theorem 8.3 *If W is a Banach space, and $1 \leq p \leq \infty$, then embedding of the space*

$$Y = \{y \mid y \in L_p(0,T;W),\ y' \in L_{p'}(0,T;W')\}$$

to $C(0,T;H)$ is continuous.

By Theorem 8.3, $y_k(t) \to y(t)$ in H for all $t \in [0,T]$ whenever $y_k \to y$ in Y.

Note the **formula of integration by parts**

$$\int_s^t \Big[\langle z'(\tau), y(\tau)\rangle + \langle z(\tau), y'(\tau)\rangle\Big] d\tau = \big(z(t), y(t)\big) - \big(z(s), y(s)\big)$$

for all y, z from the space

$$X = L_p(0,T;W) \cap L_q(0,T;H),$$

where $0 \leq p < q \leq \infty$.

Theorem 8.4 *Embedding of the space*

$$Y = \{y \mid y \in L_2(0,T;W),\ y' \in L_2(0,T;H),\ y'' \in L_2(0,T;W')\}$$

to

$$Y_1 = \{y \mid y \in C(0,T;W),\ y' \in C(0,T;H)\}$$

is continuous.

By Theorem 8.4, if $y_k \to y$ in Y, then $y_k(t) \to y(t)$ in W and $y_k'(t) \to y'(t)$ in H for all $t \in [0,T]$.

For any y, z from the space Y of Theorem 8.4 we have the following **formula of integration by parts:**

$$\int_s^t \Big[\langle z''(\tau), y(\tau)\rangle - \langle z(\tau), y''(\tau)\rangle\Big] d\tau =$$

$$\langle z'(t), y(t)\rangle - \langle z'(s), y(s)\rangle - \langle y'(t), z(t)\rangle + \langle y'(s), z(s)\rangle.$$

Remark 8.6 We shall use Theorem 8.4 and the final formula for the analysis of the control systems described by the hyperbolic equations (see Chapter 10).

Theorem 8.5 *Let W_0, W, W_1 be Banach spaces, where W_0 and W_1 are reflexive, as continuous embedding $W_0 \subset W \subset W_1$ holds, and the first of them is compact. Then embedding of the space*

$$Y = \Big\{y \mid y \in L_s(0,T;W_0),\ y' \in L_r(0,T;W_1)\Big\}$$

to $L_s(0,T;W)$ is compact, where $1 < s < \infty,\ 1 < r < \infty$.

By Theorem 8.5, if $y_k \to y$ weakly in Y, then $y_k \to y$ strongly in $L_s(0,T;W)$.

8.2 Ordinary differential equations

We used the Galerkin method for the analysis of stationary systems. The approximate solution of the problem is determined here as a linear combination of basic elements. The unknown coefficients of this value satisfy a system of algebraic equations. If we would like to apply the analogical technique for the non-stationary case, these coefficients depend upon the time and satisfy a system of ordinary differential equations. Therefore, we consider properties of these equations.

Let us have the system of the linear ordinary differential equations

$$y'(t) + A(t)y(t) = f(t), \, t > 0 \qquad (8.2)$$

where $y(t) = (y_1(t), y_2(t), ..., y_k(t))$ is a unknown vector of order k, $f(t) = (f_1(t), f_2(t), ..., f_k(t))$ is a known vector of order k, and $A(t)$ is a matrix of $k \times k$ order for all fixed t. We have also the initial condition

$$y(0) = \varphi, \qquad (8.3)$$

where φ is a vector of order k.

Theorem 8.6 *If the matrix A and vector-function f are continuous on an interval $[0, T]$, then the Cauchy problem $(8.2), (8.3)$ has a unique continuous solution on this interval.*

Remark 8.7 We shall use Theorem 8.6 for the proof of the existence of the approximate solution for the Cauchy problem for first order linear abstract evolutional equations.

Consider also the system of nonlinear differential equations

$$y' = f(t, y), \, t > 0 \qquad (8.4)$$

where $y = y(t) = (y_1(t), y_2(t), ..., y_k(t))$ is a unknown vector-function, and $f = f(t, y) = (f_1(t, y), f_2(t, y), ..., f_k(t, y))$ is a known vector-function.

Theorem 8.7 (*Peano Theorem*). *Let the vector-function f be continuous in the set*

$$Q = \left\{ (t, y) \in \mathbb{R}^{n+1} \, \middle| \, t \in [0, a], \, \|y - \varphi\| \le b \right\},$$

and the following inequality holds: $\|f(t, y)\| \le c$ for $(t, y) \in Q$. Then Cauchy problem $(8.3), (8.4)$ has a continuous solution on the interval $[0, T]$, where $T = \min \{a, b/c\}$.

Remark 8.8 Theorem 8.6 guarantees the existence of the *global solution* of the Cauchy problem, i.e., its solution exists on each time interval. However, Peano Theorem guarantees the solvability of this problem on the small enough time interval only. Therefore, we obtain the *local solution* of the problem. Note that this solution can be non-unique.

Remark 8.9 We shall apply the Peano Theorem for proving the existence of the existence of the approximate solution of the Cauchy problem for the nonlinear abstract evolutional equation (see next chapter).

Consider general linear evolutional equations.

8.3 Linear first order evolutional equations

Consider Hilbert spaces W and H such that embedding of W to H is continuous and dense. We identify the space H with its adjoint space H'. Then embedding $W \subset H \subset W'$ is continuous and dense. Suppose the space W is separable.

Let us have a linear continuous operator $A : W \to W'$ such that

$$\langle Ay, y \rangle \geq \chi \|y\|_W^2 \ \forall y \in W,$$

where $\chi > 0$. Consider the linear **first order evolutional equation**

$$y'(t) + Ay(t) = f(t), \ t \in (0, T) \tag{8.5}$$

with the initial condition

$$y(0) = \varphi, \tag{8.6}$$

where f and φ are known. The system (8.5), (8.6) is a Cauchy problem for the given evolutional equation.

Remark 8.10 Of course, the equation (8.2) is the partial case of (8.5). A non-trivial example of the first order linear evolutional equation is heat equation.

Remark 8.11 We shall consider also the nonlinear evolutional equations (see next chapter) and second order evolutional equations (see Chapter 10).

Determine the space

$$Y = \{y \mid y \in L_\infty(0, T; W), \ y' \in L_\infty(0, T; H), \ y' \in L_2(0, T; W')\}.$$

Theorem 8.8 *For any $f \in L_2(0, T; W')$, $\varphi \in H$ there exists a unique solution $y \in Y$ of the problem* (8.5), (8.6).

Proof. 1. We use the **Faedo–Galerkin method**. The space W is separable. Therefore, there exists a complete system of linear independent elements $\{\mu_m\}$ there. Then each element of this space is the sum of a convergent series with respect to these elements. Determine an approximate solution $y_k = y_k(t)$ of the problem (8.5), (8.6) by the formula

$$y_k(t) = \sum_{m=1}^{k} \xi_{mk}(t)\mu_m, \ k = 1, 2, ..., \tag{8.7}$$

where the functions $\xi_{1k}, ..., \xi_{kk}$ satisfy the equalities

$$\langle y_k'(t), \mu_j \rangle + \langle Ay_k(t), \mu_j \rangle = \langle f(t), \mu_j \rangle, \ j = 1, ..., k. \tag{8.8}$$

272 *Optimization and Differentiation*

Put here the value y_k from the formula (8.7). We get

$$\sum_{m=1}^{k} \langle \mu_m, \mu_j \rangle \xi'_{mk}(t) + \sum_{m=1}^{k} \langle A\mu_m, \mu_j \rangle \xi_{mk}(t) = \langle f(t), \mu_j \rangle, \ j = 1, ..., k.$$

Thus, (8.8) is the system of linear ordinary differential equations. The determinant of the matrix with elements $\langle \mu_m, \mu \rangle$ is not equal to zero. Therefore, we can find its solution, i.e., the derivatives $\xi'_{mk}(t)$.

The element φ of H is the limit of a sequence of elements from the space W because of the density of embedding of the space W to H. Then there exists a number sequence $\{\alpha_k\}$ such that

$$\varphi = \sum_{m=1}^{\infty} \alpha_m \mu_m.$$

Determine the sum

$$\varphi_k = \sum_{m=1}^{k} \alpha_m \mu_m.$$

We obtain the convergence

$$\varphi_k \to \varphi \ \text{in} \ H \tag{8.9}$$

because of the convergence of the series. Determine the initial values for the functions ξ_{mk} by the equality

$$\xi_{mk}(0) = \alpha_m, \ m = 1, 2, ...$$

Therefore, we have

$$y_k(0) = \varphi_k. \tag{8.10}$$

By Theorem 8.6, the Cauchy problem for the system of linear ordinary differential equations (8.8), (8.10) has a unique solution on the arbitrary interval $(0, T)$. Then we can determine the value $y_k = y_k(t)$ by the formula (8.7).

2. Multiply j-th equality (8.8) by $\xi_{jk}(t)$ and sum the result by j. Using the formula (8.7), we get

$$\langle y'_k(t), y_k(t) \rangle + \langle Ay_k(t), y_k(t) \rangle = \langle f(t), y_k(t) \rangle. \tag{8.11}$$

We have the relations

$$\frac{1}{2} \frac{d}{dt} \|y_k(t)\|_H^2 = \langle y'_k(t), y_k(t) \rangle,$$

$$\langle Ay_k(t), y_k(t) \rangle \geq \chi \|y_k(t)\|_W^2,$$

$$|\langle f(t), y_k(t) \rangle| \leq \|f(t)\|_{W'} \|y_k(t)\|_W.$$

Note the obvious inequality

$$ab \leq \frac{1}{2\chi} a^2 + \frac{\chi}{2} b^2$$

Linear first-order evolution systems

for all numbers a and b. From the equality (8.11) it follows that

$$\frac{1}{2}\frac{d}{dt}\|y_k(t)\|_H^2 + \chi\|y_k(t)\|_W^2 \le \frac{1}{2\chi}\|f(t)\|_{W'}^2 + \frac{\chi}{2}\|y_k(t)\|_W^2.$$

Then we have the inequality

$$\frac{d}{dt}\|y_k(t)\|_H^2 + \chi\|y_k(t)\|_W^2 \le \frac{1}{\chi}\|f(t)\|_{W'}^2.$$

Integrating it by t by using the condition (8.10), we have

$$\|y_k(t)\|_H^2 + \chi\int_0^t \|y_k(\tau)\|_W^2 d\tau \le \|y_{0k}\|_H^2 + \frac{1}{\chi}\int_0^t \|f(\tau)\|_{W'}^2 d\tau.$$

3. Using the condition (8.9) and the properties of the function f, we obtain the boundedness of the sequence $\{y_k\}$ in the space $L_2(0,T;W)$. By Banach–Alaoglu theorem, we can extract a subsequence with initial denotation such that

$$y_k \to y \ \text{ weakly in } L_2(0,T;W). \tag{8.12}$$

Multiply the equality (8.8) by an arbitrary function $\eta \in C^1[0,T]$ such that $\eta(T) = 0$. After integration we get

$$\int_0^T \langle y_k'(t), \eta_j(t)\rangle dt + \int_0^T \langle Ay_k(t), \eta_j(t)\rangle dt = \int_0^T \langle f(t), \eta_j(t)\rangle dt,$$

where $\eta_j(t) = \eta(t)\mu_j$. Using the formula of integration by parts and the condition (8.10), we have

$$\int_0^T \langle y_k'(t), \eta_j(t)\rangle dt = -\int_0^T \langle \eta_j'(t), y_k(t)\rangle dt - \left(\varphi_k, \eta_j(0)\right)_H.$$

Then we obtain the equality

$$-\int_0^T \langle \eta_j'(t), y_k(t)\rangle dt + \int_0^T \langle Ay_k(t), \eta_j(t)\rangle dt =$$

$$\int_0^T \langle f(t), \eta_j(t)\rangle dt + \left(\varphi_k, \eta_j(0)\right)_H, \ j = 1,...,k. \tag{8.13}$$

From the condition (8.12) and the continuity of embedding of the space W to H it follows that

$$\int_0^T \langle \eta_j'(t), y_k(t)\rangle dt \to \int_0^T \langle \eta_j'(t), y(t)\rangle dt.$$

Using the definition of the adjoint operator, we have the equality

$$\langle Ay_k(t), \eta_j(t) \rangle = \langle A^*\eta_j(t), y_k(t) \rangle.$$

By reflexivity of Hilbert space W the adjoint operator A^* is the linear continuous operator from the space W to W'. The function φ is continuous. Therefore, we have $A^*\eta_j \in L_2(0,T;W')$. Then from (8.12) the convergence follows:

$$\int_0^T \langle A^*\eta_j(t), y_k(t) \rangle dt \to \int_0^T \langle A^*\eta_j(t), y(t) \rangle dt.$$

We have

$$\int_0^T \langle Ay_k(t), \eta_j(t) \rangle dt \to \int_0^T \langle Ay(t), \eta_j(t) \rangle dt.$$

From the equality (8.9) we get

$$\left(y_{0k}, \eta_j(0) \right)_H \to \left(\varphi, \eta_j(0) \right)_H.$$

Passing to the limit in (8.13), we obtain

$$-\int_0^T \langle \eta_j'(t), y(t) \rangle dt + \int_0^T \langle Ay(t), \eta_j(t) \rangle dt =$$

$$\int_0^T \langle f(t), \eta_j(t) \rangle dt + \left(\varphi, \eta_j(0) \right)_H, \ j = 1, 2, \dots .$$

This equality is true for all functions η with given properties. Suppose it is equal to an arbitrary element of the space $D(0,T)$ of infinite differentiable finite functions on the interval $(0,T)$. Using the definition of the generalized derivative, we have the equality

$$\int_0^T \langle \eta_j'(t), y(t) \rangle dt = \int_0^T \langle y(t), \mu_j \rangle \eta_j'(t) dt = -\int_0^T \frac{d}{dt} \langle y(t), \mu_j \rangle \eta(t) dt.$$

The generalized derivative in the right-hand side here is an element of the space of distributions $D'(0,T)$. Each element of the set $D(0,T)$ is equal to zero on the boundary of this interval. Therefore, we transform the previous equality

$$\int_0^T \left[\frac{d}{dt} \langle \mu_j, y(t) \rangle + \langle Ay(t), \mu_j \rangle - \langle f(t), \mu_j \rangle \right] \eta(t) dt = 0, \ j = 1, 2, \dots . \quad (8.14)$$

Linear first-order evolution systems

The function η is arbitrary. Therefore, we get

$$\langle y'(t), \mu_j \rangle + \langle Ay(t), \mu_j \rangle = \langle f(t), \mu_j \rangle, \ j = 1, 2, \dots.$$

Each element $\mu \in W$ can be determined as a sum

$$\mu = \sum_{j=1}^{\infty} b_j \mu_j,$$

where b_j are numbers. Multiply j-th previous equality by b_j. Summing by j, we get

$$\langle y'(t) + Ay(t) - f(t), \mu \rangle = 0 \ \forall \mu \in W.$$

Then we have the equality

$$y'(t) + Ay(t) = f(t).$$

Thus the function is the solution of the equation (8.5). Finding the derivative $y' = Ay + f$, we have $y' \in L_2(0, T; W')$. Therefore y is the point of the space Y.

Consider the equality (8.14). After integration by parts by using the equality (8.5), we have

$$\left(y(0), \mu_j \right)_H \eta(0) = (\varphi \mu_j)_H \, \eta(0), \ j = 1, 2, \dots.$$

Therefore, we get

$$\left(y(0), \mu_j \right)_H = (\varphi \mu_j)_H, \ j = 1, 2, \dots.$$

Multiply j-th equality by b_j and sum the result by j. We obtain

$$\left(y(0), \mu \right)_H = (\varphi \mu)_H \ \forall \mu \in W.$$

Then the initial condition (8.6) $y(0) = \varphi$ is true. We proved the solvability of the problem (8.5), (8.6).

4. Suppose there exists two solutions y_1 and y_2 of this problem. The difference $y = y_1 - y_2$ satisfies the equation

$$y'(t) + Ay(t) = 0$$

with homogeneous initial condition. We have the equality

$$\langle y'(t) + Ay(t), y(t) \rangle = 0.$$

After integration by t we obtain

$$\int_0^T \langle y'(t), y(t) \rangle dt + \int_0^T \langle Ay(t), y(t) \rangle dt = 0.$$

Using the relations

$$\int\limits_0^T \langle y'(t), y(t) \rangle dt = \frac{1}{2} \int\limits_0^T \frac{d}{dt} \|y(t)\|_H^2 dt = \frac{1}{2} \|y(T)\|_H^2,$$

$$\langle Ay(t), y(t) \rangle \geq \chi \|y(t)\|_W^2,$$

we obtain the inequality

$$\|y(T)\|_H^2 + \chi \int\limits_0^T \|y(t)\|_W^2 dt \leq 0.$$

Then we have $y = 0$ and the equality $y_1 = y_2$. Therefore, the solution of the problem (8.5), (8.6) is unique. $\square$

Remark 8.12 This theorem has four steps. There are the determination of the approximate solution, obtaining a priori estimate, passing to the limit, and proof of the uniqueness of the solution (see also Theorem 3.5 and Theorem 4.4). The difference between the Faedo–Galerkin method and the Galerkin method for the analysis of the stationary systems is the use of ordinary differential equations for the determination of the approximate solution.

Prove also the continuity of the solution of the considered Cauchy problem with respect to the initial state φ and the absolute term f.

Lemma 8.1 *The map* $(\varphi, f) \to y$ *for the problem* (8.5), (8.6) *is a continuous and weakly continuous operator* $H \times L_2(0, T; W') \to Y$, *as the following inequality holds:*

$$\|y_1 - y_2\|_{L_2(0,T;W)}^2 \leq \alpha^{-1} \|\varphi_1 - \varphi_2\|_H^2 + \alpha^{-2} \|f_1 - f_2\|_{L_2(0,T;W')}^2, \qquad (8.15)$$

where y_i *is the solution of the problem* (8.5), (8.6) *for* $\varphi_i \in H$, $f_i \in L_2(0, T; W')$, $i = 1, 2$.

Proof. 1. The difference $y = y_1 - y_2$ satisfies the equation

$$y'(t) + Ay(t) = f(t)$$

with initial condition

$$y(0) = \varphi,$$

where $\varphi = \varphi_1 - \varphi_2$, $f = f_1 - f_2$. Then we have the equality

$$\langle y'(t) + Ay(t), y(t) \rangle = \langle f(t), y(t) \rangle.$$

After integration we get

$$\int\limits_0^T \langle y'(t), y(t) \rangle dt + \int\limits_0^T \langle Ay(t), y(t) \rangle dt = \int\limits_0^T \langle f(t), y(t) \rangle dt.$$

Using the equality

$$\int_0^T \langle y'(t), y(t)\rangle dt = \frac{1}{2}\int_0^T \frac{d}{dt}\|y(t)\|_H^2 dt = \frac{1}{2}\|y(T)\|_H^2 - \frac{1}{2}\|\varphi\|_H^2$$

and the property of the operator A, we obtain the inequality

$$\|y(T)\|_H^2 - \|\varphi\|_H^2 + 2\alpha \int_0^T \|y(t)\|_W^2 dt \le 2\int_0^T |\langle f(t), y(t)\rangle| dt \le$$

$$2\int_0^T \|f(t)\|_{W'}\|y(t)\|_W dt \le \alpha \int_0^T \|y(t)\|_W^2 + \frac{1}{\alpha}\int_0^T \|f(t)\|_{W'}^2.$$

Therefore, we get

$$\alpha \int_0^T \|y(t)\|_W^2 dt \le \|\varphi\|_H^2 + \frac{1}{\alpha}\int_0^T \|f(t)\|_{W'}^2.$$

Then we have the inequality

$$\|y\|_{L_2(0,T;W)}^2 \le \alpha^{-1}\|\varphi\|_H^2 + \alpha^{-1}\|f\|_{L_2(0,T;W')}^2,$$

which can be transformed to (8.15). Thus, $(\varphi, f) \to y$ is the continuous operator from the space $H \times L_2(0,T;W')$ to $L_2(0,T;W)$.

2. We have the inequality

$$\|y'\|_{L_2(0,T;W')}^2 \le \|Ay\|_{L_2(0,T;W')}^2 + \|f\|_{L_2(0,T;W')}^2.$$

We have also

$$\|Ay\|_{L_2(0,T;W')}^2 = \int_0^T \|Ay(t)\|_{W'}^2 dt \le \int_0^T \|A\|^2\|y(t)\|_W^2 dt \le \|A\|^2 \int_0^T \|y(t)\|_W^2 dt.$$

Therefore, we obtain the inequality

$$\|y'\|_{L_2(0,T;W')}^2 \le \|A\|^2\|y\|_{L_2(0,T;W)}^2 + \|f\|_{L_2(0,T;W')}^2.$$

Using (8.15), we prove that the map $(\varphi, f) \to y$ is a continuous operator $H \times L_2(0,T;W') \to Y$.

3. Consider now the Cauchy problem

$$y_i' + Ay_i = f_i, \quad y_i(0) = \varphi_i, \quad i = 1, 2.$$

278   *Optimization and Differentiation*

Multiply by arbitrary number a_i and sum the result. We obtain Cauchy problem

$$\big(a_1 y_1 + a_2 y_2\big)' + A\big(a_1 y_1 + a_2 y_2\big) = \big(a_1 f_1 + a_2 f_2\big),$$

$$a_1 y_1(0) + a_2 y_2(0) = a_1 \varphi_1 + a_2 \varphi_2.$$

Thus, the linear combination of the initial state and the absolute term of the equation conforms to the linear combination of solution of the problem (8.5), (8.6). Hence, the map $(\varphi, f) \to y$ is a linear operator. We know that a linear continuous operator is weakly continuous. Therefore, the dependence of the solution of a given Cauchy problem from the initial state and the absolute term is weakly continuous. $\square$

We use these results for the analysis of the optimization control problem for the system, described by first order linear evolutional equations.

8.4 Optimization control problems for linear evolutional equations

Let us have the spaces H, W and the operator A as before. We have also Hilbert space V and a linear continuous operator $B : V \to L_2(0, T; W')$. Consider the equation

$$y'(t) + Ay(t) = (Bv)(t) + f(t), \ t \in (0, T) \tag{8.16}$$

with the initial condition

$$y(0) = \varphi, \tag{8.17}$$

where v is a control, y is the state system, and $\varphi \in H$ and $f \in L_2(0, T; W')$ are known values. From Theorem 8.8 it follows that for all control $v \in V$ the problem (8.16), (8.17) has a unique solution $y = y[v]$ from the space Y. Let us have also a non-empty convex closed subset U of the space V and the functional

$$I(v) = \frac{\alpha}{2}\|v\|_V^2 + \frac{1}{2}\big\|C\big(y[v] - y_d\big)\big\|_Z^2,$$

where $\alpha > 0$, $C : L_2(0, T; W) \to Z$ is a linear continuous operator, Z is a Hilbert space, and $y_d \in L_2(0, T; W)$ is known.

Problem 8.1 *Find a control $u \in U$ that minimizes the functional I on the set U.*

Prove the existence and the uniqueness of this problem.

Lemma 8.2 *The functional I is continuous.*

Linear first-order evolution systems 279

Proof. If $v_k \to v$ in V, then we have the convergence $y[v_k] \to y[v]$ in Y by Lemma 8.1. Then $C(y[v_k] - y_d) \to C(y[v] - y_d)$ in Z. Using the continuity of the norm of Hilbert space, we prove that the functional I is continuous. $\square$

Lemma 8.3 *The functional I is strictly convex.*

Proof. The sum of convex and strictly convex functionals is strictly convex. The square of norm of a Hilbert space is strictly convex (see Chapter 2). Therefore, it will be sufficient to prove the convexity of the functional

$$J(v) = \left\| C(y[v] - y_d) \right\|_Z^2.$$

Let y_1 and y_2 be the solutions of the problem (8.16), (8.17) for the controls v_1 and v_2, and σ is a constant from the unit interval $[0,1]$. Then we have

$$y_i'(t) + Ay_i(t) = (Bv_i)(t) + f(t), \ t \in (0,T), \ y_i(0) = \varphi, \ i = 1,2.$$

Multiply the first equality by $1 - \sigma$, and the second one by σ. Summing the results, we have

$$y'(t) + Ay(t) = (Bv)(t) + f(t), \ t \in (0,T), \ y(0) = \varphi,$$

where $v = (1 - \sigma)v_1 + \sigma v_2$, $y = (1 - \sigma)y_1 + \sigma y_2$. We have the equality

$$y[(1 - \sigma)v_1 + \sigma v_2] = (1 - \sigma)y[v_1] + \sigma y[v_2].$$

Therefore, the operator $v \to y[v]$ is affine. Then we get

$$Cy[(1 - \sigma)v_1 + \sigma v_2] = (1 - \sigma)Cy[v_1] + \sigma Cy[v_2].$$

Now the convexity of the functional J follows from the convexity of the square of norm for Hilbert spaces. $\square$

Using Theorem 2.4, Theorem 2.5, and these lemmas, we prove the following result.

Theorem 8.9 *Problem 8.1 has a unique solution.*

We can prove an additional result. The square of norm of a Hilbert space is strictly uniformly convex (see Chapter 2). Then the functional I is strictly uniformly convex because it is the sum of the strictly uniformly convex square of norm and the convex functional J. Using Theorem 2.6, we get the following result.

Theorem 8.10 *Problem 8.1 is Tihonov well-posed.*

Now we determine the optimality condition for Problem 8.1. We know (see Theorem 1.8 and Theorem 2.1) that the necessary condition of minimum of Gâteaux differentiable functional I on a convex subset U of a Banach space at a point u is the variational inequality

$$\langle I'(u), v - u \rangle \geq 0 \ \forall v \in U, \tag{8.18}$$

as this is a sufficient condition of minimum if the functional is convex.

Prove the differentiability of the functional for Problem 8.1.

280 *Optimization and Differentiation*

Lemma 8.4 *The functional I is Gâteaux differentiable at the arbitrary point $u \in V$, as we have the equality*

$$I'(u) = \alpha \Lambda_V^{-1} u + B^* p, \tag{8.19}$$

where Λ_V is the canonic isomorphism of the spaces V' and V, and p is the solution of the adjoint Cauchy problem

$$-p'(t) + A^* p(t) = C^* \Lambda_Z^{-1} C\big(y[u] - y_d\big)(t), \ t \in (0, T), \tag{8.20}$$

$$p(T) = 0. \tag{8.21}$$

Proof. The adjoint operator A^* is the linear continuous operator from W to W'. The space W is reflexive. Hence, we have

$$\langle A^* y, y \rangle = \langle Ay, y \rangle \geq \chi \|y\|_Y^2 \ \forall y \in Y.$$

Thus, the operator A^* has the same properties as the initial operator A of Theorem 8.3. Note that C^* is the operator from the space Z to $L_2(0, T; W')$. Then we have

$$C^* \Lambda_Z^{-1} C\big(y[u] - y_d\big) \in L_2(0, T; W').$$

Therefore, the term at the right-hand side of the equality (8.20) has the same properties as the absolute term of the equation (8.2). Determine the abstract function q by the equality $q(t) = p(T - t)$; we transform the problem (8.20), (8.21) to

$$q'(t) + A^* q(t) = C^* \Lambda_Z^{-1} C\big[y(T - t) - y_d(T - t)\big], \ t \in (0, T),$$

$$q(0) = 0,$$

where $y = y[u]$. Using Theorem 8.3, we prove that the problem (8.20), (8.21) has a unique solution from the space Y.

For all $h \in V$ and number σ we find the difference

$$I(u + \sigma h) - I(u) = \frac{\alpha}{2}\Big[\|u + \sigma h\|_V^2 - \|u\|_V^2\Big] +$$

$$\frac{1}{2}\Big[\big\|C\big(y[u + \sigma h] - y_d\big)\big\|_Z^2 - \big\|C\big(y - y_d\big)\big\|_Z^2\Big].$$

Using the property of the square of norm for Hilbert space (see Chapter 2), we get

$$I(u + \sigma h) - I(u) = \frac{\alpha \sigma}{2}\Big[2(u, h)_V + \sigma \|h\|_V^2\Big] + \frac{1}{2}\Big[\big(C(y - y_d), C\zeta\big)_Z + \|C\zeta\|_Z^2\Big],$$

where $\zeta = y[u + \sigma h] - y[u]$. Using the definition of the adjoint operator $C^* : Z' \to L_2(0, T; W')$ and canonic isomorphism Λ_Z, we obtain

$$(\eta, C\zeta)_Z = \big\langle \Lambda_Z^{-1} \eta, C\zeta \big\rangle_Z = \big\langle C^* \Lambda_Z^{-1} \eta, \zeta \big\rangle_{L_2(0,T;W)} = \int_0^T \big\langle C^* \Lambda_Z^{-1} \eta(t), \zeta(t) \big\rangle_W dt.$$

Linear first-order evolution systems

Thus, we get the equality

$$I(u + \sigma h) - I(u) = \alpha(u, h)_V + \frac{\alpha\sigma}{2}\|h\|_V^2 +$$

$$\int_0^T \Big\langle (C^*\Lambda_Z^{-1}C(y - y_d))(t), \zeta(t) \Big\rangle_W dt + \frac{1}{2}\|C\zeta\|_Z^2. \qquad (8.22)$$

The function ζ satisfies the equation

$$\zeta'(t) + A\zeta(t) = \sigma(Bh)(t) \qquad (8.23)$$

with homogeneous initial condition. Then for all $p \in Y$ we have

$$\int_0^T \Big\langle \zeta'(t) + A\zeta(t), p(t) \Big\rangle dt = \sigma \int_0^T \langle (Bh)(t), p(t) \rangle dt.$$

Using the formula of integration by parts and the definition of the adjoint operator, we get

$$(\zeta(T), p(T)) + \int_0^T \Big\langle -p'(t) + A^*p(t), \zeta(t) \Big\rangle dt = \sigma \langle B^*p, h \rangle.$$

Suppose p is the solution of the problem (8.20), (8.21). We have

$$\int_0^T \Big\langle (C^*\Lambda_Z^{-1}C(y - y_d))(t), \zeta(t) \Big\rangle_W dt = \sigma \langle B^*p, h \rangle.$$

Transform the equality (8.22) to

$$I(u + \sigma h) - I(u) = \sigma\big(\alpha u + \Lambda_V B^*p, h\big)_V + \frac{\sigma^2}{2}\|h\|_V^2 + \frac{1}{2}\|C\zeta\|_Z^2. \qquad (8.24)$$

From (8.23) it follows that

$$\int_0^T \langle \zeta'(t), \zeta(t) \rangle dt + \int_0^T \langle A\zeta(t), \zeta(t) \rangle dt = \int_0^T \langle (Bh)(t), \zeta(t) \rangle dt.$$

Using the formula of integration by parts and the property of the operator A, we have the inequality

$$\chi \int_0^T \|\zeta(t)\|_W^2 dt \leq \sigma \int_0^T \big| \langle (Bh)(t), \zeta(t) \rangle \big| dt \leq \sigma \int_0^T \|(Bh)(t)\|_{W'} \|\zeta(t)\|_W^2 dt \leq$$

Optimization and Differentiation

$$\frac{\chi}{2} \int_0^T \|\zeta(t)\|_W^2 dt + \frac{\sigma^2}{2\chi} \int_0^T \|(Bh)(t)\|_{W'}^2 dt.$$

Therefore, we have

$$\|\zeta\|_{L_2(0,T;W)}^2 \le \frac{\sigma^2}{\chi^2} \|Bh\|_{L_2(0,T;W')}^2.$$

Using the definition of the norm of operators, we obtain the inequality

$$\|C\zeta\|_Z \le \|C\| \|z\|_{L_2(0,T;W)}.$$

After dividing the equality (8.24) by σ and passing to the limit as $\sigma \to 0$ we have

$$\langle I'(u), h \rangle = \left(\alpha u + \Lambda_V B^* p, h \right) = \left(\Lambda_V \left(\alpha \Lambda_V^{-1} u + B^* p \right), h \right) = \left\langle \alpha \Lambda_V^{-1} u + B^* p, h \right\rangle_V.$$

The element h is arbitrary here. Therefore, the formula (8.19) is true. $\square$

Put the value of the functional derivative in to the condition (8.18). We have the following result.

Theorem 8.11 *The necessary and sufficient condition of the optimality for Problem 8.1 is the variational inequality*

$$\left\langle \alpha \Lambda_V^{-1} u + B^* p, v - u \right\rangle_V \ge 0 \ \ \forall v \in U. \tag{8.25}$$

Thus, the system of optimality conditions involves the initial system on the optimal control

$$y'(t) + Ay(t) = Bu(t) + f(t), \ t \in (0, T), \tag{8.26}$$

$$y(0) = \varphi, \tag{8.27}$$

the adjoint system (8.20), (8.21), and the variational inequality (8.25).

Remark 8.13 Note that the state function is known at the initial time. However, the function p is given at the finite time. Therefore, we cannot solve the equations (8.20) and (8.26) simultaneously. We can solve the system (8.20), (8.21), (8.25)–(8.27) by the *successive approximations method*. Suppose we know the value of the control at the previous iteration. We solve the Cauchy problem (8.26), (8.27) by standard technique. Then we solve the Cauchy problem (8.20), (8.21) with inverse time. Finally, the new iteration of the control can be obtained by variational inequality (8.25).

Now consider the optimization control problem for the system described by the heat equation as an example.

8.5 Optimization control problems for the heat equation

Let Ω be an open bounded set of Euclid space $\mathbb{R}^n$ with the boundary Γ, $T > 0$, $Q = \Omega \times (0, T)$, $\Sigma = \Gamma \times (0, T)$. Consider the **heat equation**

$$\frac{\partial y(x, t)}{\partial t} = \Delta y(x, t) + f(x, t), \ (x, t) \in Q \tag{8.28}$$

with homogeneous boundary condition

$$y(x, t) = 0 \ (x, t) \in \Sigma \tag{8.29}$$

and initial condition

$$y(x, 0) = \varphi(x), \ x \in \Omega, \tag{8.30}$$

where $y = y(x, t)$ is the state function, and the functions $f = f(x, t)$ and $\varphi = \varphi(x)$ are known. We have the **first boundary problem** for the heat equation.

Determine the spaces $W = H_0^1(\Omega)$ and $H = L_2(\Omega)$. We have the equality $W' = H^{-1}(\Omega)$. Note continuous and dense embedding $W \subset H \subset W'$. Determine the operator $A = -\Delta$. It is the linear continuous operator from the space W to W'. Using Green's formula, we have the equality

$$\langle Ay, y \rangle = - \int_\Omega y \Delta y dx = - \int_\Omega \sum_{i=1}^n \left(\frac{\partial y}{\partial x_i} \right)^2 dx = \|y\|_W^2 \ \forall y \in W.$$

Therefore, the relation

$$\langle Ay, y \rangle \geq \chi \|y\|_W^2 \ \forall y \in W$$

is true as the equality with the constant $\chi = 1$, Thus, the boundary problem (8.28)–(8.30) is transformed to Cauchy problem (8.5), (8.6). From Theorem 8.3 we have the following result.

Corollary 8.1 *For all $\varphi \in L_2(\Omega)$, $f \in L_2(0, T; H^{-1}(\Omega))$ there exists a unique solution of the problem* (8.28)–(8.30) *from the space*

$$Y = \left\{ y \mid y \in L_2(0, T; H_0^1(\Omega)), \ y' \in L_2(0, T; H^{-1}(\Omega)) \right\}.$$

Using Lemma 8.1, we prove the continuous dependence of the solution of this boundary problem from the initial state and the absolute term for the heat equation.

Corollary 8.2 *The map $(\varphi, f) \rightarrow y$ for the problem* (8.28)–(8.30) *is the continuous and weakly continuous operator from the space $L_2(\Omega) \times L_2(0, T, H^{-1}(\Omega))$ to Y, as the estimate* (8.15) *is true.*

284 *Optimization and Differentiation*

Now determine the optimization problem. We have the equation

$$\frac{\partial y(x,t)}{\partial t} = \Delta y(x,t) + v(x,t) + f(x,t), \quad (x,t) \in Q \tag{8.31}$$

with boundary conditions

$$y(x,t) = 0, \quad (x,t) \in \Sigma, \tag{8.32}$$

$$y(x,0) = \varphi(x), \quad x \in \Omega. \tag{8.33}$$

The functions $f = f(x,t)$ and $\varphi = \varphi(x)$ are known elements of the corresponding spaces (see Corollary 8.1), and the function $v = v(x,t)$ is the control. We choose it from the convex closed subset U of the space $V = L_2(Q)$. By Corollary 8.1 and Corollary 8.2 for all control $v \in V$ the problem (8.31)–(8.33) has a unique solution $y = y[v]$ from the set Y, as the operator $y[\cdot] : V \to Y$ is continuous and weakly continuous. Let $B : V \to L_2(0,T;W')$ be the embedding operator of the space $L_2(Q)$ to $L_2(0,T;H^{-1}(\Omega))$. Now we transformed the problem (8.31)–(8.33) to the form (8.16), (8.17).

Determine the functional

$$I(v) = \frac{\alpha}{2} \int_Q v^2 dQ + \frac{1}{2} \int_Q \sum_{i=1}^n \left(\frac{\partial y[v]}{\partial x_i} - \frac{\partial y_d}{\partial x_i} \right)^2 dQ,$$

where $\alpha > 0$, $y_d \in L_2(0,T;H_0^1(\Omega))$ is a known function.

Problem 8.2 *Find the control $u \in U$ for the system* (8.31)–(8.33) *that minimizes the functional I on the set U.*

Determine the space $Z = [L_2(Q)]^n$ and the linear operator $C : L_2(0,T;W) \to Z$ by the equality

$$Cy = \nabla y = \left\{ \frac{\partial y}{\partial x_1}, ..., \frac{\partial y}{\partial x_n} \right\}.$$

We transform Problem 8.2 to Problem 8.1. By Theorem 9.9 and Theorem 8.10, we have the following result.

Corollary 8.3 *Problem 8.2 is Tihonov well-posed.*

Use Theorem 8.11 for obtaining the optimality conditions.

Corollary 8.4 *The necessary and sufficient condition of optimality of the control u for Problem 8.2 is the variational inequality*

$$\int_Q (\alpha u + p)(v - u) dQ \geq 0 \quad \forall v \in U, \tag{8.34}$$

Linear first-order evolution systems

where p is the solution of the boundary problem

$$-\frac{\partial p(x,t)}{\partial t} = \Delta p(x,t) + \Delta y_d(x,t) - \Delta y(x,t), \ (x,t) \in Q, \tag{8.35}$$

$$p(x,t) = 0, \ (x,t) \in \Sigma, \tag{8.36}$$

$$p(x,T) = 0, \ x \in \Omega, \tag{8.37}$$

and y is the solution of the boundary problem

$$\frac{\partial y(x,t)}{\partial t} = \Delta y(x,t) + u(x,t) + f(x,t), \ (x,t) \in Q, \tag{8.38}$$

$$y(x,t) = 0, \ (x,t) \in \Sigma, \tag{8.39}$$

$$y(x,0) = \varphi(x), \ x \in \Omega. \tag{8.40}$$

Proof. The space $V = L_2(Q)$ is self-adjoint. Therefore, the operator Λ_V is a unit. The operator $B^* : L_2(0,T;W) \to V'$ is determined by the equality

$$\langle B^*y, v \rangle = \langle Bv, y \rangle = \int_Q vy dQ \ \forall v \in V, \ y \in L_2(0,T;W).$$

Then the variational inequality (8.31) is transformed to (8.34).

Using Green's formula, we have

$$\langle A^*p, y \rangle = \langle Ay, p \rangle = -\int_Q p \Delta y dQ = -\int_Q y \Delta p dQ \ \forall y, p \in L_2(0,T;W).$$

Hence, $A^* - \Delta$. Determine the operator

$$C^* : \left\{ [L_2(Q)]^n \right\}' \to L_2\big(0,T;H_0^1(\Omega)\big)$$

by the equality

$$\langle C^*z, y \rangle = \langle Cy, z \rangle = \int_Q \sum_{i=1}^n \left(z_i \frac{\partial y}{\partial x_i} \right) dQ = -\int_Q y \sum_{i=1}^n \frac{\partial z_i}{\partial x_i} dQ = -\int_Q y \mathrm{div} z dQ$$

for all $y \in L_2\big(0,T;H_0^1(\Omega)\big)$, $z \in [L_2(Q)]^n$. The space Z is self-adjoint. Therefore, the operator Λ_Z is a unit too. Then

$$C^*z = -\mathrm{div} z, \ C^*Cy = -\mathrm{div} \triangledown y = -\Delta y.$$

Thus, the equalities (8.26), (8.27) are transformed to the boundary problem (8.35)–(8.37). $\square$

Suppose the set of admissible controls is determined by the local constraints. Particularly, consider the set

$$U = \big\{ v \in L_2(Q) \big| \ a \le v(x,t) \le b, \ (x,t) \in Q \big\},$$

where a and b are constants, as $a < b$. This is a convex closed bounded set (see Chapter 3). Consider the partial case of Problem 8.2.

286 *Optimization and Differentiation*

Problem 8.3 *Find the control for the system* (8.31)–(8.33) *that minimizes the functional I on the determined set U.*

Theorem 8.12 *The solution of Problem 8.3 is*

$$u(x,t) = \begin{cases} a, & \text{if } -p(x,t)/\alpha < a, \\ -p(x,t)/\alpha, & \text{if } a \leq -p(x,t)/\alpha \leq b, \\ b, & \text{if } -p(x,t)/\alpha > b. \end{cases} \qquad (8.41)$$

Proof. By Corollary 8.4, the unique solution of Problem 8.2 satisfies the variational inequality (8.34). Consider the set Q_* of all Lebesgue points of the functions $\alpha u + p$ and $(\alpha u + p)u$ on Q, i.e., the points, where we can use the Lebesgue Theorem for passing to the limit (see the next equalities). For arbitrary point (x_*, t_*) consider the sequence of neighborhood $\{O_k\}$ such that $m_k \to 0$ as $k \to \infty$, where m_k is the measure of the set O_k. By the Lebesgue Theorem we have

$$\lim_{k\to\infty} \frac{1}{m_k} \int_{O_k} [\alpha u(x,t) + p(x,t)] dQ = \alpha u(x_*, t_*) + p(x_*, t_*),$$

$$\lim_{k\to\infty} \frac{1}{m_k} \int_{O_k} [\alpha u(x,t) + p(x,t)] u(x,t) dQ = \alpha [u(x_*, t_*) + p(x_*, t_*)] u(x_*, t_*).$$

Determine the needle variation

$$v_k = (1 - \chi_k)u + \chi_k w,$$

where w is an arbitrary point of the interval $[a, b]$ and χ_k is the characteristic function of the set O_k.

Put $v = v_k$ in to the inequality (8.34). We get

$$\int_{O_k} [\alpha u(x,t) + p(x,t)] [w - u(x,t)] dQ \geq 0.$$

After dividing by m_k and passing to the limit we obtain

$$[\alpha u(x_*, t_*) + p(x_*, t_*)] [w - u(x_*, t_*)] \geq 0.$$

This inequality is true for all $w \in [a, b]$ and $(x_*, t_*) \in Q_*$. Therefore, almost for all $(x, t) \in Q$ we have

$$[\alpha u(x,t) + p(x,t)] [w - u(x,t)] \geq 0 \ \forall w \in [a, b]. \qquad (8.42)$$

If for a point (x, t) we have the inequality $\alpha u(x,t) + p(x,t) > 0$ then from the condition (8.42) it follows that $w - u(x,t) \geq 0$ for all $w \in [a, b]$. Then $u(x,t)$ is not greater than all numbers $w \in [a, b]$. Hence, it cannot be greater than a. However, it cannot be less than a because of the definition of the set

Linear first-order evolution systems

of admissible controls. Therefore, we have $u(x,t) = a$. Thus, if the number $-p(x,t)$ is greater than $\alpha u(x,t)$ that is equal to αa or if we have the inequality $-p(x,t)/\alpha < a$ then $u(x,t) = a$.

Analogically, if $\alpha u(x,t) + p(x,t) < 0$, then we have the inequality $w - u(x,t) \leq 0$ for all $w \in [a,b]$ because of (8.42). Then the value $u(x,t)$ is not less than all numbers from the interval $[a,b]$. Therefore it is not less than maximal of them, i.e., the number b. However, the inequality $u(x,t) \leq b$ is true because of the definition of the set U. Hence, we have the equality $u(x,t) = b$. Thus, if $-p(x,t)$ is less than the value αb that is equal to αb or $-p(x,t)/\alpha > b$ then $u(x,t) = b$.

Finally, if $\alpha u(x,t) + p(x,t) = 0$, then we find $u(x,t) = -p(x,t)/\alpha$. This value is admissible, if $a \leq -p(x,t)/\alpha \leq b$. Thus, the solution of the variational inequality (8.34) satisfies the formula (8.41). $\square$

We can put the control to the equality (8.38). We obtain the boundary problem with respect to the unknown functions y and p. If we find its solution, particularly, the function p, we find the optimal control by the formula (8.41).

Remark 8.14 We could extend these results by using the known technique. Particularly, we could consider optimization control problems by replacing the Laplace operator by the general elliptic one (see Section 3.11), equations with general nonlinearity (see Section 4.10), nonhomogeneous boundary conditions (see Section 3.11), Neumann problems (see Section 5.5), general integral functional (see Section 4.9), and boundary control (see Section 5.5).

Consider two special cases. The minimized functional can be independent directly from the control or we do not have any constraints with respect to the control.

8.6 Optimization control problems with the functional that is not dependent from the control

Consider the heat equation

$$\frac{\partial y(x,t)}{\partial t} = \Delta y(x,t) + u(x,t) + f(x,t), \ (x,t) \in Q \tag{8.43}$$

with the boundary conditions

$$y(x,t) = 0, \ (x,t) \in \Sigma, \tag{8.44}$$

$$y(x,0) = \varphi(x), \ x \in \Omega. \tag{8.45}$$

We suppose the inclusions $\varphi \in L_2(\Omega)$ and $f \in L_2(0,T; H^{-1}(\Omega))$. The control v belongs to a convex closed subset U of the space $V = L_2(Q)$. By Corollary 8.1 and Corollary 8.2 for all $v \in V$ the problem (8.43)–(8.45) has a unique solution $y = y[v]$ from the space

$$Y = \left\{ y \mid y \in L_2\big(0, T; H_0^1(\Omega)\big), \ y' \in L_2\big(0, T; H^{-1}(\Omega)\big) \right\},$$

as the operator $y[\cdot] : V \to Y$ is continuous and weakly continuous. Determine the functional

$$I(v) = \frac{1}{2} \int\limits_\Omega \sum_{i=1}^n \left(\frac{\partial y[v]}{\partial x_i} - \frac{\partial y_d}{\partial x_i} \right)^2 dQ,$$

where $y_d \in L_2\big(0, T; H_0^1(\Omega)\big)$ is a known function.

Problem 8.4 *Find the control for the system (8.43)–(8.45) that minimizes the functional I on the determined set U.*

This is Problem 8.2 for $\alpha = 0$. We can prove continuity and strict continuity of this functional (see Lemma 8.2 and Lemma 8.3). Then we obtain the following result.

Theorem 8.13 *Problem 8.4 has a unique solution.*

Remark 8.15 For this case the functional is not uniformly strictly convex. Therefore, we cannot guarantee Tihonov well-posedness of the problem.

We have the following result that is an analog of Corollary 8.4.

Theorem 8.14 *The necessary and sufficient condition of optimality of the control u for Problem 8.4 is the variational inequality*

$$\int\limits_Q p(v - u)dQ \geq 0 \ \forall v \in U, \tag{8.46}$$

where p is the solution of the boundary problem

$$-\frac{\partial p(x,t)}{\partial t} = \Delta p(x,t) + \Delta y_d(x,t) - \Delta y(x,t), \ (x,t) \in Q, \tag{8.47}$$

$$p(x,t) = 0, \ (x,t) \in \Sigma, \tag{8.48}$$

$$p(x,T) = 0, \ x \in \Omega. \tag{8.49}$$

The relation (8.46) is the corollary of the variational inequality (8.34) for $\alpha = 0$. However, these optimality conditions have different properties. Suppose there exists control $u \in U$ such that we have the equality $y[u] = y_d$. Then our functional is equal to zero, i.e., this control is optimal, as the function satisfies the homogeneous equation (8.47) with homogeneous boundary conditions (8.48), (8.49). The unique solution of this boundary problem is equal to zero. Then the relation (8.46) is transformed to the trivial inequality $0 \geq 0$. Thus, the ***optimality condition degenerates***. We cannot use it for solving our problem. This optimal control is called ***singular***.

Linear first-order evolution systems 289

We shall use **Tihonov regularization method** for finding singular optimal control. Determine the functional

$$I_k(v) = I(v) + \frac{\alpha_k}{2} \int_Q v^2 dQ,$$

where $\alpha_k > 0$ and $\alpha_k \to 0$ as $k \to \infty$.

Problem 8.5 *Find the control for the system* (8.43)–(8.45) *that minimizes the functional I_k on the set U.*

This problem is equivalent to Problem 8.2. By Corollary 8.4, we have the following result.

Corollary 8.5 *Problem 8.5 is Tihonov well-posed.*

Let u_k be the solution of Problem 8.5. Prove the convergence of the regularization method.

Theorem 8.15 $I(u_k) \to \min I(U)$ *as $k \to \infty$.*

Proof. We have

$$I_k(u_k) = \min I_k(U) \le I_k(u) = I(u) + \frac{\alpha_k}{2}\|u\|_V^2 = \min I(U) + \frac{\alpha_k}{2}\|u\|_V^2,$$

where u is the solution of Problem 8.4. We have the inequality

$$\lim_{k\to\infty} I_k(u_k) \le \min I(U). \tag{8.50}$$

Using the boundedness of the set U, we determine the boundedness of the sequence $\{u_k\}$ in the space V. Extracting a subsequence, we have the convergence $u_k \to v$ weakly in V, as $v \in V$. By the weak continuity of the operator $y[\cdot] : V \to Y$, we get $y[u_k] \to y[v]$ weakly in Y. Then we have

$$\frac{\partial y[u_k]}{\partial x_i} \to \frac{\partial y[v]}{\partial x_i} \text{ weakly in } L_2(Q), \quad i = 1,...,n.$$

Using the weak lower semicontinuity of the square of the norm for the space $L_2(Q)$, we obtain

$$I(v) \le \inf_{k\to\infty} \lim I(u_k).$$

By inequality $I(u_k) \le I_k(u_k)$ and the inequality (8.50), we have $I(v) \le \min I(U)$. Therefore, the control v is the solution of Problem 8.4. From the inequalities

$$\min I(U) = I(v) \le \inf_{k\to\infty} \lim I(u_k) \le \lim_{k\to\infty} I_k(u_k) \le \min I(U)$$

it follows that $I(u_k) \to \min I(U)$. $\square$

290 *Optimization and Differentiation*

Thus, for large enough number k the value of the minimized functional for Problem 8.4 at the solution u_k of Problem 8.5 is close enough to the minimum of the functional I on the set of admissible controls. Therefore, the control u_k can be chosen as an approximate solution of Problem 8.4.

Remark 8.16 How can we determine the approximate solution for the problem of minimization of the functional I on the set U? We can choose an admissible control that is close enough to the point minimum of this functional on the given set. However, it can be a control such that its value of the functional is close enough to the minimum of the functional. If the problem is Tihonov well-posed, then both forms of the approximate solution are equivalent. However, the non-closed controls can have closed enough values of the functional for the ill-posed problems.

Use Corollary 8.4 for finding the function u_k.

Corollary 8.6 *The control u_k is the solution of Problem 8.5 if and only if it satisfies the variational inequality. The necessary and sufficient condition of optimality of the control u for Problem 8.4 is the variational inequality*

$$\int_Q (\alpha_k u_k + p_k)(v - u_k)dQ \geq 0 \ \forall v \in U, \tag{8.51}$$

where p_k is the solution of the boundary problem

$$-\frac{\partial p_k(x,t)}{\partial t} = \Delta p_k(x,t) + \Delta y_d(x,t) - \Delta y[u_k](x,t), \ (x,t) \in Q, \tag{8.52}$$

$$p_k(x,t) = 0, \ (x,t) \in \Sigma, \tag{8.53}$$

$$p_k(x,T) = 0, \ x \in \Omega. \tag{8.54}$$

Using Theorem 8.15 and Corollary 8.6, we can choose the solution of the variational inequality (8.51) for large enough k as an approximate solution of Problem 8.4.

Remark 8.17 Using Theorem 8.15, we can pass to the limit in the formula (8.51) and obtain the variational inequality (8.46).

8.7 Non-conditional optimization control problems

Consider Problem 8.1 for the non-conditional case.

Problem 8.6 *Find a control $u \in V$ that minimizes the functional I for the abstract system (8.16), (8.17) on the space V.*

Linear first-order evolution systems

Now the variational inequality (8.25) is equal to the stationary condition (see Chapter 1)

$$\alpha \Lambda_V^{-1} u + B^* p = 0.$$

Find the control

$$u = -\alpha^{-1} \Lambda_V^{-1} B^* p. \tag{8.55}$$

Thus, the solution of Problem 8.6 satisfies the following system:

$$y' + Ay = -\alpha^{-1} B \Lambda_V^{-1} B^* p + f, \tag{8.56}$$

$$y(0) = \varphi, \tag{8.57}$$

$$-p' + A^* p = C^* \Lambda_Z^{-1} Cy - C^* \Lambda_Z^{-1} C y_d, \tag{8.58}$$

$$p(T) = 0. \tag{8.59}$$

Lemma 8.5 *There exists a linear continuous operator* $R : Y \to Y$ *and a function* $r \in Y$ *such that*

$$p = Ry + r. \tag{8.60}$$

Proof. From Theorem 8.1 it follows that for all $\varphi \in H$ and $f \in L_2(0, T; W')$ the problem (8.5), (8.6) has a unique solution $y \in Y$. Fix the value φ. Then there exists an operator $\Phi : L_2(0, T; W') \to Y$ such that $y = \Phi f$. By Lemma 8.1 the operator Φ is continuous. The adjoint problem

$$-p' + A^* p = f, \ p(T) = 0$$

has the analogical properties. Therefore, there exists a continuous operator $\Psi : L_2(0, T; W') \to Y$ such that the solution of this problem is equal to $p = \Psi f$ for all $f \in L_2(0, T; W')$. Prove that the operator Ψ is affine, i.e., it satisfies the equality

$$\Psi \big[\sigma f_1 + (1 - \sigma) f_2 \big] = \sigma \Psi f_1 + (1 - \sigma) \Psi f_2 \ \forall f_1, f_2 \in L_2(0, T; W'), \ \sigma \in [0, 1].$$

Let p_i be the solution of the adjoint problem for $f = f_i, \ i = 1, 2$. We have

$$-\big[\sigma p_1' + (1 - \sigma) p_2' \big] + \big[\sigma A^* p_1 + (1 - \sigma) A^* p_2 \big] = \sigma f_1 + (1 - \sigma) f_2,$$

$$\sigma p_1(T) + (1 - \sigma) p_2(T) = 0.$$

Hence, the function $p_1 + (1 - \sigma) p_2$ is the solution of the adjoint problem for the absolute term $\sigma f_1 + (1 - \sigma) f_2$. Therefore, we have the equality

$$\Psi \big[\sigma f_1 + (1 - \sigma) f_2 \big] = \sigma \Psi f_1 + (1 - \sigma) \Psi f_2.$$

By definition of the affine operator there exists a linear continuous operator $\Psi_0 : L_2(0, T; W') \to Y$ and a function ψ such that $p = \Psi f = \Psi_0 f + \psi$. Hence, the solution of the problem (8.58), (8.59) is

$$p = \Psi_0 C^* \Lambda_Z^{-1} C(y - y_d) + \psi.$$

292 *Optimization and Differentiation*

Determine $R = \Psi_0 C^* \Lambda_Z^{-1} C$ and $r = \psi - \Psi_0 C^* \Lambda_Z^{-1} C y_d$. This completes the proof of the lemma. $\square$

Find the equations with respect to the operator R and the function r.

Remark 8.18 Our next transformations are formal. We do not give its strict substantiation because this is a well-known result that is not related to our general problem—application of differentiation theory to the optimization control problems.

Write formally the equality (8.60)

$$p(t) = R(t)y(t) + r(t), \ t \in (0, T).$$

After differentiation we have

$$p' = R'y + Ry' + r'.$$

Put the values of the derivatives from the equalities (8.56), (8.58). We get

$$A^* p - C^* \Lambda_Z^{-1} C y + C^* \Lambda_Z^{-1} C y_d = R'y - R\big(Ay + \alpha^{-1} \Lambda_V^{-1} B^* p - f\big) + r.$$

Using (8.60), we obtain

$$\big(R' - A^* R - RA - \alpha^{-1} RB\Lambda_V B^* R + C^* \Lambda_Z^{-1} C\big)y +$$

$$\big(r' - A^* r - C^* \Lambda_Z^{-1} C y_d - \alpha^{-1} RB\Lambda_V B^* r - Rf\big) = 0.$$

Let the values in the brackets be equal to zero. We have the equations

$$R' - A^* R - RA - \alpha^{-1} RB\Lambda_V B^* R + C^* \Lambda_Z^{-1} C = 0, \qquad (8.61)$$

$$r' - A^* r - C^* \Lambda_Z^{-1} C y_d - \alpha^{-1} RB\Lambda_V B^* r - Rf = 0. \qquad (8.62)$$

For $t = T$ we have

$$p(T) = R(T)y(T) + r(T)$$

because of the equality (8.59). Let the coefficient before $y(T)$ and absolute term be equal to zero here. We obtain the equalities

$$R(T) = 0, \qquad (8.63)$$

$$r(T) = 0. \qquad (8.64)$$

Remark 8.19 The equation (8.61) with final condition (8.63) is a Cauchy problem with respect to the operator R. Of course, it is necessary to prove its solvability.

Remark 8.20 The equation (8.61) is nonlinear. The fourth term here is square with respect to R. Differential equations with square nonlinearity are called Riccati equations. Therefore, we call (8.61) a **Riccati equation**. Note that the equation (8.62) with known value R is linear.

Linear first-order evolution systems 293

Thus, we have the following algorithm for finding the solution of Problem 8.6 (see Figure 8.1). At first, we find the operator R from the equation (8.61) with condition (8.63). Then we find the function r from the equation (8.62) with condition (8.64). We obtain the direct dependence of the function p from y. Put it in to the equality (8.56). Solve the equation

$$y' + Ay = -\alpha^{-1}B\Lambda_V B^*(Ry + r) + f \tag{8.65}$$

with initial condition (8.57). We determine the function y. Then find the function p by formula (8.60). Finally, the solution of Problem 8.5 can be determined by the formula (8.55).

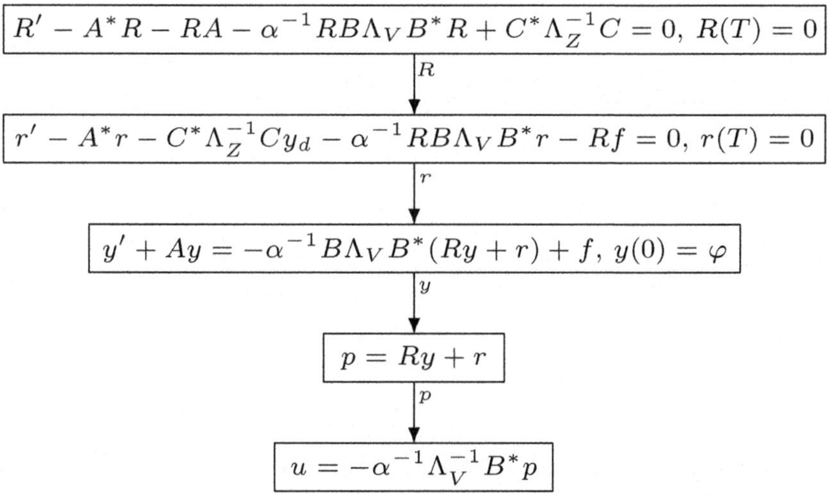

$$\boxed{R' - A^*R - RA - \alpha^{-1}RB\Lambda_V B^*R + C^*\Lambda_Z^{-1}C = 0, \ R(T) = 0}$$

$$\downarrow R$$

$$\boxed{r' - A^*r - C^*\Lambda_Z^{-1}Cy_d - \alpha^{-1}RB\Lambda_V B^*r - Rf = 0, \ r(T) = 0}$$

$$\downarrow r$$

$$\boxed{y' + Ay = -\alpha^{-1}B\Lambda_V B^*(Ry + r) + f, \ y(0) = \varphi}$$

$$\downarrow y$$

$$\boxed{p = Ry + r}$$

$$\downarrow p$$

$$\boxed{u = -\alpha^{-1}\Lambda_V^{-1}B^*p}$$

FIGURE 8.1: The solution of the program control problem.

We find the control for each time. This is the solution of the program control problem. However, we can consider the synthesis problem. The dependence of the optimal control with respect to the state function is an object of finding for this case. We find the feedback system here (see Figure 8.2). Indeed, from the equalities (8.55) and (8.60) we find the optimal control

$$u = -\alpha^{-1}\Lambda_V B^*(Ry + r) \tag{8.66}$$

as a function of o. The values R and r are known here by the equalities (8.61)–(8.64).

Remark 8.21 The program control is known for each time in advance. However, we determine the values R and r only for the synthesis problem. Then we model the dependence of the control from the state function by the formula (8.66). That is in reality a feedback system. The state function is the object of experimental measuring here. Then we can determine the optimal control by the formula (8.66) with known state y.

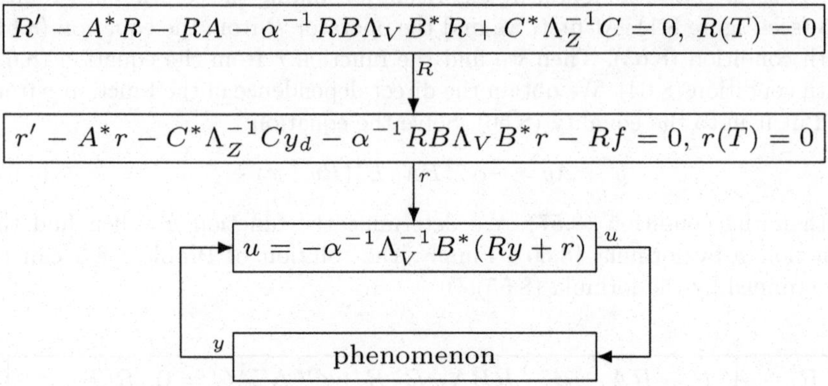

294 Optimization and Differentiation

$$R' - A^*R - RA - \alpha^{-1}RB\Lambda_V B^*R + C^*\Lambda_Z^{-1}C = 0, \; R(T) = 0$$

$\big\downarrow R$

$$r' - A^*r - C^*\Lambda_Z^{-1}Cy_d - \alpha^{-1}RB\Lambda_V B^*r - Rf = 0, \; r(T) = 0$$

$\big\downarrow r$

$$u = -\alpha^{-1}\Lambda_V^{-1}B^*(Ry + r) \quad u$$

phenomenon

FIGURE 8.2: The solution of the synthesis problem.

Remark 8.22 The equalities (8.61)–(8.64), (8.66) do not depend upon the initial state φ. Therefore, we can use the solution of the synthesis problem for the analysis of the control systems with different initial states.

8.8 Non-conditional optimization control problems for the heat equation

Consider the non-conditional optimization control problem for the heat equation as an application of the obtained results. We shall consider an easier functional

$$I(v) = \frac{\alpha}{2}\int_Q v^2 dQ + \int_Q \big(y[v] - y_d\big)dQ,$$

where $\alpha > 0$, $y_d \in L_2(Q)$ is a known function.

Problem 8.7 *Find a control that minimizes the functional I for the abstract system (8.31)–(8.33) on the space V.*

Using known results, determine the solution of this problem by the formula

$$u = -p/\alpha, \tag{8.67}$$

where p is the solution of the boundary problem

$$-\frac{\partial p(x,t)}{\partial t} = \Delta p(x,t) + y(x,t) - y_d(x,t), \; (x,t) \in Q, \tag{8.68}$$

$$\text{Linear first-order evolution systems} \qquad 295$$

$$p(x,t) = 0 \ (x,t) \in \Sigma, \tag{8.69}$$

$$p(x,T) = 0, \ x \in \Omega. \tag{8.70}$$

Put the control from the formula (8.67) in to the equality (8.38). We have

$$\frac{\partial y}{\partial t} = \Delta y - \alpha^{-1} p + f. \tag{8.71}$$

The system (8.39), (8.40), (8.67)–(8.70) with respect to the functions y and p is the problem (8.56)–(8.59) for the equation heat. By Lemma 8.5, these functions satisfy the equality function (8.60). Try to find the relation between these functions by the equality

$$p(x,t) = \int_Q R(x,\xi,t) y(\xi,t) d\xi + r(x,t), \tag{8.72}$$

where R and r are unknown functions.

Put p from the formula (8.72) to the equality (8.68). We have

$$-\int_Q \left(\frac{\partial R}{\partial t} y + R \frac{\partial y}{\partial t} \right) d\xi - \frac{\partial r}{\partial t} = \int_Q \Delta_x R y d\xi + \Delta r + y - y_d.$$

The term $\Delta_x R$ is the value of the Laplace operator with respect to the variable x at the function $R = R(x,\xi,t)$. Put here the time derivative of the function y from the equality (8.71). We get

$$-\int_Q \left[\frac{\partial R}{\partial t} y + R(\Delta y - \alpha^{-1} p + f) \right] d\xi - \frac{\partial r}{\partial t} = \int_Q \Delta_x R y d\xi + \Delta r + y - y_d.$$

Using Green's formula, we obtain

$$-\int_Q R \Delta y d\xi + \int_S R \frac{\partial y}{\partial n} d\xi = -\int_Q \Delta_\xi R y d\xi + \int_S \frac{\partial R}{\partial n} y d\xi.$$

The function y is equal to zero on the boundary because of the condition (8.39). Choose the function R such that the following equality holds:

$$R(x,\xi,t) = 0, \ (\xi,t) \in \Sigma, \ x \in \Omega. \tag{8.73}$$

We get

$$\int_Q R \Delta y d\xi = \int_Q \Delta_\xi R y d\xi.$$

Using the formula (8.72), find the integral

$$\int_\Omega R(x,\xi,t) p(\xi,t) d\xi = \int_\Omega R(x,\xi,t) \left[\int_\Omega R(\xi,\eta,t) y(\eta,t) d\eta + r(\xi,t) \right] d\xi =$$

296   *Optimization and Differentiation*

$$\int_\Omega \Big[\int_\Omega R(x,\eta,t)R(\eta,\xi,t)d\eta \Big] y(\xi,t)d\xi + \int_\Omega R(x,\xi,t)r(\xi,t)d\xi.$$

Determine δ-***function*** that is the distribution, satisfying the equality

$$\psi(x) = \int_\Omega \delta(x-\xi)\psi(\xi)d\xi, \ x \in \Omega$$

for all functions ψ. Then we have

$$y(x,t) = \int_\Omega \delta(x-\xi)y(\xi,t)d\xi.$$

Thus, we determine the equality

$$\int_\Omega \Big[\frac{\partial R(x,\xi,t)}{\partial t} + \Delta_\xi R(x,\xi,t) + \Delta_x R(x,\xi,t) -$$

$$\alpha^{-1} \int_\Omega R(x,\eta,t)R(\eta,\xi,t)d\eta + \delta(x-\xi) \Big] y(\xi,t)d\xi +$$

$$\int_\Omega \Big[\frac{\partial r}{\partial t} + \Delta r - \alpha^{-1} \int_\Omega R(x,\xi,t)r(\xi,t)d\xi + \int_\Omega R(x,\eta,t)f(\eta,t)d\xi - y_d \Big] = 0.$$

Suppose the values in the square brackets are equal to zero. We obtain the equations with respect to the functions R and r

$$\frac{\partial R(x,\xi,t)}{\partial t} + \Delta_\xi R(x,\xi,t) + \Delta_x R(x,\xi,t) -$$

$$\alpha^{-1} \int_\Omega R(x,\eta,t)R(\eta,\xi,t)d\eta + \delta(x-\xi) = 0, \ x,\xi \in \Omega, \ t \in (0,T); \qquad (8.74)$$

$$\frac{\partial r}{\partial t} + \Delta r - \alpha^{-1} \int_\Omega R(x,\xi,t)r(\xi,t)d\xi +$$

$$\int_\Omega R(x,\eta,t)f(\eta,t)d\xi = y_d, \ (x,t) \in Q. \qquad (8.75)$$

Suppose x belongs to the boundary Γ. The equality in (8.72) is true if the functions R and r are equal to zero on the surface Σ. Therefore, we have the following boundary condition:

$$R(x,\eta,t) = 0, \ (x,t) \in \Sigma, \ \xi \in \Omega; \qquad (8.76)$$

$$r(x,t) = 0, \ (x,t) \in \Sigma. \qquad (8.77)$$

Linear first-order evolution systems 297

Determine $t = T$ in the equality (8.72). Using the equality (8.70), we determine the final conditions

$$R(x, \eta, T) = 0, \ x, \xi \in \Omega; \tag{8.78}$$

$$r(x, T) = 0, \ x \in \Omega. \tag{8.79}$$

Now we can use the known algorithm. At first, we solve the nonlinear integro-differential equation (8.74) with boundary conditions (8.73), (8.76), (8.78). Then we find the function r from the linear integro-differential equation (8.75) with conditions (8.77), (8.79). The function y can be determined from the linear integro-differential equation

$$\frac{\partial y(x, t)}{\partial t} = \Delta y(x, t) - \alpha^{-1} \left[\int_\Omega R(x, \xi, t) y(\xi, t) d\xi + r(x, t) \right] + f(x, t), \ (x, t) \in Q$$

with boundary conditions (8.39), (8.40). The function p is determined by the formula (8.72). Finally, the solution to Problem 8.7 is determined by the formula (8.67).

Remark 8.23 Of course, it is necessary to prove the solvability of all boundary problems for substantiation of these results.

Remark 8.24 We could find the solution of the synthesis problem for Problem 8.7.

8.9 Hamilton–Jacobi equation

Try to determine the equation with respect to the minimal value of the functional for the system, described by the linear evolutional equation without any constraints. Consider a number $s \in (0, T)$. Determine Hilbert spaces Θ and $V_s = L_2(s, T; \Theta)$. Suppose the operator A and the spaces H and W satisfy the same properties as before, and B is a linear continuous operator from the space Θ to W'. Consider the equation

$$y'(t) + Ay(t) = Bv(t), \ t \in (s, T) \tag{8.80}$$

with initial condition

$$y(s) = \psi, \tag{8.81}$$

where $\psi \in H$. For all $v \in V_s$ the problem (8.80), (8.81) has a unique solution $y = y[v]$ from the space

$$Y_s = \left\{ y \mid y \in L_2(s, T; W), \ y' \in L_2(s, T; W') \right\}.$$

298 *Optimization and Differentiation*

Determine the functional

$$I_{s\psi} = \frac{1}{2} \int_s^T \left[\alpha \|v(t)\|_\Theta^2 + \|Cy[v](t)\|_Z^2 \right] dt,$$

where $\alpha > 0$, Z is a Hilbert space, and $C : W \to Z$ is a linear continuous operator.

Problem 8.8 *Find a control $u \in V_s$ that minimizes the functional $I_{s\psi}$ on the space V_s.*

This problem is Tihonov well-posed. The necessary and sufficient condition of optimality for it is the stationary condition

$$\alpha \Lambda_\Theta^{-1} u(t) + B^* p(t) = 0,$$

where Λ_Θ is the canonic isomorphism between the spaces Θ' and Θ. The abstract function p here is the solution of the system

$$-p'(t) + A^* p(t) = C^* \Lambda_Z^{-1} C y(t), \ t \in (s, T) \tag{8.82}$$

$$p(T) = 0. \tag{8.83}$$

Then find the optimal control

$$u(t) = -\alpha^{-1} \Lambda_\Theta B^* p(t). \tag{8.84}$$

Put it in to the equation (8.80); we have

$$y'(t) + Ay(t) = -\alpha^{-1} B \Lambda_\Theta B^* p(t), \ t \in (s, T). \tag{8.85}$$

Thus, the solution of Problem 8.8 satisfies the system (8.81)–(8.83), (8.85), which is the analog (8.56)–(8.59).

Determine

$$J(s, \psi) = \min I_{s\psi}(V_s).$$

Lemma 8.6 *The following equality holds:*

$$J(s, \psi) = \frac{1}{2} \big(p(s), \psi \big)_H. \tag{8.86}$$

Proof. Find the value of the minimizing functional at the optimal control. Using the equality (8.82), we have

$$\int_s^T \|Cy(t)\|_Z^2 dt = \int_s^T \big(Cy(t), Cy(t) \big)_Z = \int_s^T \Big\langle C^* \Lambda_Z^{-1} C y(t), y(t) \Big\rangle dt =$$

$$\int_s^T \Big\langle -p'(t) + A^* p(t), y(t) \Big\rangle dt.$$

Linear first-order evolution systems 299

By the equalities (8.81), (8.85), we find

$$\int_s^T \|Cy(t)\|_Z^2 dt = \big(p(s), \psi\big)_H + \int_s^T \Big\langle y'(t) + Ay(t), p(t)\Big\rangle dt = \big(p(s), \psi\big)_H -$$

$$\alpha^{-1}\int_s^T \Big\langle B\Lambda_\Theta B^* p(t), p(t)\Big\rangle dt = \big(p(s), \psi\big)_H - \alpha^{-1}\int_s^T \Big\|\Lambda_\Theta B^* p(t)\Big\|_\Theta^2 dt.$$

Then we get

$$\int_s^T \|u(t)\|_\Theta^2 dt = \alpha^{-2}\int_s^T \Big\|\Lambda_\Theta B^* p(t)\Big\|_\Theta^2 dt.$$

Thus, the minimal value of the functional for Problem 8.8 is determined by the formula (8.86). $\square$

We have the following analog of Lemma 8.5.

Lemma 8.7 *There exists a linear continuous operator $R = R(t)$ on the set H such that*

$$p(t) = R(t)y(t), \ t \in (s, T). \tag{8.87}$$

Indeed, the problem (8.82), (8.83) has a unique solution that is continuous with respect to the absolute term of the equation. We do not have any absolute terms of the equation and boundary condition. Therefore, the relation between the function y and p is linear.

Remark 8.25 We determine $f = 0$ and $y_d = 0$ for Problem 8.3 for obtaining easier transformations.

Theorem 8.16 *The minimal value of the functional for Problem 8.8 satisfies the **Hamilton–Jacobi equation***

$$\frac{\partial J(s, \psi)}{\partial s} + H_0\Big(\psi, \frac{\partial J(s, \psi)}{\partial \psi}\Big) = 0, \ \psi \in V, \ s \in (0, T) \tag{8.88}$$

with final condition

$$J(T, \psi) = 0, \ \psi \in V, \tag{8.89}$$

where

$$H_0(\psi, q) = \frac{1}{2}\|C\psi\|_Z^2 - \langle A\psi, q\rangle + \frac{1}{2\alpha}\|B^* q\|_V^2. \tag{8.90}$$

Proof. Repeat the technique of obtaining the system (8.61)–(8.64). We have the problem

$$R'(t) - A^* R(t) - R(t)A - \alpha^{-1}R(t)B\Lambda_\Theta B^* R(t) + C^*\Lambda_Z^{-1} C = 0, \tag{8.91}$$

300 *Optimization and Differentiation*

$$R(T) = 0. \tag{8.92}$$

Using (8.90), we obtain the equality

$$\frac{d}{dt}\langle R(t)g, \psi \rangle - \langle A^* R(t)g, \psi \rangle - \langle R(t)Ag, \psi \rangle -$$

$$\alpha^{-1}\langle R(t)B\Lambda_\Theta B^* R(t)g, \psi \rangle + \langle C^* \Lambda_Z^{-1} Cg, \psi \rangle = 0 \; \forall g, \psi \in H.$$

Then we have

$$\frac{d}{dt}\langle g, R(t)^* \psi \rangle - \langle g, R(t)^* A\psi \rangle - \langle g, A^* R(t)^* \psi \rangle -$$

$$\alpha^{-1}\langle g, R(t)^* B\Lambda_\Theta B^* R(t)^* \psi \rangle + \langle g, C^* \Lambda_Z^{-1} C\psi \rangle = 0.$$

The points g and h are arbitrary. Therefore, we have the equality

$$\left(R(t)^*\right)' - R(t)^* A - A^* R(t)^* - \alpha^{-1} R(t)^* B\Lambda_\Theta B^* R(t)^* + C^* \Lambda_Z^{-1} C = 0, \; t \in (s, T).$$

Analogically, from the equality (8.91) it follows that

$$0 = \left(R(T)g, \psi\right)_H = \left(g, R(T)^* \psi\right)_H \; \forall g, \psi \in H.$$

Therefore, $R(T)^* = 0$. Thus, the operators $R(t)$ and $R(t)^*$ satisfy the same equalities. Using the uniqueness of the solution of the problem (8.91), (8.92), we get $R(t)^* = R(t)$.

Using the equality (8.87), we transform the formula (8.74). We have

$$J(s, \psi) = \frac{1}{2}\langle R(s)\psi, \psi \rangle. \tag{8.93}$$

Find the partial derivatives

$$\frac{\partial J}{\partial \psi} = R(s)\psi, \quad \frac{\partial J}{\partial s} = \frac{1}{2}\left\langle \frac{\partial R(s)}{\partial s}\psi, \psi \right\rangle.$$

Using the equality (8.90), we get

$$2\frac{\partial J}{\partial s} = \left\langle \frac{\partial R(s)}{\partial s}\psi, \psi \right\rangle = \langle A^* R(s)\psi, \psi \rangle + \langle R(s)A\psi, \psi \rangle +$$

$$\alpha^{-1}\langle R(s)B\Lambda_\Theta B^* R(t)\psi, \psi \rangle - \langle C^* \Lambda_Z^{-1} C\psi, \psi \rangle =$$

$$2\langle A\psi, R(s)\psi \rangle + \alpha^{-1}\|B^* R(t)\psi\|_\Theta^2 - \|C\psi\|_Z^2.$$

Determine the functional H_0 by the formula (8.90). We obtain the equation (8.88). Choose $t = T$ for the equality (8.93). Using (8.92), we have the boundary condition (8.79). $\square$

Remark 8.26 The Hamilton–Jacobi equation involves the derivatives of the function J only. If a function $J(s, \psi)$ satisfies this equation, then the function $J(s, \psi) + c$ with arbitrary constant is its solution too. By equality (8.89), this constant is equal to zero.

Linear first-order evolution systems

The equation (8.88) with final condition (8.89) gives us the minimal value of the functional for Problem 8.8. Now determine the functional

$$H(\psi, q, w) = \frac{\alpha}{2}\|w\|_\Theta^2 + \frac{1}{2}\|C\psi\|_Z^2 - \langle p, A\psi + Bv \rangle$$

that is called the **Hamiltonian**.

Lemma 8.8 *The following equality holds:*

$$H_0(\psi, q) = \min_{w \in \Theta} H(\psi, q, w).$$

Proof. Indeed, the derivative of I with respect to w is $\alpha w - B^*p$. Let this value be zero. We find $w = \alpha^{-1}B^*p$. The second derivative of H is positive. Therefore, this is the point of minimum. Put the last value of w in to the formula for H. We obtain the equality (8.90). $\square$

Now we have the following result.

Corollary 8.7 *The minimal value of the functional for Problem 8.8 satisfies Bellman equation*

$$\frac{\partial J(s, \psi)}{\partial s} = \min_{w \in \Theta} H\left(\psi, \frac{\partial J(s, \psi)}{\partial \psi}, w\right) \tag{8.94}$$

with final condition (8.89).

Consider the optimization problem for the heat equation as an application of these results.

8.10 Bellman method for an optimization problem for the heat equation

Consider the heat equation

$$\frac{\partial y(x, t)}{\partial t} = \Delta y(x, t) + v(x, t) + f(x, t), \quad (x, t) \in Q \tag{8.95}$$

with boundary conditions

$$y(x, t) = 0 \ (x, t) \in \Sigma, \tag{8.96}$$

$$y(x, 0) = \varphi(x), \ x \in \Omega, \tag{8.97}$$

where $\varphi \in L_2(\Omega)$, $f \in L_2(0, T; H^{-1}(\Omega))$. We choose the control v from the set

$$U = \left\{ v \in L_2(Q) \middle| \ v(t) \in G(t), \ t \in (0, T) \right\},$$

302   *Optimization and Differentiation*

where $G(t)$ is a convex closed subset of $L_2(\Omega)$ and $v(t)$ is the value of the function $v = v(x, t)$ for the fixed t. By Corollary 8.1 for all $v \in U$ the problem (8.95)–(8.97) has a unique solution $y = y[v]$ from the space

$$Y = \left\{ y \mid y \in L_2(0, T; H_0^1(\Omega)), \ y' \in L_2(0, T; H^{-1}(\Omega)) \right\}.$$

Consider the functional

$$I(v) = \frac{\alpha}{2} \int_Q v^2 dQ + \frac{1}{2} \int_Q \sum_{i=1}^n \left(\frac{\partial y[v]}{\partial x_i} - \frac{\partial y_d}{\partial x_i} \right)^2 dQ,$$

where $\alpha > 0$, $y_d \in L_2(0, T; H_0^1(\Omega))$ is a known function.

Problem 8.9 *Find the control that minimizes the functional I on the set U.*

This is the partial case of Problem 8.2. By Corollary 8.4 it has a unique solution u that satisfies the variational inequality

$$\int_Q (\alpha u + p)(v - u) dQ \geq 0 \ \forall v \in U, \tag{8.98}$$

where p is the solution of the boundary problem

$$-\frac{\partial p(x, t)}{\partial t} = \Delta p(x, t) + \Delta y_d(x, t) - \Delta y(x, t), \ (x, t) \in Q, \tag{8.99}$$

$$p(x, t) = 0 \ (x, t) \in \Sigma, \tag{8.100}$$

$$p(x, T) = 0, \ x \in \Omega, \tag{8.101}$$

and $y = y[u]$. Determine the following analog of Theorem 8.11.

Theorem 8.17 *The solution of Problem 8.9 satisfies the inequality*

$$\int_\Omega [\alpha u(x, t) + p(x, t)] [w(x) - u(x, t)] dx \geq 0 \ \forall w \in G(t), \ t \in [0, T]. \tag{8.102}$$

Proof. For almost all values $t \in [0, T]$ and all $w \in G(t)$ consider a sequence of neighborhoods $\{O_k\}$ of the point t such that the measure m_k of the set O_k satisfies the convergence $m_k \to 0$ as $k \to \infty$ and the following equalities hold:

$$\lim_{k \to \infty} \frac{1}{m_k} \int_{O_k} \int_\Omega [\alpha u(x, t) + p(x, t)] w(x) dx dt = \int_\Omega [\alpha u(x, t_*) + p(x, t_*)] w(x) dx,$$

$$\lim_{k \to \infty} \frac{1}{m_k} \int_{O_k} \int_\Omega [\alpha u(x, t) + p(x, t)] u(x, t) dx dt = \int_\Omega [\alpha u(x, t_*) + p(x, t_*)] u(x, t_*) dx.$$

Linear first-order evolution systems 303

Determine the control $v_k = (1 - \chi_k)u + \chi_k w$, where χ_k is the characteristic function of the set O_k. Choose $v = v_k$ in the inequality (8.98). We get

$$\int\limits_{O_k} \int\limits_{\Omega} \big[\alpha u(x,t) + p(x,t)\big]\big[w(x) - u(x,t)\big]dxdt \geq 0.$$

Dividing by m_k and passing to the limit, we have the variational inequality (8.102). $\square$

Determine another form of the optimality conditions for Problem 8.9. We use the **Bellman method** from the previous section. For all $s \in (0,T)$, $\psi \in L_2(\Omega)$ we determine the functional

$$I_{s\psi}(v) = \frac{1}{2}\int\limits_s^T \Big\{\alpha \int\limits_\Omega v^2 dx + \int\limits_\Omega \sum_{i=1}^n \Big(\frac{\partial z[v]}{\partial x_i} - \frac{\partial y_d}{\partial x_i}\Big)^2 dx\Big\}dt,$$

where $z[v]$ is the solution z of the boundary problem. Consider the heat equation

$$\frac{\partial z(x,t)}{\partial t} = \Delta z(x,t) + v(x,t) + f(x,t), \ x \in \Omega, \ t \in (s,T) \tag{8.103}$$

with boundary conditions

$$z(x,t) = 0 \ (x,t) \in S, \ t \in (s,T), \tag{8.104}$$

$$z(x,s) = \psi(x), \ x \in \Omega. \tag{8.105}$$

Determine the set

$$U_s = \Big\{v \in L_2\big((s,T) \times \Omega\big) \mid v(t) \in G(t), \ t \in (s,T)\Big\}.$$

Try to minimize the functional $I_{s\psi}$ on the set U_s. This problem has a unique solution because of Theorem 8.6.

Denote

$$J(s,\psi) = \min I_{s,\psi}(U_s).$$

We use the **Bellman optimality principle**: if the function u_s minimizes the functional $I_{s,\psi}$ on the set U_s, then the function $u_{s'}$ determined by the equality $u_{s'}(x,t) = u_s(x,t)$ for $x \in \Omega$ and $t \in (s',T)$ minimizes the functional $I_{s',\psi}$ on the set $U_{s'}$ for all $s' \in (s,T)$.

Remark 8.27 By the Bellman optimality principle, each final part of the optimal trajectory is optimal. If we choose the part of the optimal trajectory from an arbitrary point I, then it is the optimal trajectory for the problem with point M being the beginning of the time (see Figure 8.3).

Remark 8.28 This idea is the basis of the dynamic programming that is the serious method of solving optimization problems for multiphase phenomenon. By Bellman principle, the optimal control does not depend upon the prehistory of the system. Therefore, we begin to find optimal control from the final stage of the phenomenon. The dynamic programming is an effective enough method of solving optimization problems for discrete systems.

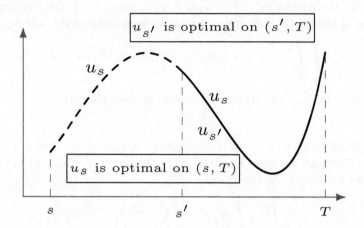

FIGURE 8.3: Bellman optimality principle.

Remark 8.29 The value J is called the **Bellman functional** or **Bellman function**.

Theorem 8.18 *The functional J satisfies Bellman the equation*

$$\frac{\partial J(s,\psi)}{\partial s} + H_1\left(s,\psi, \frac{\partial J(s,\psi)}{\partial \psi}\right) + \min_{w \in G(s)} H\left(\frac{\partial J(s,\psi)}{\partial \psi}, w\right) = 0 \qquad (8.106)$$

with $s \in (0,T)$, $\psi \in L_2(\Omega)$, and the final condition

$$J(T,\psi) = 0, \ \psi \in L_2(\Omega), \qquad (8.107)$$

where

$$H_1(s,\psi,\eta) = \int_\Omega \left\{ \frac{1}{2} \sum_{i=1}^n \left[\frac{\partial \psi(x)}{\partial x_i} - \frac{\partial y_d(x,s)}{\partial x_i}\right]^2 + \eta[\Delta\psi(x) + f(x,s)] \right\} dx,$$

$$H(\eta,w) = \int_\Omega \left(\frac{\alpha w^2}{2} + \eta w\right) dx.$$

Proof. The condition (8.107) is the direct corollary of the definition of functional J. Prove that this functional satisfies the Bellman equation. Using the Bellman optimality principle, we get

$$J(s,\psi) = \min_{v \in U_s} \int_s^T F_v(t)dt = \min_{v \in U_s}\left\{\int_s^{s+\tau} F_v(t)dt + \int_{s+\tau}^T F_v(t)dt\right\} =$$

Linear first-order evolution systems

$$\min_{v \in U_s} \left\{ \int_s^{s+\tau} F_v(t)dt + J\left(s+\tau, z[v](s+\tau)\right) \right\},$$

where

$$F_v(t) = \frac{1}{2} \int_\Omega \left\{ \alpha v^2 + \sum_{i=1}^n \left[\frac{\partial \psi(x)}{\partial x_i} - \frac{\partial y_d(x,s)}{\partial x_i} \right]^2 \right\} dx.$$

Then we have the equality

$$0 = \min_{v \in U_s} \left[\frac{1}{2} \int_s^{s+\tau} F_v(t)dt + r_v(t) \right], \tag{8.108}$$

where

$$r_v(t) = J\left(s+\tau, z[v](s+\tau)\right) - J\left(s, z[v](s)\right).$$

Using the mean value theorem and the condition (8.105), we get

$$\lim_{\tau \to 0} \frac{1}{\tau} \int_s^{s+\tau} F_v(t)dt = F_v(s) = \frac{1}{2} \int_\Omega \left\{ \alpha [v(x,s)]^2 + \sum_{i=1}^n \left[\frac{\partial \psi(x)}{\partial x_i} - \frac{\partial y_d(x,s)}{\partial x_i} \right]^2 \right\} dx.$$

Analogically, we have

$$\lim_{\tau \to 0} \frac{r_v(\tau)}{\tau} = \frac{\partial J(s,\psi)}{\partial s} + \left\langle \frac{\partial J(s,\psi)}{\partial \psi}, \frac{\partial z}{\partial t} \Big|_{t=s} \right\rangle =$$

$$\frac{\partial J(s,\psi)}{\partial s} + \int_\Omega \frac{\partial J(s,\psi)}{\partial \psi} (\Delta \psi + v + f) \Big|_{t=s} dx.$$

After dividing the equality (8.108) by τ and passing to the limit as $\tau \to 0$ we have

$$0 = \min_{v \in U_s} \left\{ \frac{1}{2} \int_\Omega \left\{ \alpha [v(x,s)]^2 + \sum_{i=1}^n \left[\frac{\partial \psi(x)}{\partial x_i} - \frac{\partial y_d(x,s)}{\partial x_i} \right]^2 \right\} dx + \right.$$

$$\left. \frac{\partial J(s,\psi)}{\partial s} + \int_\Omega \frac{\partial J(s,\psi)}{\partial \psi} \left[\Delta \psi(x) + v(x,s) + f(x,s) \right] dx \right\}.$$

Then we obtain

$$\frac{\partial J}{\partial s} + \int_\Omega \left[\frac{1}{2} \sum_{i=1}^n \left(\frac{\partial \psi}{\partial x_i} - \frac{\partial y_d}{\partial x_i} \right)^2 + \frac{\partial J}{\partial \psi} (\Delta \psi + f) \right] dx + \min_{v \in U_s} \int_\Omega \left(\frac{\alpha v^2}{2} + \frac{\partial J}{\partial \psi} v \right) dx = 0,$$

where all terms are considered for $t = s$. Denote the value of the function v for $t = s$ by w, and using its belonging to the set $G(s)$ we have the equality (8.106). $\square$

The practical using of this theorem is determin at first the control that minimizes H and finding the functional J from the system (8.106), (8.107).

306   *Optimization and Differentiation*

Corollary 8.8 *The solution of Problem 8.9 satisfies the inequality*

$$\int_\Omega \left[\alpha u(x,t) + \frac{\partial J(t,\psi)}{\partial \psi} \right] [w(x) - u(x,t)] \ \forall w \in G(t), \ t \in [0,T]. \quad (8.109)$$

Proof. By Bellman equation, the optimal control at the time t is the function $u = u(x,t)$ that minimizes on the set $G(t)$ the functional

$$K = K(w) = H\left(\frac{\partial J(t,\psi)}{\partial \psi}, w \right) = \int_\Omega \left[\frac{\alpha w^2}{2} + \frac{\partial J(t,\psi)}{\partial \psi} w \right] dx$$

for fixed value of the first argument of H. The functional is convex and differentiable. This control minimizes this functional if and only if it satisfies the variational inequality

$$\langle K'[u(t)], w - u(t) \rangle \geq 0 \ \forall w \in G(t),$$

where $u(t)$ is the function $u = u(x,t)$ for the fixed t. Now we obtain variational inequality (8.109) after the calculation of the Gâteaux derivative of the functional K. $\square$

The variational inequalities (8.102) and (8.109) give the necessary and sufficient optimality conditions for Problem 8.9. Comparing these relations, we obtain the equality

$$p(t) = \frac{\partial J(t,\psi)}{\partial \psi},$$

where $p(t)$ is the function $p = p(x,t)$ for the fixed t.

The Bellman equation is the **sufficient condition optimality**.

Theorem 8.19 *Suppose there exists Fréchet differentiable functional J that satisfies the problem* (8.106), (8.107). *Then the function u that is determined by the equality*

$$H\left(\frac{\partial J(s,\psi)}{\partial \psi}, u(s) \right) = \min_{w \in G(s)} \left(\frac{\partial J(s,\psi)}{\partial \psi}, w \right), \ s \in (0,T), \ \psi \in L_2(\Omega), \quad (8.110)$$

is the solution of Problem 8.9, as $I(u) = J(0,\varphi)$.

Proof. Find the derivative

$$\frac{d}{dt} J[t, y(t)] = \frac{\partial J}{\partial t} + \int_\Omega \frac{\partial J}{\partial y} \frac{\partial y}{\partial t} dx,$$

where $y(t)$ is the function $y = y(x,t)$ that is the solution of the boundary problem (8.95)–(8.97) for an admissible control v with fixed t. Using the equation (8.95), transform the previous equality

$$\frac{d}{dt} J[t, y(t)] = \frac{\partial J}{\partial t} + \int_\Omega \frac{\partial J}{\partial y} (\Delta y + v + f) dx.$$

Linear first-order evolution systems 307

Integrating this equality by t by using condition (8.107) and (8.97), we get

$$-J(0,\varphi) = \int\limits_0^T \left[\frac{\partial J}{\partial t} + \int\limits_\Omega \frac{\partial J}{\partial y}(\Delta y + v + f)\,dx\right]dt.$$

From the equality (8.106) and the definition of the minimizing functional after integration, the inequality follows:

$$\int\limits_0^T \left[\frac{\partial J}{\partial t} + \int\limits_\Omega \frac{\partial J}{\partial y}(\Delta y + v + f)\,dx\right]dt + I(v) \geq 0.$$

This condition is true as the equality for $v = u$ because of (8.110). Then we have

$$I(v) \geq J(0,\varphi)\ \forall v \in U;\quad I(u) \geq J(0,\varphi).$$

Therefore, $I(u) \leq I(v)$ for all $v \in U$, i.e., the control u is the solution of the considered optimization problem. $\square$

Remark 8.30 This optimality condition uses the supposition of the existence of the functional J.

The optimality conditions (8.109) and (8.110) are equivalent. The variational inequality (8.109) is the necessary and sufficient condition of minimum of the functional with respect to the second argument for the fixed value of its first argument and some number s. Note that (8.109) is equivalent to the variational inequality (8.102) that is the sufficient optimality condition. Therefore, now we have the necessary optimality condition too. Thus, the functional J for the considered problem exists in reality.

Remark 8.31 Of course, solving of the Bellman equation for more difficult optimization problem is very difficult.

We considered optimization control problems for linear evolutional systems. Therefore, we did not have any necessity to use the serious differentiation theory. However, the situation changes if we have a nonlinear system.

8.11 Comments

The short introduction on the abstract functions theory is given, for example, by H. Gajewski, K. Gröger, and K. Zacharias [200]. The classic spaces of the differential and integrable functions are considered in each standard course of the functional analysis. The distributions theory is considered by P. Antosic, J. Mikusinski, and R. Sikorski [18], H. Bremermann [90], F.G. Friedlander [185], I.M. Gelfand and G.E. Shilov [206], J.I. Richards, H.K. Joun [430], L. Schwartz [448], S.L. Sobolev [508], and V.S. Vladimirov [542]. The

308 *Optimization and Differentiation*

different Sobolev spaces are considered in the following books: R. Adams [4], O.V. Besov, V.P. Ilyin, and S.M. Nikolsky [67], H. Gajewski, K. Gröger, and K. Zacharias [200], J.L. Lions and E. Magenes [319], and S.L. Sobolev [508]. The Bochner integral was proposed by S. Bochner (see K. Iosida [256]). The different forms of integration are described by F. Burk [100]. L. Schwartz analyzes the distributions with values in Banach spaces [448]. We base on the book H. Gajewski, K. Gröger, K. Zacharias [200] for describing the functional spaces. The proof of the Theorems 8.1–8.4 is given by H. Gajewski, and K. Gröger, K. Zacharias [200], and the proof of the Theorem 8.5 is given by J.L. Lions [316].

There exist many books where the theory of the ordinary differential equations (particularly, Theorem 8.6 and Theorem 8.7) is described (see, for example, V.I. Arnold [24] and S. Lefschetz [305]). We use the properties of the abstract linear differential equations from the book by J.L. Lions [315], where the proof of Theorem 8.8 is given (see also S.G. Krein [283], and J.L. Lions and E. Magenes [319]).

Optimal control problems for the abstract linear evolutional equations are considered by J.L. Lions [314], [315], P. Neittaanmäki and D. Tiba [384], L.P. Pan and K.L. Teo [392], and others. L.P. Pan and K.L. Teo [392] use the penalty method for obtaining conditions of optimality. G. Knowles [272] analyzes a time optimal problem for these systems. P. Neittaanmäki and D. Tiba [384] consider extremum problems with state constraints and non-smooth functionals.

The theory of the linear parabolic equations is considered, for example, by A. Friedman [186], O.A. Ladyzhenskaya, V.A. Solonnikov, and N.N. Uraltseva [293], J.L. Lions [315], and J.L. Lions and E. Magenes [319]. Optimal control problems for the linear parabolic equations including Problem 8.2 are examined by different authors (see, for example, N.U. Ahmed and K.L. Teo [10], A.I. Egorov [154], [155], Yu.V. Egorov [156], H.O. Fattorini [170], A.V. Fursikov [197], J.L. Lions [314], [315], V.I. Plotnikov [403], [404], T.I. Seidman [453], T.K. Sirazetdinov [503], F.P. Vasiliev [538], [539], and Z.S. Wu and K.L. Teo [557].

Tihonov regularization method for extremum problems is applied by Yu.V. Egorov [156], H.W. Engl [166], R.P. Fedorenko [172], T. Kobayashi [275], M.M. Potapov [410], S.Ya. Serovajsky [456], [482], A.N. Tihonov [523], A.N. Tihonov and V.Ya. Arsenin [524], and A.N. Tihonov and F.P. Vasiliev [525]. Other regularization methods are considered by J. Baranger [47], M. Edelstein [151], [152], J.L. Lions [315], [317], and S.Ya. Serovajsky [464].

The general theory of the singular control is described by R. Gabasov and F.M. Kirillova [199]. The singular controls for the ordinary differential equations are considered by Ya.M. Bershchansky [66], S.Ya. Serovajsky [482], and B.P. Yeo [561]. Particularly, B.P. Yeo [561] applies Tihonov regularization method for finding the singular control. O.V. Vasiliev [540] consider the singular control for the system described by the Goursat problem.

We consider the determination of the integro-differential Riccati equation for Problem 8.6 on the basis of the book by J.L. Lions [315] (see also A. Bensoussan [61] and P.K.C. Wang [550]). The direct analysis of this equation is given, for example, by G. Da Prato [135]. The synthesis problem for the system described by linear parabolic equations is considered by A.I. Egorov [154], [155], T. Banks and C. Wang [46], Da Prato [135], T. Kobayashi [275], I. Lasiecka and R. Triggiani [299], and T.K. Sirazetdinov [503].

Hamilton–Jacobi equation is the classic result of the calculus of variations (see, for example, N.I. Ahiezer [7], G.A. Bliss [75], O. Bolza [82], G. Buttazzo, G. Mariano, and S. Hildebrandt [104], R. Courant [130], E. Elsgolts [164], I.M. Gelfand and S.V. Fomin [205], M.R. Hestens [241], M.A. Lavrentiev and L.A. Lusternik [300], G. Leitmann [307], T.K. Surazhetdinov [503], L. Young [563], and L. Zlaf [570]). It was obtained by W. Hamilton and C. Jacobi, although some results in this direction have been set previously by P. Fermat, C. Huygens, J. Bernulli, and others. Hamilton–Jacobi equation for the optimization problems for the systems described by the ordinary differential equations is considered by G.Q. Chen and B. Su [123], and others. The analogical results for the distributed parameter systems are obtained, for example, by V. Barbu [50], [51], P. Cannarsa and G. Da Prato [108], P. Cannarsa and M.E. Tessitore [109], and J.L. Lions [315]. H. Frankowska [183], J.-P. Penot, and C. Zălinescu [399] analyze the Hamilton–Jacobi equations.

The Bellman method and dynamical programming are the natural extension of the Hamilton–Jacobi theory to the extremum problems with constraints (see R. Bellman [57],

Linear first-order evolution systems 309

R. Bellman and R. Kalaba [58], and R. Gabasov and F.M. Kirillova [198]). The applications of this theory to the optimization problems for the systems described by the parabolic equations are considered by V. Barbu [51], P. Cannarsa and O. Cârjua [107], C.-W. Chen, J. S.-H. Tsai, and L.-S. Shieh [123], A.I. Egorov [154], [155], A. Just [264], J.L. Lions [315], F. Masiero [342], A.J. Pritchard and M.J.E. Mayhew [413], and T.K. Sirazetdinov [503]. The analogical results were obtained by T.K. Sirazetdinov [503] for hyperbolic equations; by M. Bardi and S. Bottacin [54] for elliptic equations; by V. Barbu [51] and G. Da Prato [135] for abstract evolutional equations; by R. Buckdahn and J. Li [96] for differential games. S. Chaumont [122] and Z. Liu [322] analyze the Bellman equation directly.

Using ideas of dynamical programming, V.F Krotov obtains the sufficient conditions of optimality (see V.F Krotov [285], V.F Krotov, V.Z. Bukreev, and V.I. Gurman [286], V.F. Krotov and V.I. Gurman [287]). The theory of the sufficient conditions of optimality is analyzed in the review of D. Peterson and J.H. Zalkind [401]. The sufficient conditions of optimality for the systems described by the ordinary differential equations are considered by V.I. Blagodatskikh [73], [74], V.G. Boltyansky [77], V.N. Brandin [88], F.P. Vasiliev [538], [539], D. Wachsmuth [545], and V.I. Yakubovich [559]. The different sufficient conditions of optimality for the elliptic equations are obtained by Yu.P. Krivenkov [284], A. Rösch and F. Tröltzsch [433], S. Russak [440], and S.Ya. Serovajsky [460]. The analogical results were obtained by E. Casas, F. Tröltzsch [531], and Z.S. Wu and K.L. Teo [557] for parabolic equations; by M.T. Dzhenaliev [150] for hyperbolic equations; by E. Casas, M. Mateos, and J.-P. Raymond [113], J.C. Reyes and R. Griesse [428], J.C. Reyes and F. Tröltzsch [429], and D. Wachsmuth [545] for Navier–Stokes equations.

K.-S. Chou, G.-X. Li, and C. Qu [127] apply group theory for analysis of optimization problems for systems described by linear parabolic equations. J.L. Lions [315] uses duality theory for it. The approximate solution of these problems can be found by using the Fourier method (see A.I. Egorov [154], [155], and V.I. Plotnikov [403]), Ritz and Galerkin methods (see I. Lasiecka [297], I. Lasiecka and K. Malanowski P. [298], and Neittaanmäki and D. Tiba [384]), the finite elements method (see W. Alt and U. Mackenroth [16], G. Knowles [274], I. Lasiecka [297], U. Mackenroth [336], D. Meidner and B. Vexler [349], and Neittaanmäki and D. Tiba [383]), the gradient methods (see J.W. He and R. Glowinski [238], P. Neittaanmäki and D. Tiba [384], and Z.S. Wu and K.L. Teo [556]), and the finite difference method (see W. Alt and U. Mackenroth [16]).

J.L. Lions [315] considers optimization problems for the high order parabolic equations; W. Kotarsky [279] considers parabolic equations with an infinite quantity of variables; A.V. Fursikov [197] solves the optimization problems with inverse time. V.I. Plotnikov [404] considers the system with an isoperimetric condition. H.O. Fattorini [170] and A.I. Egorov [154], [155] analyze systems with fixed final state. W. Alt and U. Mackenroth [16], P. Neittaanmäki and D. Tiba [384], and T.I. Seidman [453] solve optimization problems with state constraints. G. Knowles [273], W. Kotarsky [279], J.L. Lions [315], F. Masiero [342], K.L. Teo [517], and P.K.C. Wang [550] solve optimization problems for the systems with delay. P. Neittaanmäki and D. Tiba [384] consider optimization problems with non-smooth functionals. The time optimal problems for these systems are considered by V. Barbu [51], [52], I. Lasiecka [297], J.L. Lions [315], G. Knowles [273], [274], and V. Krabs and U. Lamp [281]. The pulse control problems are analyzed by J.L. Lions [315]. F. Masiero [342] and T.K. Sirazetdinov [503] solve optimization problems for stochastic systems.

Chapter 9

Nonlinear first order evolutional systems

9.1	Nonlinear evolutional equations with monotone operators	312
9.2	Optimization control problems for evolutional equations with monotone operator	321
9.3	Optimization control problems for the nonlinear heat equation	323
9.4	Necessary optimality conditions for the nonlinear heat equation	326
9.5	Optimization control problems with the functional that is not dependent upon the control	334
9.6	Sufficient optimality conditions for the nonlinear heat equation	337
9.7	Coefficient optimization problems for linear parabolic equations	339
9.8	Coefficient optimization problems for nonlinear parabolic equations	346
9.9	Initial optimization control problems for nonlinear parabolic equations	353
9.10	Optimization control problems for nonlinear parabolic equations with final functional	358
9.11	Comments	361

We considered optimization control problems for linear first order evolutional equations. The example is the classic heat equation. The natural continuation of these results is the analysis of the analogical problems for the nonlinear equations (see Figure 9.1). At first, we consider nonlinear evolutional equations with monotone operators that are the non-stationary analogs of the equations from Chapter 4. We prove the one-valued solvability of the Cauchy problem and weak continuity of its solution with respect to the absolute term. We consider a parabolic equation with nonlinearity power as an example of this system. This is the analog of the stationary equations of Chapter 4 and Chapter 5 (see Figure 9.1).

We know (see previous results) the differentiability of the state function with respect to the control. We prove that this dependence is not Gâteaux differentiable for the concrete example. Hence, we use the extended differentiability for obtaining the necessary conditions of optimality as before. We analyze also optimization problems with functionals that do not depend upon the control and the sufficient optimality conditions. We consider these problems for the linear evolutional systems (see Figure 9.1).

311

We consider also optimization control problems for the linear and nonlinear heat equations for the case of the presence of the control in the coefficients of the state operator. The one-valued solvability of the boundary problems is true here on the given sets admissible controls only. We cannot of use the Gâteaux derivative or the extended one for this situation. We replace it by the derivatives with respect to the convex set here (see Chapter 6). This is the analog of the optimization control problems for the stationary systems with the control in the coefficients (see Figure 9.1). Our final results are the optimization control problems with the initial control and with the functional that depends upon the state function at the final time.

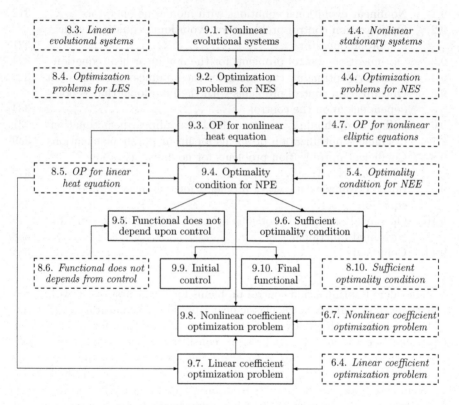

FIGURE 9.1: The structure of Chapter 9.

9.1 Nonlinear evolutional equations with monotone operators

Consider Hilbert space H, a separable Hilbert space W_1 and a reflexive Banach space W_2 such that embedding of the spaces W_1 and W_2 to H is

Nonlinear first order evolutional systems

313

continuous and dense. Identify the space H with its adjoint space H'. Then we have continuous and dense embedding $W_i \subset H \subset W_i'$, $i = 1, 2$. Let us have a linear continuous operator $A_1 : W_1 \to W_1'$ and a nonlinear continuous operator $A_2 : W_2 \to W_2'$ such that

$$\langle A_1 y, y \rangle \geq \chi_1 \|y\|_{W_1}^2 \ \forall y \in W_1, \tag{9.1}$$

$$\langle A_2 y, y \rangle \geq \chi_2 \|y\|_{W_2}^q \ \forall y \in W_2, \tag{9.2}$$

$$\|A_2 y\|_{W_2'} \leq c \|y\|_{W_2}^{q-1} \ \forall y \in W_2, \tag{9.3}$$

$$\langle A_2 y - A_2 z, y - z \rangle \geq 0 \ \forall y, z \in W_2, \tag{9.4}$$

where $\chi_1 > 0$, $\chi_2 > 0$, $c > 0$, $q > 1$. The inequalities (9.1) and (9.2) determine the coercivity of the operators. The inequality (9.3) characterizes the velocity of the nonlinearity increase. The operator with property (9.4) is monotone.

Determine the spaces $W = W_1 \cap W_2$, $W' = W_1' + W_2'$, and the operator $A : W \to W'$ by the formula $A = A_1 + A_2$.

Remark 9.1 Obviously, the operator A is monotone.

Consider the nonlinear first order evolutional equation

$$y'(t) + A(t)y(t) = f(t), \ t \in (0, T) \tag{9.5}$$

with the initial condition

$$y(0) = \varphi, \tag{9.6}$$

where f and φ are known.

Remark 9.2 The equation (9.5) has the same form as (8.5). However, it is nonlinear. This equation is equivalent to (8.5) whenever the operator A_2 has the zero values only.

Remark 9.3 The heat equation with power nonlinearity is the example of the equation (9.5) with considered properties (see Section 9.3).

Determine the space

$$X = L_2(0, T; W_1) \cap L_q(0, T; W_2).$$

It has the adjoint space

$$X' = L_2(0, T; W_1') + L_{q'}(0, T; W_2'),$$

where $1/q + 1/q' = 1$. Determine the solution of the problem (9.5), (9.6) on the set

$$Y = \{ y \mid y \in X, \ y' \in X') \}.$$

This is a Banach space because of Theorem 8.1.

Theorem 9.1 *Under the inequalities (9.1)–(9.4) for any $f \in X'$, $\varphi \in H$ there exists a unique solution $y \in Y$ of the problem (9.5), (9.6).*

314 *Optimization and Differentiation*

Proof. Determine an approximate solution of the problem. Let $\{\mu_m\}$ be a complete set of the linear independent elements of the space W. Using the Faedo–Galerkin method, determine the approximate solution $y_k = y_k(t)$ of the problem (9.5), (9.6) by the formula

$$y_k(t) = \sum_{m=1}^{k} \xi_{mk}(t)\mu_m, \ k = 1, 2, ..., \tag{9.7}$$

where the functions $\xi_{1k}, ..., \xi_{kk}$ satisfy the equalities

$$\langle y_k'(t), \mu_j \rangle + \langle Ay_k(t), \mu_j \rangle = \langle f(t), \mu_j \rangle, \ j = 1, ..., k. \tag{9.8}$$

Denote by M_k the set of the linear combinations of $\mu_1, ..., \mu_k$ with scalar product of H. Determine the operator $B_k : M_k \to M_k'$ and the element $f_k(t) \in M_k'$ by equalities

$$\langle B_k y, z \rangle = \langle Ay, z \rangle \ \forall y, z \in M_k,$$

$$\langle f_k(t), z \rangle = \langle f(t), z \rangle \ \forall z \in M_k.$$

The arbitrary element $z \in M_k$ is equal to a sum $\mu_1 \beta_1 + ... + \mu_k \beta_k$ with coefficients $\beta_1, ..., \beta_k$. Multiply j-th equality (9.8) by the number β_j. After summing by j we get

$$\langle y_k'(t) + B_k y_k(t) - f_k(t), z \rangle = 0 \ \forall z \in M_k.$$

Thus, we have

$$y_k'(t) + B_k y_k(t) = f_k(t), \ t \in (0, T). \tag{9.9}$$

This is the system of nonlinear ordinary differential equations with respect to the functions $\xi_{1k}, ..., \xi_{kk}$.

For all functions φ of H there exists a sequence $\{\alpha_k\}$ such that

$$\varphi = \sum_{m=1}^{\infty} \alpha_m \mu_m.$$

Determine the sum

$$\varphi_k = \sum_{m=1}^{k} \alpha_m \mu_m.$$

We obtain the convergence

$$\varphi_k \to \varphi \ \text{in} \ H. \tag{9.10}$$

Determine the initial values for the functions ξ_{mk} by the equality

$$\xi_{mk}(0) = \alpha_m, \ m = 1, 2, ... \ .$$

Nonlinear first order evolutional systems

Therefore, we have

$$\varphi_k(0) = \varphi_k. \tag{9.11}$$

Thus, we have Cauchy problem (9.9), (9.11) for the system of nonlinear ordinary differential equations. Its solvability follows from the Peano Theorem (see Theorem 8.2). The operator B_k is continuous and bounded because of the analogical properties of the operators A_1 and A_2. Therefore, the suppositions of Theorem 8.2 are true. Hence, the problem (9.9), (9.11) is solvable on an interval $(0, T_k)$. Thus, we can determine the abstract function y_k at this interval by the formula (9.7).

2. Determine properties of the approximate solution. Multiply j-th equality (9.8) by ξ_{jk} and sum the result by j. We have

$$\langle y'_k(t), y_k(t) \rangle + \langle Ay_k(t), y_k(t) \rangle = \langle f(t), y_k(t) \rangle.$$

Consider the equality

$$\langle y'_k(t), y_k(t) \rangle = \frac{1}{2} \frac{d}{dt} \|y_k(t)\|_H^2.$$

Then we get

$$\frac{1}{2} \frac{d}{dt} \|y_k(t)\|_H^2 + \langle Ay_k(t), y_k(t) \rangle = \langle f(t), y_k(t) \rangle. \tag{9.12}$$

Integrating by t by using (9.11), we have

$$\|y_k(t)\|_H^2 + 2 \int_0^t \langle Ay_k(\tau), y_k(\tau) \rangle d\tau = \|\varphi_k\|_H^2 + 2 \int_0^t \langle f(\tau), y_k(\tau) \rangle d\tau \leq$$

$$\|\varphi_k\|_H^2 + 2\|f\|_{X'}\|y_k\|_{X_t},$$

where $X_t = L_2(0, t; W_1) \cap L_q(0, t; W_2)$. From (9.1), (9.2) it follows the coercivity of the operator such that

$$\lim_{\|y_k\|_{X_t} \to \infty} \frac{1}{\|y_k\|_{X_t}} \int_0^t \langle Ay_k(\tau), y_k(\tau) \rangle d\tau = \infty.$$

Using the convergence (9.10), we have

$$\|y_k\|_{X_t} \leq c_1, \quad \|y_k\|_H \leq c_2,$$

where the positive constant c_1 and c_2 do not depend upon k and t. The value t is arbitrary; therefore, we can determine $t = T$. Thus, the sequence $\{y_k\}$ is bounded in the space X, and the sequence $\{y_k(T)\}$ is bounded in H.

316 *Optimization and Differentiation*

Consider a sequence $\{\psi_k\}$ of the space X that tends to zero. Suppose the sequence $\{A\psi_k\}$ is not bounded in the space X', i.e., $\|A\psi_k\|_{X'} \to \infty$. Determine the numbers

$$\beta_k = 1 + \|A\psi_k\|_{X'}\|\psi_k\|_X, \ k = 1, 2, \dots .$$

The operator A is monotone. Therefore, for all $\psi \in X$ we get

$$(\beta_k)^{-1} \int_0^T \langle A\psi_k(t), \psi(t)\rangle dt \leq$$

$$(\beta_k)^{-1}\left(\int_0^T \langle A\psi_k(t), \psi_k(t)\rangle dt + \int_0^T \langle A\psi(t), \psi(t) - \psi_k(t)\rangle dt\right) \leq$$

$$1 + (\beta_k)^{-1}\|A\psi\|_{X'}\big(\|\psi\|_X + \|\psi_k\|_X\big).$$

The sequence $\{\beta_k\}$ does not tend to zero. Then the right-hand side of the previous inequality is bounded by a constant that does not depend upon k. Thus, we obtain

$$\sup \lim_{k\to\infty} \int_0^T \langle A\psi_k(t), \psi(t)\rangle dt < \infty \ \forall \psi \in X.$$

Using the Banach–Steinhaus theorem, we determine the existence of a constant $c' > 0$ such that

$$(\beta_k)^{-1}\|A\psi\|_{X'} \leq c'.$$

Then we have the inequality

$$\|A\psi_k\|_{X'} \leq c'\big(1 + \|A\psi_k\|_{X'}\|\psi_k\|_X\big).$$

Choose the number k so large that $2c'\|\psi_k\|_X \leq 1$. Therefore, we have the inequality $\|A\psi_k\|_{X'} \leq 2c'$ that contradicts the condition $\|A\psi_k\|_{X'} \to \infty$. Thus, from the convergence $\psi_k \to 0$ in X the boundedness of the sequence $\{A\psi_k\}$ follows. Hence, there exists a constant $c_1 > 0$ such that $\|A\psi\|_{X'} \leq c_1$ if $\|\psi\|_X \leq 1$.

Using the monotony of the operator A, we have

$$\|Ay_k\|_{X'} = \sup_{\|\psi\|_X=1} \int_0^T |\langle Ay_k(t), \psi(t)\rangle| dt \leq$$

$$\sup_{\|\psi\|_X=1} \int_0^T \Big[\langle Ay_k(t), y_k(t)\rangle + \langle A\psi(t), \psi(t)\rangle - \langle A\psi(t), y_k(t)\rangle\Big] dt.$$

Then the sequence $\{Ay_k\}$ is bounded in the space X'.

3. Prove the convergence of the sequence of the approximate solutions to the solution of the initial problem. Using the boundedness of the sequences

Nonlinear first order evolutional systems

and the Banach–Alaoglu theorem, we extract subsequences by saving its denotations such that

$$y_k \to y \text{ weakly in } X, \tag{9.13}$$

$$Ay_k \to z \text{ weakly in } X', \tag{9.14}$$

$$y_k(T) \to z \text{ weakly in } H. \tag{9.15}$$

Multiply the equality (9.8) by the arbitrary function $\eta \in D(0,T)$ and integrate the result. We obtain equality

$$\int_0^T \langle y_k'(t), \eta_j(t) \rangle dt + \int_0^T \langle Ay_k(t), \eta_j(t) \rangle dt = \int_0^T \langle f(t), \eta_j(t) \rangle dt, \tag{9.16}$$

where $\eta_j(t) = \eta(t)\mu_j$, $j = 1, ..., k$. Using the formula of integration by parts, we have

$$\int_0^T \langle y_k'(t), \eta_j(t) \rangle dt = - \int_0^T \langle \eta_j'(t), y_k(t) \rangle dt$$

because the function η as the element of $D(0,T)$ is equal to zero on the boundary of the given interval. After passing to the limit by using (9.13), we get

$$\lim_{k\to\infty} \int_0^T \langle y_k'(t), \eta_j(t) \rangle dt = - \lim_{k\to\infty} \int_0^T \langle \eta_j'(t), y_k(t) \rangle dt = \int_0^T \langle y'(t), \eta_j(t) \rangle dt.$$

Using (9.14), we have

$$\lim_{k\to\infty} \int_0^T \langle Ay_k(t), \eta_j(t) \rangle dt = \int_0^T \langle z(t), \eta_j(t) \rangle dt.$$

Thus, after passing to the limit in the equality (9.16) we obtain

$$\int_0^T \Big\langle y'(t) + z(t) - f(t), \eta_j(t) \Big\rangle dt = 0, \ j = 1, 2, ... \ .$$

Each element $w \in W$ can be transformed to the series with respect to the basis $\{\mu_j\}$ and coefficients $\{\beta_j\}$. Multiplying the j-th previous equality β_j and sum by j, we get

$$\int_0^T \Big\langle y'(t) + z(t) - f(t), \eta(t)w \Big\rangle dt = 0.$$

318   *Optimization and Differentiation*

The functions η and w are arbitrary here. Therefore, we get

$$y'(t) + z(t) = f(t), \ t \in (0, T). \tag{9.17}$$

If the function $\eta = \eta(t)$ is continuously differentiable in the interval $[0, T]$, then we can use the formula of the integration by parts. We obtain

$$\int_0^T \langle y_k'(t), \eta_j(t) \rangle dt = \left(y_k(T), \eta_j(T)\right)_H - \left(\varphi_k, \eta_j(0)\right)_H - \int_0^T \langle \eta_j'(t), y_k(t) \rangle dt$$

because of the equality (9.11). Passing to the limit at the equality (9.16) by using the conditions (9.10), (9.13)–(9.15), we get

$$\left(h, \eta_j(T)\right)_H - \left(\varphi, \eta_j(0)\right)_H - \int_0^T \langle \eta_j'(t), y(t) \rangle dt + \int_0^T \langle z(t) - f(t), \eta_j(t) \rangle dt = 0.$$

Integrating by parts, we have

$$\left(h - y(T), \eta_j(T)\right)_H - \left(\varphi - y(0), \eta_j(0)\right)_H + \int_0^T \langle y'(t) + z(t) - f(t), \eta_j(t) \rangle dt = 0.$$

Using (9.17), we obtain

$$\left(h - y(T), \eta_j(T)\right)_H - \left(\varphi - y(0), \eta_j(0)\right)_H = 0, \ j = 1, 2, \dots . \tag{9.18}$$

Choose $\eta(T) = 0$ at (9.18). We have the equality

$$\left(\varphi - y(0), \mu_j\right)_H = 0, \ j = 1, 2, \dots .$$

Then we obtain the equality (9.6). Determine $\eta(0) = 0$ at (9.18). We get

$$\left(h - y(T), \mu_j\right)_H = 0, \ j = 1, 2, \dots .$$

Therefore, $y(T) = h$.

Using (9.12), we have

$$\sup \lim_{k \to \infty} \int_0^T \langle A y_k(t), y_k(t) \rangle dt =$$

$$\sup \lim_{k \to \infty} \int_0^T \langle f(t), y_k(t) \rangle dt - \sup \lim_{k \to \infty} \int_0^T \langle y_k'(t), y_k(t) \rangle dt =$$

$$\sup \lim_{k \to \infty} \int_0^T \langle f(t), y_k(t) \rangle dt - \frac{1}{2} \sup \lim_{k \to \infty} \left(\|y_k(T)\|_H^2 - \|y_{0k}\|_H^2 \right).$$

From the conditions (9.10), (9.13), (9.15) and the weak semicontinuity of the norm of Banach space (see Section 2.3) it follows that

$$\sup \lim_{k \to \infty} \int_0^T \langle Ay_k(t), y_k(t) \rangle dt \le \int_0^T \langle f(t), y(t) \rangle dt +$$

$$\frac{1}{2} \sup \lim_{k \to \infty} \left(\|\varphi\|_H^2 - \|y(T)\|_H^2 \right) = \int_0^T \langle f(t) - y'(t), y(t) \rangle dt = \int_0^T \langle z(t), y(t) \rangle dt$$

because of the equality (9.17).

Using the last relations and the monotony of the operator, we find

$$\int_0^T \left\langle z(t) - A\psi(t), y(t) - \psi(t) \right\rangle dt = \int_0^T \langle z(t), y(t) \rangle dt - \int_0^T \langle z(t), \psi(t) \rangle dt -$$

$$\int_0^T \langle A\psi(t), y(t) \rangle dt + \int_0^T \langle A\psi(t), \psi(t) \rangle dt \ge \sup \lim_{k \to \infty} \int_0^T \langle Ay_k(t), y_k(t) \rangle dt -$$

$$\sup \lim_{k \to \infty} \int_0^T \langle Ay_k(t), \psi(t) \rangle dt - \sup \lim_{k \to \infty} \int_0^T \langle A\psi(t), y_k(t) \rangle dt +$$

$$\int_0^T \langle A\psi(t), \psi(t) \rangle dt = \sup \lim_{k \to \infty} \int_0^T \left\langle Ay_k(t) - A\psi(t), y_k(t) - \psi(t) \right\rangle dt \ge 0$$

for all $\psi \in Y$. Determine $\psi = y - \sigma h$, where $h \in Y$, $\sigma > 0$. We get

$$\sigma \int_0^T \left\langle z(t) - A[y(t) - \sigma h(t)], h(t) \right\rangle dt \ge 0.$$

Dividing by σ and passing to the limit as $\sigma \to 0$ by using the continuity of the operator, we get

$$\int_0^T \langle z(t) - Ay(t), h(t) \rangle dt \ge 0 \ \forall h \in Y.$$

Replace the arbitrary element h by $-h$ here. We have

$$\int_0^T \langle z(t) - Ay(t), h(t) \rangle dt = 0 \ \forall h \in Y.$$

320　　　　　　　　*Optimization and Differentiation*

Then $z(t) = Ay(t)$. From the equality (9.17), it follows that the abstract function y satisfies the equation (9.5). It is sufficient to find the derivative from the equality (9.5) for obtaining the inclusion $y' \in X'$. Hence, $y \in Y$.

4. Prove the uniqueness of the solution of the problem (9.5), (9.6). Suppose the existence of the solutions y_1 and y_2 of this problem. We have

$$y_1'(t) - y_2'(t) + Ay_1(t) - Ay_2(t) = 0, \ t \in (0, T).$$

Therefore, we get the equality

$$\left\langle y_1'(t) - y_2'(t) + Ay_1(t) - Ay_2(t), y_1(t) - y_2(t) \right\rangle = 0, \ t \in (0, T).$$

Then for all $t \in (0, T)$ we have

$$\int_0^t \left\langle y_1'(\tau) - y_2'(\tau) + Ay_1(\tau) - Ay_2(\tau), y_1(\tau) - y_2(\tau) \right\rangle d\tau = 0. \qquad (9.19)$$

The functions y_1 and y_2 are equal for $t = 0$. Therefore, we obtain

$$\int_0^t \langle y_1'(\tau) - y_2'(\tau), y_1(\tau) - y_2(\tau) \rangle d\tau = \frac{1}{2} \left\| y_1(t) - y_2(t) \right\|_H^2.$$

By the monotony of the operator A we get

$$\int_0^t \left\langle Ay_1(\tau) - Ay_2(\tau), y_1(\tau) - y_2(\tau) \right\rangle d\tau \geq 0.$$

Hence, from the equality (9.19) it follows that

$$\left\| y_1(t) - y_2(t) \right\|_H^2 \leq 0.$$

Then $y_1 = y_2$, and the solution of the problem is unique. $\square$

Remark 9.4 Theorem 9.1 is a natural extension of Theorem 4.4 of the one-value solvability of the equation with the monotone operator for the case of the evolutional system.

Remark 9.5 We cannot the necessity to the consideration the operator as the sum the linear operator and monotone one here. However, we have this form of the operator for the nonlinear heat equation.

Now consider an optimization control problem for the system, described by the evolutional equation with monotone operator.

9.2 Optimization control problems for evolutional equations with monotone operator

Consider Hilbert space V and a linear continuous operator $B : V \to L_2(0,T;W_1')$. Let us have a system, described by the equation

$$y'(t) + Ay(t) = (Bv)(t) + f(t), \ t \in (0,T), \tag{9.20}$$

with the initial condition

$$y(0) = \varphi. \tag{9.21}$$

The abstract function v from the space V is the control here. Fix $\varphi \in H$ and $f \in L_2(0,T;W')$. From Theorem 9.1 it follows that for all $v \in V$ the problem (9.20), (9.21) has a unique solution $y = y[v]$ from the space Y. Consider a non-empty convex closed subset U of the space V and the functional

$$I(v) = \frac{\alpha}{2}\|v\|_V^2 + \frac{1}{2}\|C(y[v] - y_d)\|_Z^2,$$

where $\alpha > 0$, $C : L_2(0,T;W_1) \to Z$ is a linear continuous operator, Z is Hilbert space, and $y_d \in L_2(0,T;W_1)$ is known.

Problem 9.1 *Find a control $u \in U$ that minimizes the functional I on the set U.*

Remark 9.6 If A_2 is the zero operator, i.e., $A = A_2$, then Problem 9.1 can be transformed to Problem 8.1 for the linear evolutional equation.

Lemma 9.1 *The map $y[\cdot] : V \to Y$ for the problem (9.20), (9.21) is weakly continuous.*

Proof. Suppose the convergence $v_k \to v$ weakly in V. Then $Bv_k \to Bv$ weakly in $L_2(0,T;W')$. The solution $y_k = y[v_k]$ of the problem (9.20), (9.21) for $v = v_k$ satisfies the equality

$$y_k'(t) + Ay_k(t) = (Bv_k)(t) + f(t), \ t \in (0,T). \tag{9.22}$$

Then we get

$$\langle y_k'(t), y_k(t) \rangle + \langle Ay_k(t), y_k(t) \rangle = \langle (Bv_k)(t) + f(t), y_k(t) \rangle. \tag{9.23}$$

We have

$$\langle y_k'(t), y_k(t)\rangle = \frac{1}{2}\frac{d}{dt}\big\|y_k(t)\big\|_H^2,$$

$$\langle Ay_k(t), y_k(t)\rangle = \sum_{i=1}^{2}\langle A_i y_k(t), y_k(t)\rangle \geq \chi_1\big\|y_k(t)\big\|_{W_1}^2 + \chi_2\big\|y_k(t)\big\|_{W_2}^q,$$

$$\big|\langle (Bv_k)(t) + f(t), y_k(t)\rangle\big| \leq \big\|(Bv_k)(t) + f(t)\big\|_{W_1'}\big\|y_k(t)\big\|_{W_1} \leq$$

$$\frac{1}{2\chi_1}\big\|(Bv_k)(t) + f(t)\big\|_{W_1'}^2 + \frac{\chi_1}{2}\big\|y_k(t)\big\|_{W_1}^2.$$

From the equality (9.23) it follows that

$$\frac{1}{2}\frac{d}{dt}\big\|y_k(t)\big\|_H^2 + \frac{\chi_1}{2}\big\|y_k(t)\big\|_{W_1}^2 + \chi_2\big\|y_k(t)\big\|_{W_2}^q \leq \frac{1}{2\chi_1}\big\|(Bv_k)(t) + f(t)\big\|_{W_1'}^2.$$

Integrating this inequality by t by using the initial condition (9.21), we get

$$\chi_1\int_0^T \big\|y_k(t)\big\|_{W_1}^2 dt + 2\chi_2\int_0^T \big\|y_k(t)\big\|_{W_2}^q dt \leq \|\varphi\|_H^2 + \frac{1}{\chi_1}\int_0^T \big\|(Bv_k)(t) + f(t)\big\|_{W_1'}^2.$$

Then the sequence $\{y_k\}$ is bounded in the space X.

From the equality (9.22) it follows that

$$\big\|y_k'\big\|_{X'} \leq \big\|Ay_k\big\|_{X'} + \big\|Bv_k\big\|_{L_2(0,T;W_1')} + \|f\|_{L_2(0,T;W_1')}.$$

By the inequality (9.3) the sequence $\{Ay_k\}$ is bounded in the space X'. The sequence $\{y_k'\}$ is bounded in the same space because of the previous inequality.

Thus, sequence $\{y_k\}$ is bounded in the space Y. After extracting a subsequence we have the convergence $y[v_k] \to y$ weakly in Y. We pass to the limit at the equality (9.22) as an analogical action of Theorem 9.1. Therefore, $y = y[v]$. $\square$

From Lemma 9.1 we obtain the following result.

Corollary 9.1 *The functional I for Problem 9.1 is weakly lower semicontinuous.*

Using Theorem 8.7, we have the following result.

Theorem 9.2 *Problem 9.1 has a unique solution.*

Remark 9.7 The map $v \to y[v]$ for the problem (9.20), (9.21) is not affine because of the nonlinearity of the operator A_2. Therefore, we cannot to determine the convexity of the minimizing functional. Hence, we cannot prove the uniqueness of the solution of this optimization control problem and its well-posedness. The analogical results are true for the stationary control systems.

9.3 Optimization control problems for the nonlinear heat equation

Consider an example of the nonlinear first order evolutional system. Let Ω be an open bounded set of Euclid space $\mathbb{R}^n$ with the boundary Γ, $T > 0$, $Q = \Omega \times (0,T)$, $\Sigma = \Gamma \times (0,T)$. We have the nonlinear heat equation

$$\frac{\partial y}{\partial t} - \Delta y + |y|^\rho y = f \tag{9.24}$$

in the set Q with the boundary condition

$$y(x,t) = 0,\ (x,t) \in \Sigma \tag{9.25}$$

and the initial condition

$$y(x,0) = \varphi(x),\ x \in \Omega, \tag{9.26}$$

where $\rho > 0$.

Remark 9.8 The problem (9.24)–(9.26) is the evolutional analog of the boundary problem for the elliptic equation with power nonlinearity from Chapter 4 and Chapter 5.

Prove that the problem (9.24)–(9.26) can be transformed to (9.5), (9.6). Determine the spaces $H = L_2(\Omega)$, $W_1 = H_0^1(\Omega)$, $W_2 = L_q(\Omega)$ and $W = W_1 \cap W_2 = H_0^1(\Omega) \cap L_q(\Omega)$, where $q = \rho + 2$. Consider also its adjoint spaces $H' = L_2(\Omega)$, $W' = H^{-1}(\Omega)$, $W_2' = L_q(\Omega)$, where $1/q + 1/q' = 1$. We have continuous and dense embedding $W_i \subset H \subset W_i'$. Determine the following functional spaces:

$$X = L_2\big(0,T;H_0^1(\Omega)\big) \cap L_q(Q),\ X' = L_2\big(0,T;H^{-1}(\Omega)\big) + L_{q'}(Q),$$

$$Y = \{y\mid y \in X,\ y' \in X'\}.$$

Determine the linear operator $A_1 : H_0^1(\Omega) \to H^{-1}(\Omega)$ by the equality $A_1 y = -\Delta y$. Using Green's formula, we get

$$\langle A_1 y, y\rangle = -\int_\Omega y \Delta y dx = \int_\Omega \sum_{i=1}^n \left(\frac{\partial y}{\partial x_i}\right)^2 dx = \|y\|^2\ \forall y \in H_0^1(\Omega).$$

Thus, we have the relation (9.1) as the equality with the constant $\chi_1 = 1$.

Determine the power operator $A_2 : L_q(\Omega) \to L_{q'}(\Omega)$ by the equality $A_2 y = |y|^\rho$. Then we have

$$\langle A_2 y, y\rangle = \int_\Omega |y|^{\rho+2} dx = \|y\|_q^q\ \forall y \in L_q(\Omega).$$

324   *Optimization and Differentiation*

Therefore, the condition (9.2) is true as the equality with the constant $\chi_1 = 2$. Using Hölder inequality, we have

$$\|A_2 y\|_{q'} = \left(\int_\Omega ||y|^p y|^{q'} dx \right)^{1/q'} = \left(\int_\Omega |y|^{p+2} dx \right)^{(q-1)/q} = \|y\|_q^{q-1} \ \forall y \in L_q(\Omega).$$

Then the condition (9.3) is true as the equality with the constant $c = 1$. Finally, we have

$$\langle A_2 y - A_2 z, y - z \rangle = \int_\Omega (|y|^p y - |z|^p z)(y - z)dx \geq 0 \ \forall y, z \in L_q(\Omega).$$

This is the monotony condition (9.4).

Thus, the boundary problem (9.24)–(9.26) is transformed to the system (9.5), (9.6), as the conditions of Theorem 9.1 are true. Then we have the following result.

Theorem 9.3 *For all functions $\varphi \in L_2(\Omega)$ and $f \in X'$ the problem (9.24)– (9.26) has a unique solution $y \in Y$.*

We can prove also Theorem 9.3 by using compact embedding of the functional spaces.

Second proof of Theorem 9.3. Let $\{\mu_m\}$ be again a complete set of the linear independent elements of the space W. Determine the approximate solution y_k of the problem (9.24)–(9.26) by the formula

$$y_k = \sum_{m=1}^{k} \xi_{mk} \mu_m, \ k = 1, 2, \dots ,$$

where the functions $\xi_{1k} = \xi_{1k}(t), ..., \xi_{kk} = \xi_{kk}(t)$ satisfy the equalities

$$\int_\Omega \frac{\partial y_k}{\partial t} \mu_j dx + \int_\Omega \sum_{1=1}^{n} \frac{\partial y_k}{\partial x_i} \frac{\partial \mu_j}{\partial x_i} dx + \int_\Omega |y_k|^p y_k \mu_j dx = \int_\Omega f \mu_j dx \qquad (9.27)$$

with the initial conditions

$$y_k(x, 0) = \varphi_k(x), \ x \in \Omega, \qquad (9.28)$$

where $j = 1, ..., k$, and $\varphi_k \to \varphi$ in H. Cauchy problem (9.27), (9.28) has a solution $\xi_{mk} = \xi_{mk}(t)$ on an interval $(0, T_k)$.

We prove the boundedness of the sequence $\{y_k\}$ in the space Y by using the technique from Theorem 9.1. Then we can extract a subsequence such that $y_k \to y$ weakly in Y. Now we use Theorem 8.5 of compact embedding of the functional spaces. Determine there $W_0 = H_0^1(\Omega)$, $W = L_2(\Omega)$, $W_1 = H^{-1}(\Omega)$,

Nonlinear first order evolutional systems 325

$r = s = 2$. Then embedding of the space Y to $L_2(Q)$ is compact. Extracting a subsequence, we have $y_k \to y$ strongly in $L_2(Q)$ and a.e. on Q. Then $|y_k|^p y_k \to |y|^p y$ a.e. on Q. From the boundedness of the sequence $\{y_k\}$ in the space $L_q(Q)$ it follows that sequence $\{|y_k|^p y_k\}$ is bounded in the space $L_{q'}(Q)$. Using Lemma 4.6, we have the convergence $|y_k|^p y_k \to |y|^p y$ weakly in $L_{q'}(Q)$. We can finish now passing to the limit at the equalities (9.27), (9.28) without any difficulties. $\square$

Consider now the control system, described by the equation (9.24)

$$\frac{\partial y}{\partial t} - \Delta y + |y|^p y = v + f \tag{9.29}$$

in the set Q with the boundary conditions

$$y(x, t) = 0, \ (x, t) \in \Sigma, \tag{9.30}$$

$$y(x, 0) = \varphi(x), \ x \in \Omega, \tag{9.31}$$

where the control v belongs to the space $V = L_2(Q)$. Choose the operator $B : V \to L_2(0, T; W_1')$ equal to embedding of these spaces. We transform the equation (9.29) with the boundary condition (9.30) to the operator equation (9.20). From Theorem 9.3 it follows that for all $v \in V$ the problem (9.29)–(9.31) has a unique solution $y = y[v]$ from the space Y. Consider the functional

$$I(v) = \frac{\alpha}{2} \int_Q v^2 dQ + \frac{1}{2} \int_Q \sum_{i=1}^n \left(\frac{\partial y[v]}{\partial x_i} - \frac{\partial y_d}{\partial x_i} \right)^2 dQ,$$

where $\alpha > 0$, $y_d \in L_2(0, T; H_0^1(\Omega))$ is known. Determine $Z = L_2(0, T; H_0^1(\Omega))$. Let C be the unit operator. We have the functional I from Problem 9.1. Consider a non-empty convex closed subset U of the space V.

Problem 9.2 *Find the control $u \in U$ for the system* (9.29)–(9.31) *that minimizes the functional I on the set U.*

This is the partial case of Problem 9.1. From Theorem 9.2 it follows the theorem.

Theorem 9.4 *Problem 9.2 is solvable.*

We would like to determine a necessary condition of optimality for the considered problem. If our functional is differentiable, then the optimal control satisfies the variational inequality that is an analog of (8.25) and (8.34). However, it is necessary to analyze the differentiability of the solution of the problem (9.29)–(9.31) with respect to the control. This is the general difficulty of the substantiation of the optimization methods for the nonlinear infinite dimensional systems.

9.4 Necessary optimality conditions for the nonlinear heat equation

Consider again the boundary problem (9.24)–(9.26). For all $\varphi \in L_2(\Omega)$ and $f \in X'$ it has a unique solution $y \in Y$ because of Theorem 9.1. Let the initial state be fixed. Choose the absolute term $f = v$ of the equation as the control. Denote by $y[v]$ the solution of this problem for this control.

Lemma 9.2 *The map $y[\cdot] : X' \to Y$ for the problem (9.24)–(9.26) is not Gâteaux differentiable for large enough values ρ and n.*

Proof. Suppose this map is Gâteaux differentiable at the arbitrary point $u \in X'$. Then there exists a linear continuous operator $y'[u] : X' \to Y$ such that for all $h \in X'$ we have the convergence $(y_\sigma - y_0)/\sigma \to y'[u]h$ in Y as $\sigma \to 0$, where y_σ and y_0 are the solutions of the problem (9.24)–(9.26) for the controls $u + \sigma h$ and u. From (9.24) the equality follows:

$$\frac{\partial(y_\sigma - y_0)}{\partial t} - \Delta(y_\sigma - y_0) + |y_\sigma|^\rho y_\sigma - |y_0|^\rho y_0 = \sigma h.$$

After dividing by σ and passing to the limit as $\sigma \to 0$ we have the equality

$$\frac{\partial y'[u]h}{\partial t} - \Delta y'[u]h + (\rho + 1)|y_0|^\rho y'[u]h = h.$$

Thus, for all $h \in X'$ the equation

$$\frac{\partial \eta}{\partial t} - \Delta \eta + (\rho + 1)|y_0|^\rho \eta = h \tag{9.32}$$

with homogeneous boundary conditions has the solution $\eta = y'[u]h$ from the space Y. Obviously, this solution is unique.

Choose the functions $\varphi \in L_2(\Omega)$ and $u \in X'$ so smooth that the solution $y[u]$ of the problem (9.24)–(9.26) is continuous. Determine the values ρ and n so large that the inclusion $W_{q'}^{2,1}(Q) \subset L_q(Q)$ gets broken, where $W_{q'}^{2,1}(Q)$ is the set of the functions that belong to the space $L_{q'}(Q)$ with its generalized derivatives before second order with respect to the space variables and the first order with respect to the time. Then there exists a point $\eta_* \in W_{q'}^{2,1}(Q) \backslash L_q(Q)$ that satisfies the homogeneous boundary conditions. Determine the function

$$h_* = \frac{\partial \eta_*}{\partial t} - \Delta \eta_* + (\rho + 1)|y_0|^\rho \eta_*.$$

We have the inclusion $h_* \in L_{q'}(Q)$. Then $h_* \in X'$. Thus, there exists a point $h = h_*$ from X' such that the solution η_* of the homogeneous boundary problem for the equation (9.32) does not have any solutions of the space $L_q(Q)$. However, this contradicts the previous result. Thus, our supposition

Nonlinear first order evolutional systems 327

about Gâteaux differentiability at this point of the solution of the problem (9.24)–(9.26) with respect to the absolute term is false. $\square$

Thus, the parabolic equation with nonlinearity power has the same properties as the analogical elliptic equation (see Chapter 5). Therefore, we try to apply the methods of the analysis of the nonlinear stationary systems.

Consider the linearized equation (9.32). If we prove its one-value solvability in the space for all $h \in X'$, then we can find derivative η' from this equality, and this is the point from the space X'. Then we have $\eta \in Y$. This is an analog of the suppositions of the Inverse function theorem and the Implicit function theorem. We could prove the differentiability of the considered dependence for this situation. However, this result contradicts Lemma 9.2. But we can determine positive results for the linearized equation.

Multiply the equality (9.32) by the function η and integrate the result by using the homogeneous boundary conditions. We obtain the inequality

$$\int_Q \sum_{i=1}^n \left|\frac{\partial \eta}{\partial x_i}\right|^2 dQ + (\rho+1)\int_Q |y[u]|^\rho \eta^2 dQ \le \int_Q \eta h dQ. \tag{9.33}$$

The first term here is the square of the norm in the space $Y_* = L_2\big(0,T; H_0^1(\Omega)\big)$. The second term is non-negative. The considered equation is linear. Using the previous technique (see Chapter 8), we can probe the one-valued solvability of the homogeneous boundary problem for the equation (9.32) in the space Y_* for all h from the the adjoint space $Y_*' = L_2\big(0,T; H^{-1}(\Omega)\big)$.

We can prove a stronger proposition if we interpret the second term of the inequality (9.33) as a square of the norm of a Hilbert space. Determine the space

$$X_0 = \left\{y \in Y_* \big|\ |y[u]|^{\rho/2} y \in L_2(Q)\right\}.$$

We can determine the norm here by the equality

$$\|\eta\|_{X_0}^2 = \|\eta\|_{Y_*}^2 + (\rho+1)\left\||y[u]|^{\rho/2}\eta\right\|_{L_2(Q)}^2.$$

It has the adjoint space

$$X_0' = \left\{h \big|\ h = h_1 + (\rho+1)|y[u]|^{\rho/2} h_2,\ h_1 \in Y_*',\ h_2 \in L_2(Q)\right\}.$$

Indeed, for all $\eta \in X_0$, $h \in X_0'$ we have

$$\left|\int_Q \eta h dQ\right| \le \left|\int_Q \eta h_1 dQ\right| + (\rho+1)\left|\int_Q \eta|y[u]|^{\rho/2} h_2 dQ\right| < \infty.$$

Using the definition of the norm of the adjoint space, we get

$$\|h\|_{X_0'} = \sup_\eta \left[\left|\int_Q \eta h dQ\right| \le \left|\int_Q \eta h_1 dQ\right| + (\rho+1)\left|\int_Q \eta|y[u]|^{\rho/2} h_2 dQ\right|\right] \le$$

$$\sup_{\eta} \left[\|\eta\|_{Y_*} \|h_1\|_{Y'_*} + (\rho+1) \left\| |\eta|y[u]|^{\rho/2} \right\|_{L_2(Q)} \|h_2\|_{L_2(Q)} \right] \le \|h_1\|_{Y'_*} + \|h_2\|_{L_2(Q)},$$

where we have sup with respect to all η with unit norm in X_0. From the inequality (9.33) the a priori estimate follows of the solution for the homogeneous boundary problem for the equation (9.32) in the space X_0 as from equality (9.32) it follows:

$$\left\| \frac{\partial \eta}{\partial t} \right\|_{X'_0} \le \|\Delta\eta\|_{Y'_*} + (\rho+1) \left\| |y[u]|^{\rho/2} \right\|_{L_{2q/\rho}(Q)} \left\| |y[u]|^{\rho/2}\eta \right\|_{L_2(Q)} + \|h\|_{X'_0}.$$

Thus, the homogeneous boundary problem for the equation (9.32) has a priori estimate of the solution in the space

$$Y_0 = \{y| \ y \in X_0, \ y' \in X'_0\}.$$

Then for all $h \in X'_0$ this problem has a unique solution in the space Y_0.

Thus, the linearized equation is solvable in the larger space for the narrower set of the absolute terms as these sets depend upon the point of the differentiation. We had the analogical properties for the stationary systems. We considered extended differentiability of the state function with respect to the control there (see Chapter 5). Prove that the map $y[\cdot] : V \to Y$ is now extended differentiably at the arbitrary point $u \in V$.

Lemma 9.3 *The map $y[\cdot] : V \to Y$ for the problem (9.29)–(9.31) has $(V, Y_0; V, Y^w_*)$-extended derivative $y'[u]$ at the arbitrary point $u \in V$, where Y^w_* is the space Y_* with the weak topology, and the spaces Y_0 and Y_* are determined before as there exists a linear continuous operator $y'_T[u] : V \to L_2(\Omega)$ such that for all $h \in V$ we have the convergence*

$$\{y[u + \sigma h](T) - y[u](T)\}/\sigma \to y'_T[u]h \text{ weakly in } L_2(\Omega). \tag{9.34}$$

Aditionally, the following equality holds:

$$\int_Q \mu_Q y'[u]h \, dQ + \int_\Omega \mu_\Omega y'_T[u]h \, dQ = \int_Q p_\mu[u]h \, dQ \ \forall h \in V, \tag{9.35}$$

where $p_\mu[u]$ is the solution of the equation

$$-\frac{\partial p_\mu[u]}{\partial t} - \Delta p_\mu[u] + (\rho+1)|y[u]|^\rho p_\mu[u] = \mu_Q \tag{9.36}$$

in the set Q with boundary conditions

$$p_\mu[u](x,t) = 0, \ (x,t) \in \Sigma, \tag{9.37}$$

$$p_\mu[u](x,T) = \mu_\Omega(x), \ x \in \Omega, \tag{9.38}$$

and $\mu = (\mu_Q, \mu_\Omega)$.

$$\text{Nonlinear first order evolutional systems} \qquad 329$$

Proof. 1. The function $\eta[v] = y[v] - y[u]$ is the solution of the equation

$$\frac{\partial \eta[v]}{\partial t} - \Delta\eta[v] + \big(g[v]\big)^2 \eta[v] = v - u \qquad (9.39)$$

with homogeneous boundary conditions, where

$$\big(g[v]\big)^2 = (\rho+1)\Big|y[u] + \varepsilon\big(y[v] - y[u]\big)\Big|^\rho, \ \varepsilon \in [0,1].$$

Note that $\big(g[u]\big)^2 = (\rho+1)|y[u]|^\rho$. Consider also the boundary problem

$$-\frac{\partial p}{\partial t} - \Delta p + \big(g[v]\big)^2 p = \mu_Q \qquad (9.40)$$

in the set Q with boundary conditions

$$p(x,t) = 0, \ (x,t) \in \Sigma, \qquad (9.41)$$

$$p(x,T) = \mu_\Omega(x), \ x \in \Omega. \qquad (9.42)$$

The system (9.36)–(9.38) is its partial case for $v = u$.

Determine the space

$$X[v] = \Big\{ y\big|\ y \in Y_*, \ g[v]y \in L_2(Q) \Big\}.$$

This is a Hilbert space with scalar product

$$(y,\lambda)_{X[v]} = \int_Q \nabla y \nabla \lambda dQ + \int_Q \big(g[v]\big)^2 y\lambda dQ.$$

It has the adjoint space

$$\big(X[v]\big)' = \Big\{ \varphi + g[v]\psi\big|\ \varphi \in Y_*', \ \psi \in L_2(Q) \Big\}.$$

Determine the space

$$Y[v] = \Big\{ y\big|\ y \in X[v], \ y' \in \big(X[v]\big)' \Big\}.$$

Note that it is equal to the space Y_0 for $v = u$.

Multiply the equality (9.40) by the function p and integrate the result by using the conditions (9.41), (9.42). We have

$$\|p\|_{X[v]}^2 \le \|\mu_Q\|_{(X[v])'}^2 + \frac{1}{2}\|\mu_\Omega\|_2^2 \le \frac{1}{2}\|p\|_{X[v]}^2 + \frac{1}{2}\|\mu_Q\|_{(X[v])'}^2 + \frac{1}{2}\|\mu_\Omega\|_2^2.$$

From the equality (9.40) it follows that

$$\|p'\|_{(X[v])'} \le \|\Delta p\|_{(X[v])'} + \|y[v]p\|_{L_2(Q)}\|y[v]\|_{L_{2q/\rho}(Q)} + \|\mu_Q\|_{(X[v])'}.$$

330   *Optimization and Differentiation*

The solution of the problem (9.29)–(9.31) belongs to the space $L_q(Q)$. Then we have $g[v] \in L_{2q/\rho}(Q)$. The Laplace operator is the linear continuous operator from the space Y_* to Y_*'. From the previous inequalities it follows that for all functions $\mu_Q \in \big(X[v]\big)'$, $\mu_\Omega \in L_2(\Omega)$ the problem (9.40)–(9.42) has a priori estimate of the solution in the space $Y[v]$. Using the classic theory of the linear parabolic equation (see Theorem 8.7), we prove that this problem under considered suppositions has a unique solution $p = p_\mu[v]$ from the space $Y[v]$.

2. Now suppose the convergence $\sigma \to 0$. By weak continuity of the map $y[\cdot] : V \to Y$ we determine the convergence $y[u + \sigma h] \to y[u]$ for all $h \in V$. Therefore, the set $\{y[u + \sigma h]\}$ is bounded in the space $L_q(Q)$. Then the sequence $\{g_\sigma\}$ is bounded in $L_{2q/\rho}(Q)$, where $g_\sigma = g[u + \sigma h]$.

Denote by $p_\mu^\sigma[h]$ the solution of the problem (9.40)–(9.42) for $v = u + \sigma h$. For all number σ and functions $\mu_Q \in \big(X[v]\big)'$, $\mu_\Omega \in L_2(\Omega)$ we determine the inequalities

$$\left\|p_\mu^\sigma[h]\right\|_{Y_*}^2 + \left\|g_\sigma p_\mu^\sigma[h]\right\|_{L_2(Q)}^2 \le \left\|\mu_Q\right\|_{Y_*'}\left\|p_\mu^\sigma[h]\right\|_{Y_*} + \frac{1}{2}\|\mu_\Omega\|_2^2,$$

$$\left\|\big(p_\mu^\sigma[h]\big)'\right\|_{P_1} \le \left\|p_\mu^\sigma[h]\right\|_{Y_*} + \left\|g_\sigma p_\mu^\sigma[h]\right\|_{L_2(Q)}\|g_\sigma\|_{L_{2q/\rho}(Q)} + \left\|\mu_Q\right\|_{Y_*'},$$

where $P_1 = Y_*' + L_{q'}(Q)$. Then the set $\{p_\mu^\sigma[h]\}$ is bounded in the space

$$P = \big\{p\, |\, p \in Y_*, \, p' \in P_1\big\},$$

and the set $\{g_\sigma p_\mu^\sigma[h]\}$ is bounded in $L_2(Q)$. Using Hölder inequality, we determine the boundedness of the set $\{(g_\sigma)^2 p_\mu^\sigma[h]\}$ in the space $L_{q'}(Q)$. Using the Banach–Alaoglu Theorem, after extracting of a subsequence we have the convergence $p_\mu^\sigma[h] \to \varphi_\mu$ weakly in P. Using compact embedding of the spaces Y and P in $L_2(Q)$ (see Theorem 8.5), we obtain $y[u+\sigma h] \to y[u]$ and $p_\mu^\sigma[h] \to \varphi_\mu$ strongly in $L_2(Q)$ and a.e. on Q. Then $p_\mu^\sigma[h] \to (\rho + 1)|y[u]|^\rho \varphi_\mu$ a.e. on Q. Thus, the previous convergence is realized also in the weak topology of the space $L_{q'}(Q)$. Passing to the limit in the problem (9.40)–(9.42) for $p = p_\mu^\sigma[h]$ and $\sigma \to 0$ we have $\varphi_\mu = p_\mu[u]$. Using the continuous embedding of P to $C\big(0, T; L_2(\Omega)\big)$ (see Theorem 8.4), we have $p_\mu^\sigma[h](T) \to p_\mu[u](T)$ weakly in $L_2(\Omega)$.

Multiplying now the equality (9.39) for $v = u + \sigma h$ by the function $p_\mu^\sigma[h]$ and integrating the result, we get

$$\int_Q \mu_Q \eta[u + \sigma h]dQ + \int_\Omega \mu_\Omega \eta[u + \sigma h](T)dx =$$

$$\int_Q p_\mu^\sigma[h]h\,dQ \,\, \forall h \in V, \,\, \mu_Q \in X_\sigma', \,\, \mu_\Omega \in L_2(\Omega). \tag{9.43}$$

Nonlinear first order evolutional systems

Note that for $\mu_\Omega = 0$ the equality (9.35) determines a linear continuous operator $y'[u] : V \to X_0$. Then

$$\int_Q \mu_Q \Big\{ \big(y[u + \sigma h] - y[u] \big) / \sigma - y'[u]h \Big\} \mu_Q dQ = \int_Q \Big\{ p_\mu^\sigma[h] - p_\mu[h] \Big\} h dQ.$$

Thus, we have $\big(y[u + \sigma h] - y[u] \big) / \sigma \to y'[u]h$ weakly in Y_*. Thus, the operator $y'[\cdot] : V \to Y$ is $\big(V, X_0; V, Y_*^w \big)$-extended differentiable at the point u.

3. Using the known technique, we determine that for all $h \in V$ the set of the solutions $\{\eta[u + \sigma h]\}$ of the homogeneous boundary problem for the equation (9.39) is bounded in the space P as $\sigma \to 0$. After extracting a subsequence, we have the convergence $\eta[u + \sigma h] \to \eta$ weakly in P. Repeat the previous transformation. Pass to the limit at the equation (9.39). We determine that the function η satisfies the linearized equation (9.32). However, $y'[u]h$ is the limit of $\{\eta[u + \sigma h]\}$. Find the derivative of η with respect to t from the equality (9.32). This is the element of X_0'. Thus, $y'[u]h \in Y_0$ and the operator $y'[u]h$ is $\big(V, Y_0; V, Y_*^w \big)$-extended derivative of the dependence of the state function of the system (9.29)–(9.31) from the control at the point u.

4. For $\mu_Q = 0$ the equality (9.35) determines the linear continuous operator $y_T'[u]$. Then we have

$$\int_\Omega \Big\{ \big(y[u + \sigma h](T) - y[u](T) \big) / \sigma - y_T'[u]h \Big\} \mu_\Omega dx = \int_Q \Big\{ p_\mu^\sigma[h] - p_\mu'h \Big\} h dQ.$$

Thus, the convergence (9.34) is true. We obtain the equality (9.35) after passing to the limit at the equality (9.43). $\square$

Using the extended differentiability of the control–state mapping, we could obtain the necessary optimality conditions. Determine the differentiability of the functional for Problem 9.2.

Lemma 9.4 *If $u_k \to u$ in V, then $y[u_k] \to y[u]$ in $L_2\big(0, T; H_0^1(\Omega)\big)$.*

Proof. Multiply the equality (9.39) for $v = u_k$ by the function $\eta[u_k]$. After integration we have

$$\int_Q \frac{\partial \eta[u_k]}{\partial t} \eta[u_k] dQ - \int_Q \Delta \eta[u_k] \eta[u_k] dQ + \int_Q \big(g[u_k] \eta[u_k] \big)^2 dQ = \int_Q (u_k - u) \eta[u_k] dQ.$$

The boundary conditions for the functions $\eta[u_k]$ are homogeneous. Then we get

$$\big\| \eta[u_k](T) \big\|_2^2 + \big\| \eta[u_k] \big\|_{L_2(0,T;H_0^1(\Omega))}^2 + \big\| g[u_k] \eta[u_k] \big\|_{L_2(Q)}^2 =$$

$$\int_Q (u_k - u) \eta[u_k] dQ \leq \big\| \eta[u_k] \big\|_{L_2(0,T;H_0^1(\Omega))} \big\| u_k - u \big\|_{L_2(0,T;H^{-1}(\Omega))}.$$

If $u_k \to u$ in V, then $\eta[u_k] \to 0$. This completes the proof of the lemma. $\sqcup$

332　　　　　　　　　*Optimization and Differentiation*

Lemma 9.5 *The functional I for Problem 9.2 has Gâteaux derivative*

$$I'(u) = \alpha u + p$$

at the arbitrary point u, where p is the solution of the equation

$$-\frac{\partial p}{\partial t} - \Delta p + (\rho + 1)|y[u]|^{\rho} p = \Delta y_d - \Delta y[u] \qquad (9.44)$$

in the set Q with the boundary conditions

$$p(x,t) = 0, \ (x,t) \in \Sigma, \qquad (9.45)$$

$$p(x,T) = 0, \ x \in \Omega. \qquad (9.46)$$

Proof. We have the equality

$$I(u + \sigma h) - I(u) = \frac{\alpha}{2} \int_Q (2\sigma u h + \sigma^2 h^2) dQ +$$

$$\frac{1}{2} \int_Q \sum_{i=1}^n \frac{\partial\big(y[u + \sigma h] + y[u] - 2y_d\big)}{\partial x_i} \frac{\partial\big(y[u + \sigma h] - y[u]\big)}{\partial x_i} dQ.$$

From Lemma 9.4 and Lemma 9.5 it follows that

$$(y[u + \sigma h] - y[u])/\sigma \to y'[u]h \ \text{weakly in} \ L_2\big(0,T; H_0^1(\Omega)\big),$$

$$y[u + \sigma h] \to y[u] \ \text{strongly in} \ L_2\big(0,T; H_0^1(\Omega)\big).$$

Thus, we have

$$\frac{1}{\sigma} \frac{\partial\big(y[u + \sigma h] - y[u]\big)}{\partial x_i} \to y'[u]h \ \text{weakly in} \ L_2(Q),$$

$$\frac{\partial y[u + \sigma h]}{\partial x_i} \to \frac{\partial y[u]}{\partial x_i} \ \text{strongly in} \ L_2(Q)$$

for all $i = 1, ..., n$. Dividing the previous equality by σ and passing to the limit as $\sigma \to 0$ we have

$$\langle I'(u), h \rangle = \alpha \int_Q u h dQ + \int_Q \sum_{i=1}^n \frac{\partial(y[u] - y_d)}{\partial x_i} \frac{\partial y'[u]h}{\partial x_i} dQ =$$

$$\alpha \int_Q u h dQ + \int_Q \big(\Delta y_d - \Delta y[u]\big) h dQ.$$

Choose $\mu_\Omega = 0$, $\mu_Q = \Delta y_d - \Delta y[u]$ at the equality (9.35). We have

$$\langle I'(u), h \rangle = \int_Q (\alpha u + p) h dQ \ \forall h \in V.$$

Nonlinear first order evolutional systems 333

This completes the proof of the lemma. □

Using Lemma 9.5 and the standard variational inequality, we obtain the following result.

Theorem 9.5 *If the control u is the solution of Problem 9.2, it satisfies the variational inequality*

$$\int_Q (\alpha u + p)(v - u)dQ \geq 0 \quad \forall v \in U. \tag{9.47}$$

Thus, the system of the optimality conditions consists of the boundary problem (9.29)–(9.31) for $v = u$, the adjoint system (9.44)–(9.46), and the variational inequality (9.47).

We can find the solution of this variational inequality, if we have the following set of admissible control:

$$U = \{v \in L_2(Q) \mid a \leq v(x,t) \leq b, \ (x,t) \in Q\},$$

where a and b are constants, as $a < b$. Using the coincidence of the variational inequalities (8.34) and (9.47) and Theorem 8.12, we obtain the following result.

Theorem 9.6 *The solution of Problem 9.2 with given set U is*

$$u(x,t) = \begin{cases} a, & \text{if } -p(x,t)/\alpha < a, \\ -p(x,t)/\alpha, & \text{if } a \leq -p(x,t)/\alpha \leq b, \\ b, & \text{if } -p(x,t)/\alpha > b. \end{cases}$$

Remark 9.9 We have the coincidence of the optimality condition for the linear and nonlinear cases because we have the same set of admissible control and the same form of presence of the control to the state equations and functionals for Problem 8.2 and Problem 9.2. We had the analogical results for the stationary control systems.

Consider a partial case of the set of admissible control. Suppose the control is admissible if it has non-negative values only, i.e.,

$$U = \{v \in L_2(Q) \mid v(x,t) \geq 0 \ (x,t) \in Q\}.$$

Corollary 9.2 *The optimal control for Problem 9.2 with non-negative control satisfies the conditions*

$$u(x,t) \geq 0, \ [\alpha u(x,t) + p(x,t)] \geq 0, \ u(x,t)[\alpha u(x,t) + p(x,t)] = 0. \tag{9.48}$$

Proof. The condition (9.48) is the corollary of the set of admissible controls. From the relation (9.47) the pointwise variational inequality follows:

$$[\alpha u(x,t) + p(x,t)][w - u(x,t)] \geq \quad \forall w \geq 0.$$

334   *Optimization and Differentiation*

If $w = u(x,t) + z$ with $z \geq 0$, then we have the inequality

$$[\alpha u(x,t) + p(x,t)]z \geq 0 \ \forall z \geq 0.$$

Therefore, we obtain second condition (9.48). Finally, determine $w = 0$ at the previous inequality. We have

$$[\alpha u(x,t) + p(x,t)]u(x,t) \leq 0.$$

Thus, the product of two non-negative multipliers is non-positive. This is possible if one of them is equal to zero only, i.e., the final condition (9.48) is true. $\square$

Remark 9.10 By third condition (9.48) set Q can be divided into three parts. We have the equality $u(x,t) = 0$ for $\alpha u(x,t) + p(x,t) \geq 0$ in the first of them. The conditions $\alpha u(x,t) + p(x,t) = 0$ and $u(x,t) \geq 0$ are true for the second of them. Finally, it is possible that a subset of Q can be empty such that we have both equality $u(x,t) = 0$ and $\alpha u(x,t) + p(x,t) = 0$. Thus, solving of the optimization control problem can be transformed to finding of these sets.

Remark 9.11 We extend our results to the general integral functionals and to the equations with general nonlinearity (see Chapter 3). We shall consider also the optimization control problems with the initial control (see Section 9.9) and with the functional that depends upon the state function at the final time (see Section 9.10).

9.5 Optimization control problems with the functional that is not dependent upon the control

Consider again the system described by the boundary problem (9.24)–(9.26)

$$\frac{\partial y}{\partial t} - \Delta y + |y|^\rho y = u(x,t) + f(x,t), \ (x,t) \in Q \tag{9.49}$$

in the set Q with the boundary conditions

$$y(x,t) = 0, \ (x,t) \in \Sigma, \tag{9.50}$$

$$y(x,0) = \varphi(x), \ x \in \Omega. \tag{9.51}$$

Determine the space

$$Y = \left\{ y \big| \ y \in X, \ y' \in X' \right\},$$

where

$$X = L_2\big(0,T; H_0^1(\Omega)\big) \cap L_q(Q), \ X = L_2\big(0,T; H^{-1}(\Omega)\big) + L_{q'}(Q).$$

Nonlinear first order evolutional systems

Suppose $\varphi \in L_2(\Omega)$ and $f \in X'$. The control v belongs to a convex closed bounded subset U of the space $V = L_2(Q)$. For all control $v \in V$ this problem has a unique solution $y = y[v]$ from the set Y.

Consider the functional

$$I(v) = \frac{1}{2} \int_Q \sum_{i=1}^{n} \left(\frac{\partial y[v]}{\partial x_i} - \frac{\partial y_d}{\partial x_i} \right)^2 dQ,$$

where $y_d \in L_2(0, T; H_0^1(\Omega))$ is a known function.

Problem 9.3 *Find the control for the system (9.49)–(9.51) that minimizes the functional I on the determined set U.*

This is the analog of Problem 8.4 and the partial case of Problem 9.2 for $\alpha = 0$. Using the standard technique, we determine its solvability.

Theorem 9.7 *Problem 9.3 has a solution.*

By Lemma 9.3 the solution of the problem (9.49)–(9.51) is extended differentiably with respect to the control. We have the following analog of Lemma 9.5.

Lemma 9.6 *Functional I for Problem 9.3 at the arbitrary point u has the Gâteaux derivative $I'(u) = p$, where p is the solution of the equation*

$$-\frac{\partial p}{\partial t} - \Delta p + (\rho + 1)|y|^\rho p = \Delta y_d - \Delta y \tag{9.52}$$

in the set Q with the boundary conditions

$$p(x, t) = 0, \ (x, t) \in \Sigma, \tag{9.53}$$

$$p(x, T) = 0, \ x \in \Omega. \tag{9.54}$$

Put the functional derivative in to the standard variational inequality. We obtain the following analog of Theorem 9.6.

Theorem 9.8 *If the control u is the solution of Problem 9.3, then it satisfies the variational inequality*

$$\int_Q p(v - u)dQ \geq 0 \ \forall v \in U. \tag{9.55}$$

This result follows from the variational inequality (9.47) for $\alpha = 0$. However, it can degenerate as the analogical condition (8.46). Indeed, if there exists a control $u \in U$ such that $y[u] = y_d$, then the value of the given functional is equal to zero, i.e., this control is optimal. Then the term at the right-hand side

336 *Optimization and Differentiation*

of the equation (9.52) is equal to zero too. Therefore, the solution of the problem (9.52)–(9.54) is zero, and the inequality (9.55) is transformed to $0 \geq 0$. Thus, the optimal control is singular. It cannot be found from the obtained optimality condition. We can find it by using the Tihonov regularization method. Determine the functional

$$I_k(v) = I(v) + \frac{\alpha_k}{2} \int_Q v^2 dQ,$$

where $\alpha_k > 0$ and $\alpha_k \to 0$ as $k \to \infty$.

Problem 9.4 *Find the control for the system (9.49)–(9.51) that minimizes the functional I_k on the set U.*

From Theorem 9.4 its solvability follows.

Corollary 9.3 *Problem 9.4 is solvable.*

Using Theorem 9.5, determine necessary conditions of optimality.

Corollary 9.4 *If the control u_k is the solution of Problem 9.4, then the following variational inequality holds:*

$$\int_Q (\alpha_k u_k + p_k)(v - u_k)dQ \geq 0 \ \forall v \in U, \tag{9.56}$$

where p_k is the solution of the boundary problem

$$-\frac{\partial p_k}{\partial t} - \Delta p_k + (\rho + 1)\big|y[u_k]\big|^\rho p = \Delta y_d - \Delta y[u_k] \tag{9.57}$$

in the set Q with the boundary conditions

$$p_k(x,t) = 0, \ (x,t) \in \Sigma, \tag{9.58}$$

$$p_k(x,T) = 0, \ x \in \Omega. \tag{9.59}$$

Determine the convergence of the regularization method. The following result is the analog of Theorem 8.15.

Theorem 9.9 $I(u_k) \to \min I(U)$ *as $k \to \infty$.*

Proof. We have

$$I_k(u_k) = \min I_k(U) \leq I_k(u) = I(u) + \frac{\alpha_k}{2}\|u\|_V^2 = \min I(U) + \frac{\alpha_k}{2}\|u\|_V^2,$$

where u is the solution of Problem 9.3. Thus, we get the inequality

$$\lim_{k \to \infty} I_k(u_k) \leq \min I(U). \tag{9.60}$$

Nonlinear first order evolutional systems 337

By the boundedness of the set U, we determine the boundedness of the sequence $\{u_k\}$ in the space V. Extracting a subsequence, we have the convergence $u_k \to v$ weakly in V, as $v \in V$. Using the weak continuity of the map $y[\cdot] : V \to Y$, we get $y[u_k] \to y[v]$ weakly in Y. Then we have the convergence

$$\frac{\partial y[u_k]}{\partial x_i} \to \frac{\partial y[v]}{\partial x_i} \text{ weakly in } L_2(Q), \quad i = 1, ..., n.$$

Using the weak lower semicontinuity of the square of the norm for the space $L_2(Q)$, we obtain

$$I(v) \leq \inf \lim_{k \to \infty} I(u_k).$$

By inequality $I(u_k) \leq I_k(u_k)$ and the inequality (9.60), we have $I(v) \leq \min I(U)$. Therefore, the control v is the solution of Problem 9.3. From the inequalities

$$\min I(U) = I(v) \leq \inf \lim_{k \to \infty} I(u_k) \leq \lim_{k \to \infty} I_k(u_k) \leq \min I(U)$$

it follows that $I(u_k) \to \min I(U)$. $\square$

Thus, for the large enough numbers k the value of the minimizing functional for Problem 9.3 at the solution u_k of Problem 9.4 is close enough to the minimum of the functional I on the set of admissible controls. Therefore, we can choose the solution u_k from the system (9.49)–(9.51) for $v = u_k$, and (9.56)–(9.59) as an approximate solution to Problem 9.3.

Remark 9.12 In reality, the proofs of Theorem 8.15 and Theorem 9.9 coincide because we use weak continuity of the solution of the boundary problem with respect to the control only. This result is true for linear and nonlinear cases. However, the regularized problem is Tihonov well-posed for the linear system only.

Remark 9.13 We can pass to the limit in the relation (9.56) and obtain the variational inequality (9.55).

9.6 Sufficient optimality conditions for the nonlinear heat equation

Consider again the system, described by the equalities (9.29)–(9.31). The control v belongs to the set

$$U = \left\{ v \in L_2(Q) \middle| v(t) \in G(t), \ t \in (0, T) \right\},$$

where $G(t)$ is a convex closed subset of the space $L_2(\Omega)$. We have the functional

$$I(v) = \frac{\alpha}{2} \int_Q v^2 dQ + \frac{1}{2} \int_Q \sum_{i=1}^{n} \left(\frac{\partial y[v]}{\partial r_i} - \frac{\partial y_d}{\partial x_i} \right)^2 dQ,$$

338   *Optimization and Differentiation*

where $\alpha > 0$, $y_d \in L_2(0, T; H_0^1(\Omega))$ is the known function.

Problem 9.5 *Find the control that minimizes the functional I on the set U.*

This problem is solvable because of Theorem 9.4. Determine a sufficient optimality condition that is an analog of Theorem 8.18.

Theorem 9.10 *Suppose there exists Fréchet differentiable functional J that satisfies the Bellman equation*

$$\frac{\partial J(s, \psi)}{\partial s} + H_1\left(s, \psi, \frac{\partial J(s, \psi)}{\partial \psi}\right) + \min_{w \in G(s)} H\left(\frac{\partial J(s, \psi)}{\partial \psi}, w\right) = 0 \qquad (9.61)$$

with $s \in (0, T)$, $\psi \in L_2(\Omega)$, and the final condition

$$J(T, \psi) = 0, \ \psi \in L_2(\Omega), \qquad (9.62)$$

where

$$H_1(s, \psi, \eta) = \int_\Omega \left\{ \frac{1}{2} \sum_{i=1}^n \left[\frac{\partial \psi(x)}{\partial x_i} - \frac{\partial y_d(x, s)}{\partial x_i}\right]^2 + \eta\left[\Delta\psi(x) + f(x, s)\right] \right\} dx,$$

$$H(\eta, w) = \int_\Omega \left(\frac{\alpha w^2}{2} + \eta w\right) dx.$$

Then the function u that satisfies the equality

$$H\left(\frac{\partial J(s, \psi)}{\partial \psi}, u(s)\right) = \min_{w \in G(s)} \left(\frac{\partial J(s, \psi)}{\partial \psi}, w\right), \ s \in (0, T), \ \psi \in L_2(\Omega), \quad (9.63)$$

is the solution of Problem 9.5, as $I(u) = J(0, \varphi)$.

Proof. We have the equality

$$\frac{d}{dt} J\big[t, y(t)\big] = \frac{\partial J}{\partial t} + \int_\Omega \frac{\partial J}{\partial y} \frac{\partial y}{\partial t} dx,$$

where $y(t)$ is the function $y = y(x, t)$ that is the solution of the boundary problem (9.29)–(9.31) for an admissible control v with fixed t. Using the equation (9.29), transform the previous equality

$$\frac{d}{dt} J\big[t, y(t)\big] = \frac{\partial J}{\partial t} + \int_\Omega \frac{\partial J}{\partial y} (\Delta y - |y|^p y + v + f) dx.$$

Integrating this equality by t by using condition (9.62), we have

$$-J(0, \varphi) = \int_0^T \left[\frac{\partial J}{\partial t} + \int_\Omega \frac{\partial J}{\partial y} (\Delta y - |y|^p y + v + f) dx\right] dt.$$

Nonlinear first order evolutional systems 339

Then we obtain inequality

$$\int_0^T \left[\frac{\partial J}{\partial t} + \int_\Omega \frac{\partial J}{\partial y} (\Delta y - |y|y^\rho + v + f) dx \right] dt + I(v) \geq 0.$$

This relation is realized as the equality for $v = u$ because of (9.61). Then we get

$$I(v) \geq J(0, \varphi) \ \forall v \in U; \quad I(u) \geq J(0, \varphi).$$

Now we have $I(u) \leq I(v)$ for all $v \in U$, i.e., the control u is the solution of the considered optimization problem. $\square$

This proof is not different in reality from the analogical proof of Theorem 8.18. We do not have any additional difficulties for the determination of the sufficient optimality condition for the nonlinear systems. However, the practical using of this result is bounded enough because of the absence of effective methods for the analysis of the Bellman equation (9.61).

9.7 Coefficient optimization problems for linear parabolic equations

We considered optimization problems for the evolutional equations with the control at the absolute term. Now we analyze the system with the control that is the coefficient of the state operator. We have an open bounded set Ω of $\mathbb{R}^n$ with the boundary Γ, $T > 0$, $Q = \Omega \times (0, T)$, $\Sigma = \Gamma \times (0, T)$. Consider the linear equation

$$\frac{\partial y}{\partial t} - \Delta y + vy = f \tag{9.64}$$

on the set of Q with the boundary conditions

$$y(x, t) = 0 \ (x, t) \in \Sigma, \tag{9.65}$$

$$y(x, 0) = \varphi(x), \ x \in \Omega, \tag{9.66}$$

where f and φ are known functions. Determine the control space $V = L_2(Q)$ and the set of admissible controls

$$U = \left\{ v \in V \middle| a \leq v(x) \leq b, \ x \in \Omega \right\},$$

where a and b are constants, as $b > a > 0$.

340   *Optimization and Differentiation*

Multiply the equality (9.64) by the function y and integrate the result. We have

$$\int_Q \left(\frac{\partial y}{\partial t} - \Delta y + vy \right) y \, dQ = \int_Q f y \, dQ.$$

Using boundary conditions, we get

$$\frac{1}{2} \int_\Omega |y(x,T)|^2 dx + \int_Q \sum_{i=1}^n \left(\frac{\partial y}{\partial x_i} \right)^2 dQ + \int_Q vy^2 dQ = \frac{1}{2} \int_\Omega \varphi^2 dx + \int_Q f y \, dQ.$$

By the definition of the set of admissible control, we obtain inequality

$$\|y\|_{Y_*}^2 \le \frac{1}{2} \|\varphi\|_2^2 + \|y\|_{Y_*} \|f\|_{Y_*'},$$

where

$$Y_* = L_2\big(0,T; H_0^1(\Omega)\big), \quad Y_*' = L_2\big(0,T; H^{-1}(\Omega)\big).$$

Thus, we get

$$\|y\|_{Y_*}^2 \le \|\varphi\|_2^2 + \|f\|_{Y_*'}^2. \tag{9.67}$$

Thus, for $\varphi \in L_2(\Omega)$ the problem (9.64)–(9.66) has an a priori estimate of the solution in the space Y_*. If $y \in Y_*$, then from the equality (9.64) the inequality follows:

$$\left\| \frac{\partial y}{\partial t} \right\|_{Y_*'} \le \|\Delta y\|_{Y_*'} + \|vy\|_{Y_*'} + \|f\|_{Y_*'}. \tag{9.68}$$

Using the boundedness of the set of admissible control and the continuity of the operator $\Delta : Y_* \to Y_*'$, we determine that the time derivative of the function y belongs to the space Y_*'. Therefore, the solution of the problem (9.64)–(9.66) is an element of the space

$$Y = \Big\{ y \big| \ y \in Y_*, \ y' \in Y_*' \Big\}.$$

Thus, this system is an analog of the boundary problem for the linear heat equation (see the previous chapter). Using Corollary 8.1, we have the following result.

Theorem 9.11 *Suppose $\varphi \in L_2(\Omega)$, $f \in Y_*'$. Then for all control $v \in U$ there exists a unique solution $y = y[u]$ of the problem (9.64)–(9.66) from the space Y.*

Determine the functional

$$I(v) = \frac{\alpha}{2} \int_Q v^2 dQ + \frac{1}{2} \int_Q \sum_{i=1}^n \left(\frac{\partial y[v]}{\partial x_i} - \frac{\partial y_d}{\partial x_i} \right)^2 dQ,$$

where $\alpha > 0$, $y_d \in L_2\big(0,T; H_0^1(\Omega)\big)$ is a known function.

Nonlinear first order evolutional systems 341

Problem 9.6 *Find the control $u \in U$ for the system (9.64)–(9.66) that minimizes the functional I on the set U.*

We cannot use the previous results for proving the solvability of the problem because we have the need to obtain the weak continuity of control–state mapping.

Lemma 9.7 *The map $y[\cdot] : V \to Y$ for the problem (9.64)–(9.66) is weakly continuous.*

Proof. Suppose for a sequence $\{v_k\}$ of the set U we have the convergence $v_k \to v$ weakly in V. Using the convexity and the closeness of the set U, we determine its weak closeness. Hence, $v \in U$. Using (9.67), we obtain the boundedness of the sequence $\{y[v_k]\}$ in the space Y_*. By the inequality (9.68) and the definition of the set of admissible control, we prove the boundedness of the time derivatives for the solutions of the problem (9.64)–(9.66) in the space Y'_*. Then the sequence $\{y[v_k]\}$ is bounded in the space Y. Extracting a subsequence, we obtain the convergence $y[v_k] \to y$ weakly in Y. Use Theorem 8.5 for $Z_0 = H_0^1(\Omega)$, $Z = L_2(\Omega)$, $Z_1 = H^{-1}(\Omega)$. Determine compact embedding of the space Y to $L_2(Q)$. Then we have the convergence $y[v_k] \to y$ strongly in $L_2(Q)$.

For all λ from the distributions set $D(Q)$ we have

$$\left| \int_Q v_k y[v_k] \lambda dQ - \int_Q vy\lambda dQ \right| \leq \left| \int_Q (v_k - v)y\lambda dQ \right| + \left| \int_Q v_k(y[v_k] - y)\lambda dQ \right|.$$

Using the condition $y\lambda \in L_2(Q)$, we obtain the boundedness of the sequence $\{v_k\lambda\}$ in the space $L_2(Q)$. The sequence $\{v_k\}$ converges weakly and $\{y[v_k]\}$ converges strongly in this space. Therefore, we obtain

$$\int_Q v_k y[v_k] \lambda dQ \to \int_Q vy\lambda dQ.$$

Multiply the equality (9.64) for $v = v_k$ by the arbitrary function $\lambda \in D(Q)$. After integration and passing to the limit as $k \to \infty$ we get $y = y[v]$. $\square$

Remark 9.14 This is a nonstationary analog of Lemma 6.4 about the weak continuity of the solution of the homogeneous Dirichlet problem for the linear elliptic equation with coefficient control with respect to the control.

Then we determine the solvability of Problem 9.6.

Theorem 9.12 *Problem 9.6 is solvable.*

We consider a nonstationary analog of Problem 6.1. The boundary problems here are one-valued solvable on the set of admissible controls. However, if we would like to determine the Gâteaux derivative of the map $y[\cdot] : V \to Y$ at

342 Optimization and Differentiation

a point $u \in U$ we consider the boundary problem (9.64)–(9.66) for the control $u + \sigma h$ with arbitrary h. Probably, this value is not an element of the set U. Therefore, we use the derivative with respect to the convex set (see Chapter 6).

By Theorem 6.2, if the functional I has the extended derivative $I_U'(u)$ with respect to the convex set U at the point u of its minimum on this set, then we get

$$\langle I_U'(u), v - u \rangle \geq 0 \ \forall v \in U. \tag{9.69}$$

Lemma 9.8 *The map* $y[\cdot] : V \to Y$ *for the problem* (9.64)–(9.66) *has* $(V, Y_*; V, Y_*^w)$*-extended derivative* $y'[u]$ *with respect to the set* U *at the arbitrary point* $u \in U$ *as there exists an affine continuous operator* $y'_{UT}[u]$ *on the set* $\delta(u) = \{v - u | v \in U\}$ *with values in the space* $L_2(\Omega)$ *such that for all* $v \in U$ *we have the convergence*

$$\{y[u + \sigma(v - u)](T) - y[u](T)\}/\sigma \to y'_{UT}[u](v - u) \ \text{weakly in} \ L_2(\Omega). \tag{9.70}$$

In addition, we have the equality

$$\int_Q \mu_Q y'[u](v - u)dQ + \int_\Omega \mu_\Omega y'_{UT}[u](v - u)dQ = \int_Q p_\mu[u](u - v)dQ \tag{9.71}$$

for all $v \in U$, $\mu_Q \in Y_*'$, $\mu_\Omega \in L_2(\Omega)$, *where* $p_\mu[u]$ *is the solution of the equation*

$$-\frac{\partial p_\mu[u]}{\partial t} - \Delta p_\mu[u] + u p_\mu[u] = \mu_Q \tag{9.72}$$

in the set Q *with boundary conditions*

$$p_\mu[u](x, t) = 0, \ (x, t) \in \Sigma, \tag{9.73}$$

$$p_\mu[u](x, T) = \mu_\Omega(x), \ x \in \Omega, \tag{9.74}$$

and $\mu = (\mu_Q, \mu_\Omega)$.

Proof. 1. For all $v \in U$ and number σ the function

$$\eta_\sigma[v] = (y[u + \sigma(v - u)] - y[u])/\sigma$$

is the solution of the equation

$$\frac{\partial \eta_\sigma[v]}{\partial t} - \Delta \eta_\sigma[v] + u \eta_\sigma[v] + (v - u)y[u + \sigma(v - u)] = 0 \tag{9.75}$$

with homogeneous boundary conditions. Then for all functions $\lambda \in Y$ we have

$$\int_Q \left(-\frac{\partial \lambda}{\partial t} - \Delta \lambda + u\lambda \right) \eta_\sigma[v]dQ + \int_\Omega \lambda(x, T)\eta_\sigma[v](T)dx =$$

$$\int_Q \lambda(u-v)y[u+\sigma(v-u)]dQ.$$

The equation (9.72) coincides to (9.64). Then for all $\mu_Q \in Y'_*$, $\mu_\Omega \in L_2(\Omega)$ there exists its unique solution in the space Y. Choose $\lambda = p_\mu[u]$ in the previous equality. We have

$$\int_Q \mu_Q \eta_\sigma[v]dQ + \int_\Omega \mu_\Omega \eta_\sigma[v](T)dx = \int_Q p_\mu[u](u-v)y[u+\sigma(v-u)]dQ. \quad (9.76)$$

Obviously, the equality (9.71) for $\mu_\Omega = 0$ determines an affine continuous operator $y'_U[u]$ on the set of differences $\delta(u) = \{v-u|\ v \in U\}$ with values from Y_*. By (9.71) and (9.76) we get

$$\int_Q \mu_Q \Big[\{y[u+\sigma(v-u)] - y[u]\}/\sigma - y'_U[u](v-u)\Big]dQ =$$

$$\int_Q p_\mu[u](u-v)\{y[u+\sigma(v-u)] - y[u]\}dQ.$$

Using the definition of the set of admissible control, we get $p_\mu[u](u-v) \in L_2(Q)$. By Lemma 9.7 we have the convergence $y[u+\sigma(v-u)] \to y[u]$ weakly in Y as $\sigma \to 0$. Then after passing to the limit at the previous equality we have the convergence

$$\{y[u+\sigma(v-u)] - y[u]\}/\sigma \to y'_U[u](v-u) \ \text{ weakly in } Y_* \qquad (9.77)$$

for all $v \in U$.

2. Multiply the equality (9.75) by an arbitrary function $\lambda \in Y$ that is equal to zero for $t = T$. After integration we get

$$\int_Q \Big(-\frac{\partial \lambda}{\partial t} - \Delta\lambda + u\lambda \Big)\eta_\sigma[v]dQ + \int_Q \lambda(v-u)y[u+\sigma(v-u)]dQ = 0.$$

Passing to the limit, we have

$$\int_Q \Big(-\frac{\partial \lambda}{\partial t} - \Delta\lambda + u\lambda \Big)y'_U[u](v-u)dQ + \int_Q \lambda(v-u)y[u]dQ = 0.$$

Then we obtain

$$\int_Q \frac{\partial y'_U[u](v-u)}{\partial t}\lambda dQ = \int_Q \Big\{ \Delta y'_U[u](v-u) - uy'_U[u](v-u) - (v-u)y[u] \Big\}\lambda dQ.$$

344   *Optimization and Differentiation*

Determine the derivative $\partial \eta_\sigma[v]/\partial t$ from the equality (9.75). After passing to the limit by using the condition (9.77), we obtain the convergence

$$\frac{\partial \eta_\sigma[v]}{\partial t} \to \left\{ \Delta y'_U[u](v-u) - u y'_U[u](v-u) - (v-u)y[u] \right\} \text{ weakly in } Y'_*.$$

Therefore, we get

$$\frac{\partial \eta_\sigma[v]}{\partial t} \to \frac{\partial y'_U[u](v-u)}{\partial t} \text{ weakly in } Y'_*.$$

Thus, we obtain convergence

$$\{y[u+\sigma(v-u)] - y[u]\}/\sigma \to y'_U[u](v-u) \text{ weakly in } Y.$$

3. For $\mu_Q = 0$ the equality (9.71) determines the affine continuous operator $y'_{UT} : \delta(u) \to L_2(\Omega)$. From (9.71) and (9.76) we get

$$\int_\Omega \mu_\Omega \Big[\{y[u+\sigma(v-u)](T) - y[u](T)\}/\sigma - \{y'_{UT}[u](v-u)\}(T) \Big] dx =$$

$$\int_Q p_\mu[u](u-v)\{y[u+\sigma(v-u)] - y[u]\}dQ.$$

Pass to the limit he; we obtain the condition (9.74). $\square$

Consider the function $\psi_\sigma[v] = \sigma \eta_\sigma[v]$, where $\eta_\sigma[v]$ was determined before.

Lemma 9.9 *For all $v \in U$ we have the convergence $\psi_\sigma[v] \to 0$ in Y_* as $\sigma \to 0$.*

Proof. The function $\psi_\sigma[v]$ satisfies the equation

$$\frac{\partial \psi_\sigma[v]}{\partial t} - \Delta \psi_\sigma[v] + u\psi_\sigma[v] + \sigma(v-u)y[u+\sigma(v-u)] = 0.$$

Multiply it by $\psi_\sigma[v]$ and integrate. We have

$$\frac{1}{2}\int_\Omega \left(\psi_\sigma[v](x,T)\right)^2 dx + \left\|\psi_\sigma[v]\right\|_{Y_*}^2 +$$

$$\int_Q u\left(\psi_\sigma[v]\right)^2 dQ = \sigma \int_Q (u-v)y[u+\sigma(v-u)]\psi_\sigma[v]dQ \le \frac{1}{2}\left\|\psi_\sigma[v]\right\|_{Y_*}^2 +$$

$$\frac{\sigma^2}{2}\left\|(u-v)y[u+\sigma(v-u)]\right\|_{Y'_*}^2 \le \frac{1}{2}\left\|\psi_\sigma[v]\right\|_{Y_*}^2 + b\sigma^2\left\|y[u+\sigma(v-u)]\right\|_{Y'_*}^2$$

by the definition of the set U. Thus, we have inequality

$$\left\|\psi_\sigma[v]\right\|_{Y_*}^2 \le 2b\sigma^2\left\|y[u+\sigma(v-u)]\right\|_{Y'_*}^2.$$

$$\text{Nonlinear first order evolutional systems} \qquad 345$$

From Lemma 9.7 the boundedness of the set of the functions $y[u + \sigma(v - u)]$ follows as $\sigma \to 0$ in the space Y_*, and in Y'_* too. Using the previous inequality, we complete the proof of the lemma. $\square$

Using Lemma 9.8 and Lemma 9.9, find the derivative of the minimizing functional for Problem 9.6.

Lemma 9.10 *The functional I for Problem 9.6 has the derivative with respect to the set U at the arbitrary point $u \in U$ such that*

$$\langle I'_U(u), v - u \rangle = \int_Q (\alpha u + py)(v - u)dQ \quad \forall v \in U, \qquad (9.78)$$

where $y = y[u]$, and p is the solution of the equation

$$-\frac{\partial p}{\partial t} - \Delta p + up = \Delta y_d - \Delta y[u] \qquad (9.79)$$

in the set Q with the boundary conditions

$$p(x, t) = 0, \ (x, t) \in \Sigma, \qquad (9.80)$$

$$p(x, T) = 0, \ x \in \Omega. \qquad (9.81)$$

Proof. We have the equality

$$I[u + \sigma(v - u) - I(u)] = \frac{\alpha}{2} \int_Q \left[2\sigma u(v - u) + \sigma^2(v - u)^2 \right]dQ +$$

$$\frac{1}{2} \int_Q \sum_{i=1}^n \frac{\partial}{\partial x_i} \left\{ y[u + \sigma(v - u)] + y[u] - 2y_d \right\} \frac{\partial}{\partial x_i} \left\{ y[u + \sigma(v - u)] - y[u] \right\} dQ.$$

From Lemma 9.8 the convergence follows:

$$\frac{1}{\sigma} \left\{ \frac{\partial y[u + \sigma(v - u)]}{\partial x_i} - \frac{\partial y[u]}{\partial x_i} \right\} \to \frac{\partial y'_U[u](v - u)}{\partial x_i} \quad \text{weakly in } L_2(Q).$$

Using Lemma 9.9, we have the convergence

$$\frac{\partial}{\partial x_i} \left\{ y[u + \sigma(v - u)] - y[u] \right\} \to 0 \quad \text{strongly in } L_2(Q).$$

Divide the previous equality by σ and pass to the limit as $\sigma \to 0$. We get

$$\langle I'_U(u), v - u \rangle = \int_Q [\alpha u(v - u)dQ + \int_Q \sum_{i=1}^n \frac{\partial(y - y_d)}{\partial x_i} \frac{\partial y'_U[u](v - u)}{\partial x_i} dQ =$$

346 Optimization and Differentiation

$$\int_Q [\alpha u(v-u)dQ + \int_Q (\Delta y_d - \Delta y)y'_U[u](v-u)dQ.$$

Put $\mu_Q = \Delta y_d - \Delta y$ and $\mu_\Omega = 0$ in the problem (9.71)–(9.73). We obtain the boundary problem (9.79)–(9.81). Using (9.70), we transform the previous equality to (9.78). □

Theorem 9.13 *The solution of Problem 9.6 with given set U is*

$$u(x,t) = \begin{cases} a, & \text{if } -p(x,t)/\alpha < a, \\ -p(x,t)/\alpha, & \text{if } a \le -p(x,t)/\alpha \le b, \\ b, & \text{if } -p(x,t)/\alpha > b. \end{cases} \tag{9.82}$$

Indeed, from the condition (9.78) it follows that the variational inequality

$$\int_Q (\alpha u + yp)(v-u)dQ \ge 0 \ \forall v \in U$$

is the necessary optimality condition for Problem 9.3. Now we determine the formula (9.82) by the standard method.

Remark 9.15 Theorem 9.13 is the nonstationary analog of Theorem 6.5.

9.8 Coefficient optimization problems for nonlinear parabolic equations

Now we consider the optimization control problems for the nonlinear heat equation. We have the equation

$$\frac{\partial y}{\partial t} - \Delta y + |y|^p y + vy = f \tag{9.83}$$

in the set of Q with the boundary conditions

$$y(x,t) = 0 \ (x,t) \in \Sigma \tag{9.84}$$

$$y(x,0) = \varphi(x), \ x \in \Omega, \tag{9.85}$$

where $\rho > 0$, $\varphi \in L_2(\Omega)$, $f \in L_2(0,T; H^{-1}(\Omega))$. Determine the control space $V = L_2(Q)$ and the set of admissible controls

$$U = \left\{ v \in V \mid a \le v(x) \le b, \ x \in \Omega \right\},$$

where a and b are constants, as $b > a > 0$.

Nonlinear first order evolutional systems

347

Multiply the equality (9.83) by the function y and integrate the result. After simple transformations, we obtain inequality

$$\int_Q \sum_{i=1}^n \left(\frac{\partial y}{\partial x_i}\right)^2 dQ + \int_Q |y|^{\rho+2} dQ + \int_Q v y^2 dQ = \frac{1}{2} \int_\Omega \varphi^2 dx + \int_Q f y dQ.$$

Then we have

$$\|y\|_{Y_*}^2 + \|y\|_{L_2(Q)}^q \leq \frac{1}{2}\|\varphi\|_2^2 + \|y\|_{Y_*}\|f\|_{Y_*'} \leq \frac{1}{2}\|\varphi\|_2^2 + \frac{1}{2}\|y\|_{Y_*}^2 + \frac{1}{2}\|f\|_{Y_*'}^2,$$

where

$$q = \rho + 2, \quad Y_* = L_2\big(0, T; H_0^1(\Omega)\big), \quad Y_*' = L_2\big(0, T; H^{-1}(\Omega)\big).$$

The problem (9.83)–(9.85) has a priori estimate of the solution in the space

$$X = Y_* \cap L_q(Q).$$

If $y \in X$, then from the equality (9.83) the inequality follows:

$$\left\|\frac{\partial y}{\partial t}\right\|_{Y_*'} \leq \|\Delta y\|_{Y_*'} + \||y|^\rho y\|_{L_{q'}(Q)} \|vy\|_{Y_*'} + \|f\|_{Y_*'}.$$

Then the time derivative of the functions belongs to the space

$$X' = Y_*' + L_{q'}(Q).$$

Therefore, the function y is an element of the space

$$Y = \big\{y \,\big|\, y \in X, \; y' \in X'\big\}.$$

Thus, the boundary problem (9.83)–(9.85) has the same properties as the system (9.29)–(9.31). We prove the solvability of the considered problem as the analog of Theorem 9.3.

Theorem 9.14 *Suppose $\varphi \in L_2(\Omega)$, $f \in Y_*'$. Then for all control $v \in U$ there exists a unique solution $y = y[u]$ of the problem (9.83)–(9.85) from the space Y.*

Consider the functional

$$I(v) = \frac{\alpha}{2} \int_Q v^2 dQ + \frac{1}{2} \int_Q \sum_{i=1}^n \left(\frac{\partial y[v]}{\partial x_i} - \frac{\partial y_d}{\partial x_i}\right)^2 dQ,$$

where $\alpha > 0$, $y_d \in Y_*$ is a known function.

Problem 9.7 *Find the control $u \in U$ for the system (9.83)–(9.85) that minimizes the functional I on the set U.*

348 *Optimization and Differentiation*

This problem has properties of Problem 9.2 by nonlinearity of the equation and of Problem 9.4 because of the form of the control. This is the nonstationary analog of Problem 6.5. Prove the weak continuity of the solution of the boundary problem with respect to the control that is the general step of the substantiation of the problem solvability.

Lemma 9.11 *The map $y[\cdot] : V \to Y$ for the problem* (9.83)–(9.85) *is weakly continuous.*

Proof. Consider a sequence $\{v_k\}$ of the set U such that $v_k \to v$ weakly in V. Using the obtained estimates, we determine the boundedness of the sequence $\{y[v_k]\}$ in the space Y. Then extracting a subsequence, we obtain convergence $y[v_k] \to y$ weakly in Y. Using compact embedding of the space Y in $L_2(Q)$ we have $y[v_k] \to y$ strongly in $L_2(Q)$ and a.e. on Q. Using the technique from Lemma 9.7, we obtain the convergence

$$\int\limits_Q v_k y[v_k] \lambda dQ \to \int\limits_Q v y \lambda dQ.$$

By the boundedness of the sequence $\{|y[v_k]|^\rho y[v_k]\}$ in the space $L_{q'}(Q)$ and the convergence $|y[v_k]|^\rho y[v_k] \to |y[v]|^\rho y[v]$ a.e. on Q, we have $|y[v_k]|^\rho y[v_k] \to |y[v]|^\rho y[v]$ weakly in $L_{q'}(Q)$. Determine $v = v_k$ for the problem (9.83)–(9.85) and pass to the limit as $k \to \infty$. Obtain the equality $y = y[v]$. $\square$

Remark 9.16 Lemma 9.11 is a nonstationary analog of Lemma 6.8.

Using the previous result, we prove the solvability of the considered problem.

Theorem 9.15 *Problem* 9.7 *is solvable.*

Prove the extended differentiability with respect to the set U of the control–state mapping at the arbitrary point $u \in U$. Determine the space

$$X_0 = \left\{ p \big|\ p \in Y_*,\ |y[u]|^{\rho/2} p \in L_2(Q) \right\}$$

and the equation

$$-\frac{\partial p_\mu[u]}{\partial t} - \Delta p_\mu[u] + (\rho + 1)|y[u]|^\rho p_\mu[u] + u p_\mu[u] = \mu_Q \tag{9.86}$$

in the set Q with boundary conditions

$$p_\mu[u](x,t) = 0,\ (x,t) \in \Sigma, \tag{9.87}$$

$$p_\mu[u](x,T) = \mu_\Omega(x),\ x \in \Omega, \tag{9.88}$$

where $\mu = (\mu_Q, \mu_\Omega)$.

Nonlinear first order evolutional systems

Lemma 9.12 *The map $y[\cdot] : V \to Y$ for the problem (9.83)–(9.85) has $(V, X_0; V, X_0^w)$-extended derivative $y'[u]$ with respect to the set U at the arbitrary point $u \in U$ as there exists an affine continuous operator $y'_{UT}[u]$ on the set $\delta(u) = \{v - u|\ v \in U\}$ with values in the space $L_2(\Omega)$ such that for all $v \in U$ we have the convergence*

$$\{y[u + \sigma(v - u)](T) - y[u](T)\}/\sigma \to y'_{UT}[u](v - u) \text{ weakly in } L_2(\Omega).$$

In addition, we have the equality

$$\int_Q \mu_Q y'[u](v - u)dQ + \int_\Omega \mu_\Omega y'_{UT}[u](v - u)dQ = \int_Q p_\mu[u](u - v)dQ \quad (9.89)$$

for all $v \in U$, $\mu_Q \in X'_0$, $\mu_\Omega \in L_2(\Omega)$.

Proof. 1. Consider the equality

$$\frac{\partial \eta_\sigma[v]}{\partial t} - \Delta \eta_\sigma[v] + \big(g_\sigma[v]\big)^2 \eta_\sigma[v] + u\eta_\sigma[v] + (v - u)y[v_\sigma] = 0$$

that is the analog of (9.39) and (9.75), where

$$v_\sigma = u + \sigma(v - u), \quad \eta_\sigma[v] = \big(y[u + \sigma(v - u)] - y[u]\big)/\sigma,$$

$$\big(g_\sigma[v]\big)^2 = (\rho + 1)\big|y[u] + \varepsilon\eta_\sigma[v]\big|^\rho, \quad \varepsilon \in [0, 1].$$

Multiply by smooth enough function λ that is equal to zero on the surface Σ and for $t = T$. We have

$$\int_Q \Big[-\frac{\partial \lambda}{\partial t} - \Delta\lambda + \big(g_\sigma[v]\big)^2\lambda + u\lambda\Big]\eta_\sigma[v]dQ +$$

$$\int_\Omega \lambda(T)\eta_\sigma[v](T)dx = \int_Q \lambda(u - v)y[u + \sigma(v - u)]dQ. \quad (9.90)$$

Consider the equation

$$-\frac{\partial p}{\partial t} - \Delta p + \big(g_\sigma[v]\big)^2 p + up = \mu_Q \quad (9.91)$$

in the set Q with boundary conditions

$$p(x, t) = 0, \ (x, t) \in \Sigma, \quad (9.92)$$

$$p(x, T) = \mu_\Omega(x), \ x \in \Omega. \quad (9.93)$$

We have the system (9.86)–(9.88) here for $\sigma = 0$. Multiply the equality (9.91) by the function p and integrate the result. We obtain inequality

$$\int_Q \sum_{i=1}^n \Big(\frac{\partial p}{\partial x_i}\Big)^2 dQ + \int_Q \big(g_\sigma[v]p\big)^2 dQ + \int_Q up^2 dQ \le \int_Q \mu_Q p dQ + \frac{1}{2}\|\mu_\Omega\|_2^2.$$

350   *Optimization and Differentiation*

Determine the space

$$X_\sigma = \Big\{ y \big| \, y \in Y_*, \; g_\sigma[v] y \in L_2(Q) \Big\}.$$

The third integral at the left-hand side of the previous equality is non-negative. Then we obtain

$$\|p\|^2_{X_\sigma} \leq \|\mu_Q\|^2_{X'_\sigma} + \|\mu_\Omega\|^2_2.$$

From the equality (9.91) it follows that

$$\|p'\|_{X'_\sigma} \leq \|p'\|_{Y'_*} + \big\|g_\sigma[v]p\big\|_{L_2(Q)} \big\|g_\sigma[v]\big\|_{L_{2q/\rho}(Q)} + b\|p\|_{X_\sigma} + \|\mu_Q\|_{X'_\sigma}.$$

Thus, the boundary problem (9.91)–(9.93) for all numbers σ and the functions $\mu_Q \in X'_\sigma$, $\mu_\Omega \in L_2(\Omega)$ has a priori estimate of the solution in the space

$$Y_\sigma = \Big\{ y \big| \, y \in X_\sigma, \; y' \in X'_\sigma \Big\}.$$

Then this problem has a unique solution $p = p^\sigma_\mu[v]$ from the Y_σ. Note that for $\sigma = 0$ the equation (9.91) is transformed to (9.86), and X_σ is the space X_0. Then for all functions $\mu_Q \in X'_\sigma$, $\mu_\Omega \in L_2(\Omega)$ the problem (9.86)–(9.88) has a unique solution from space Y_0. Therefore, for $\mu_\Omega = 0$ the equality (9.89) determines an affine continuous operator $y'_U[u]$ on the set of differences $\delta(u)$ with values from X_0.

Choose $\lambda = p^\sigma_\mu[v]$ at the equality (9.90). We have

$$\int\limits_Q \mu_Q \eta_\sigma[v] dQ + \int\limits_\Omega \mu_\Omega \eta_\sigma[v](T) dx = \int\limits_Q p_\mu[u](u-v)y[v_\sigma] dQ.$$

Using this equality and the condition (9.89) for $\mu_\Omega = 0$, we get

$$\int\limits_Q \mu_Q \Big[\big\{ y[u+\sigma(v-u)] - y[u] \big\}/\sigma - y'_U[u](v-u) \Big] dQ =$$

$$\int\limits_Q (u-v) \Big\{ p^\sigma_\mu[v]y[v_\sigma] - p_\mu[u]y[u] \Big\} dQ. \tag{9.94}$$

2. Now suppose the convergence $\sigma \to 0$. Then by Lemma 9.11 we have $y[u+\sigma(v-u)] \to y[u]$ weakly in Y. Therefore, the set $\{y[v_\sigma]\}$ is bounded in the space $L_q(Q)$. We obtain the boundedness of $\{g_\sigma[v]\}$ in the space $L_{2q/\rho}(Q)$. Using the properties of the functions $p^\sigma_\mu[v]$ we obtain the boundedness of the set $\{p^\sigma_\mu[v]\}$ in the space X_σ and the boundedness of $\big\{ \big(g_\sigma[v]\big)^2 p^\sigma_\mu[v] \big\}$ in the space $L_{q'}(Q)$. From the equality (9.91) follows the boundedness of set $\big(\{p^\sigma_\mu[v]\}'\big)$ in the space $P_1 = Y'_* + L_{q'}(Q)$. Then the set $\{p^\sigma_\mu[v]\}$ is bounded in the space

$$P = \Big\{ p \big| \, p \in Y_*, \; p' \in P_1 \Big\}.$$

Nonlinear first order evolutional systems

Extracting subsequence we have the convergence $p_\mu^\sigma[v] \to p_\mu$ weakly in P. Using compact embedding of the spaces P and Y to $L_2(Q)$, determine $p_\mu^\sigma[v] \to p_\mu$ and $y[v_\sigma] \to y[u]$ strongly in $L_2(Q)$ and a.e. on Q. Then $\left(g_\sigma[v]\right)^2 p_\mu^\sigma[v] \to (\rho+1)|y[u]|^\rho p_\mu$ a.e. on Q. Using the boundedness of $\left\{\left(g_\sigma[v]\right)^2 p_\mu^\sigma[v]\right\}$ in the space $L_{q'}(Q)$, we get $\left(g_\sigma[v]\right)^2 p_\mu^\sigma[v] \to (\rho+1)|y[u]|^\rho p_\mu$ weakly in $L_{q'}(Q)$. Then after passing to the limit at the equation (9.91) for $p = p_\mu^\sigma[v]$ and $\sigma \to 0$ we have $p_\mu = p_\mu[u]$.

3. Return to the equality (9.94). Using the definition of the set of admissible controls, we get

$$\left| \int_Q (u-v)\left\{ p_\mu^\sigma[v]y[v_\sigma] - p_\mu[u]y[u] \right\} dQ \right| \le$$

$$2b \int_Q \left| p_\mu^\sigma[v]\{y[v_\sigma] - y[u]\} \right| dQ + 2b \int_Q \left| \{p_\mu^\sigma[v] - p_\mu[u]\}y[v_\sigma] \right| dQ \le$$

$$\le 2b \left\| p_\mu^\sigma[v] \right\|_{L_2(Q)} \left\| y[v_\sigma] - y[u] \right\|_{L_2(Q)} + 2b \left\| p_\mu^\sigma[v] - p_\mu[u] \right\|_{L_2(Q)} \left\| y[v_\sigma] \right\|_{L_2(Q)}.$$

By the convergence $p_\mu^\sigma[v] \to p_\mu[u]$ and $y[v_\sigma] \to y[u]$ strongly in $L_2(Q)$ we obtain

$$\left| \int_Q (u-v)\left\{ p_\mu^\sigma[v]y[v_\sigma] - p_\mu[u]y[u] \right\} dQ \right| \to 0.$$

From (9.94) the convergence follows:

$$\{y[u + \sigma(v-u)] - y[u]\}/\sigma \to y_U'[u](v-u) \quad \text{weakly in } Y_*$$

for all $v \in U$. Thus, the affine continuous operator $y_U'[u]$ that is determined by the equality (9.89) is the extended derivative with respect to the given set of the considered math at the point u. Repeat the final transformations from Lemma 9.3 and Lemma 9.8. This completes the proof of Lemma 9.12. $\square$

Now consider an analog of Lemma 9.4 and Lemma 9.9. Determine the function $\psi_\sigma[v] = \sigma \eta_\sigma[v]$, where $\eta_\sigma[v]$ was definite before.

Lemma 9.13 *For all* $v \in U$ *we have the convergence* $\psi_\sigma[v] \to 0$ *in* Y_* *as* $\sigma \to 0$.

Proof. The function $\psi_\sigma[v]$ satisfies the equation

$$\frac{\partial \psi_\sigma[v]}{\partial t} - \Delta \psi_\sigma[v] + \left(g_\sigma[v]\right) + u\psi_\sigma[v] + \sigma(v-u)y[v_\sigma] = 0.$$

Multiply it by $\psi_\sigma[v]$ and integrate. We have

$$\frac{1}{2} \int_\Omega \left(\psi_\sigma[v](x,T)\right)^2 dx + \left\| \psi_\sigma[v] \right\|_{Y_*}^2 + \left\| g_\sigma[v]\psi_\sigma[v] \right\|_{L_2(Q)}^2 +$$

Optimization and Differentiation

$$\int_Q u(\psi_\sigma[v])^2 dQ = \sigma \int_Q (u-v)y[v_\sigma]\psi_\sigma[v]dQ \leq \frac{1}{2}\|\psi_\sigma[v]\|_{Y_*}^2 +$$

$$\frac{\sigma^2}{2}\|(u-v)y[v_\sigma]\|_{Y_*'}^2 \leq \frac{1}{2}\|\psi_\sigma[v]\|_{Y_*}^2 + b\sigma^2\|y[v_\sigma]\|_{Y_*'}^2.$$

Thus, we have inequality

$$\|\psi_\sigma[v]\|_{Y_*}^2 \leq 2b\sigma^2\|y[v_\sigma]\|_{Y_*'}^2.$$

From Lemma 9.11 the boundedness of the set of the functions $y[v_\sigma]$ follows in the space Y^*, and in Y_* too as Y_*'. Now we use the previous inequality. This completes the proof of the lemma. $\square$

Prove the differentiability of the state functional.

Lemma 9.14 *The functional I for Problem 9.6 has the derivative with respect to the set U at the arbitrary point $u \in U$ such that*

$$\langle I_U'(u), v-u \rangle = \int_Q (\alpha u + py)(v-u)dQ \quad \forall v \in U, \tag{9.95}$$

where $y = y[u]$, and p is the solution of the equation

$$-\frac{\partial p}{\partial t} - \Delta p + (\rho+1)|y|^\rho + up = \Delta y_d - \Delta y[u] \tag{9.96}$$

with homogeneous conditions on the surface Σ and for $t = T$.

Proof. We have the equality

$$I[u + \sigma(v-u)] - I(u)] = \frac{\alpha}{2}\int_Q [2\sigma u(v-u) + \sigma^2(v-u)^2]dQ +$$

$$\frac{1}{2}\int_Q \sum_{i=1}^n \frac{\partial}{\partial x_i}\Big\{y[u + \sigma(v-u)] + y[u] - 2y_d\Big\}\frac{\partial}{\partial x_i}\Big\{y[u + \sigma(v-u)] - y[u]\Big\}dQ.$$

From Lemma 9.12 the convergence follows:

$$\frac{1}{\sigma}\Big\{\frac{\partial y[u + \sigma(v-u)]}{\partial x_i} - \frac{\partial y[u]}{\partial x_i}\Big\} \to \frac{\partial y_U'[u](v-u)}{\partial x_i} \quad \text{weakly in } L_2(Q).$$

Using Lemma 9.13, we have the convergence

$$\frac{\partial}{\partial x_i}\Big\{y[u + \sigma(v-u)] - y[u]\Big\} \to 0 \quad \text{strongly in } L_2(Q).$$

Divide the previous equality by σ and pass to the limit as $\sigma \to 0$. We get

$$\langle I_U'(u), v - u \rangle = \int_Q [\alpha u(v-u)dQ + \int_Q \sum_{i=1}^n \frac{\partial(y-y_d)}{\partial x_i} \frac{\partial y_U'[u](v-u)}{\partial x_i} dQ =$$

$$\int_Q [\alpha u(v-u)dQ + \int_Q (\Delta y_d - \Delta y)y_U'[u](v-u)dQ.$$

Using the final transformations of Lemma 9.9, we finish the proof of lemma. $\square$

Remark 9.17 We have the extended differentiability with respect to the convex set for the control–state mapping and the usual differentiability with respect to the convex set for the minimizing functional. Therefore, the extended differentiation is the technique for obtaining the result in the standard form.

Now we can find the optimal control as the analog of Theorem 9.13.

Theorem 9.16 *The solution of Problem 9.7 with given set U is*

$$u(x,t) = \begin{cases} a, & \text{if } -p(x,t)/\alpha < a, \\ -p(x,t)/\alpha, & \text{if } a \le -p(x,t)/\alpha \le b, \\ b, & \text{if } -p(x,t)/\alpha > b. \end{cases}$$

Indeed, put the functional derivative from the equality (9.95) to the relation (9.69). We obtain the variational inequality

$$\int_Q (\alpha u + yp)(v-u)dQ \ge 0 \; \forall v \in U$$

that can be transformed to the previous formula by using the standard technique.

9.9 Initial optimization control problems for nonlinear parabolic equations

We considered the optimization control problems with the distributed and the boundary controls. Let Ω be an open bounded set of Euclid space $\mathbb{R}^n$ with the boundary Γ, $T > 0$, $Q = \Omega \times (0,T)$, $\Sigma = \Gamma \times (0,T)$. We have the nonlinear equation

$$\frac{\partial y}{\partial t} - \Delta y + |y|^\rho y = f \tag{9.97}$$

354   *Optimization and Differentiation*

in the set Q with the boundary condition

$$y(x,t) = 0, \ (x,t) \in \Sigma \tag{9.98}$$

and the initial condition

$$y(x,0) = v(x), \ x \in \Omega, \tag{9.99}$$

where $\rho > 0$, $f = f(x,t)$ is a known function, and v is the control that is an element of a non-empty convex closed set U of the space $V = L_2(\Omega)$. This is the boundary problem (9.24)–(9.26) with **initial control**. Determine the spaces

$$X = L_2\big(0,T;H_0^1(\Omega)\big) \cap L_q(Q), \ X' = L_2\big(0,T;H^{-1}(\Omega)\big) + L_{q'}(Q),$$

where $1/q + 1/q' = 1$. Now we determine the state space

$$Y = \big\{y \big| \ y \in X, \ y' \in X'\big\}.$$

Suppose f belongs to the space X'. By Theorem 9.3, for all $v \in V$ the problem (9.97)–(9.99) has a unique solution $y = y[v]$ from the space Y.

Consider the functional

$$I(v) = \frac{\alpha}{2} \int_\Omega v^2 dQ + \frac{1}{2} \int_\Omega \sum_{i=1}^n \Big(\frac{\partial y[v]}{\partial x_i} - \frac{\partial y_d}{\partial x_i}\Big)^2 dQ,$$

where $y_d \in L_2\big(0,T;H_0^1(\Omega)\big)$ is a known function.

Problem 9.8 *Find the control $u \in U$ for the system (9.97)–(9.99) that minimizes the functional I on the set U.*

Determine the weak continuity of the control–state mapping.

Lemma 9.15 *The operator $y[\cdot] : V \to Y$ for the problem (9.97)–(9.99) is weakly continuous.*

Proof. 1. Suppose the convergence $v_k \to v$ weakly in V. Multiply the equality (9.97) for $v = v_k$ by the function $y_k = y[v_k]$. After the integration we have

$$\frac{1}{2}\frac{d}{dt} \int_\Omega |y_k|^2 dx + \int_\Omega \sum_{i=1}^n \Big|\frac{\partial y_k}{\partial x_i}\Big|^2 dx + \int_\Omega |y_k|^q dx = \int_\Omega f y_k dx.$$

Then

$$\frac{1}{2}\frac{d}{dt}\|y_k(t)\|^2 + \|y_k(t)\|^2 + \|y_k(t)\|_q^q \le \|y_k(t)\|\|f(t)\|_* \le \frac{1}{2}\|y_k(t)\|^2 + \frac{1}{2}\|f(t)\|_*^2.$$

Integrate this inequality by t by using the condition (9.99). We get

$$\|y_k(t)\|_2^2 + \int_0^t \|y_k(\tau)\|^2 d\tau + 2 \int_0^t \|y_k(\tau)\|_q^q d\tau \le \|v_k\|_V^2 + \|f\|_{L_2(0,T;H^{-1}(\Omega))}^2.$$

Nonlinear first order evolutional systems

Then the sequence $\{y_k\}$ is bounded in X. From the equality (9.97), it follows that

$$\|y'\|_{X'} \le \|\Delta y\|^2_{L_2(0,T;H^{-1}(\Omega))} + \||y_k|^\rho y_k\|_{L_{q'}(Q)} + \|f\|^2_{L_2(0,T;H^{-1}(\Omega))}.$$

The Laplace operator is the continuous operator from $H_0^1(\Omega)$ to $H^{-1}(\Omega)$. Consider the equality

$$\|y_k\|^{q'}_{L_{q'}(Q)} = \int_Q \||y_k|^q y_k\|^{q'} dQ = \int_Q |y_k|^{(\rho+1)q'} dQ = \int_Q |y_k|^q dQ = \|y_k\|^q_{L_q(Q)}.$$

From the previous inequality, it follows that the sequence $\{y_k'\}$ is bounded in X'. Then $\{y_k\}$ is bounded in Y.

2. Extracting a subsequence of $\{y_k\}$, we have $y_k \to y$ weakly in Y. From Theorem 8.5, compact embedding of Y to $L_2(Q)$ follows. Then $y_k \to y$ strongly in $L_2(Q)$ and a.e. on Q. Therefore, $|y_k|^\rho y_k \to |y|^\rho y$ a.e. on Q. The sequence $\{y_k\}$ is bounded in $L_q(Q)$. Therefore, $\{|y_k|^\rho y_k\}$ is bounded in $L_{q'}(Q)$. Using Lemma 4.6, we obtain $|y_k|^\rho y_k \to |y|^\rho y$ weakly in $L_{q'}(Q)$. Multiply the equality (9.97) for $v = v_k$ by a smooth enough function λ. After the integration and passing to the limit we have $y = y[v]$. This completes the proof of the lemma. $\square$

From Lemma 9.15 it follows that the considered functional is weakly lower semicontinuous. Then we obtain the solvability of the optimization problem.

Theorem 9.17 *Problem* 9.8 *is solvable.*

Determine the differentiability of the control–state mapping at an arbitrary point $u \in V$. Consider the spaces

$$Y_* = L_2\big(0,T;H_0^1(\Omega)\big), \quad X_0 = \Big\{y \in Y_* \big|\, |y[u]|^{\rho/2} y \in L_2(Q)\Big\}$$

and the boundary problem $p_\mu[u]$ is the solution of the equation

$$-\frac{\partial p_\mu[u]}{\partial t} - \Delta p_\mu[u] + (\rho+1)|y[u]|^\rho p_\mu[u] = \mu_Q \qquad (9.100)$$

in the set Q with boundary conditions

$$p_\mu[u](x,t) = 0, \ (x,t) \in \Sigma, \qquad (9.101)$$

$$p_\mu[u](x,T) = \mu_\Omega(x), \ x \in \Omega, \qquad (9.102)$$

where $\mu = \big(\mu_Q, \mu_\Omega\big)$.

Lemma 9.16 *The map* $y[\cdot] : V \to Y$ *for the problem* (9.97)–(9.99) *has* $\big(V, X_0; V, Y_*^w\big)$*-extended derivative* $y'[u]$ *at the arbitrary point* $u \in V$, *where* Y_*^w *is the space* Y_* *with the weak topology. Besides, there exists a linear continuous operator* $y'_T[u] : V \to L_2(\Omega)$ *such that for all* $h \in V$ *we have the*

356   *Optimization and Differentiation*

convergence $\{y[u+\sigma h](T) - y[u](T)\}/\sigma \to y'_T[u]h$ weakly in $L_2(\Omega)$. We have also the equality

$$\int_Q \mu_Q y'[u]hdQ + \int_\Omega \mu_\Omega y'_T[u]hdQ = \int_\Omega p_\mu[u](x,0)h(x)dx \qquad (9.103)$$

for all $h \in V$, $\mu_Q \in X_0$, $\mu_\Omega \in L_2(\Omega)$.

Proof. For all value $n \in V$ and the number σ the function

$$\eta_\sigma[h] = (y[u+\sigma h] - y[u])/\sigma$$

satisfies the equation

$$\frac{\partial \eta_\sigma[h]}{\partial t} - \Delta \eta_\sigma[h] + |y[u+\sigma h]|^\rho y[u+\sigma h] - |y[u]|^\rho y[u] = 0,$$

the homogeneous boundary condition, and the initial condition

$$\eta_\sigma[h]\big|_{t=0} = h.$$

Multiply the first equality by an arbitrary function $\lambda \in Y$. After integration we have

$$\int_Q \left[\frac{\partial \lambda}{\partial t} - \Delta \lambda + \left(g_\sigma[h]\right)^2 \lambda\right]\eta_\sigma[h]dQ + \int_\Omega \eta_\sigma[h](x,T)\lambda(x,T)dx, \qquad (9.104)$$

where

$$\left(g_\sigma[h]\right)^2 = (\rho+1)\big|y[u] + \varepsilon\big(y[u+\sigma h] - y[u]\big)\big|^\rho, \ \varepsilon \in [0,1].$$

Consider the boundary problem

$$-\frac{\partial p}{\partial t} - \Delta p + \left(g_\sigma[h]\right)^2 p = \mu_Q \ \text{ in } Q,$$

$$p = 0, \ \text{ in } \Sigma,$$

$$p(x,T) = \mu_\Omega(x), \ x \in \Omega.$$

This is the system (9.40)–(9.42) in reality, and this is equal to the problem (9.100)–(9.102) for $\sigma = 0$. Determine the space

$$X_\sigma = \left\{y\big|\ y \in Y_*, \ g_\sigma[h]y \in L_2(Q)\right\}$$

that is equal to X_0 for $\sigma = 0$. Consider also its adjoint space

$$X'_\sigma = \left\{\varphi + g_\sigma[h]\psi\big|\ \varphi \in Y'_*, \ \psi \in L_2(Q)\right\}.$$

Nonlinear first order evolutional systems

Determine the space

$$Y_\sigma = \left\{ y \mid y \in X_\sigma, \ y' \in X'_\sigma \right\}.$$

By Lemma 9.3, for all number σ and for all function $h \in V$, $\mu_Q \in X'_\sigma$, $\mu_\Omega \in L_2(\Omega)$ this problem has a unique solution $p = p^\sigma_\mu[v]$ from the set Y_σ, as $p^\sigma_\mu[h] \to p_\mu[u]$ weakly in the space

$$P = \left\{ p \mid p \in Y_*, \ p' \in Y'_* + L_{q'}(Q) \right\}$$

as $\sigma \to 0$. For $\sigma = 0$ the problem (9.100)–(9.102) has a unique solution from the space Y_0. Embedding of this space to $C\big(0, T; L_2(\Omega)\big)$ is compact. Then $p_\mu[u]$ for $t = 0$ belongs to $L_2(\Omega)$. Therefore, the equality (9.100) for $\mu_\Omega = 0$ determines a linear continuous operator $y'[u] : V \to X_0$ in reality.

Choose $\lambda = p^\sigma_\mu[h]$ at the equality (9.104). We have

$$\int_Q \eta_\sigma[h] \mu_Q dQ + \int_\Omega \eta_\sigma[h](T) \mu_\Omega dx = \int_\Omega h p^\sigma_\mu[h](0) dx.$$

Using (9.103), we obtain

$$\int_Q \big(\eta_\sigma[h] - y'[u]h\big) \mu_Q dQ + \int_\Omega \big(\eta_\sigma[h](T) - y'_T[u]h\big) \mu_\Omega dx =$$

$$\int_\Omega h\big(p^\sigma_\mu[h](0) - p_\mu[u](0)\big) dx. \tag{9.105}$$

Suppose $\mu_\Omega = 0$. Using the convergence $p^\sigma_\mu[h](0) \to p_\mu[u](0)$ weakly in $L_2(\Omega)$, we have $\big(y[u + \sigma h] - y[u]\big)/\sigma \to y'[u]h$ weakly in Y_*. We complete the proof of the lemma, if we choose $\mu_Q = 0$ at the equality (9.105) and pass to the limit. $\square$

Using Lemma 9.16, we prove the differentiability of the minimizing functional.

Lemma 9.17 *The functional I for Problem 9.8 has Gâteaux derivative $I'(u) = \alpha u + p(0)$ at the arbitrary point $u \in V$, where p is the solution of the equation*

$$-\frac{\partial p}{\partial t} - \Delta p + (\rho + 1)\big|y[u]\big|^\rho p = \Delta y_d - \Delta y[u]$$

that is equal to zero on the surface Σ and for $t = T$.

Proof. We have the equality

$$I(u + \sigma h) - I(u) = \frac{\alpha}{2} \int_\Omega (2\sigma u h + \sigma^2 h^2) dx +$$

Optimization and Differentiation

$$\frac{1}{2} \int\limits_Q \sum_{i=1}^n \frac{\partial \big(y[u + \sigma h] + y[u] - 2y_d\big)}{\partial x_i} \frac{\partial \big(y[u + \sigma h] - y[u]\big)}{\partial x_i} dQ.$$

Using Lemma 9.16, we have

$$\frac{1}{\sigma} \frac{\partial \big(y[u + \sigma h] - y[u]\big)}{\partial x_i} \to \frac{\partial y'[u]h}{\partial x_i} \quad \text{weakly in } L_2(Q).$$

Besides,

$$\frac{\partial y[u + \sigma h]}{\partial x_i} \to \frac{\partial y[u]}{\partial x_i} \quad \text{strongly in } L_2(Q).$$

Divide the previous equality by σ and pass to the limit as $\sigma \to 0$. We obtain

$$\langle I'(u), h \rangle = \alpha \int\limits_\Omega uh dx + \int\limits_Q \sum_{i=1}^n \frac{\partial \big(y[u] - y_d\big)}{\partial x_i} \frac{\partial y'[u]h}{\partial x_i} dQ =$$

$$\alpha \int\limits_\Omega uh dx + \int\limits_Q \big(\Delta y_d - \Delta y[u]\big) y'[u]h dQ.$$

Choose $\mu_\Omega = 0$, $\mu_Q = \Delta y_d - \Delta y[u]$ at the equality (9.103). We have

$$\langle I'(u), h \rangle = \int\limits_\Omega \big[\alpha u + p(x, 0)\big] h(x) dx \ \forall h \in V.$$

This completes the proof of the lemma. $\square$

Putting the value of the functional derivative in to the standard necessary condition of optimality, we obtain the following result.

Theorem 9.18 *The solution u of Problem 9.8 satisfies the variational inequality*

$$\int\limits_\Omega \big[\alpha u + p(x, 0)\big] \big[v(x) - u(x)\big] dx \geq 0 \ \forall v \in U.$$

9.10 Optimization control problems for nonlinear parabolic equations with final functional

We considered before the optimization control problems for the minimizing functionals that depend upon the state function at each point of its domain. Now the functional depends upon the state function at the final time only. Let

Nonlinear first order evolutional systems 359

Ω be an open bounded set of Euclid space $\mathbb{R}^n$ with the boundary Γ, $T > 0$, $Q = \Omega \times (0, T)$, $\Sigma = \Gamma \times (0, T)$. We have the nonlinear heat equation

$$\frac{\partial y}{\partial t} - \Delta y + |y|^\rho y = v + f \tag{9.106}$$

in the set Q with the boundary condition

$$y(x, t) = 0, \ (x, t) \in \Sigma \tag{9.107}$$

and the initial condition

$$y(x, 0) = \varphi(x), \ x \in \Omega, \tag{9.108}$$

where $\rho > 0$. Determine the spaces

$$X = L_2\big(0, T; H_0^1(\Omega)\big) \cap L_q(Q), \ X' = L_2\big(0, T; H^{-1}(\Omega)\big) + L_{q'}(Q),$$

where $q = \rho + 2$. Suppose the conditions $\varphi \in L_2(\Omega)$, $f \in X'$. The control v belongs to the non-empty convex closed set U of the space $V = L_2(Q)$. From Theorem 9.3 it follows that for all $v \in V$ the problem (9.106)–(9.108) has a unique solution $y = y[v]$ from the space

$$Y = \big\{ y \big| \ y \in X, \ y' \in X' \big\}.$$

We considered Problem 9.2 with minimizing functional that depends upon the state function on the whole of its domain Q. Now we have the functional

$$I(v) = \frac{\alpha}{2} \int_Q v^2 dQ + \frac{1}{2} \int_\Omega \big(y[v]_{t=T} - y_d\big)^2 dx,$$

where $\alpha > 0$, $y_d \in L_2(\Omega)$ is a known function.

Problem 9.9 *Find the control $u \in U$ for the system (9.106)–(9.108) that minimizes the functional I on the set U.*

Prove the existence of the optimal control.

Theorem 9.19 *Problem 9.9 is solvable.*

Proof. Suppose $v_k \to v$ weakly in V. Using Lemma 9.1, we have the convergence $y[v_k] \to y[v]$ weakly in Y. Therefore, we obtain $y[v_k]\big|_{t=T} \to y[v]\big|_{t=T}$ weakly in $L_2(\Omega)$ because embedding of the space Y to $C\big(0, T; L_2(\Omega)\big)$ is continuous. Hence, the given functional is weakly lower semicontinuous. Then we determine the solvability of the optimization problem by using the standard technique. $\square$

360 *Optimization and Differentiation*

It is necessary to analyze the differentiability of control–state mapping for obtaining the conditions of optimality. By Lemma 9.2, this dependence is not Gâteaux differentiable for large enough values of the parameters ρ and n. However, by Lemma 9.3, there exists its $(V, Y_0; V, Y_*^w)$-extended derivative $y'[u]$ at the arbitrary point $u \in V$. We use the following spaces here:

$$Y_* = L_2(0, T; H_0^1(\Omega)), \ Y_0 = \{y| \ y \in X_0, \ y' \in X_0'\},$$

where

$$X_0 = \left\{y \in Y_*| \ |y[u]|^{\rho/2} y \in L_2(Q)\right\},$$

$$X_0' = \left\{h| \ h = h_1 + (\rho + 1)|y[u]|^{\rho/2} h_2, \ h_1 \in Y_*', \ h_2 \in L_2(Q)\right\},$$

and Y_*^w is the space Y_* with the weak topology. Besides there exists a linear continuous operator $y_T'[u] : V \to L_2(\Omega)$ such that for all $h \in V$ we have the convergence $\{y[u + \sigma h](T) - y[u](T)\}/\sigma \to y_T'[u]h$ weakly in $L_2(\Omega)$. These operators satisfy the equality

$$\int_Q \mu_Q y'[u]hdQ + \int_\Omega \mu_\Omega y_T'[u]hdQ = \int_Q p_\mu[u]hdQ \ \forall h \in V, \tag{9.109}$$

where $p_\mu[u]$ is the solution of the equation

$$-\frac{\partial p_\mu[u]}{\partial t} - \Delta p_\mu[u] + (\rho + 1)|y[u]|^\rho p_\mu[u] = \mu_Q \tag{9.110}$$

in the set Q with boundary conditions

$$p_\mu[u](x, t) = 0, \ (x, t) \in \Sigma, \tag{9.111}$$

$$p_\mu[u](x, T) = \mu_\Omega(x), \ x \in \Omega, \tag{9.112}$$

and $\mu = (\mu_Q, \mu_\Omega)$. We have the following result that is the analog of Lemma 9.5.

Lemma 9.18 *The functional I for Problem 9.9 has Gâteaux derivative*

$$I'(u) = \alpha u + p$$

at the arbitrary point u, where p is the solution of the equation

$$-\frac{\partial p}{\partial t} - \Delta p + (\rho + 1)|y[u]|^\rho p = 0 \tag{9.113}$$

in the set Q with the boundary conditions

$$p(x, t) = 0, \ (x, t) \in \Sigma, \tag{9.114}$$

$$p(x, T) = y[u](x, T) - y_d(x), \ x \in \Omega. \tag{9.115}$$

Nonlinear first order evolutional systems

Proof. Divide the equality

$$I(u + \sigma h) - I(u) = \frac{\alpha}{2} \int_Q (2\sigma uh + \sigma^2 h^2) dQ +$$

$$\frac{1}{2} \int_\Omega \left\{ \left(y[u + \sigma h]\big|_{t=T} - y_d \right)^2 - \left(y[u]\big|_{t=T} - y_d \right)^2 \right\} dx$$

by σ and pass to the limit as $\sigma \to 0$. We have

$$\langle I'(u), h \rangle = \alpha \int_Q uh \, dQ + \int_\Omega \left(y[u]\big|_{t=T} - y_d \right) y'[u] h\big|_{t=T} dx. \qquad (9.116)$$

Choose $\mu_Q = 0$, $\mu_\Omega = y[u]\big|_{t=T} - y_d$ at the equalities (9.110)–(9.112). We obtain the system (9.113)–(9.115). Then the equality (9.109) can be transformed to

$$\langle I'(u), h \rangle = \int_Q (\alpha u + p) h \, dQ \ \ \forall h \in V.$$

This completes the proof of the lemma. $\square$

Putting the value of the functional derivative in to the standard optimality condition, we obtain the following result.

Theorem 9.20 *If the control u is the solution of Problem 9.9, it satisfies the variational inequality*

$$\int_Q (\alpha u + p)(v - u) dQ \geq 0 \ \ \forall v \in U.$$

Remark 9.18 This equals to the variational inequality (9.47) for Problem 9.2 because the control belongs identically to the state equations and the cost functionals for both optimization problems. Therefore, necessary conditions of optimality differ only in the adjoint systems.

Our next step is the analysis of optimization control problems for second order evolutional equations.

9.11 Comments

The theory of the evolutional equations with monotone operators including the proof of Theorem 9.1 is given in the books of H. Gajewski, K. Gröger, and K. Zacharias [200] and J.L. Lions [316]. The optimal control problems for the abstract evolutional system with

362   *Optimization and Differentiation*

analysis of Problem 9.1 is considered by R. Cannarsa and G. Da Prato [108], H.O. Fattorini [169], X. Li and J. Yong [313], N. Papageorgiou [393], and D. Tiba [521]. N.U. Ahmed [9] and N. Papageorgiou [393] solve time optimal problems for these systems. N.U. Ahmed and K.L. Teo [12] consider the problems with state constraints. R. Cannarsa and G. Da Prato analyze the synthesis problem [108]. N. Papageorgiou [394] considers optimization problems for the systems with non-monotone operators. R. Cannarsa and G. Da Prato [108] consider the system with infinite horizon.

The theory of the nonlinear parabolic equations is described by O.A. Ladyzhenskaya, V.A. Solonnikov, and N. N. Uraltseva [293], J.L. Lions [316], S. Mizohata [363], and D. Henry [240]. Lemma 9.3 is considered, for example, by J.L. Lions [316]. The optimal control problems for nonlinear second order parabolic equations are considered by N.U. Ahmed [9], N.U. Ahmed and K.L. Teo [10], [12], V. Barbu [51], A. Friedmann [188], A.V. Fursikov [197], L. Lei and G. Wang [306], J.L. Lions [315], Z. Liu [322] and H.W. Lou [329], V.I. Maksimov [337], A.S. Matveev and V.A. Yakubovich [343], S. Nababan and E.S. Noussair [377], P. Neittaanmäki and D. Tiba [384], N. Papageorgiou [393], T. Seidman and H.X. Zhou [452], S.Ya. Serovajsky [456], [458], [469], [488], [493], J. Sokolowsky [509], D. Tiba [518], [521], F. Tröltzsch [531], [532], M.D. Voisie [544], L. Wolfersdorf [554], J. Yong [562], and others. The uniqueness of the optimal control for these systems is proved by T. Seidman and H.X. Zhou [452]. The regularity of its solution is determined by J.C. Reyes and F. Tröltzsch [429]. N. Papageorgiou [395] and T. Seidman and H.X. Zhou [453] analyze Hadamard well-posedness of these problems. H.O. Fattorini [169] applies extension methods there.

The Hamilton–Jacobi equation for optimization problems described by nonlinear parabolic equations is considered by R. Cannarsa and G. Da Prato [108] and R. Cannarsa and M.E. Tessitore [109]. A. Just [264] and R. Cannarsa and O. Cârjua [107] apply the dynamic programming for these systems. H. Goldberg and F. Tröltzsch [218] and F. Tröltzsch [531] obtain the sufficient conditions of optimality. S.H. Farag and M.H. Farag [167] apply the penalty method. The different approximate methods for the systems are considered by A. Sage and P. Chaudhuri [443], S.Ya. Serovajsky [456], [492], and F. Tröltzsch [532].

Optimal control problems for the parabolic equations with nonlinear boundary conditions are considered by V. Barbu [51], H.O. Fattorini and T. Murphy [171], H. Goldberg, andJ.L. Lions [315], and F. Tröltzsch [531]. N. Papageorgiou [393] analyzes optimization problems for the strongly nonlinear parabolic equations. J. Dronion and J.P. Raymond [146] consider a pulse control problem. R. Cannarsa and G. Da Prato [108] consider an optimal control problem with infinite time interval. S. Nababan and K.L. Teo [378] and L. Pan and J. Yong [392] analyze optimization problems for nonlinear parabolic equations with delay.

Optimization problems for the different nonlinear singular parabolic equations are considered by A.V. Fursikov [197], J.L. Lions [317], S.Ya. Serovajsky [487], G. Wang and C. Liu [548], and C. Zhao, M. Wang, and P. Zhao [567]. N. Papageorgiou [394] considers high order nonlinear parabolic equations. V. Barbu [49], [51], P. Neittaanmäki and D. Tiba [384], and D. Tiba [519] solve optimization problems for the parabolic differential inclusions.

Optimization problems for the nonlinear parabolic equations with state constraints are considered by N. Arada and J. P. Raymond [22], M. Bergounioux and F. Tröltzsch [64], E. Casas [111], H.O. Fattorini [169], H. Goldberg and F. Tröltzsch [218], A.S. Matveev and V.A. Yakubovich [343], S. Nababan and E.S. Noussair [377], P. Neittaanmäki and D. Tiba [384], M.M. Novozhenov and V.I. Plotnikov [388], A. Rösch and Tröltzsch [433], G. Wang and C. Liu [548], and others. V. Barbu and G. Wang [53] and S.Ya. Serovajsky [487] analyze the analogical problems for the systems with fixed final state. V. Barbu [50], V. Barbu and G. Wang [53], and R. Cannarsa and G. Da Prato [108] solve the synthesis problems. V. Barbu [51], P. Cannarsa and O. Cârjua [107], G. Knowles [272], and N. Papageorgiou [393], analyze time optimal problems. G. Knowles [272] solves a problem with a non-convex set of admissible controls. V. Barbu [51], P. Neittaanmäki and D. Tiba [384], S.Ya. Serovajsky [487], and D. Tiba [521] consider non-smooth extremum problems for these systems. A vector optimization problem for these systems is considered by S.Ya. Serovajsky [462].

Control–state mapping for the nonlinear parabolic system is frequently non-differentiable. This case is not realized if the velocity of the increase of the nonlinear term

Nonlinear first order evolutional systems

and the dimension of the given set are small enough. The system is weakly nonlinear for this situation (see the previous chapter). Therefore, these suppositions are often used (see, for example, A.V. Fursikov [197], Chapter 2, Theorem 8.1, 8.2; J.L. Lions [317], Chapter 1, Theorem 5.5, Chapter 2, Theorem 2.1; [384], Chapter 4, p. 184). The analogical difficulty is realized also if the derivative of the minimized functional is not determined directly, and the state equation is interpreted as the constraint. If we apply the penalty method here, then it is necessary to obtain the strong enough a priori estimates for the solution of the adjoint system (see J.L. Lions [317], Chapter 1, Section 3.5). The application of the Lagrange principle or analogical ideas here (see, for example, A.V. Fursikov [197]) is to conform to using the results of the general infinite dimensional extremum theory (see, for example, V.M. Alekseev, V.M. Tihomirov, and S.V. Fomin [14], E.R. Avakov [34], V. Boltyansky [80], A.V. Dmitruk, A.A. Milutin, and N.P. Osmolovsky [142], A.D. Dubovitsky and A.A. Milutin [147], X. Guo and J. Ma [228], H. Halkin [235], D. Ioffe and V.M. Tihomirov [253], W. Kotarsky [279], A.A. Milutin [359], L. Neustadt [385], [386], M.M. Novozhenov and V.I. Plotnikov [388], B.N. Pshenichny [414], V.M. Tihomirov [522], and V.A. Yakubovich [558]). However, these results are based on the linearized theorems (the Inverse function theorem, the Implicit function theorem, Lusternik theorem, and others) that are non-applicable for the strongly nonlinear systems.

There exist results of the analysis of the optimization control problems for the nonlinear equations with infinite increase of velocity (see D. Tiba [518], p. 35 for the parabolic case and J.J. Alibert and J.P. Raymond [15] for the elliptic case). However, the solutions of these equations are chosen from the subspace of L_∞. Lemma 9.2 is not applicable for this situation. Moreover, it is possible to prove the Gâteaux differentiability of control–state mapping here. Then the necessary conditions of optimality can be obtained on the basis of the standard operator derivatives. V. Barbu [51] and D. Tiba [518] choose the control from Hilbert spaces. Therefore, the Gâteaux derivative of the state function with respect to the control exists too. For all these cases, the solution of the boundary problem is smooth enough. However, this is true for additional suppositions with respect to parameters of the system (the additional regularity of the coefficients of the state operator, boundary data, and given set). We prefer to analyze the boundary problem in the natural spaces that corresponds to the easiest a priori estimates. Note that the additional regularity is not necessary for obtaining the solvability of the problem.

Optimal control problems for the systems described by the linear parabolic equations with coefficient controls are considered by N.U. Ahmed and K.L. Teo [12], W.H. Fleming [178], A. Friedmann [188], A.D. Iskenderov and R.K. Tagiev [258], G. Knowles [272], J.L. Lions [315], F. Murat [373], [374], P. Neittaanmäki and D. Tiba [384], T.I. Seidman [453], J. Sokolowsky [509], K.L. Teo [517], Z.S. Wu and K.L. Teo [557], T. Zolezzi [571], and others. These problems can be non-solvable (see F. Murat [373] and K.L. Teo [517] for the system with delay, N.U. Ahmed and K.L. Teo [12] for Cauchy problem, A. Belmiloudi [59] for a minimax problem). The optimization problems for the nonlinear parabolic equations with coefficient control are considered, for example, by S.Ya. Serovajsky [469], [488], and J. Sokolowsky [509].

Chapter 10

Second order evolutional systems

10.1 Linear second order evolutional equations 366

10.2 Optimization control problem for linear second order evolutional equations 372

10.3 Non-conditional optimization control problems 377

10.4 Optimization control problem for the wave equation 380

10.5 Nonlinear wave equation 383

10.6 Optimization control problem for the nonlinear wave equation . 390

10.7 Optimality conditions for the optimization optimal problem for the nonlinear wave equation 392

10.8 Non-differentiability of the solution of the nonlinear wave equation with respect to the absolute term 397

10.9 Optimization control problem for the linear hyperbolic equation with coefficient control 398

10.10 Optimization control problem for the nonlinear wave equation with coefficient control 403

10.11 Comments 409

We considered before the control systems for first order evolutional systems. Now we consider second order evolutional equations. At first, we analyze linear second order evolutional equations and optimization control problems for these systems. We solve also the synthesis problem for the unconditional case. The wave equation is considered as an example.

The nonlinear hyperbolic equation has additional difficulties in comparison with the nonlinear parabolic one. Particularly, we prove the uniqueness of the boundary problem by using constraints on the dimension of the set and the degree of the increase of the nonlinear term. We obtain the extended differentiability of control–state mapping. This dependence can be Gâteaux non-differentiable.

We consider also linear and nonlinear systems with a coefficient control. The analogical boundary problems for elliptic and parabolic equations are one-value solvable on the set of admissible controls only. However, this result is true for the hyperbolic case for the whole control space. Therefore, we do not use the derivatives with respect to the convex set here. The standard derivatives are applicable for this situation.

366 Optimization and Differentiation

10.1 Linear second order evolutional equations

Consider Hilbert spaces W and H such that embedding of W to H is continuous and dense. We identify the space H with its adjoint space H'. Then embedding $W \subset H \subset W'$ is continuous and dense. Suppose the space W is separable. Determine a linear continuous operator $A : W \to W'$ that is self-adjoint and coercive, i.e., we have the conditions

$$\langle Ay, z \rangle = \langle Az, y \rangle \ \forall y, z \in W$$

and

$$\langle Ay, y \rangle \geq \chi \|y\|_W^2 \ \forall y \in W,$$

where $\chi > 0$. Consider the linear **second order evolutional equation**

$$y''(t) + Ay(t) = f(t), \ t \in (0, T) \tag{10.1}$$

with the initial conditions

$$y(0) = \varphi, \ y'(0) = \psi, \tag{10.2}$$

where f, φ, and ψ are known functions. We have the Cauchy problem for the given second order evolutional equation.

Remark 10.1 The wave equation is the natural example of the given equation.

Determine the space

$$Y = \{y \mid y \in L_\infty(0, T; W), \ y' \in L_\infty(0, T; H), \ y'' \in L_2(0, T; W')\}.$$

Theorem 10.1 *For any $f \in L_2(0, T; H)$, $\varphi \in W$, $\psi \in H$ there exists a unique solution $y \in Y$ of the problem* (10.1), (10.2).

We shall use the following result.

Lemma 10.1 (***Gronwall lemma***) *Suppose for non-negative measurable functions $\eta = \eta(t)$ and $a = a(t)$ such that the product ηa is integrable on the interval $(0, T)$, and the following inequality holds:*

$$\eta(t) \leq \int_0^t a(\tau)\eta(\tau)d\tau + b, \ t \in (0, T),$$

where b is a positive constant. Then we get

$$\eta(t) \leq \exp\left[\int_0^t a(\tau)d\tau\right], \ t \in (0, T).$$

Second order evolutional systems

Proof of Theorem 10.1. 1. We use the Faedo–Galerkin method for proving the solvability of the problem. Consider a complete system of linear independent elements $\{\mu_m\}$ in the space W. Determine an approximate solution $y_k = y_k(t)$ of the problem (10.1), (10.2) by the formula

$$y_k(t) = \sum_{m=1}^{k} \xi_{mk}(t)\mu_m, \ k = 1, 2, ..., \tag{10.3}$$

where the functions $\xi_{1k}, ..., \xi_{kk}$ satisfy the equalities

$$\langle y_k''(t), \mu_j \rangle + \langle Ay_k(t), \mu_j \rangle = \langle f(t), \mu_j \rangle, \ j = 1, ..., k. \tag{10.4}$$

Put here the value y_k from the formula (10.3). We get

$$\sum_{m=1}^{k} \langle \mu_m, \mu_j \rangle \xi_{mk}''(t) + \sum_{m=1}^{k} \langle A\mu_m, \mu_j \rangle \xi_{mk}(t) = \langle f(t), \mu_j \rangle, \ j = 1, ..., k.$$

Thus, (10.4) is the system of linear second order ordinary differential equations.

Determine the initial values of the functions φ and ψ as a series with respect to μ_m,

$$\varphi = \sum_{m=1}^{\infty} \alpha_m \mu_m, \ \psi = \sum_{m=1}^{\infty} \beta_m \mu_m,$$

where $\{\alpha_k\}$ and $\{\beta_k\}$ are numerical sequences. Consider the sums

$$\varphi_k = \sum_{m=1}^{k} \alpha_m \mu_m, \ \psi_k = \sum_{m=1}^{k} \beta_m \mu_m.$$

We have the convergence

$$\varphi_k \to \varphi \text{ in } W, \ \psi_k \to \psi \text{ in } H. \tag{10.5}$$

Choose the initial conditions for the functions ξ_{mk} by the equalities

$$\xi_{mk}(0) = \alpha_m, \ \xi_{mk}'(0) = \beta_m, \ m = 1, 2, ...$$

Therefore, we have

$$y_k(0) = \varphi_k, \ y_k'(0) = \psi_k. \tag{10.6}$$

Thus, we have the Cauchy problem for the system of linear ordinary differential equations with respect to the functions ξ_{mk}. Choose the first derivatives of the unknown functions as new variables. We have the system of the first order linear differential equations. It has a unique solution on an interval $(0, T)$. Then we can determine the function $y_k = y_k(t)$ by the formula (10.3). Prove that this is in reality the approximate solution of the problem (10.1), (10.2).

Optimization and Differentiation

2. Multiply j-th equality (10.3) by $\xi'_{jk}(t)$ and sum the result by j. Using the formula (10.3), we get

$$\langle y''_k(t), y'_k(t) \rangle + \langle A y_k(t), y'_k(t) \rangle = \langle f(t), y'_k(t) \rangle. \tag{10.7}$$

We have the equality

$$\frac{1}{2}\frac{d}{dt}\|y'_k(t)\|^2_H = \langle y''_k(t), y'_k(t) \rangle.$$

Using the properties of the operator A, we have

$$\frac{d}{dt}\langle A y_k(t), y_k(t) \rangle \ge \langle A y'_k(t), y_k(t) \rangle + \langle A y_k(t), y'_k(t) \rangle = 2\langle A y_k(t), y'_k(t) \rangle.$$

Now we obtain the inequality

$$\left|\langle f(t), y'_k(t) \rangle\right| \le \|f(t)\|_H \|y'_k(t)\|_H \le \frac{1}{2}\|f(t)\|^2_H + \frac{1}{2}\|y'_k(t)\|^2_H.$$

From the equality (10.7), it follows that

$$\frac{1}{2}\frac{d}{dt}\Big[\|y'_k(t)\|^2_H + 2\langle A y_k(t), y_k(t) \rangle\Big] \le \|f(t)\|^2_H + \frac{1}{2}\|y'_k(t)\|^2_H.$$

Using the inequality

$$\langle Ay, y \rangle \ge \chi\|y\|^2_W \ \forall y \in W,$$

we get

$$\frac{1}{2}\frac{d}{dt}\Big[\|y'_k(t)\|^2_H + 2\chi\|y_k(t)\|^2_W\Big] \le \|f(t)\|^2_H + \frac{1}{2}\|y'_k(t)\|^2_H.$$

Integrating this result by using (10.6), we obtain

$$\|y'_k(t)\|^2_H + 2\chi\|y_k(t)\|^2_W \le \|\psi_k(t)\|^2_H + 2\chi\|\varphi_k(t)\|^2_W +$$

$$\int_0^t \|f(\tau)\|^2_H d\tau + \int_0^t \|y'_k(\tau)\|^2_H d\tau. \tag{10.8}$$

Thus, we have the inequality

$$\|y'_k(t)\|^2_H \le \|\psi_k(t)\|^2_H + 2\chi\|\varphi_k(t)\|^2_W + \int_0^t \|f(\tau)\|^2_H d\tau + \int_0^t \|y'_k(\tau)\|^2_H d\tau.$$

Three terms at the right-hand side are bounded because of the conditions (10.5) and the properties of the abstract function f. Using the Gronwall lemma, we prove the existence of a constant $c_1 > 0$ such that $\|y'_k(t)\|_H \le c_1$. Then from the inequality (10.8) it follows that $\|y_k(t)\|_W \le c_2$ for a constant $c_2 > 0$.

Second order evolutional systems

3. Thus, the sequence $\{y_k\}$ is bounded in the space $L_\infty(0, T; W)$ and the sequence $\{y_k'\}$ is bounded in $L_2(0, T; H)$. Using the Banach–Alaoglu theorem after extracting a subsequence we have the convergence

$$y_k \to y \text{ weakly in } L_\infty(0, T; W), \tag{10.9}$$

$$y_k' \to y' \text{ weakly in } L_2(0, T; H). \tag{10.10}$$

Using continuous embedding of the space

$$Y = \left\{ y \mid y \in L_\infty(0, T; W), \ y' \in L_2(0, T; H) \right\}$$

to $C(0, T; H)$, from the formulas (10.9), (10.10) the convergence

$$y_k(0) \to y(0) \text{ weakly in } H$$

follows. Then from (10.5) it follows that $y(0) = \varphi$.

Multiply the equality (10.4) by the arbitrary function $\eta \in C^1[0, T]$ such that $\eta(T) = 0$. After integration we obtain

$$\int_0^T \langle y_k''(t), \eta_j(t) \rangle dt + \int_0^T \langle Ay_k(t), \eta_j(t) \rangle dt = \int_0^T \langle f(t), \eta_j(t) \rangle dt,$$

where $\eta_j(t) = \eta(t)\mu_j$. Using the formula of the integration by parts and second condition (10.6), we obtain

$$\int_0^T \langle y_k''(t), \eta_j(t) \rangle dt = - \int_0^T \langle \eta_j'(t), y_k'(t) \rangle dt - (\psi_k, \eta_j(0)).$$

Then we get the equality

$$- \int_0^T \langle \eta_j'(t), y_k'(t) \rangle dt + \int_0^T \langle A\eta_j(t), y_k(t) \rangle dt =$$

$$\int_0^T \langle f(t), \eta_j(t) \rangle dt + (\psi_k, \eta_j(0)), \ j = 1, ..., k.$$

Passing to the limit by using the conditions (10.5), (10.9), (10.10), we obtain

$$- \int_0^T \langle \eta_j'(t), y'(t) \rangle dt + \int_0^T \langle A\eta_j(t), y(t) \rangle dt =$$

$$\int_0^T \langle f(t), \eta_j(t) \rangle dt + (\psi, \eta_j(0)). \tag{10.11}$$

Suppose η is an arbitrary function of the space $D(0,T)$ infinite differentiable functions on the interval $(0,T)$. Using the definition of the generalized derivative, we obtain the equality

$$\int_0^T \langle \eta_j'(t), y'(t) \rangle dt = \int_0^T \langle \mu_j(t), y'(t) \rangle \eta'(t) dt = -\int_0^T \frac{d^2}{dt^2} \langle \mu_j(t), y(t) \rangle \eta(t) dt.$$

We have the generalized derivative at the right-hand side here. This is the element of the distribution space $D'(0,T)$. Therefore, we transform the previous equality

$$\int_0^T \left[\frac{d^2}{dt^2} \langle y(t), \mu_j \rangle + \langle Ay(t) - f(t), \mu_j \rangle \right] \eta(t) dt = 0, \ j = 1, 2, \dots . \tag{10.12}$$

The function η is arbitrary. Hence, we get

$$\langle y''(t), \mu_j \rangle + \langle Ay(t), \mu_j \rangle = \langle f(t), \mu_j \rangle, \ j = 1, 2, \dots .$$

Each element $\mu \in W$ can be determined as a sum

$$\mu = \sum_{j=1}^{\infty} b_j \mu_j,$$

where b_j are numbers. Multiply j-th previous equality by b_j. Summing by j, we get

$$\langle y''(t) + Ay(t) - f(t), \mu \rangle = 0 \ \forall \mu \in W.$$

Then we have the equality

$$y''(t) + Ay(t) = f(t).$$

Thus the function o is the solution of the equation (10.1). Finding the derivative $y'' = Ay + f$, we have $y'' \in L_2(0,T;W')$. Therefore y is the point of the space Y.

Consider the equality (10.11). After integration by parts by using the equality (10.1), we have

$$\left(y'(0), \mu_j \right)_H \eta(0) = (\psi, \mu_j)_H \, \eta(0), \ j = 1, 2, \dots .$$

Therefore, we get

$$\left(y'(0), \mu_j \right)_H = (\psi, \mu_j)_H, \ j = 1, 2, \dots .$$

Multiply j-th equality by b_j and sum the result by j. We obtain

$$\left(y'(0), \mu \right)_H = (\psi, \mu)_H \ \forall \mu \in W.$$

Second order evolutional systems

Then the second initial condition (10.2) $y'(0) = \psi$ is true. We proved the solvability of the problem (10.1), (10.2).

4. Suppose there exists two solutions y_1 and y_2 of this problem. The difference $y = y_1 - y_2$ satisfies the equation

$$y''(t) + Ay(t) = 0$$

with homogeneous initial condition.

Fix the value $s \in (0, T)$. Determine the abstract function

$$z(t) = \begin{cases} -\int_t^s y(\tau)d\tau, & t \leq s, \\ 0, & t > s. \end{cases}$$

It satisfies the problem

$$z' = y, \quad z(s) = 0.$$

Using the equation with respect to y, we have

$$\int_0^s \langle y''(t), z(t) \rangle dt + \int_0^s \langle Ay(t), z(t) \rangle dt = 0. \tag{10.13}$$

After integration by parts we get

$$\int_0^s \langle y''(t), z(t) \rangle dt = -\int_0^s \langle y'(t), z'(t) \rangle dt = -\int_0^s \langle y'(t), y(t) \rangle =$$

$$-\frac{1}{2} \int_0^s \frac{d}{dt} \|y(t)\|_H^2 dt = -\frac{1}{2} \|y(s)\|_H^2.$$

Then

$$\frac{d}{dt} \langle Az(t), z(t) \rangle = \langle Az'(t), z(t) \rangle + \langle Az(t), z'(t) \rangle = 2 \langle Az'(t), z(t) \rangle.$$

Therefore, we get

$$\langle Ay(t), z(t) \rangle = \langle Az'(t), z(t) \rangle = \frac{d}{dt} \langle Az(t), z(t) \rangle.$$

Transform the second integral of the equality (10.13). We have

$$\int_0^s \langle Ay(t), z(t) \rangle dt = \frac{1}{2} \int_0^s \frac{d}{dt} \langle Az(t), z(t) \rangle dt = -\frac{1}{2} \langle Az(0), z(0) \rangle$$

because $z(s) = 0$. Therefore, we transform the equality (10.13) to the form

$$\|y(s)\|_H^2 + \langle Az(0), z(0) \rangle = 0.$$

372   *Optimization and Differentiation*

Using the coercive condition, we obtain the inequality

$$\|y(s)\|_H^2 + \chi\|z(0)\|_H^2 \leq 0.$$

Then we have $y = 0$ and the equality $y_1 = y_2$. Therefore, the solution of the problem (10.1), (10.2) is unique. $\square$

Remark 10.2 This result is the analog of Theorem 8.3 about the one-value solvability of the Cauchy problem for the first order evolutional equation.

10.2   Optimization control problem for linear second order evolutional equations

Now consider an optimization control problem for the system described by the second order linear evolutional equation. Let V be a Hilbert space, and $B : V \to L_2(0, T; H)$ be a linear continuous operator. Consider the system described by the equation

$$y''(t) + Ay(t) = (Bv)(t) + f(t), \ t \in (0, T) \tag{10.14}$$

with the initial conditions

$$y(0) = \varphi, \ y'(0) = \psi, \tag{10.15}$$

where v is a control, y is the state system, and $\varphi \in W$, $\psi \in H$, and $f \in L_2(0, T; H)$ are known values. From Theorem 10.1 it follows that for all control $v \in V$ the problem (10.14), (10.15) has a unique solution $y = y[v]$ from the space Y. Let us have also a non-empty convex closed subset U of the space V and the functional

$$I(v) = \frac{\alpha}{2}\|v\|_V^2 + \frac{1}{2}\|C(y[v] - y_d)\|_Z^2,$$

where $\alpha > 0$, $C : L_2(0, T; H) \to Z$ is a linear continuous operator, Z is a Hilbert space, and $y_d \in L_2(0, T; H)$ is known.

Problem 10.1 *Find a control $u \in U$ that minimizes the functional I on the set U.*

Prove the existence and the uniqueness of this problem. It is known that the minimization problem of the lower bounded strict convex continuous coercive functional on the nonempty convex closed subset of a Hilbert space has a unique solution. We need to prove the continuity of the given functional only.

At first, prove the following result.

$$\text{Second order evolutional systems} \qquad 373$$

Lemma 10.2 *The map* $y[\cdot] : V \to L_2(0, T; H)$ *is continuous.*

Proof. Suppose y_1 and y_2 are solutions of the problem (10.1), (10.2) for controls v_1 and v_2. The function $y = y_1 - y_2$ satisfies the equation

$$y''(t) + Ay(t) = (Bv)(t), \ t \in (0, T) \qquad (10.16)$$

with homogeneous initial conditions, where $v = v_1 - v_2$. Fix a point $s \in (0, T)$. Determine the abstract function

$$z(t) = \begin{cases} -\int\limits_t^s y(\tau)d\tau, & t \le s, \\ 0, & t > s. \end{cases}$$

From (10.16) the equality follows:

$$\int\limits_0^s \langle y''(t), z(t) \rangle dt + \int\limits_0^s \langle Ay(t), z(t) \rangle dt = \int\limits_0^s \langle (Bv)(t), z(t) \rangle dt \qquad (10.17)$$

that is the analog of (10.13). We transform two integrals by using the technique from Theorem 10.1. Determine the function

$$w(t) = \int\limits_0^t (Bv)(\tau)d\tau.$$

It satisfies the equalities $w(0) = 0$, $w'(t) = Bv(t)$. Then we transform the term at the right-hand side of the equality (10.17):

$$\int\limits_0^s \langle (Bv)(t), z(t) \rangle dt = \int\limits_0^s \langle w'(t), z(t) \rangle dt = -\int\limits_0^s \langle w(t), y(t) \rangle dt.$$

From the equality (10.17), it follows that

$$\|y(s)\|_H^2 + \langle Az(0), z(0) \rangle = 2 \int\limits_0^s \langle w(t), y(t) \rangle dt.$$

Using coercivity, we get

$$\|y(s)\|_H^2 \le 2 \int\limits_0^s \|w(t)\|_H \|y(t)\|_H dt \le \int\limits_0^s \|w(t)\|_H^2 dt + \int\limits_0^s \|y(t)\|_H^2 dt.$$

By the Gronwall lemma, we obtain

$$\|y(s)\|_H^2 \le c \int\limits_0^T \|w(t)\|_H^2 dt,$$

374   *Optimization and Differentiation*

where $c > 0$. From the definition of w the inequality follows:

$$\|w(t)\|_H \le \int_0^t d\tau \le \sqrt{t}\Big[\|(Bv)(\tau)\|_H^2 d\tau\Big]^{1/2} \le \sqrt{t}\|Bv\|_{L_2(0,T;H)} \le \sqrt{t}\|B\|\|v\|_V.$$

Therefore, we get the inequality

$$\|y(s)\|_H^2 \le cT^2\|B\|\|v\|_V.$$

This completes the proof of the lemma. $\square$

Using Lemma 10.2 and the definition of the given functional, we have the following result.

Lemma 10.3 *The functional I is continuous.*

Now we can prove one-valued solvability of the problem.

Theorem 10.2 *Problem 10.1 has a unique solution.*

Using the strict uniform convexity of the square of the norm of Hilbert space and Theorem 2.6, we obtain the following result.

Theorem 10.3 *Problem 10.1 is Tihonov well-posed.*

Remark 10.3 These results are the analog of the properties of the linear stationary systems and first order linear evolutional systems.

Determine optimality conditions for Problem 10.1. The necessary condition of minimum for Gâteaux differentiable functional I at a point u on a convex subset U of a Banach space is the variational inequality

$$\langle I'(u), v - u \rangle \ge 0 \ \forall v \in U. \tag{10.18}$$

If this functional is convex, then (10.18) is the necessary and sufficient condition of minimum. Find the derivative of the given functional.

Lemma 10.4 *The functional I is Gâteaux differentiable at the arbitrary point $u \in V$, as we have the equality*

$$I'(u) = \alpha \Lambda_V^{-1} u + B^* p, \tag{10.19}$$

where p is the solution of the adjoint Cauchy problem

$$p''(t) + A^* p(t) = C^* \Lambda_Z^{-1} C\big(y[u] - y_d\big)(t), \ t \in (0, T), \tag{10.20}$$

$$p(T) = 0, \ p'(T) = 0, \tag{10.21}$$

$\Lambda_V : V' \to V$ *and* $\Lambda_Z : Z' \to Z$ *are canonic isomorphisms.*

Second order evolutional systems 375

Proof. The operator C^* maps the space Z to $L_2(0,T;H)$. Then we have

$$C^*\Lambda_Z^{-1}C\big(y[u]-y_d\big) \in L_2(0,T;W').$$

Hence, the term at the right-hand side of the equality (10.20) is the element of the same space as the absolute term of the equation (10.1). Determine the abstract function q by the equality $q(t) = P(T-t)$; we transform the problem (10.20), (10.21)

$$q''(t) + A^*q(t) = C^*\Lambda_Z^{-1}C\big(y[u]-y_d\big)(T-t), \ t \in (0,T);$$

$$q(0) = 0, \quad q'(0) = 0.$$

This has the same form as the problem (10.1), (10.2). From Theorem 10.1 one-valued solvability of the problem (10.20), (10.21) in the space Y follows.

For all $h \in V$ and number σ we find the difference

$$I(u+\sigma h) - I(u) = \frac{\alpha}{2}\Big[\|u+\sigma h\|_V^2 - \|u\|_V^2\Big] +$$

$$\frac{1}{2}\Big[\|C(y[u+\sigma h]-y_d)\|_Z^2 - \|C(y-y_d)\|_Z^2\Big] =$$

$$\frac{\alpha\sigma}{2}\Big[2(u,h)_V + \sigma\|h\|_V^2\Big] + \frac{1}{2}\Big[(C(y-y_d),C\zeta)_Z + \|C\zeta\|_Z^2\Big],$$

where $\zeta = y[u+\sigma h] - y[u]$. We have the equality

$$(\eta,C\zeta)_Z = \langle\Lambda_Z^{-1}\eta,C\zeta\rangle_Z = \big(C^*\Lambda_Z^{-1}\eta,\zeta\big)_{L_2(0,T;H)} = \int_0^T \big(C^*\Lambda_Z^{-1}\eta(t),\zeta(t)\big)_H dt.$$

Thus, we get

$$I(u+\sigma h) - I(u) = \alpha(u,h)_V + \frac{\alpha\sigma}{2}\|h\|_V^2 +$$

$$\int_0^T \big(C^*\Lambda_Z^{-1}C(y-y_d)(t),\zeta(t)\big)_H dt + \frac{1}{2}\|C\zeta\|_Z^2. \qquad (10.22)$$

Obviously, the function ζ satisfies the equation

$$\zeta''(t) + A\zeta(t) = \sigma(Bh)(t) \qquad (10.23)$$

with homogeneous initial conditions. Then for all $p \in Y$ we have

$$\int_0^T \big\langle\zeta''(t) + A\zeta(t),p(t)\big\rangle dt = \sigma\int_0^T \langle(Bh)(t),p(t)\rangle dt.$$

Optimization and Differentiation

Using the formula of integration by parts, we get

$$\big(\zeta'(T), p(T)\big) - \big(\zeta(T), p'(T)\big) + \int\limits_0^T \Big\langle p''(t) + A^* p(t), \zeta(t) \Big\rangle dt \, = \, \sigma \big\langle B^* p, h \big\rangle_V.$$

Let p be the solution of the problem (10.20), (10.21). Then

$$\int\limits_0^T \Big(C^* \Lambda_Z^{-1} C(y[u] - y_d)(t), \zeta(t) \Big)_H dt \, = \, \sigma \big\langle B^* p, h \big\rangle_V.$$

Therefore, the equality (10.22) can be transformed to

$$I(u + \sigma h) - I(u) \, = \, \sigma \big(\alpha u + \Lambda_V B^* p, h \big)_V + \frac{\sigma^2}{2} \|h\|_V^2 + \frac{1}{2} \|C\zeta\|_Z^2. \qquad (10.24)$$

From (10.23) it follows that

$$\int\limits_0^t \big\langle \zeta''(t), \zeta'(t) \big\rangle dt + \int\limits_0^t \big\langle A\zeta(t), \zeta'(t) \big\rangle dt \, = \, \sigma^2 \int\limits_0^t \big\langle (Bh)(t), \zeta'(t) \big\rangle dt.$$

Determine the inequality

$$\big\| \zeta'(t) \big\|_H^2 + 2\chi \big\| \zeta(t) \big\|_W^2 \, \le \, \sigma^2 \int\limits_0^t \big\| (Bh)(\tau) \big\|_H^2 d\tau + \int\limits_0^t \big\| \zeta'(\tau) \big\|_H^2 d\tau$$

as the analog of (10.16). Using Gronwall the lemma, we get

$$\big\| \zeta'(t) \big\|_H^2 \, \le \, c_1 \sigma^2 \big\| Bh \big\|_{L_2(0,T;H)}^2,$$

where $c_1 > 0$. From the previous inequality it follows that

$$\big\| \zeta(t) \big\|_W^2 \, \le \, c_2 \sigma^2 \big\| Bh \big\|_{L_2(0,T;H)}^2,$$

where $c_2 > 0$. Using continuous embedding of the space $L_\infty(0,T;W)$ to $L_2(0,T;H)$, we obtain

$$\big\| \zeta \big\|_{L_2(0,T;H)}^2 \, \le \, c_3 \sigma^2,$$

where $c_3 > 0$. Then

$$\big\| C\zeta \big\|_Z^2 \, \le \, \|C\|^2 \big\| \zeta \big\|_{L_2(0,T;H)}^2 \, \le \, c_3 \|C\|^2 \sigma^2.$$

After dividing the equality (10.24) by σ and passing to the limit as $\sigma \to 0$ we have

$$\langle I'(u), h \rangle = \left(\alpha u + \Lambda_V B^* p, h \right) = \left(\Lambda_V (\alpha \Lambda_V^{-1} u + B^* p), h \right) = \langle \alpha \Lambda_V^{-1} u + B^* p, h \rangle_V.$$

The element h is arbitrary here. Therefore, the formula (10.19) is true. $\square$

Put the value of the functional derivative in to the condition (10.18). We have the following result.

Theorem 10.4 *The necessary and sufficient condition of the optimality for Problem 10.1 is the variational inequality*

$$\left\langle \alpha \Lambda_V^{-1} u + B^* p, v - u \right\rangle_V \geq 0 \ \forall v \in U. \tag{10.25}$$

Thus, we have the state system

$$y''(t) + Ay(t) = Bu(t) + f(t), \ t \in (0, T), \tag{10.26}$$

$$y(0) = \varphi, \ y'(0) = \psi, \tag{10.27}$$

the adjoint system (10.20), (10.21), and the variational inequality (10.25).

10.3 Non-conditional optimization control problems

Consider the partial case of Problem 10.1. Suppose the constraints with respect to the control are absent.

Problem 10.2 *Find a control $u \in V$ that minimizes the functional I for the abstract system* (10.14), (10.15) *on the space V.*

For the considered case the variational inequality (10.25) can be transformed to the stationary condition

$$\alpha \Lambda_V^{-1} u + B^* p = 0.$$

Find the control

$$u = -\alpha^{-1} \Lambda_V^{-1} B^* p. \tag{10.28}$$

Thus, the solution of Problem 10.2 satisfies the following system:

$$y'' + Ay = -\alpha^{-1} B \Lambda_V^{-1} B^* p + f, \tag{10.29}$$

$$y(0) = \varphi, \ y'(0) = \psi, \tag{10.30}$$

$$p'' + A^* p = C^* \Lambda_Z^{-1} Cy - C^* \Lambda_Z^{-1} Cy_d, \tag{10.31}$$

378   *Optimization and Differentiation*

$$p(T) = 0, \ p'(T) = 0. \tag{10.32}$$

Determine new variables $z = (z_1, z_2) = (y, y')$, $w = (w_1, w_2) = (p, p')$. Then we transform the equation (10.29) to the system

$$\begin{cases} z_1' - z_2 = 0, \\ z_2' + Az_1 = -\alpha^{-1} B \Lambda_V^{-1} B^* w_1 + f. \end{cases}$$

Determine the matrix operators

$$\Theta = \begin{pmatrix} 0 & -E \\ A & 0 \end{pmatrix}, \quad \Phi = \begin{pmatrix} 0 & 0 \\ -\alpha^{-1} B \Lambda_V^{-1} B^* & 0 \end{pmatrix},$$

where A is the unit operator, and 0 is the zero operator that maps the arbitrary element of the space to the number 0.

Remark 10.4 Of course, it is necessary to determine the domains and the codomains of these operators. However, our transformations will be formal here.

Now the equation (10.29) can be transformed to

$$z' + \Theta z = \Phi w + \zeta, \tag{10.33}$$

where $\zeta = (0, f)$. Denote $z_0 = (z_{01}, z_{02})$. Then we transform the initial condition (10.30).

$$z(0) = z_0. \tag{10.34}$$

The equation (10.31) is transformed to the system

$$\begin{cases} w_1' - w_2 = 0, \\ w_2' + Aw_1 = C^* \Lambda_Z^{-1} C z_1 - C^* \Lambda_Z^{-1} C y_d. \end{cases}$$

Determine the matrix

$$\Xi = \begin{pmatrix} 0 & 0 \\ C^* \Lambda_Z^{-1} C & 0 \end{pmatrix}$$

and the vector $\xi = (0, -C^* \Lambda_Z^{-1} C y_d)$. We obtain the equation

$$w' + \Theta w = \Xi z + \xi \tag{10.35}$$

with final condition

$$w(T) = 0. \tag{10.36}$$

Thus, the system of the optimality condition (10.29)–(10.32) is transformed to (10.33)–(10.36). This is the analog of the system (8.55)–(8.59). Therefore, we have the analog of Lemma 8.5.

Lemma 10.5 *There exists a linear continuous operator R and a function r such that*

$$w(t) = R(t)z(t) + r(t), \ t \in (0, T). \tag{10.37}$$

Second order evolutional systems

Remark 10.5 Of course, the properties of the operators of the considered system are different. We cannot transform the hyperbolic equation to the system of parabolic equations by using a change of variables. However, we analyze Problem 10.2 formally.

Differentiate formally the equality (10.37).

$$w'(t) = R'(t)z(t) + R(t)z'(t) + r'(t), \ t \in (0, T).$$

Put in the values of the derivatives from the equalities (10.33), (10.35). We obtain

$$R'z - R\Theta z + R\Phi w + R\zeta + \Theta w - \Xi z - \xi + r' = 0.$$

Using the equality (10.37), we get

$$\left(R' + R\Phi R - R\Theta + \Theta R - \Xi\right)z + \left(r' + R\Phi r + R\zeta r - \xi + \Theta r\right) = 0.$$

Suppose the terms under the brackets are equal to zero. We obtain the following equations.

$$R' + R\Phi R - R\Theta + \Theta R - \Xi = 0, \tag{10.38}$$

$$r' + R\Phi r + R\zeta r - \xi + \Theta r = 0. \tag{10.39}$$

For $t = T$ from the equality (10.37) it follows that

$$w(T) = R(T)z(T) + r(T) = 0.$$

This equality is true, if we have the final conditions

$$R(T) = 0, \tag{10.40}$$

$$r(T) = 0. \tag{10.41}$$

Remark 10.6 The equation (10.38) has the square nonlinearity with respect to the unknown value R. Therefore, we can call it Riccati equations.

Thus, we have formally the following algorithm of solving Problem 10.2 (see Figure 10.1). At first, we determine the operator R from the equation (10.38) with condition (10.40). Then we find r from the linear equation (10.39) with condition (10.41). After putting w with known R and r in to the equality (10.33) we have the equation

$$z' + \Theta z = \Phi R z + \Phi r + \zeta. \tag{10.42}$$

Solve it with initial condition (10.34); we find z. Then we determine w by the formula (10.37). Its first component is p. Finally, the solution u of Problem 10.2 can be found by the formula (10.28).

We can find also the solution of the synthesis problem. Determine the dependence of the optimal control from the state system (see Figure 10.2). Indeed, using the equality $z = (y, y')$ from the equalities (10.28) and (10.37) we find the following dependence

$$u = -\alpha^{-1} \Lambda_V B^* (Rz + r). \tag{10.43}$$

R and r are known here by the equalities (10.38)–(10.41).

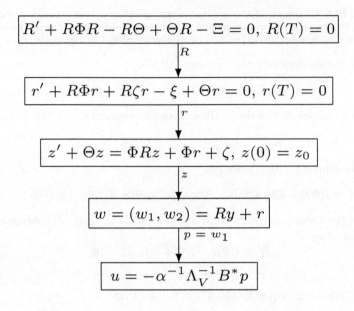

FIGURE 10.1: The solution of the program control problem.

10.4 Optimization control problem for the wave equation

The standard examples of the second order evolutional equations are hyperbolic equations. Suppose Ω is an open bounded set of Euclid space $\mathbb{R}^n$ with boundary Γ, $T > 0$, $Q = \Omega \times (0,T)$, $\Sigma = \Gamma \times (0,T)$. Consider the **wave equation**

$$\frac{\partial^2 y(x,t)}{\partial t^2} = \Delta y(x,t) + v(x,t), \ (x,t) \in Q \qquad (10.44)$$

with homogeneous boundary condition

$$y(x,t) = 0 \ (x,t) \in \Sigma \qquad (10.45)$$

and initial condition

$$y(x,0) = \varphi(x), \quad \frac{\partial y(x,0)}{\partial t} = \psi(x), \ x \in \Omega, \qquad (10.46)$$

where $y = y(x,t)$ is the state function, and φ and ψ are known functions.

Determine the spaces $W = H_0^1(\Omega)$ and $H = L_2(\Omega)$. Then we have the equality $W' = H^{-1}(\Omega)$ and continuous and dense embedding $W \subset H \subset W'$.

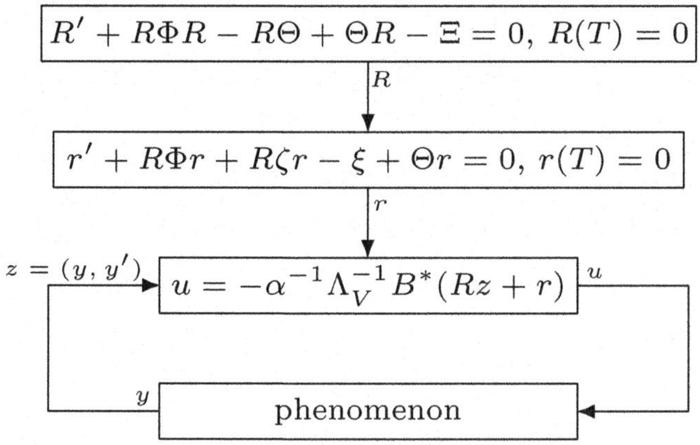

$$\boxed{R' + R\Phi R - R\Theta + \Theta R - \Xi = 0, \ R(T) = 0}$$

$\downarrow R$

$$\boxed{r' + R\Phi r + R\zeta r - \xi + \Theta r = 0, \ r(T) = 0}$$

$\downarrow r$

$z = (y, y')$ $\boxed{u = -\alpha^{-1}\Lambda_V^{-1}B^*(Rz + r)}$ u

y $\boxed{\text{phenomenon}}$

FIGURE 10.2: The solution of the synthesis problem.

Determine the operator $A = -\Delta$ that is a linear continuous mapping from the space W to W'. Using Green's formula, we have the equality

$$\langle Ay, z \rangle = -\int_\Omega y\Delta z dx = -\int_\Omega z\Delta y dx = \langle y, Az \rangle \ \forall y, z \in W,$$

i.e., the operator A is self-adjoint. Then we have

$$\langle Ay, y \rangle = -\int_\Omega y\Delta y dx = -\int_\Omega \sum_{i=1}^n \left(\frac{\partial y}{\partial x_i}\right)^2 dx = \|y\|_W^2 \ \forall y \in W.$$

Thus, the coercivity condition

$$\langle Ay, y \rangle \geq \chi \|y\|_W^2 \ \forall y \in W$$

is true as the equality with the constant $\chi = 1$. Determine the control space $V = L_2(Q)$. Let B be a unit operator. Therefore, we transform the boundary problem (10.44)–(10.46) to (10.13), (10.14). From Theorem 10.1 the one-valued solvability of the system follows.

Corollary 10.1 *For all $\varphi \in H_0^1(\Omega)$, $\psi \in L_2(\Omega)$ there exists a unique solution of the problem* (10.44) $-$ (10.46) *from the space*

$$Y = \left\{y \middle| \ y \in L_\infty(0, T; H_0^1(\Omega)), \ y' \in L_\infty(0, T; L_2(\Omega)), \ y'' \in L_2(0, T; H^{-1}(\Omega))\right\}.$$

Using Lemma 10.2, we obtain continuity of the control–state mapping.

Corollary 10.2 *The operator for the problem* (10.44)–(10.46) *is continuous.*

382 *Optimization and Differentiation*

Consider a convex closed subset U of the space V and the functional

$$I(v) = \frac{\alpha}{2} \int_Q v^2 dQ + \frac{1}{2} \int_Q (y[v] - y_d)^2 dQ,$$

where $\alpha > 0$, $y_d \in L_2(Q)$ is a known function.

Problem 10.3 *Find the control $u \in U$ for the system (10.44)–(10.46) that minimizes the functional I on the set U.*

Determine the space $Z = L_2(Q)$. Let C be a unit operator. Then Problem 10.3 is transformed to Problem 10.1. From Theorem 10.2 and Theorem 10.3 the result follows.

Corollary 10.3 *Problem 10.3 is Tihonov well-posed.*

Using Theorem 10.4 we determine the necessary conditions of optimality.

Corollary 10.4 *The necessary and sufficient condition of optimality of the control u for Problem 10.3 is the variational inequality*

$$\int_Q (\alpha u + p)(v - u) dQ \geq 0 \ \ \forall v \in U, \tag{10.47}$$

where p is the solution of the equation

$$-\frac{\partial^2 p(x,t)}{\partial t^2} = \Delta p(x,t) + y(x,t) - y_d(x,t), \ (x,t) \in Q \tag{10.48}$$

in the set Q with boundary conditions

$$p(x,t) = 0, \ (x,t) \in \Sigma, \tag{10.49}$$

$$p(x,T) = 0, \ \frac{\partial p(x,T)}{\partial t} = 0, x \in \Omega, \tag{10.50}$$

and y is the solution of the boundary problem

$$\frac{\partial^2 y}{\partial t^2} = \Delta y + u \tag{10.51}$$

in the set Q with boundary conditions

$$y(x,t) = 0, \ (x,t) \in \Sigma, \tag{10.52}$$

$$y(x,0) = \varphi(x), \ \frac{\partial y(x,0)}{\partial t} = \psi(x), \ x \in \Omega. \tag{10.53}$$

Second order evolutional systems

Proof. By Theorem 10.4 the solution of Problem 10.3 satisfies the system (10.18), (10.21), (10.25)–(10.27). Obviously, Cauchy problem (10.26), (10.27) for the second order evolutional equation now is the boundary problem (10.51)–(10.53) for the wave equation. By the equality $V = L_2(Q)$ the canonic isomorphism Λ_V is the unit operator. The operator B is a unit to. Therefore, it is equal to its adjoint operator. Then the variational inequality (10.25) can be transformed to (10.47). We have $Z = L_2(Q)$; then C is a unit. Hence, Λ_Z and C^* are units too. Therefore, the adjoint equation (10.20) is (10.48) here. The boundary conditions (10.59), (10.50) are obvious. Hence, the solution of Problem 10.3 satisfies the equalities (10.47)–(10.53). $\square$

Suppose the set of admissible controls is determined by the local constraints. Particularly, consider the set

$$U = \{v \in L_2(Q) \mid a \leq v(x,t) \leq b, \ (x,t) \in Q\},$$

where a and b are constants, as $a < b$.

Theorem 10.5 *The solution of Problem* 10.3 *with a given set of admissible controls is determined by the formula*

$$u(x,t) = \begin{cases} a, & \text{if } -p(x,t)/\alpha < a, \\ -p(x,t)/\alpha, & \text{if } a \leq -p(x,t)/\alpha \leq b, \\ b, & \text{if } -p(x,t)/\alpha > b. \end{cases} \tag{10.54}$$

Indeed, we can prove this result by using Theorem 8.7 because of the equality of the variational inequality (8.34) and (10.47).

We can put this control in to the equality (10.51). Then we have the boundary problem with respect to the unknown functions y and p. If we find now the function p, we can determine the optimal control by the formula (10.54).

Remark 10.7 We could consider the optimization problem for the wave equation without any constraints. It will be an analog of results of Chapter 8.

10.5 Nonlinear wave equation

Now consider nonlinear second order evolutional equations. Suppose Ω is an open Euclid bounded set of the space $\mathbb{R}^n$ with the boundary Γ, $T > 0$, $Q = \Omega \times (0,T)$, $\Sigma = \Gamma \times (0,T)$. We have the nonlinear heat equation

$$\frac{\partial^2 y}{\partial t^2} - \Delta y + |y|^p y - f \tag{10.55}$$

Optimization and Differentiation

in the set Q with the boundary condition

$$y(x,t) = 0 \ (x,t) \in \Sigma \tag{10.56}$$

and the initial conditions

$$y(x,0) = \varphi(x), \quad \frac{\partial y(x,0)}{\partial t} = \psi(x), \ x \in \Omega, \tag{10.57}$$

where $\rho > 0$, and functions $f = f(x,t)$, $\varphi = \varphi(x)$, $\psi = \psi(x)$ are known.

Multiply the equality (10.55) by the time derivative of the function y and integrate the result. We have equality

$$\int_\Omega \left(\frac{\partial^2 y}{\partial t^2} - \Delta y + |y|^\rho \right) \frac{\partial y}{\partial t} dx = \int_\Omega f \frac{\partial y}{\partial t} dx. \tag{10.58}$$

Remark 10.8 These transformations are formal, because we do not know the existence of these integrals.

We have

$$\frac{\partial^2 y}{\partial t^2} \frac{\partial y}{\partial t} = \frac{1}{2} \frac{\partial}{\partial t} \left(\frac{\partial y}{\partial t} \right)^2, \quad |y|^\rho y \frac{\partial y}{\partial t} = \frac{1}{q} \frac{\partial}{\partial t} (|y|^q),$$

$$-\int_\Omega \Delta y \frac{\partial y}{\partial t} dx = \int_\Omega \sum_{i=1}^n \frac{\partial y}{\partial x_i} \frac{\partial}{\partial t} \frac{\partial y}{\partial x_i} dx = \frac{1}{2} \|y(t)\|^2,$$

where $q = \rho + 2$. Therefore, from the equality (10.58) it follows that

$$\frac{1}{2} \frac{d}{dt} \left[\|y'(t)\|_2^2 + \frac{1}{q} \|y(t)\|_q^q + \frac{1}{2} \|y(t)\|^2 \right] \le \int_\Omega f \frac{\partial y}{\partial t} dx \le \frac{1}{2} \|y'(t)\|_2^2 + \frac{1}{2} \|f(t)\|_2^2.$$

Integrating this inequality by using the initial condition (10.57) we get

$$\|y'(t)\|_2^2 + \frac{2}{q} \|y(t)\|_q^q + \|y(t)\|^2 \le$$

$$\|\psi\|_2^2 dx + \frac{2}{q} \|\varphi\|_q^q + \|\varphi\|^2 + \int_0^t \|y'(t)\|_2^2 dt + \int_0^t \|f(t)\|_2^2 dt. \tag{10.59}$$

From (10.59) and the conditions $\varphi \in H_0^1(\Omega) \cap L_q(\Omega)$, $\psi \in L_2(\Omega)$, $f \in L_2(Q)$ the inequality follows:

$$\|y'(t)\|_2^2 \le c + \int_0^t \|y'(t)\|_2^2 dt,$$

where c is a positive constant. Using the Gronwall lemma, we obtain the

Second order evolutional systems 385

estimate of the time derivative of the function y in the space $L_\infty(0, T; L_q(\Omega))$. Then from the previous inequality the estimate of the function y in the space $L_\infty(0, T; H_0^1(\Omega) \cap L_q(\Omega))$ follows. Determine the second time derivative from the equality (10.55). This is an element of the space

$$W = L_\infty(0, T; H^{-1}(\Omega)) + L_{q'}(Q).$$

Determine the space

$$Y = \{y \mid y \in L_\infty(0, T; H_0^1(\Omega) \cap L_q(\Omega)), \ y' \in L_\infty(0, T; L_2(\Omega)), \ y'' \in W\}.$$

Theorem 10.6 *For all functions $\varphi \in H_0^1(\Omega) \cap L_q(\Omega)$, $\psi \in L_2(\Omega)$, $f \in L_2(Q)$ the problem (10.55)–(10.57) has a unique solution $y \in Y$.*

Proof. 1. Let $\{\mu_m\}$ be a complete set of linear independent elements of the space $H_0^1(\Omega) \cap L_q(\Omega)$. Determine an approximate solution of the problem (10.55)–(10.57) by the formula

$$y_k = \sum_{m=1}^{k} \xi_{mk} \mu_m, \ k = 1, 2, \dots, \tag{10.60}$$

where the functions $\xi_{1k} = \xi_{1k}(t), ..., \xi_{kk} = \xi_{kk}(t)$ satisfy the equalities

$$\int_\Omega \frac{\partial^2 y_k}{\partial t^2} \mu_j dx + \int_\Omega \sum_{1=1}^{n} \frac{\partial y_k}{\partial x_i} \frac{\partial \mu_j}{\partial x_i} dx + \int_\Omega |y_k|^p y_k \mu_j dx = \int_\Omega f \mu_j dx. \tag{10.61}$$

We have the system of the second order nonlinear ordinary differential equations. Add the following initial conditions

$$y_k(x, 0) = \varphi_k(x), \ y_k'(x, 0) = \psi_k(x), \ x \in \Omega, \tag{10.62}$$

where

$$\varphi_k \to \varphi \ \text{in} \ H_0^1(\Omega) \cap L_q(\Omega), \ \psi_k \to \psi \ \text{in} \ L_2(\Omega). \tag{10.63}$$

This Cauchy problem has a solution on an interval $(0, T_k)$. Then we can determine the function y_k by the formula (10.60).

2. Multiply j-th equality (10.61) by ξ_{jk}' and sum the result by j. We get the equality

$$\int_\Omega \left(\frac{\partial^2 y_k}{\partial t^2} - \Delta y_k + |y_k|^p y_k \right) \frac{\partial y_k}{\partial t} dx = \int_\Omega f \frac{\partial y_k}{\partial t} dx$$

that is the analog of (10.58). Using the known technique, we have inequality

$$\|y_k'(t)\|_2^2 + \frac{2}{q} \|y_k(t)\|_q^q + \|y_k(t)\|^2 \le$$

$$\|\psi_k\|_2^2 + \frac{2}{q}\|\varphi_k\|_q^q + \|\varphi_k\|^2 + \int\limits_0^t \|y_k'(t)\|_2^2 dt + \int\limits_0^t \|f(t)\|_2^2 dt$$

that is an analog of (10.59). From (10.62) it follows that the first three integrals at the right-hand side of this inequality are bounded. Therefore, using the Gronwall lemma, we get boundedness of the sequence of the derivatives $\{y_k'\}$ in the space $L_\infty(L_2(\Omega))$. Then we have boundedness of the sequence $\{y_k\}$ in the space $L_\infty(H_0^1(\Omega) \cap L_q(\Omega))$. Thus, the sequence $\{y_k\}$ is bounded in the space

$$Y_1 = \big\{y \big| \, y \in L_\infty(H_0^1(\Omega) \cap L_q(\Omega)), \, y' \in L_\infty(L_2(\Omega))\big\}.$$

3. Using the Banach–Alaoglu Theorem, after extracting a subsequence we get the convergence $y_k \to y$ *-weakly in Y_1. Then

$$\lim_{k\to\infty} \langle \lambda, y_k \rangle = \langle \lambda, y \rangle \; \forall \lambda \in \Lambda,$$

where $\Lambda' = Y_1$. We have $y_k \to y$ weakly in $H^1(Q)$. Using the Rellich–Kondrashov Theorem, after extracting a subsequence we have $y_k \to y$ strongly in $L_2(Q)$ and a.e. on Q. Therefore, we get the convergence $|y_k|^p y_k \to |y|^p y$ a.e. on Q. By the boundedness of the sequence $\{y_k\}$ in the space $L_q(Q)$ the sequence $\{|y_k|^p y_k\}$ is bounded in $L_{q'}(Q)$. Therefore, $|y_k|^p y_k \to |y|^p y$ weakly in $L_{q'}(Q)$. Multiply the equality (10.61) by the arbitrary function $\eta \in C^1[0,T]$ such that $\eta(T) = 0$. After integration we obtain

$$\int\limits_Q \Big(\frac{\partial^2 y_k}{\partial t^2} - \Delta y_k + |y_k|^p y_k\Big)\eta_j dQ = \int\limits_Q f\eta_j dQ,$$

where $\eta_j(t) = \eta(t)\mu_j$. After integration by parts we get

$$\int\limits_Q \frac{\partial^2 y_k}{\partial t^2}\eta_j dQ = -\int\limits_Q \frac{\partial y_k}{\partial t}\frac{\partial \eta_j}{\partial t} dQ - \int\limits_\Omega \psi_k \eta_j(x,0)dx.$$

Now transform the previous equality.

$$-\int\limits_Q \frac{\partial y_k}{\partial t}\frac{\partial \eta_j}{\partial t} dQ - \int\limits_\Omega \psi_k \eta_j(x,0)dx + \int\limits_Q \sum_{1=1}^n \frac{\partial y_k}{\partial x_i}\frac{\partial \eta_j}{\partial x_i} dQ +$$

$$\int\limits_Q |y_k|^p y_k \eta_j dQ = \int\limits_Q f\eta_j dQ.$$

Passing to the limit, we obtain

$$-\int\limits_Q \frac{\partial y}{\partial t}\frac{\partial \eta_j}{\partial t} dQ - \int\limits_\Omega \psi\eta_j(x,0)dx + \int\limits_Q \sum_{1=1}^n \frac{\partial y}{\partial x_i}\frac{\partial \eta_j}{\partial x_i} dQ + \int\limits_Q |y|^p y\eta_j dQ = \int\limits_Q f\eta_j dQ.$$

Second order evolutional systems 387

Let η be an arbitrary element of the space $D(0,T)$. After integration by parts by using Green's formula, we get the equality

$$\int_0^T \eta \int_\Omega \left(\frac{\partial^2 y}{\partial t^2} - \Delta y + |y|^p y - f\right)\mu_j dx dt = 0, \ j = 1, 2, \dots .$$

Then we have

$$\int_\Omega \left(\frac{\partial^2 y}{\partial t^2} - \Delta y + |y|^p y - f\right)\mu_j dx = 0, \ j = 1, 2, \dots$$

because η is arbitrary. The set $\{\mu_j\}$ is complete. Then the function y satisfies in reality the equality (10.55). Find the second time derivative of y from this function. This is an element of the W. Thus, we have $y \in Y$. $\square$

Remark 10.9 By the equality (10.55) we can determine the second time derivative of y. This belongs to some space. However, we shall not use this property.

The uniqueness of the solution can be determined by using additional conditions.

Theorem 10.7 *The solution of the problem* (10.55)–(10.57) *is unique, if we have the additional condition*

$$n = 2 \ \ or \ \ \rho \le 2/(n-2) \ \ for \ \ n > 2. \tag{10.64}$$

Proof. Suppose there exist two solutions y_1 and y_2 of the given problem. Then its difference $y = y_1 - y_2$ satisfies the equality

$$\frac{\partial^2 y}{\partial t^2} - \Delta y + |y_1|^p y_1 - |y_2|^p y_2 = 0 \tag{10.65}$$

with homogeneous the boundary condition. Fix the value $s \in (0,T)$. Determine the functions

$$z(x,t) = \begin{cases} -\int_t^s y(x,\tau)d\tau, \ t \le s, \\ 0, \ t > s. \end{cases} \qquad w(x,t) = \int_0^t y(x,\tau)d\tau.$$

We have the equalities

$$\frac{\partial z}{\partial t} = y, \ \ \frac{\partial w}{\partial t} = y, \ \ z(x,s) = 0, \ \ w(x,0) = 0, \ \ z(x,t) = w(x,t) - w(x,s).$$

Multiply the equality (10.65) by the function z and integrate the result. We get

$$\int_0^s \int_\Omega \frac{\partial^2 y}{\partial t^2} z \, dx dt - \int_0^s \int_\Omega \Delta y z \, dx dt = \int_0^s \int_\Omega \left(|y_1|^p y_1 - |y_2|^p y_2\right) z \, dx dt. \tag{10.66}$$

We have the equality

$$\int\limits_0^s \int\limits_\Omega \frac{\partial^2 y}{\partial t^2} z\,dxdt = -\int\limits_0^s \int\limits_\Omega \frac{\partial y}{\partial t}\frac{\partial z}{\partial t}\,dxdt =$$

$$-\int\limits_0^s \int\limits_\Omega \frac{\partial y}{\partial t} y\,dxdt = -\frac{1}{2}\int\limits_0^s \frac{d}{dt}\int\limits_\Omega y^2 dx = -\frac{1}{2}\|y(t)\|_2^2,$$

where we use our previous denotations of the norm. Using Green's formula, we obtain

$$-\int\limits_0^s \int\limits_\Omega \Delta y z\,dxdt = \int\limits_0^s \int\limits_\Omega \sum_{i=1}^n \frac{\partial y}{\partial x_i}\frac{\partial z}{\partial x_i}\,dxdt = \int\limits_0^s \int\limits_\Omega \sum_{i=1}^n \frac{\partial}{\partial t}\frac{\partial w}{\partial x_i}\frac{\partial z}{\partial x_i}\,dxdt =$$

$$\int\limits_0^s \int\limits_\Omega \sum_{i=1}^n \frac{\partial w}{\partial x_i}\frac{\partial}{\partial x_i}\frac{\partial z}{\partial t}\,dxdt = -\int\limits_0^s \int\limits_\Omega \sum_{i=1}^n \frac{\partial w}{\partial x_i}\frac{\partial}{\partial x_i}\frac{\partial w}{\partial t}\,dxdt =$$

$$-\frac{1}{2}\int\limits_0^s \frac{\partial}{\partial t}\int\limits_\Omega \sum_{i=1}^n \left(\frac{\partial w}{\partial x_i}\right)^2 = -\frac{1}{2}\|w(s)\|^2.$$

Therefore, the equality (10.66) can be transformed to

$$\|y(s)\|_2^2 + \|w(s)\|^2 = K, \qquad (10.67)$$

where

$$K = 2\int\limits_0^s \int\limits_\Omega \left(|y_1|^\rho y_1 - |y_2|^\rho y_2\right) z\,dxdt \le$$

$$2(\rho+1)\int\limits_0^s \int\limits_\Omega \left(|y_1|^\rho + |y_2|^\rho\right)|y|\left[|w(x,t)| + |w(x,s)|\right]dxdt.$$

From the Hölder inequality it follows that

$$\left|\int\limits_\Omega \xi\eta\zeta dx\right| \le c\|\xi\|_p\|\eta\|_r\|\zeta\|_s \;\forall \xi \in L_p(\Omega),\, \eta \in L_r(\Omega),\, \zeta \in L_s(\Omega),$$

where $1/p+1/r+1/s = 1$. For $n = 2$ by the Sobolev theorem we have continuous embedding $H^1(\Omega) \subset L_\pi(\Omega)$, where π is an arbitrary number. Particularly, we choose $\pi = 2(\rho+1)$. Then we have

$$y \in L_\infty\big(0,T;L_2(\Omega)\big),\, w \in L_\infty\big(0,T;L_\pi(\Omega)\big),\, |y_i|^\rho \in L_\infty\big(0,T;L_{\pi/\rho}(\Omega)\big),\, i = 1,2.$$

Second order evolutional systems

Using Hölder inequality, we get

$$K \le c \int_0^s \Big\{ \sum_{i=1}^2 \|y_i(t)\|_{\pi/\rho} \big[\|w(t)\|_\pi + \|w(s)\|_\pi \big] \|y_i(t)\|_2 \Big\} dt \le$$

$$c_1 \int_0^s \big[\|w(t)\|_\pi + \|w(s)\|_\pi \big] \|y_i(t)\|_2 \Big\} dt,$$

where c_1 is a positive constant. From Sobolev theorem, it follows that

$$\|\xi\|_\pi \le c' \|\xi\| \ \forall h \in H_0^1(\Omega).$$

Then we transform the previous inequality

$$K \le c_2 \int_0^s \big[\|w(t)\|_H + \|w(s)\|_H \big] \|y_i(t)\|_2 dt, \tag{10.68}$$

where $c_2 > 0$.

For $n > 2$ determine the parameter r from the equality $1/r + 1/n = 1/2$, i.e., $r = 2n/(n-2)$. From Sobolev theorem by using (10.64) continuous embedding $H^1(\Omega) \subset L_r(\Omega)$ and $Y \subset L_\infty(0, T; L_r(\Omega))$ follows. Then we have $y_i \in L_\infty(0, T; L_r(\Omega))$ and $|y_i|^\rho \in L_\infty(0, T; L_{r/\rho}(\Omega))$, $i = 1, 2$. Using the inequality $\rho n \le r$ that is the corollary of the condition (10.64), determine $|y_i|^\rho \in L_\infty(0, T; L_n(\Omega))$. Using Hölder inequality, we get

$$K \le c \int_0^s \sum_{i=1}^2 \|y_i(t)\|_2 \big[\|w(t)\|_r + \|w(s)\|_r \big] \|y_i(t)\|_2 dt \le$$

$$c_3 \int_0^s \big[\|w(t)\|_r + \|w(s)\|_r \big] \|y_i(t)\|_2 dt,$$

where $c_3 > 0$. Using embedding $H^1(\Omega) \subset L_r(\Omega)$, obtain again the condition (10.68) with another constant c_2.

Thus, from the conditions (10.64) and (10.67) it follows that

$$\|y(s)\|_2^2 + \|w(s)\|^2 \le c_2 \int_0^s \big[\|w(t)\| + \|w(s)\| \big] \|y(t)\|_2^2 dt \le$$

$$c_2 \int_0^s \|w(t)\| \|y(t)\|_2 dt + \frac{1}{2} \|w(s)\|^2 + c_2^2 \int_0^s \|w(t)\|^2 dt.$$

390   *Optimization and Differentiation*

Then we have the inequality

$$\|y(s)\|_2^2 + \|w(s)\|^2 \le c_4 \int\limits_0^s \|y(t)\|_2^2 + \|w(t)\|^2 dt,$$

where $c_4 > 0$. Using the Gronwall lemma, we have

$$\|y(s)\|_2^2 + \|w(s)\|^2 \le 0.$$

Therefore, $y = 0$, i.e., $y_1 = y_2$. $\square$

Thus, we can guarantee the uniqueness of the solution of the boundary problem for small enough values of the dimension set and the nonlinearity degree.

10.6 Optimization control problem for the nonlinear wave equation

Consider the optimization control problem for the system described by the wave equation with nonlinearity power. We have the equation

$$\frac{\partial^2 y}{\partial t^2} - \Delta y + |y|^p y = v \tag{10.69}$$

in the set Q with the boundary conditions

$$y(x, t) = 0, \ (x, t) \in \Sigma, \tag{10.70}$$

$$y(x, 0) = \varphi(x), \ \frac{\partial y(x, 0)}{\partial t} = \psi(x), \ x \in \Omega, \tag{10.71}$$

where $\rho > 0$, $\varphi \in H_0^1(\Omega) \cap L_q(\Omega)$, $\psi \in L_2(\Omega)$. Determine the control space $V = L_2(Q)$. By Theorem 10.6 for all control $v \in V$ the problem (10.69)–(10.71) has the solution $y = y[v]$ from the space

$$Y = \left\{ y \middle| y \in L_\infty\left(0, T; H_0^1(\Omega) \cap L_q(\Omega)\right), \ y' \in L_2\left(0, T; L_2(\Omega)\right), \ y'' \in W \right\},$$

where

$$W = L_\infty\left(0, T; H-1(\Omega)\right) + L_{q'}(Q).$$

From Theorem 10.7 the uniqueness of this solution follows under the condition (10.64).

Consider a convex closed subset U of the space V and the functional

$$I(v) = \frac{\alpha}{2} \int\limits_\Omega v^2 dQ + \frac{1}{2} \int\limits_\Omega \left(y[v] - y_d\right)^2 dQ,$$

where $\alpha > 0$, $y_d \in L_2(Q)$ is known.

Second order evolutional systems 391

Problem 10.4 *Find the control $u \in U$ for the system (10.69)–(10.71) that minimizes the functional I on the set U.*

Lemma 10.6 *If $v_k \to v$ weakly in V, then there exists a subsequence $\{v_s\}$ from $\{v_k\}$ such that $y[v_s] \to y[v]$ $*$-weakly in the space*

$$Y_1 = \left\{ y \,\middle|\, y \in L_\infty\big(0, T; H_0^1(\Omega)\big) \cap L_q(\Omega)\big), \; y' \in L_2\big(0, T; L_2(\Omega)\big) \right\}.$$

Proof. Suppose the convergence $v_k \to v$ weakly in V. The term at the right-hand side of the equality (10.69) belongs to the space $L_2(Q)$. Therefore, the term at the left-hand side of this equality has the analogical property. Multiply it for $v = v_k$ by the time derivative of the function $y_k = y[v_k]$. After integration we get

$$\int_\Omega \int_0^\tau \left(\frac{\partial^2 y_k}{\partial t^2} - \Delta y_k + |y_k|^p y_k \right) \frac{\partial y_k}{\partial t} \, dx dt = \int_\Omega \int_0^\tau v_k \frac{\partial y_k}{\partial t} \, dx dt,$$

where $\tau \in (0, T)$. Transform it by using the standard technique. We have inequality

$$\left\| y_k'(\tau) \right\|_2^2 + \frac{2}{q} \left\| y_k(\tau) \right\|_q^q + \left\| y_k(\tau) \right\|^2 \le$$

$$\left\| \psi \right\|_2^2 + \frac{2}{q} \left\| \varphi \right\|_q^q + \left\| \varphi \right\|^2 + \int_0^\tau \left\| y_k'(t) \right\|_2^2 dt + \int_0^\tau \left\| v_k(t) \right\|_2^2 dt$$

that is the analog of (10.59). Using the boundedness of the sequence $\{v_k\}$ in the space $L_2(Q)$, we have the existence of a constant $C > 0$ such that

$$\left\| y_k'(\tau) \right\|_2^2 \le c + \int_0^\tau \left\| y_k'(t) \right\|_2^2 dt.$$

From the Gronwall lemma it follows that the sequence of time derivative of the functions y_k is bounded in the space $L_\infty\big(0, T; L_2(\Omega)\big)$. Then from the previous inequality the boundedness of the sequence $\{y_k\}$ in the space Y_1 follows. Therefore, we can extract a subsequence $\{v_s\}$ from $\{v_k\}$ such that $y[v_s] \to y$ $*$-weakly in Y_1. Repeat the transformations of the proof of Theorem 10.6. We determine $y = y[v]$. $\square$

Remark 10.10 We do not use the condition (10.64). The value $y_k = y[v_k]$ is a solution (maybe non-unique) of the considered boundary problem.

Remark 10.11 If the solution of the boundary problem is unique, then each subsequence of $\{y_k\}$ tends to the value $y[v]$. Hence, whole sequence $\{y_k\}$ has $y[v]$ as the limit.

392 *Optimization and Differentiation*

Lemma 10.7 *If $v_k \to v$ weakly in V, then there exists a subsequence $\{v_s\}$ from $\{v_k\}$ such that $I(v_s) \to I(v)$.*

Proof. Suppose $v_k \to v$ weakly in V. By Lemma 10.6 there exists a subsequence $\{v_s\}$ from $\{v_k\}$ such that $y[v_s] \to y[v]$ *-weakly in Y_1, and weakly in $H^1(Q)$. Using the Rellich–Kondrashov theorem, we have $y[v_s] \to y[v]$ strongly in $L_2(Q)$. This completes the proof of Lemma 10.7. $\square$

Theorem 10.8 *Problem 10.4 is solvable.*

Proof. Suppose $\{v_s\}$ is a minimizing sequence for Problem 10.4. Using the coercivity of the functional, we prove its boundedness. Then after extracting a subsequence we have the convergence $v_k \to v$ weakly in V, besides $v \in U$. Using Lemma 10.7, after extracting the subsequence we get $I(v_k) \to I(v)$. The sequence $\{v_s\}$ is minimizing. Hence, the limit $\{I(v_s)\}$ is the lower bound of this functional on the set U. Then the control v is the solution of the given problem. $\square$

Remark 10.12 We prove the solvability of the optimization problem without the requirement of uniqueness of the solution.

10.7 Optimality conditions for the optimization optimal problem for the nonlinear wave equation

We use the variational inequality (10.18) as a necessary optimality condition for Problem 10.4. The general step is the proof of the differentiability of the minimizing property that is a corollary of the analogical property of the control–state mapping. Prove that this dependence is an extended differentiable at the arbitrary point $u \in V$.

Consider the equation

$$\frac{\partial^2 p_\mu[u]}{\partial t^2} - \Delta p_\mu[u] + (\rho + 1)\big|y[u]\big|^\rho p_\mu[u] = \mu_Q \tag{10.72}$$

in the set Q with boundary conditions

$$p_\mu[u](x,t) = 0, \ (x,t) \in \Sigma, \tag{10.73}$$

$$p_\mu[u](x,T) = \mu_\Omega(x), \ x\frac{\partial p_\mu[u](x,T)}{\partial t^2} = \mu_{\Omega 1}(x), \ x \in \Omega, \tag{10.74}$$

where $\mu = \big(\mu_Q, \mu_\Omega, \mu_{\Omega 1}\big)$. Determine the space

$$Z = L_2(Q) \times H_0^1(\Omega) \times L_2(\Omega).$$

Second order evolutional systems

Lemma 10.8 *Under the condition* (10.64) *the operator map* $y[\cdot] : V \to Y$ *for the problem* (10.69)–(10.71) *has* $(V, V; V, V^w)$-*extended derivative* $y'[u]$ *at the arbitrary point* $u \in V$, *where* V^w *is the space* V *with the weak topology as there exist linear continuous operators* $y'_T[u] : V \to L_2(\Omega)$ *and* $y'_{T1}[u] : V \to H^{-1}(\Omega)$ *such that for all* $h \in V$ *we have the convergence*

$$\{y[u + \sigma h](T) - y[u](T)\}/\sigma \to y'_T[u]h \text{ weakly in } L_2(\Omega), \tag{10.75}$$

$$\frac{1}{\sigma}\left\{\frac{\partial y[u + \sigma h](T)}{\partial t} - \frac{\partial y[u](T)}{\partial t}\right\} \to y'_{T1}[u]h \text{ weakly in } H^{-1}(\Omega). \tag{10.76}$$

In addition, for all $h \in V$, $\mu \in Z$ *the following equality holds:*

$$\int_Q \mu_Q y'[u]h dQ + \int_\Omega \mu_\Omega y'_{T1}[u]h dQ - \int_\Omega \mu_\Omega y'_T[u]h dQ = \int_Q p_\mu[u]h dQ. \tag{10.77}$$

Proof. 1. Under the condition (10.64) the considered boundary problem has a unique solution. For any function $h \in V$ and the number σ the function $\eta_\sigma[h] = (y[u + \sigma h] - y[u])/\sigma$ is the solution of the equation

$$\frac{\partial^2 \eta_\sigma[h]}{\partial t^2} - \Delta \eta_\sigma[h] + (g_\sigma[h])^2 \eta_\sigma[h] = h \tag{10.78}$$

with homogeneous boundary conditions, where

$$(g_\sigma[h])^2 = (\rho + 1)\Big|y[u] + \varepsilon(y[u + \sigma h] - y[u])\Big|^\rho, \ \varepsilon \in [0, 1].$$

Consider the equation

$$\frac{\partial^2 p}{\partial t^2} - \Delta p + (g_\sigma[h])^2 p = \mu_Q \tag{10.79}$$

in the set Q with boundary conditions

$$p(x, t) = 0, \ (x, t) \in \Sigma, \tag{10.80}$$

$$p(x, T) = \mu_\Omega(x), \ \frac{\partial p(x, T)}{\partial t} = \mu_\Omega(x), \ x \in \Omega. \tag{10.81}$$

The system (10.72)–(10.74) is its partial case for $v = u$. Multiply the equality (10.79) by $\partial p/\partial t$ and integrate the result.

$$\int_\tau^T \int_\Omega \left(\frac{\partial^2 p}{\partial t^2} - \Delta p\right)\frac{\partial p}{\partial t}dxdt =$$

$$-\int_\tau^T \int_\Omega (g_\sigma[h])^2 \frac{\partial p}{\partial t}dxdt + \int_\tau^T \int_\Omega \mu_Q \frac{\partial p}{\partial t}dxdt, \tag{10.82}$$

where $\tau \in (0, T)$.

We have the equality

$$\int\limits_{\tau}^{T}\int\limits_{\Omega}\left(\frac{\partial^2 p}{\partial t^2}-\Delta p\right)\frac{\partial p}{\partial t}dxdt=\frac{1}{2}\int\limits_{\tau}^{T}\frac{\partial}{\partial t}\int\limits_{\Omega}\left[\left(\frac{\partial p}{\partial t}\right)^2+\sum_{i=1}^{n}\left(\frac{\partial p}{\partial x_i}\right)^2\right]dxdt=$$

$$\frac{1}{2}\left[\left\|\mu_{\Omega 1}\right\|_2^2+\left\|\mu_{\Omega}\right\|^2\right]-\frac{1}{2}\left[\left\|\frac{\partial p(\tau)}{\partial t}\right\|_2^2+\left\|p(\tau)\right\|^2\right].$$

From $y\in Y$ it follows that $\left(g_\sigma[h]\right)^2\in L_{q/\rho}(\Omega)$. Using the technique from the proof of Theorem 10.7 with condition (10.64), we get

$$\left|\int\limits_{\tau}^{T}\int\limits_{\Omega}\left(g_\sigma[h]\right)^2 p)\frac{\partial p}{\partial t}dxdt\right|\leq c\int\limits_{\tau}^{T}\left\|p(t)\right\|\left\|\frac{\partial p(t)}{\partial t}\right\|_2 dt,$$

where $c>0$. Therefore, from (10.82) it follows that

$$\frac{1}{2}\left[\left\|\frac{\partial p(\tau)}{\partial t}\right\|_2^2+\left\|p(\tau)\right\|^2\right]\leq 2c\int\limits_{\tau}^{T}\left\|p(t)\right\|\left\|\frac{\partial p(t)}{\partial t}\right\|_2 dt+\frac{1}{2}\left[\left\|\mu_{\Omega 1}\right\|_2^2+\left\|\mu_{\Omega}\right\|^2\right]+$$

$$\int\limits_{\tau}^{T}\left\|\mu_Q(t)\right\|_2\left\|\frac{\partial p(t)}{\partial t}\right\|_2 dt\leq c_1\int\limits_{\tau}^{T}\left[\frac{1}{2}\left[\left\|\frac{\partial p(t)}{\partial t}\right\|_2^2+\left\|p(t)\right\|^2\right]dt+$$

$$\frac{1}{2}\left[\left\|\mu_{\Omega 1}\right\|_2^2+\left\|\mu_{\Omega}\right\|^2\right]+c_2\int\limits_{\tau}^{T}\left\|\mu_Q(t)\right\|_2^2 dt,$$

where $c_1>0$, $c_2>0$

Consider the following analog of the Gronwall lemma. If for a non-negative integrable function $\eta=\eta(t)$ the following inequality holds:

$$\eta(\tau)\leq\alpha+\beta\int\limits_{\tau}^{T}\eta(t)dt,\ \tau\in(0,T),$$

where α, β are positive constants, then there exists a positive constant $c=c(\beta,T)$ such that $\eta(t)\leq c\alpha$. Then from the previous inequality we have

$$\left\|\frac{\partial p(\tau)}{\partial t}\right\|_2^2+\left\|p(\tau)\right\|^2\right]\leq c_3\left[\left\|\mu_{\Omega 1}\right\|_2^2+\left\|\mu_{\Omega}\right\|^2+\left\|\mu_Q\right\|_{L_2(Q)}^2\right],\qquad(10.83)$$

where $c_1>0$. Thus, the linear boundary problem (10.79)–(10.81) has a priori estimate in the space of functions from the set $L_\infty\left(0,T;H_0^1(\Omega)\right)$ with time derivatives from the set $L_\infty\left(0,T;L_2(\Omega)\right)$. This is the space Y_1 under the condition (10.64). Find the second time derivative of the function p from the

Second order evolutional systems 395

equality (10.79) by using (10.64). We have a priori estimate of the solution of this problem in the space Y. Using Theorem 10.1, we prove that for all $h \in V$, $\mu \in Z$ and number σ this problem has a unique solution $p = p_\mu^\sigma[h]$ from the set Y.

Multiply the equality (10.75) by the function $p_\mu^\sigma[v]$ and integrate the result by using the equation (10.79) and the given boundary conditions. We obtain

$$\int_Q \mu_Q \eta_\sigma[h] dQ + \int_\Omega \mu_\Omega \frac{\partial \eta_\sigma[h](T)}{\partial t} dx - \int_\Omega \mu_{\Omega 1} \eta_\sigma[h](T) dx = \int_Q h p_\mu^\sigma[h] dQ \quad (10.84)$$

for all $h \in V$, $\mu \in Z$. Note that if $\mu_\Omega = 0$, $\mu_{\Omega 1} = 0$, then the equality (10.77) determines a linear continuous operator $y'[u] : L_2(Q) \to L_2(Q)$. Then from this equality and (10.84) it follows that

$$\int_Q \mu_Q \big\{ (y[u + \sigma h] - y[u]) / \sigma - y'[u] h \big\} dQ = \int_Q h \big(p_\mu^\sigma[h] - p_\mu[h] \big) dQ. \quad (10.85)$$

2. Suppose we have the convergence $\sigma \to 0$. Then from Lemma 10.6 the convergence $y[u + \sigma h] \to y[u]$ *-weakly follows in Y. Then the set $\{y[u + \sigma h]\}$ is bounded in the space $L_q(Q)$, and the set $\{g_\sigma[h]\}$ is bounded in $L_{2q/\rho}(Q)$. From the inequality (10.83) it follows that the set $\{p_\mu^\sigma[h]\}$ is bounded in the space Y_1. After extracting a subsequence we get the convergence $p_\mu^\sigma[h] \to p_\mu$ *-weakly in Y_1 and weakly in $H^1(Q)$. Using the Rellich–Kondrashov theorem, we have the convergence $p_\mu^\sigma[h] \to p$ and $y[u + \sigma h] \to y[u]$ strongly in $L_2(Q)$ and a.e. on Q. Then $\big(g_\sigma[h] \big)^2 p_\mu^\sigma[h] \to (\rho + 1)|y[u]|^\rho p_\mu$ a.e. on Q. Under the condition (10.64) we have continuous embedding $Y \subset L_q(Q)$. Therefore, the set $\{p_\mu^\sigma[h]\}$ is bounded in the space $L_q(Q)$. Then $\big\{ \big(g_\sigma[h] \big)^2 p_\mu^\sigma[h] \big\}$ is bounded in the space $L_{q'}(Q)$. Hence, $\big(g_\sigma[h] \big)^2 p_\mu^\sigma[h] \to (\rho + 1)|y[u]|^\rho p_\mu$ weakly in $L_{q'}(Q)$. Passing to the limit at the equality (10.76) for $p = p_\mu^\sigma[h]$ and $\sigma \to 0$ we determine $p_\mu = p_\mu[h]$. Pass to the limit at the equality (10.85). We have

$$\big(y[u + \sigma h] - y[u] \big) / \sigma \to y'[u] h \text{ weakly in } L_2(Q)$$

for all $h \in V$. Thus, $y'[u] h$ is in reality the extended derivative of the control–state mapping for the boundary problem (10.69)–(10.71).

For $\mu_Q = 0$, $\mu_\Omega = 0$ the equality (10.77) determines a linear continuous operator $y_T[u] : V \to L_2(\Omega)$. Using (10.84), we get

$$\int_\Omega \mu_{\Omega 1} \big\{ (y[u + \sigma h](T) - y[u](T)) / \sigma - y'_T[u] h \big\} \mu_\Omega dx = \int_Q h \big(p_\mu^\sigma[h] - p_\mu[h] \big) dQ.$$

Pass to the limit as $\sigma \to 0$. We have the condition (10.75).

396　　　　　　　　*Optimization and Differentiation*

For $\mu_Q = 0$, $\mu_{\Omega 1} = 0$ the equality (10.77) determines a linear continuous operator $y'_T[u] : V \to H^{-1}(\Omega)$. Using condition (10.84) we get

$$\int_\Omega \mu_\Omega \left[\frac{1}{\sigma} \left\{ \frac{\partial y[u + \sigma h](T)}{\partial t} - \frac{\partial y[u](T)}{\partial t} \right\} - y'_T[u]h \right] \mu_\Omega dx = \int_Q h \big(p^\sigma_\mu[h] - p_\mu[h] \big) dQ.$$

Then the condition (10.76) is true. We obtain the general case of the equality (10.77) after passing to the limit at the equality (10.84). $\square$

Determine the differentiability of the given functional.

Lemma 10.9 *Under the condition* (10.64) *the functional I for Problem* 10.4 *has Gâteaux derivative* $I'(u) = \alpha u + p$ *at the arbitrary point* $u \in V$, *where* p *is the solution of the equation*

$$\frac{\partial^2 p}{\partial t^2} - \Delta p + (\rho + 1)\big|y[u]\big|^\rho p = y[u] - y_d \qquad (10.86)$$

in the set Q *with the boundary conditions*

$$p(x, t) = 0, \ (x, t) \in \Sigma, \qquad (10.87)$$

$$p(x, T) = 0, \ \frac{\partial p(x, T)}{\partial t} x \in \Omega. \qquad (10.88)$$

Proof. We have the equality

$$I(u + \sigma h) - I(u) = \frac{\alpha}{2} \int_Q (2\sigma u h + \sigma^2 h^2) dQ +$$

$$\frac{1}{2} \int_Q \big(y[u + \sigma h] + y[u] - 2y_d \big) \big(y[u + \sigma h] - y[u] \big) dQ.$$

By Lemma 10.6 if $\sigma \to 0$, then we have the convergence $*$-weakly in Y_1 and strongly in $L_2(Q)$. Divide the previous equality by σ and pass to the limit by using Lemma 10.8. We obtain

$$\langle I'(u), h \rangle = \alpha \int_Q u h dQ + \int_Q \big(y[u] - y_d \big) y'[u] h dQ.$$

Using the condition (10.77), determine $\mu_Q = y[u + \sigma h] - y[u]$, $\mu_\Omega = 0$, $\mu_{\Omega 1} = 0$ at the equality (10.71)–(10.75). We get

$$\langle I'(u), h \rangle = \int_Q (\alpha u + p) h dQ \ \forall h \in V.$$

This completes the proof of the lemma. $\square$

Put the derivative of the functional to the variational inequality (10.18). We obtain the necessary conditions of the optimality.

Second order evolutional systems   397

Theorem 10.9 *Under the condition* (10.64) *the solution of Problem* 10.4 *satisfies the variational inequality*

$$\int_Q (\alpha u + p)(v - u)dQ \geq 0 \ \forall v \in U. \tag{10.89}$$

Thus, we have the state system (10.69)–(10.71) for $v = u$, the adjoint system (10.86)–(10.88), and the variational inequality (10.89) for finding the optimal control.

Remark 10.13 We cannot guarantee the uniqueness of the optimal control and the sufficiency of the optimality condition because of the nonlinearity of the state equation.

Remark 10.14 We can find the variational inequality solution (10.89) for the local conditions with respect to the control as an analog of Theorem 10.5.

Remark 10.15 We could extend our results to the general integral functionals and the general hyperbolic equations (see Chapter 3). The control could be initial, and the functional could depend upon the state function at the final time (see Chapter 9).

10.8   Non-differentiability of the solution of the nonlinear wave equation with respect to the absolute term

Theorem 10.9 gives the necessary conditions of optimality for Problem 10.4 if the degree of nonlinearity ρ and the dimension of the set n are small enough. Prove that the dependence of the state function with respect to the control is not Gâteaux differentiable for large enough values of these parameters.

Lemma 10.10 *For large enough values of the parameters ρ and n the dependence of the solutions of the problem* (10.69)–(10.71) *from the absolute term is not Gâteaux differentiable.*

Proof. Suppose this dependence is Gâteaux differentiable at the arbitrary point $u \in V$. Then there exists a linear continuous operator $y'[u] : V \to Y$ such that for all $h \in V$ we have the convergence $\big(y[u+\sigma h] - y[u]\big)/\sigma \to y'[u]h$ in Y as $\sigma \to 0$. We have the equality

$$\frac{\partial^2 (y[u+\sigma h] - y[u])}{\partial t^2} - \Delta(y[u+\sigma h] - y[u])+$$

$$|y[u+\sigma h]|^\rho y[u+\sigma h] - |y[u]|^\rho y[u] = \sigma h.$$

Divide this equality by σ and pass to the limit as $\sigma \to 0$. We have

$$\frac{\partial^2 y'[u]h}{\partial t^2} - \Delta y'[u]h + (\rho + 1)|y[u]|^\rho y'[u]h = h.$$

398   *Optimization and Differentiation*

Thus, for any $h \in V$ the equation

$$\frac{\partial^2 z}{\partial t^2} - \Delta z + (\rho + 1)|y[u]|^\rho z = h \tag{10.90}$$

with homogeneous boundary conditions has the solution $z = y'[u]h$ from the space Y.

Choose the parameters of the problem (10.69)–(10.71) such that its solution $y[u]$ is a continuous function. For example, we can choose a continuous function that belongs to the Sobolev space $H^2(Q)$ with zero values on the surface Σ; then we find the values of the absolute term and the initial data from the equalities (10.69)–(10.71). Choose the values ρ and n so large that embedding $H^2(Q) \subset L_\infty(0, T; L_q(\Omega))$ is false. Then there exists a point z_* from the set $H^2(Q) \backslash L_\infty(0, T; L_q(\Omega))$ that satisfies the conditions (10.70), (10.71). Determine the function

$$h_* = \frac{\partial^2 z_*}{\partial t^2} - \Delta z_* + (\rho + 1)|y[u]|^\rho z_*.$$

We have $h_* \in V$. Thus, there exists a function $h = h_*$ from the space V such that the solution z_* of the homogeneous boundary problem for the equation (10.90) does not belong to the space $L_\infty(0, T; L_q(\Omega))$ nor to Y too. However, this contradicts our previous result. Therefore, our supposition about Gâteaux differentiability of the dependence solution of the considered boundary problem (10.69)–(10.71) with respect to the absolute term is false. $\square$

This result is the analog of the properties of nonlinear elliptic and parabolic equations.

10.9   Optimization control problem for the linear hyperbolic equation with coefficient control

Consider the control system described by the equation

$$\frac{\partial^2 y}{\partial t^2} - \Delta y + vy = f \tag{10.91}$$

on the set of Q with the boundary conditions

$$y(x, t) = 0 \ (x, t) \in \Sigma, \tag{10.92}$$

$$y(x, 0) = \varphi(x), \quad \frac{\partial y(x, 0)}{\partial t} = \psi(x), \ x \in \Omega, \tag{10.93}$$

Second order evolutional systems

where f, φ, and ψ are known functions. Suppose $f \in L_2(Q)$, $\varphi \in H_0^1(\Omega)$, and $\psi \in L_2(\Omega)$. Determine the control space $V = L_2(Q)$ and the set of admissible controls

$$U = \left\{ v \in V \mid a \le v(x) \le b, \ x \in \Omega \right\},$$

where a and b are constants, as $b > a$.

Multiply the equality (10.91) by the time derivative of y and integrate the result. We get

$$\int_\Omega \left(\frac{\partial^2 y}{\partial t^2} - \Delta y \right) \frac{\partial y}{\partial t} dx = \int_\Omega (f - vy) \frac{\partial y}{\partial t} dx.$$

Transform this equality. We have

$$\frac{1}{2} \frac{d}{dt} \left[\int_\Omega \left(\frac{\partial y}{\partial t} \right)^2 dx + \sum_{i=1}^n \left(\frac{\partial y}{\partial x_i} \right)^2 dx \right] \le \left| \int_\Omega f \frac{\partial y}{\partial t} dx \right| +$$

$$+ \delta \left| \int_\Omega y \frac{\partial y}{\partial t} dx \right| \le \frac{1+\delta}{2} \|y'(t)\|_2^2 + \frac{\delta}{2} \|y(t)\|_2^2 + \frac{1}{2} \|f(t)\|_2^2,$$

where $\delta = \max |a|, |b|$. By continuous embedding of the space $H_0^1(\Omega)$ to $L_2(\Omega)$ we have the inequality

$$\|z\|_2 \le c\|z\| \ \forall z \in H_0^1(\Omega)$$

where is a positive constant. From the previous inequality it follows

$$\frac{d}{dt} \left[\|y'(t)\|_2^2 + \|y(t)\|^2 \right] \le c_1 \left[\|y'(t)\|_2^2 + \|y(t)\|^2 \right] + \|f(t)\|_2^2 dx,$$

where $c_1 > 0$. Integrate this inequality from zero to a value $\tau \in (0, T)$. We obtain

$$\|y'(\tau)\|_2^2 + \|y(\tau)\|^2 \le \|\psi\|_2^2 + \|\varphi\|^2 + \|f\|_{L_2(Q)}^2 + c_1 \int_0^\tau \left[\|y'(t)\|_2^2 + \|y(t)\|^2 \right] dt.$$

Using the Gronwall lemma, we have the estimate

$$\|y'(\tau)\|_2^2 + \|y(\tau)\|^2 \le c_2,$$

where $c_2 > 0$. Thus, linear boundary problem (10.91)–(10.93) has a priori estimate of the solution in the space

$$Y_1 = \left\{ y \mid y \in L_\infty(0, T; H_0^1(\Omega)), \ y' \in L_\infty(0, T; L_2(\Omega)) \right\}.$$

Using the analog of Theorem 10.1, we can determine that for all $v \in V$ the problem (10.91)–(10.93) has a unique solution $y = y[v]$ from the space

$$Y = \left\{ y \mid y \in Y_1, \ y'' \in L_2(0, T; H^{-1}(\Omega)) \right\}.$$

400   *Optimization and Differentiation*

Remark 10.16 The difference between the equations (10.91) and (10.44) is the presence of the term with a function o. However, this is not the obstacle for obtaining a priori estimates and proving the one-valued solvability of the boundary problem.

Determine the functional

$$I(v) = \frac{\alpha}{2} \int_Q v^2 dQ + \frac{1}{2} \int_Q \big(y[v] - y_d\big)^2 dQ,$$

where $\alpha > 0$, $y_d \in L_2(Q)$ is a known function.

Problem 10.5 *Find the control $u \in U$ for the system (10.91)–(10.93) that minimizes the functional I on the set U.*

Prove the following result.

Lemma 10.11 *The map $y[\cdot] : V \to Y$ for the problem (10.91)–(10.93) is $*$-weakly continuous.*

Proof. Suppose the convergence $v_k \to v$ weakly in V. Using the known technique, determine the boundedness of the sequence $\{y[v_k]\}$ in the space Y. Then after extracting a subsequence we get the convergence $y[v_k] \to y$ $*$-weakly in Y. Using compact embedding of the space Y to $L_2(Q)$ we have the convergence $y[v_k] \to y$ strongly in $L_2(Q)$. For all $\lambda \in D(Q)$ we have

$$\left| \int_Q v_k y[v_k] \lambda dQ - \int_Q vy\lambda dQ \right| \le \left| \int_Q (v_k - v)y\lambda dQ \right| + \left| \int_Q v_k (y[v_k] - y)\lambda dQ \right|.$$

Using the condition $y\lambda \in L_2(Q)$, boundedness of the sequence $\{v_k\lambda\}$ in the space $L_2(Q)$, and the convergence of the sequences $\{v_k\}$ weakly and $\{y[v_k]\}$ strongly in this space, we have

$$\int_Q v_k y[v_k] \lambda dQ \to \int_Q vy\lambda dQ.$$

Using the definition of the generalized derivative, we have

$$\lim_{k \to \infty} \int_Q \frac{\partial^2 y[v_k]}{\partial t^2} \lambda dQ = - \lim_{k \to \infty} \int_Q \frac{\partial y[v_k]}{\partial t} \frac{\partial \lambda}{\partial t} dQ = - \int_Q \frac{\partial y}{\partial t} \frac{\partial \lambda}{\partial t} dQ = \int_Q \frac{\partial^2 y}{\partial t^2} \lambda dQ.$$

Multiply the equality (10.91) for $v = v_k$ by an arbitrary function $\lambda \in D(Q)$. Integrating the result and passing to the limit as $k \to \infty$ we get $y = y[v]$. $\square$

Using Lemma 10.11, we have the following result.

Theorem 10.10 *Problem 10.5 is solvable.*

Determine the differentiability of the control–state mapping.

Second order evolutional systems 401

Remark 10.17 Unlike the elliptic and the parabolic cases we can use the possibility to determine the one-valued solvability of the boundary problem for all control. This can by non-admissible too. Then we do not have any necessity to use the derivative with respect to the convex set. The usual derivatives are sufficient for this situation.

Lemma 10.12 *The map $y[\cdot] : V \to Y$ for the problem (10.91)–(10.93) has $(V, V; V, V^w)$-extended derivative $y'[u]$ at the arbitrary point $u \in V$ as, there exist affine continuous operators $y'_T[u] : V \to L_2(\Omega)$ and $y'_{T1}[u] : V \to H^{-1}(\Omega)$ such that for all $h \in V$ we have the convergence*

$$\{y[u + \sigma h](T) - y[u](T)\}/\sigma \to y'_T[u]h \text{ weakly in } L_2(\Omega),$$

$$\frac{1}{\sigma}\left\{\frac{\partial y[u + \sigma h](T)}{\partial t} - \frac{\partial y[u](T)}{\partial t}\right\} \to y'_{T1}[u]h \text{ weakly in } H^{-1}(\Omega).$$

In addition, we have the equality

$$\int_Q \mu_Q y'[u]h\,dQ + \int_\Omega \mu_\Omega \frac{\partial y'[u]h}{\partial t}\Big|_{t=T} dx - \int_\Omega \mu_{\Omega 1} y'[u]h\big|_{t=T} dx = \int_Q p_\mu[u]h\,dQ$$

$$(10.94)$$

for all $h \in V$, $\mu \in Z$, where $\mu = (\mu_Q, \mu_\Omega, \mu_{\Omega 1})$, $Z = L_2(Q) \times H_0^1(\Omega) \times L_2(\Omega)$, and $p_\mu[u]$ is the solution of the equation

$$\frac{\partial^2 p_\mu[u]}{\partial t^2} - \Delta p_\mu[u] + u p_\mu[u] = \mu_Q \qquad (10.95)$$

in the set Q with boundary conditions

$$p_\mu[u](x, t) = 0, \ (x, t) \in \Sigma, \qquad (10.96)$$

$$p_\mu[u](x, T) = \mu_\Omega(x), \ \frac{\partial p_\mu[u](x, T)}{\partial t} = \mu_{\Omega 1}(x), \ x \in \Omega. \qquad (10.97)$$

Proof. 1. The equation (10.95) has the same form as (10.91). The equalities (10.97) can be transformed to the initial condition after the change of the variable $\tau = T - t$. Then for all $\mu \in Z$ the boundary of the problem (10.95)–(10.97) has a unique solution $p_\mu[u] \in Y$. Therefore, the equality (10.94) for $\mu_\Omega = 0$, $\mu_{\Omega 1} = 0$ in reality determines a linear continuous operator $y'[u] : V \to V$.

Consider the equality

$$\frac{\partial^2 \eta_\sigma[h]}{\partial t^2} - \Delta \eta_\sigma[h] + u \eta_\sigma[h] + h y[u + \sigma h] = 0,$$

where $\eta_\sigma[h] = (y[u + \sigma h] - y[u])/\sigma$. Multiply it by the function $p_\mu[u]$ and integrate the result by using the boundary conditions. We obtain

$$\int_Q \mu_Q \eta_\sigma[h]\,dQ + \int_\Omega \mu_\Omega \frac{\partial \eta_\sigma[h]}{\partial t}\Big|_{t=T} dx - \int_\Omega \mu_{\Omega 1} \eta_\sigma[h]\big|_{t=T} dx =$$

402   *Optimization and Differentiation*

$$-\int_Q hp_\mu[u]y[u + \sigma h]dQ.$$

Then from the condition (10.94) for $\mu_\Omega = 0$, $\mu_{\Omega 1} = 0$ it follows that

$$\int_Q \mu_Q\{(y[u + \sigma h] - y[u])/\sigma - y'[u]h\}dQ = \int_Q hp_\mu[u](y[u] - y[u + \sigma h])dQ.$$

Using the definition of the set of admissible control, we have $hp_\mu[u] \in L_2(Q)$. If $\sigma \to 0$, then we have the convergence $y[u + \sigma h] \to y[u]$ *-weakly in Y_1 because of Lemma 10.11. Therefore, after passing to the limit at the previous equality we have the convergence

$$(y[u + \sigma h] - y[u])/\sigma \to y'[u]h \text{ weakly in } L_2(Q)$$

for all $h \in V$. Thus, $y'[u]$ is the extended derivative of the operator $y[\cdot] : V \to Y$ at the point u. We find the proof by using the transformation of Lemma 10.8. $\square$

The state functional and the differential property of the operator $y[\cdot] : V \to Y$ for this optimization problem are the same as for Problem 10.4. Then we can prove the following result as the analog of Lemma 10.9 and Lemma 9.12.

Lemma 10.13 *The functional I for Problem 10.5 has Gâteaux derivative $I'(u) = \alpha u + yp$ at the arbitrary point $u \in V$, where $y = y[u]$, and p is the solution of the equation*

$$-\frac{\partial^2 p}{\partial t^2} - \Delta p + up = y[u] - y_d$$

with homogeneous boundary conditions.

Now we obtain the necessary condition of optimality.

Theorem 10.11 *The solution of Problem 10.5 with given set U is*

$$u(x,t) = \begin{cases} a, & \text{if } -p(x,t)/\alpha < a, \\ -p(x,t)/\alpha, & \text{if } a \le -p(x,t)/\alpha \le b, \\ b, & \text{if } -p(x,t)/\alpha > b. \end{cases}$$

Indeed, put the derivative of the functional to the standard variational inequality. We get

$$\int_Q (\alpha u + yp)(v - u)dQ \ge 0 \ \forall v \in U.$$

Its solution is determined by the standard transformations.

Second order evolutional systems

10.10 Optimization control problem for the nonlinear wave equation with coefficient control

Now we consider the nonlinear wave equation with coefficient control. Let us have the equation

$$\frac{\partial^2 y}{\partial t^2} - \Delta y + |y|^p y + vy = f \tag{10.98}$$

in the set of Q with the boundary conditions

$$y(x,t) = 0 \ (x,t) \in \Sigma, \tag{10.99}$$

$$y(x,0) = \varphi(x), \ \frac{\partial y(x,0)}{\partial t} x \in \Omega, \tag{10.100}$$

where $\rho > 0$, $\varphi \in H_0^1(\Omega) \cap L_q(\Omega)$, $\psi \in L_2(\Omega)$, $f \in L_2(Q)$. Determine the control space $V = L_2(Q)$ and the set of admissible controls

$$U = \Big\{ v \in V \big| \ a \le v(x) \le b, \ x \in \Omega \Big\},$$

where a and b are constants, as $b > a$.

Multiply the equality (10.98) by the time derivative of the function y and integrate the result. We get

$$\int\limits_{\Omega} \left(\frac{\partial^2 y}{\partial t^2} - \Delta y + |y|^p y \right) \frac{\partial y}{\partial t} dx = \int\limits_{\Omega} (f - vy) \frac{\partial y}{\partial t} dx.$$

Using the known transformations, we obtain the inequality

$$\frac{d}{dt} \left[\frac{1}{2} \|y'(t)\|_2^2 + \frac{1}{2} \|y(t)\|^2 + \frac{1}{q} \|y(t)\|_q^q \right] \le \frac{1+\delta}{2} \|y'(t)\|_2^2 + \frac{\delta}{2} \|y(t)\|_2^2 +$$

$$\frac{1}{2} \|f(t)\|_2^2 \le c_1 \left[\frac{1}{2} \|y'(t)\|_2^2 + \frac{1}{2} \|y(t)\|^2 \right] + \frac{1}{2} \|f(t)\|_2^2,$$

where $\delta = \max \big\{ |a|, |b| \big\}$, $c_1 > 0$. Integrating this formula zero to a number $\tau \in (0, T)$ we obtain

$$\left\| \frac{\partial y(\tau)}{\partial t} \right\|_2^2 + \|y(\tau)\|^2 + \frac{2}{q} \|y(\tau)\|_q^q \le$$

$$\|\psi\|_2^2 + \|\varphi\|^2 + \frac{2}{q} \|\varphi\|_q^q + \|f\|_{L_2(Q)}^2 + c_1 \int\limits_0^\tau \left[\left\| \frac{\partial y(t)}{\partial t} \right\|_2^2 + \|y(t)\|^2 \right] dt.$$

404   *Optimization and Differentiation*

Using the Gronwall lemma, we have a priori estimate of the solution of the boundary problem in the space

$$Y_1 = \left\{ y \,\middle|\, y \in L_\infty(0, T; H_0^1(\Omega) \cap L_q(Q)), \; y' \in L_\infty(0, T; H^{-1}(\Omega)) \right\}.$$

Find the second time derivative of the function y from the equality (10.98). It belongs to the space

$$W = L_\infty(0, T; H^{-1}(\Omega)) + L_{q'}(Q).$$

Then the function y is the element of the space

$$Y = \left\{ y \,\middle|\, y \in Y_1, \; y'' \in W \right\}.$$

Thus, the problem (10.98)–(10.100) is the analog of the boundary problem (10.55)–(10.57) for the hyperbolic equation with power nonlinearity, i.e., the presence of the fourth term with control at the equality (10.98) is not important for the properties of the boundary problem. Using the proofs of Theorem 10.6 and Theorem 10.7, we determine that for all $v \in V$ the problem (10.98)–(10.100) is solvable in the space Y, as its solution is unique under the condition (10.64).

Determine the functional

$$I(v) = \frac{\alpha}{2} \int_Q v^2 dQ + \frac{1}{2} \int_Q \left(y[v] - y_d \right)^2 dQ,$$

where $\alpha > 0$, $y_d \in L_2(Q)$ is a known function.

Problem 10.6 *Find the control $u \in U$ for the system (10.98)–(10.100) that minimizes the functional I on the set U.*

Lemma 10.14 *Under the condition (10.64) the operator $y[\cdot] : V \to Y$ for the problem (10.98)–(10.100) is $*$-weakly continuous.*

Proof. Suppose we have the convergence $v_k \to v$ weakly in V. Using the standard technique, we determine the boundedness of the sequence $\{y[v_k]\}$ in the space Y_1. After extracting a subsequence we get the convergence $y[v_k] \to y$ $*$-weakly in Y_1. Using compact embedding of the space Y_1 to $L_2(Q)$, we have the convergence $y[v_k] \to y$ strongly in $L_2(Q)$ and a.e. on Q. Repeating the transformations from Lemma 10.11, we get for all $\lambda \in D(Q)$ the convergence

$$\int_Q v_k y[v_k] \lambda dQ \to \int_Q v y \lambda dQ,$$

$$\int_Q \frac{\partial^2 y[v_k]}{\partial t^2} \lambda dQ \to \int_Q \frac{\partial^2 y}{\partial t^2} \lambda dQ.$$

Using the technique from the proof of Theorem 10.6, we obtain $|y[v_k]|^p y[v_k] \to |y|^p y$ weakly in $L_{q'}(Q)$. Multiply the equality (10.98) for $v = v_k$ by the arbitrary function $\lambda \in D(Q)$. After passing to the limit we have $y = y[v]$. $\square$

Second order evolutional systems

Remark 10.18 The condition (10.64) guarantees the uniqueness of the solution of the boundary problem. Therefore, we can determine the convergence of the whole sequence without extraction of its subsequence.

Using Lemma 10.14, we prove the existence of the optimal control for the considered optimization problem.

Theorem 10.12 *Under the condition* (10.64) *Problem* 10.6 *is solvable.*

Remark 10.19 In principle we can prove the solvability of Problem 10.6 without condition (10.64).

Try to prove the differentiability of the solution of the boundary problem with respect to the control. Suppose u is an arbitrary admissible control. Consider the equation

$$\frac{\partial^2 p_\mu[u]}{\partial t^2} - \Delta p_\mu[u] + (\rho + 1)|y[u]|^\rho p_\mu[u] + u p_\mu[u] = \mu_Q \tag{10.101}$$

in the set Q with boundary conditions

$$p_\mu[u](x, t) = 0, \ (x, t) \in \Sigma, \tag{10.102}$$

$$p_\mu[u](x, T) = \mu_\Omega(x), \ \frac{\partial p_\mu[u](x, T)}{\partial t} = \mu_{\Omega 1}(x), \ x \in \Omega, \tag{10.103}$$

where $\mu = (\mu_Q, \mu_\Omega, \mu_{\Omega 1})$. Determine the space

$$Z = L_2(Q) \times H_0^1(\Omega) \times L_2(\Omega).$$

Lemma 10.15 *Under the condition* (10.64) *the operator* $y[\cdot] : V \to Y$ *for the problem* (10.101)–(10.103) *at the arbitrary point* $u \in V$ *has* $(V, V; V, V^w)$-*extended derivative* $y'[u]$ *as there exists the linear continuous operator* $y_T'[u] : V \to L_2(\Omega)$ *and* $y_{T1}'[u] : V \to H^{-1}(\Omega)$ *such that for all* $h \in V$ *we have the convergence*

$$\{y[u + \sigma h](T) - y[u](T)\}/\sigma \to y_T'[u]h \ \text{weakly in} \ L_2(\Omega).$$

$$\frac{1}{\sigma} \left\{ \frac{\partial y[u + \sigma h](T)}{\partial t} - \frac{\partial y[u](T)}{\partial t} \right\} \to y_{T1}'[u]h \ \text{weakly in} \ H^{-1}(\Omega).$$

In addition, we have the equality

$$\int_Q \mu_Q y'[u]h \, dQ + \int_\Omega \mu_\Omega \frac{\partial y'[u]h}{\partial t} \bigg|_{t=T} dQ - \int_\Omega \mu_{\Omega 1} y'[u]h \big|_{t=T} dQ =$$

$$- \int_Q h p_\mu[u] \, dQ \ \forall h \in V, \ \mu \in Z. \tag{10.104}$$

406   *Optimization and Differentiation*

Proof. 1. Under the condition (10.64) the considered boundary problem has a unique solution. For all functions $h \in V$ and the number σ we have the equality

$$\frac{\partial^2 \eta_\sigma[h]}{\partial t^2} - \Delta \eta_\sigma[h] + \left(g_\sigma[h]\right)^2 \eta_\sigma[h] + u \eta_\sigma[h] + hy[u + \sigma h] = 0, \qquad (10.105)$$

where

$$\eta_\sigma[h] = (y[u + \sigma h)] - y[u])/\sigma,$$

$$\left(g_\sigma[h]\right)^2 = (\rho + 1)\big|y[u] + \varepsilon \eta_\sigma[h]\big|^\rho, \ \ \varepsilon \in [0,1].$$

Consider the equation

$$\frac{\partial^2 p}{\partial t^2} - \Delta p + \left(g_\sigma[v]\right)^2 p + up = \mu_Q \qquad (10.106)$$

in the set Q with boundary conditions

$$p(x,t) = 0, \ (x,t) \in \Sigma, \qquad (10.107)$$

$$p(x,T) = \mu_\Omega(x), \ \frac{\partial p(x,T)}{\partial t} = \mu_{\Omega 1}(x), \ x \in \Omega. \qquad (10.108)$$

We have the system (10.101)–(10.103) here for $\sigma = 0$. Multiply the equality (10.106) by the time derivative of the function y and integrate the result. We obtain

$$\int\limits_\tau^T \int\limits_\Omega \left(\frac{\partial^2 p}{\partial t^2} - \Delta p\right) \frac{\partial p}{\partial t} dx dt =$$

$$- \int\limits_\tau^T \int\limits_\Omega \left(g_\sigma[h]\right)^2 p \frac{\partial p}{\partial t} dx dt - \int\limits_\tau^T \int\limits_\Omega up \frac{\partial p}{\partial t} dx dt + \int\limits_\tau^T \int\limits_\Omega \mu_Q \frac{\partial p}{\partial t} dx dt. \qquad (10.109)$$

Using the transformations from the proof of Lemma 10.8, we get

$$\int\limits_\tau^T \int\limits_\Omega \left(\frac{\partial^2 p}{\partial t^2} - \Delta p\right) \frac{\partial p}{\partial t} dx dt = \frac{1}{2}\left(\|\mu_{\Omega 1}\|_2^2 + \|\mu_\Omega\|^2\right) - \frac{1}{2}\left[\left\|\frac{\partial p(\tau)}{\partial t}\right\|_2^2 + \|p(\tau)\|^2\right],$$

$$\left|\int\limits_\tau^T \int\limits_\Omega \left(g_\sigma[h]\right)^2 p \frac{\partial p}{\partial t} dx dt\right| \leq c_1 \int\limits_\tau^T \|p(t)\| \left\|\frac{\partial p(t)}{\partial t}\right\|_2 dt,$$

where $c_1 > 0$. Using continuous embedding of the space $H_0^1(\Omega)$ to $L_2(\Omega)$ we have the inequality

$$\left|\int\limits_\tau^T \int\limits_\Omega up \frac{\partial p}{\partial t} dx dt\right| \leq c_2 \int\limits_\tau^T \|p(t)\| \left\|\frac{\partial p(t)}{\partial t}\right\|_2 dt,$$

where $c_2 > 0$. Therefore, from (10.109) it follows that

$$\left\|\frac{\partial p(\tau)}{\partial t}\right\|_2^2 + \|p(\tau)\|^2 \leq c_3 \int_\tau^T \left[\left\|\frac{\partial p(t)}{\partial t}\right\|_2^2 + \|p(t)\|^2\right]dt +$$

$$\|\mu_{\Omega 1}\|_2^2 + \|\mu_\Omega\|^2 + c_4 \int_\tau^T \|\mu_Q(t)\|_2^2 dt,$$

where $c_3 > 0$, $c_4 > 0$. Using the Gronwall lemma, we get

$$\left\|\frac{\partial p(\tau)}{\partial t}\right\|_2^2 + \|p(\tau)\|^2 \leq c_5\left(\|\mu_{\Omega 1}\|_2^2 + \|\mu_\Omega\|^2 + \|\mu_Q\|_{L_2(Q)}^2\right),$$

where $c_5 > 0$. Thus, the boundary problem (10.106)–(10.108) for the linear hyperbolic equation has a priori estimate of the solution in the space Y_1. Then this problem has a unique solution $p = p_\mu^\sigma[h]$ from the set Y_1 for all numbers σ and functions $h \in V$, $\mu \in Z$. This is the analog of Theorem 10.1. Find the second time derivative of the function p from the equality (10.106). Then function p belongs to the space Y.

Multiplying the equality (10.105) by the function $p_\mu^\sigma[h]$ and integrating the result by using the boundary condition, we get

$$\int_Q \mu_Q \eta_\sigma[h]dQ + \int_\Omega \mu_\Omega \frac{\partial \eta_\sigma[h]}{\partial t}\Big|_{t=T}dx - \int_\Omega \mu_{\Omega 1}\eta_\sigma[h]\big|_{t=T}dx =$$

$$- \int_Q hp_\mu^\sigma[h]y[u + \sigma h]dQ.$$

By the condition $p_\mu[u] \in Y_1$, for $\mu_\Omega = 0$, $\mu_{\Omega 1} = 0$ the equality (10.104) determines a linear continuous operator $y'[u] : L_2(Q) \to L_2(Q)$. Then from this equality and the previous formula we obtain

$$\int_Q \mu_Q\big(\eta_\sigma[h] - y'[u]h\big)dQ = \int_Q h\big(y[u]p_\mu[u] - p_\mu^\sigma[h]y[u + \sigma h]dQ\big)dQ =$$

$$\int_Q hy[u]\big(p_\mu[u] - p_\mu^\sigma[h]\big)dQ + \int_Q h\big(y[u] - y[u + \sigma h]\big)p_\mu^\sigma[h]dQ. \qquad (10.110)$$

Suppose we have the convergence $\sigma \to 0$. From Lemma 10.14 the convergence $y[u+\sigma h] \to y[u]$ $*$-weakly in Y_1 and strongly in $L_2(Q)$ follows. Using the known technique, determine the boundedness of the set $\{p_\mu^\sigma[h]\}$ in the space Y_1. By definition of U the set $\{hp_\mu^\sigma[h]\}$ is bounded in $L_2(Q)$. Then we have the convergence

$$\int_Q h\big(y[u] - y[u + \sigma h]\big)p_\mu^\sigma[h]dQ \to 0.$$

408 *Optimization and Differentiation*

In addition, after extracting a subsequence we get the convergence $p_\mu^\sigma[h] \to p_\mu$ *-weakly in Y_1 and weakly in $H^1(Q)$. Using the Rellich–Kondrashov theorem, we have the convergence $p_\mu^\sigma[v] \to p_\mu$ and $y[u + \sigma h] \to y[u]$ strongly in $L_2(Q)$ and a.e. on Q. Then $\left(g_\sigma[h]\right)^2 p_\mu^\sigma[h] \to (\rho + 1)|y[u]|^\rho p_\mu$ a.e. on Q. By the boundedness of the set $\{y[u + \sigma h]\}$ in the space $L_q(Q)$ we have the boundedness of $\left\{\left(g_\sigma[h]\right)\right\}$ in $L_{q/\rho}(Q)$. Under the condition (10.64) we have continuous embedding $Y_1 \subset L_q(Q)$. Then the set $\{p_\mu^\sigma[h]\}$ is bounded in the space $L_q(Q)$, and the set $\left\{\left(g_\sigma[h]\right)^2 p_\mu^\sigma[h]\right\}$ is bounded in $L_{q'}(Q)$. Therefore, $\left(g_\sigma[h]\right)^2 p_\mu^\sigma[h] \to (\rho + 1)|y[u]|^\rho p_\mu$ weakly in $L_{q'}(Q)$. Pass to the limit at the equality (10.106) for $p = p_\mu^\sigma[h]$ and $\sigma \to 0$. Then $p_\mu = p_\mu[u]$. Hence, $p_\mu^\sigma[h] \to p_\mu[u]$ strongly in $L_2(Q)$. Using the condition $hy[u] \in L_2(Q)$, we get

$$\int_Q hy[u]\left(p_\mu[u] - p_\mu^\sigma[h]\right)dQ \to 0.$$

Passing to the limit at the equality (10.110), we have $\left(y[u + \sigma h] - y[u]\right)/\sigma \to y'[u]h$ weakly in $L_2(Q)$ for all $h \in L_2(Q)$. We can finish the proof by using the standard technique. $\square$

We have the following result as the analog of Lemma 10.13.

Lemma 10.16 *Under the condition* (10.64) *the functional I for Problem* 10.6 *has Gâteaux derivative* $I'(u) = \alpha u + py$ *at the arbitrary point* $u \in V$, *where* p *is the solution of the equation*

$$\frac{\partial^2 p}{\partial t^2} - \Delta p + (\rho + 1)|y|^\rho + up = y[u] - y_d.$$

Determine the necessary condition of optimality that is the analog of Theorem 10.10.

Theorem 10.13 *Under the condition* (10.64) *the solution of Problem* 10.6 *with given set U is*

$$u(x, t) = \begin{cases} a, & \text{if } -p(x, t)/\alpha < a, \\ -p(x, t)/\alpha, & \text{if } a \le -p(x, t)/\alpha \le b, \\ b, & \text{if } -p(x, t)/\alpha > b. \end{cases}$$

Remark 10.20 In reality the properties of Problem 10.6 are the composition of the properties of Problem 10.4 for the nonlinear wave equation with the control at the absolute term and Problem 10.5 for the linear wave equation with the coefficient control.

10.11 Comments

The theory of the abstract second order linear evolutional equations including Theorem 10.1 is considered in the book of J.L. Lions [315]. H. Gajewski, K. Gröger, and K. Zacharias [200] describe the abstract second order nonlinear evolutional equations. The Gronwall lemma is considered there too. The linear hyperbolic equations are analyzed, for example, by J. Leray [309] and S.L. Sobolev [507], and the nonlinear hyperbolic equations are described by J.L. Lions [315].

The optimal control problems for the linear second order evolutional equations are considered by J.L. Lions [315]. Lemma 10.5 and the determination of the Riccati equation are given by J.L. Lions [315]. The different optimization problems for the linear hyperbolic equations are considered, for example, by N.U. Ahmed [8], A.G. Butkovsky [101], [102], H. Knöpp [271], V. Krabs [281], J.L. Lions [315], J.M. Sloss, I.S. Sadek, Jr., J.C. Bruch, and S. Adali [505], V.M. Tihomirov [522], L.V. Petuhov and V.A. Troutsky [402], F.P. Vasiliev [539], and others. A.G. Butkovsky [102] and K.D. Graham and D.L. Russell [221] apply moments method for these problems. M.T. Dzhenaliev [150] obtains the sufficient conditions of optimality. The extremum problems for the linear characteristic boundary problems are considered by A.G. Butkovsky [101], [102] and G. Emmanuele and A. Villani [165]. B.S. Mordukhovich and J.-P. Raymond [368] consider the system with state constraints. I. Lasiecka and A. Tuffaha [1] analyze the synthesis problem. T.K. Sirazetdinov [503] considers the stochastic systems. The optimization problems for four order linear hyperbolic equations are considered by Jr.J.C. Bruch, S. Adali, J.M. Sloss, and I.S. Sadek [94], A.G. Butkovsky [101], and J.L. Lions [315]. L. Wolfersdorf [554] solves an optimization problem for the linear first order hyperbolic equations and systems.

J. Ha, S. Nakagiri, and H. Tanabe [232] prove the differentiabily of the solution of the nonlinear hyperbolic equation with respect to the initial data. Optimal control problems for the nonlinear hyperbolic equations are considered by A.V. Fursikov [197], J. Ha and S. Nakagiri [231], A.S. Matveev and V.A. Yakubovich [343], S.Ya. Serovajsky [464], [488], [489], [494], D. Tiba [520], [518], J. Yong [562], and others. A. Kowalewsky [280] considers the system with delay. D. Tiba [518] solves the non-smooth extremum problem. S.Ya. Serovajsky [489] considers the system with non-unique solution. A.V. Fursikov [197], J.L. Lions [317], and D. Tiba [518] analyze the singular equations. S.Ya. Serovajsky [494] and D. Tiba [518] consider regularization methods. J.L. Lions [317] and D. Tiba [518] consider the penalty method.

Optimization problems for the systems described by Goursat problems for nonlinear hyperbolic equations are considered by P. Burgmeier and H. Jasinski [99], A.I. Egorov [155], V.S. Gavrilov and M.I. Sumin [204], K.A. Lurie [333], N.I. Pogodaev [405], M.M. Potapov [410], M.E. Salukvadze and M.T. Tcutcunava [445], T.K. Sirazetdinov [503], M.B. Suryanarayana [514], F.P. Vasiliev [539], O.V. Vasiliev [540], Z.S. Wu and K.L. Teo [556], and others. Particularly, P. Burgmeier and H. Jasinski [99], M.M. Potapov [410], Z.S. Wu and K.L. Teo [556] determine the approximate solution of these problems. O.V. Vasiliev [540] analyzes the singular control. M.E. Salukvadze and M.T. Tcutcunava [445] consider the systems with delay. V.S. Gavrilov and M.I. Sumin [204] analyze the system with state constraints.

The optimization problems for the nonlinear first order hyperbolic equations are considered by M. Brokate [93], K.A. Lurie [333], T.K. Sirazetdinov [503], and others.

Coefficient control problems for the hyperbolic equations are considered by N.U. Ahmed [8], J. Baranger [47], S.I. Kabanihin [265], J.L. Lions [315], A. Lorenzi [325], and M.B. Suryanarayana [514].

Chapter 11

Navier–Stokes equations

11.1	Evolutional Navier–Stokes equations	411
11.2	Optimization control problems for the evolutional Navier–Stokes equations	419
11.3	Stationary Navier–Stokes equations	430
11.4	Optimization control problems for the stationary Navier–Stokes equations	434
11.5	System of Navier–Stokes and heat equations	440
11.6	Optimization control problems for Navier–Stokes and heat equations	448
11.7	Comments	456

The optimization control problems for the systems described by Navier–Stokes equations are the subject of the final chapter of Part III. We consider Navier–Stokes equations for the stationary and evolutional cases and with heat equations. At first, we analyze the properties of the considered boundary problems. Then we prove the existence of the optimal controls. Finally, we determine the necessary conditions of optimality by using the extended differentiability of control–state mapping.

11.1 Evolutional Navier–Stokes equations

Let Ω be an open bounded set on the plane with the boundary Γ, $T > 0$, $Q = \Omega \times (0,T)$, $\Sigma = \Gamma \times (0,T)$. Consider **Navier–Stokes equations** in the set Q

$$\frac{\partial y^1}{\partial t} - \nu \Delta y^1 + \sum_{i=1}^{2} y^i \frac{\partial y^1}{\partial x_i} + \frac{\partial q}{\partial x_1} = f^1,$$

$$\frac{\partial y^2}{\partial t} - \nu \Delta y^2 + \sum_{i=1}^{2} y^i \frac{\partial y^2}{\partial x_i} + \frac{\partial q}{\partial x_2} = f^2,$$

$$\sum_{i=1}^{2} \frac{\partial y^i}{\partial x_i} = 0,$$

411

412 *Optimization and Differentiation*

where $\nu > 0$. Determine

$$y = (y^1, y^2), \ f = (f^1, f^2), \ y' = \frac{\partial y}{\partial t} = \left(\frac{\partial y^1}{\partial t}, \frac{\partial y^2}{\partial t}\right), \ D_i = \frac{\partial}{\partial x_i}, \ \nabla = (D_1, D_2)$$

and the operator

$$\operatorname{div} y = \sum_{i=1}^{2} D_i y^i$$

that is called the ***divergence***. Then we obtain the following equations

$$y' - \nu \Delta y + \sum_{i=1}^{2} y^i D_i y + \nabla q = f, \tag{11.1}$$

$$\operatorname{div} y = 0 \tag{11.2}$$

in the set Q. Consider also the boundary conditions

$$y(x, t) = 0 \ (x, t) \in \Sigma, \tag{11.3}$$

$$y(x, 0) = \varphi(x), \ x \in \Omega, \tag{11.4}$$

where $\varphi = (\varphi^1, \varphi^2)$.

Remark 11.1 Navier–Stokes equations describe the movement of the incompressible viscous fluid. The vector y is the velocity here, q is the pressure, and ν is viscous friction. The equalities (11.1) are called the ***motion equations***, and (11.2) is called the ***continuity equation***.

Determine the set

$$D = \left\{ y \in \left[D(\Omega)\right]^2 \middle| \operatorname{div} y = 0 \right\}.$$

Denote by H the closure of the set D in the space $\left[L_2(\Omega)\right]^2$. Let W be the closure of D in $\left[H_0^1(\Omega)\right]^2$. The functions of $D(\Omega)$ are equal to zero in the boundary Γ. Then these properties are true for the elements of the sets H and W. Thus, we have the equality

$$W = \left\{ y \in \left[H_0^1(\Omega)\right]^2 \middle| \operatorname{div} y = 0 \right\}.$$

Determine the scalar products

$$(\psi, \eta)_H = \sum_{i=1}^{2} \int_{\Omega} \psi^i \eta^i dx \ \forall \psi, \eta \in H,$$

$$(\psi, \eta)_W = \sum_{i,j=1}^{2} \int_{\Omega} D_j \psi^i D_j \eta^i dx \ \forall \psi, \eta \in W.$$

Navier–Stokes equations

The spaces H and W are Hilbert, as embedding of W of H is continuous and dense. Using Riesz theorem, we can identify H with its adjoint space W'. We have continuous and dense embedding $W \subset H \subset W'$. The considered spaces are separable.

We use a special basis for proving the solvability of the boundary problem. Consider the equality

$$(\eta, \mu)_W = \lambda(\eta, \mu)_H \ \forall \eta \in W. \tag{11.5}$$

For any number λ the function $\mu = 0$ satisfies this equality. However, maybe there exists a non-zero solution of this problem for a concrete λ.

Remark 11.2 This is the *spectrum problem*. The number λ such that there exists a non-zero solution of (11.5) is called the *eigenvalue*, and the corresponding solution is called the *eigenfunction* of the operator of this problem.

Lemma 11.1 *There exists a sequence of positive eigenvalues $\{\lambda_j\}$ for the problem (11.5) and the corresponding sequence $\{\mu_j\}$ of the eigenfunctions that is the basis of the space W.*

Thus, the functions μ_j satisfy the equalities

$$(\eta, \mu_j)_W = \lambda_j(\eta, \mu_l)_H \ \forall \eta \in W. \tag{11.6}$$

Determine the functionals

$$a(\psi, \eta) = \sum_{i,j=1}^{2} \int_{\Omega} D_i \psi^j D_i \eta^j \, dx \ \forall \psi, \eta \in W,$$

$$b(\psi, \eta, \zeta) = \sum_{i,j=1}^{2} \int_{\Omega} \psi^i D_i \eta^j \zeta^j \, dx \ \forall \psi, \eta, \zeta \in W.$$

We have the equalities

$$a(\psi, \psi) = \|\psi\|_W^2 \ \forall \psi \in W, \tag{11.7}$$

$$b(\psi, \psi, \eta) = -b(\psi, \eta, \psi); \ b(\psi, \eta, \eta) = 0 \ \forall \psi, \eta \in W. \tag{11.8}$$

Determine the linear continuous operator $A : W \to W'$ and the nonlinear operator $B : W \to W'$ by the equalities

$$a(\psi, \eta) = \langle A\psi, \eta \rangle, \ b(\psi, \psi, \eta) = \langle B\psi, \eta \rangle \ \forall \psi, \eta \in W.$$

The following inequality holds:

$$\|B\psi\|_{W'} \le c\|\psi\|_W^2, \tag{11.9}$$

where c is a positive constant. Therefore, the operator B is continuous.

Note the following property by using our denotations of the norms of the concrete spaces.

414   *Optimization and Differentiation*

Lemma 11.2 *For $n = 2$ there exists a positive constant $c = c(\Omega)$ such that*

$$\|\psi\|_4 \le \|\psi\|_2 \|\psi\| \; \forall \psi \in H_0^1(\Omega).$$

Determine the spaces

$$X = L_2(0, T; W), \; X' = L_2(0, T; W'), \; Y = \Big\{ y \big| \; y \in X, \; y' \in X' \Big\}.$$

Theorem 11.1 *For any $f \in X'$, $\varphi \in H$ the problem (11.1)–(11.4) has a unique solution $y \in Y$, $q \in D'(Q)$.*

Proof. 1. Transform our boundary problem. Consider an element $\eta = (\eta^1, \eta^2)$ of the space W. Suppose the functions $y \in Y$, $q \in D'(Q)$ satisfy the equalities (11.1)–(11.4). Multiply the first equality (11.1) by η^1 and the second equality multiply by η^2. After the addition and the integration of the result we get

$$\sum_{j=1}^{2} \int_{\Omega} (y')^j \eta^j \, dx + \nu \sum_{i,j=1}^{2} \int_{\Omega} D_i y^j D_i \eta^j \, dx + \sum_{i,j=1}^{2} \int_{\Omega} y^i D_i y^j \eta^j \, dx +$$

$$\sum_{j=1}^{2} \int_{\Omega} D_j q \eta^j \, dx = \sum_{j=1}^{2} \int_{\Omega} f^j \eta^j \, dx \; \forall \eta \in W. \qquad (11.10)$$

Using Green's formula, the boundary condition and the equality $\operatorname{div} \eta = 0$ we have

$$\sum_{j=1}^{2} \int_{\Omega} D_j q \eta^j \, dx = -\int_{\Omega} q \operatorname{div} \eta \, dx = 0 \; \forall \eta \in W. \qquad (11.11)$$

By the definition of the functionals a and b, the equality (11.7) can be transformed to

$$\langle y'(t), \eta \rangle + \nu a \big(y(t), \eta \big) + b \big(y(t), y(t) \eta \big) = \langle f(t), \eta \rangle \; \forall \eta \in W. \qquad (11.12)$$

Thus, we transform the problem (11.1)–(11.4) to the equation (11.12) with the initial condition (11.4).

Prove the inverse proposition. Let the function $y \in Y$ be the solution of the problem (11.12), (11.4). Then the function

$$z = y' - \nu \Delta y + \sum_{i=1}^{2} y^i D_i y + \nabla q - f$$

belongs to the set $\big[D'(\Omega) \big]^2$. This is the system of two equalities. Multiply the first of the equalities by η^1 and the second of them by η^2. After summing and the integration we have

$$\langle z(t), \eta \rangle = \langle y'(t), \eta \rangle + \nu a \big(y(t), \eta \big) + b \big(y(t), y(t) \eta \big) - \langle f(t), \eta \rangle = 0 \; \forall \eta \in W$$

Navier–Stokes equations

because of (11.12). Thus, we obtain the equality

$$\sum_{j=1}^{2} \int_{\Omega} z^j \eta^j \, dx = 0 \quad \forall \eta \in W.$$

Compare the result with (11.11). We can determine $z^j = -D_j q$. Thus, there exists $q \in D'(\Omega)$ such that the equality (11.1) is true. The conditions (11.2), (11.3) are the corollaries of the definition of the space Y.

Thus, the problems (11.1)–(11.4) and (11.12), (11.4) are equivalent. However, the second problem does not contain the function q. Hence, this problem is easier for the analysis. The theorem will be proved if there exists a unique function that satisfies the equalities (11.12), (11.4).

2. We prove the solvability of the considered problem by the Faedo–Galerkin method with the basis $\{\mu_m\}$ that is determined by the equalities (11.6). Determine the approximate solution $y_k = y_k(t)$ of the problem (11.12), (11.4) by the equality

$$y_k(t) = \sum_{m=1}^{k} \xi_{mk}(t)\mu_m, \quad k = 1, 2, ..., \tag{11.13}$$

where the functions $\xi_{1k}, ..., \xi_{kk}$ satisfy the equalities

$$\langle y_k'(t), \mu_j \rangle + \nu a \big(y_k(t), \mu_j\big) + b\big(y_k(t), y_k(t), \mu_j\big) = \langle f(t), \mu_j \rangle, \tag{11.14}$$

where $j = 1, ..., k$. This is the system of the nonlinear ordinary differential equations.

For all functions φ of H there exists a sequence $\{\alpha_k\}$ such that

$$\varphi = \sum_{m=1}^{\infty} \alpha_m \mu_m.$$

Determine the sum

$$\varphi_k = \sum_{m=1}^{k} \alpha_m \mu_m.$$

We obtain the convergence

$$\varphi_k \to \varphi \text{ in } H. \tag{11.15}$$

Determine the initial values for the functions ξ_{mk} by the equality $\xi_{mk}(0) = \alpha_m$, $m = 1, 2, ...$ Therefore, we have

$$y_k(0) = \varphi_k. \tag{11.16}$$

This is the initial condition for the system (11.14). The corresponding Cauchy problem has a solution on an interval $(0, T_k)$.

416   *Optimization and Differentiation*

3. Determine the a priori estimates of the solutions of the problem (11.14), (11.16). Multiply the j-th equality (11.14) by ξ_{jk} and sum the result by $j = 1, ..., k$. Using the formula (11.13), we get

$$\langle y_k'(t), y_k(t) \rangle + \nu a\big(y_k(t), y_k(t)\big) + b\big(y_k(t), y_k(t), y_k(t)\big) = \langle f(t), y_k(t) \rangle.$$

We have

$$\langle y_k'(t), y_k(t) \rangle = \frac{1}{2} \frac{d}{dt} \big\|y_k(t)\big\|_H^2.$$

Using (11.7), (11.8), transform the previous equality

$$\frac{1}{2} \frac{d}{dt} \big\|y_k(t)\big\|_H^2 + \nu \big\|y_k(t)\big\|_W^2 = \langle f(t), y_k(t) \rangle.$$

Thus, we have

$$\frac{d}{dt} \big\|y_k(t)\big\|_H^2 + 2\nu \big\|y_k(t)\big\|_W^2 \le 2\|f(t)\|_{W'} \|y_k(t)\|_W \le \nu \|y_k(t)\|_W^2 + \nu^{-1} \|f(t)\|_{W'}^2.$$

Integrate this inequality by using (11.16)

$$\big\|y_k(t)\big\|_H^2 + \nu \int\limits_0^t \big\|y_k(\tau)\big\|_W^2 d\tau \le \big\|\varphi_k\big\|_H^2 + \frac{1}{\nu} \int\limits_0^t \|f(\tau)\|_{W'}^2 d\tau \le \frac{1}{\nu} \|f\|_{X'}^2.$$

$$(11.17)$$

The value at the right-hand side of this inequality does not depend upon t. Hence, we can choose $T_k = T$. Thus, from the inequality (11.17) it follows that the sequence $\{y_k\}$ is bounded in space $Z = X \cap L_\infty(0, T; H)$.

4. Find the additional a priori estimate of the solution of the problem (11.14), (11.16) by using the special property of the basis. Determine the operator P_k on the space H by the equality

$$P_k = \sum_{m=1}^k (\psi, \mu_m)_H \mu_m, \ k = 1, 2, \dots .$$

Note that P_k is the projection operator from the space H to the set M_k of the linear combinations of $\mu_1, ..., \mu_k$. Then we have the equality $P_k \psi = \psi$ for all $\psi \in M_k$. Using the property of the Fourier series, determine the equality

$$\big\|P_k \psi\big\|_H^2 = \sum_{m=1}^k \big|(\psi, \mu_m)_H\big|^2 \le \sum_{m=1}^\infty \big|(\psi, \mu_m)_H\big|^2 = \|\psi\|_H^2.$$

We have

$$\|P_k\| = \sup_{\|\psi\|_H = 1} \big\|P_k \psi\big\|_H \le \sup_{\|\psi\|_H = 1} \|\psi\|_H = 1.$$

We can interpret P_k as the operator from the space W to W. Then its adjoint operator maps W' to W' equal to P_k. Its norm of the is bounded by 1.

Navier–Stokes equations

Multiply j-th equality (11.14) by μ_j and sum by j by using the definition of the operators P_k, A, B and the inclusion $y'_k(t) \in M_k$. We have

$$y'_k = -\nu P_k A y_k - P_k B y_k + P_k f.$$

Using the boundedness of the sequence $\{y_k\}$ in the space X and the operators $A : W \to W'$, $B : W \to W'$, $P_k : W \to W'$, determine the boundedness of the sequences $\{P_k A y_k\}$ and $\{P_k B y_k\}$ in X'. From the last equality it follows that the sequence $\{y'_k\}$ is bounded in X'. Using the known estimate, determine the boundedness of the sequence $\{y_k\}$ in Y.

Using Lemma 11.2, for any functions ψ, η, we have the inequality

$$\|\psi\eta\|_2 \le \|\psi\|_4 \|\eta\|_4 \le c^2 \|\psi\|^{1/2} \|\psi\|_2^{1/2} \|\eta\|^{1/2} \|\eta\|_2^{1/2}.$$

Then for all $i, m = 1, 2$ we have

$$\left\| y_k^i(t) y_k^m(t) \right\|_2^2 \le c \|y_k^i(t)\| \|y_k^i(t)\|_2 \|y_k^m(t)\| \|y_k^m(t)\|_2 \le c_1 \|y_k^i(t)\| \|y_k^m(t)\|$$

because of the boundedness of the sequence $\{y_k^i\}$ in space $L_\infty(0, T; L_2(\Omega))$, where $c_1 > 0$. Now we obtain the inequality

$$\left\| y_k^i y_k^m \right\|_{L_2(Q)}^2 \le c_1 \int_0^T \|y_k^i(t)\| \|y_k^m(t)\| dt \le c_1 \|y_k^i\|_{L_2(0,T;H_0^1(\Omega))} \|y_k^m\|_{L_2(0,T;H_0^1(\Omega))}.$$

Thus, sequence $\{y_k^i y_k^m\}$ is bounded in the space $L_2(Q)$.

5. After extracting a subsequence determine the convergence $y_k \to y$ weakly in Y and $y_k^i y_k^m \to \psi^{im}$ weakly in $L_2(Q)$. Use Theorem 8.5 about compact embedding. Determine $Z_0 = W$, $Z = H$, $Z_1 = W'$, $r = s = 2$ there. Then embedding of the space Y to $L_2(0, T; H)$ is compact. Thus, $y_k \to y$ strongly in $L_2(0, T; H)$ and a.e. on Q. Then $y_k^i y_k^m \to y^i y^m$ a.e. on Q, $i, m = 1, 2$. Using the boundedness of sequence $\{y_k^i y_k^m\}$ in the space $L_2(Q)$ we have $y_k^i y_k^m \to y^i y^m$ weakly in $L_2(Q)$.

Multiply the equality (11.14) by a function $\omega \in C^1[0, T]$. After integration we have

$$\int_0^T \langle y'_k(t), \xi_j(t) \rangle dt + \nu \int_0^T a(y_k(t), \xi_j(t)) dt +$$

$$\int_0^T b(y_k(t), y_k(t), \xi_j(t)) dt = \int_0^T \langle f(t), \xi_j(t) \rangle dt, \tag{11.18}$$

where $\xi_j(t) = \omega\mu_j$, $j = 1, ..., k$. Using the convergence $y'_k \to y'$ weakly in X' we have

$$\lim_{k \to \infty} \int_0^T \langle y'_k(t), \xi_j(t) \rangle dt = \int_0^T \langle y'(t), \xi_j(t) \rangle.$$

418   *Optimization and Differentiation*

Using the convergence $y_k \to y$ weakly in X, we obtain

$$\lim_{k \to \infty} \int_0^T a\big(y_k(t), \xi_j(t)\big)\, dt = \int_0^T a\big(y(t), \xi_j(t)\big)\, dt.$$

Using the first equality (11.8), we get

$$\int_0^T b\big(y_k(t), y_k(t), \xi_j(t)\big)\, dt = -\int_0^T b\big(y_k(t), \xi_j(t), y_k(t)\big)\, dt =$$

$$\sum_{i,m=1}^2 \int_Q y_k^i y_k^m D_i \xi_j^m\, dQ.$$

From the convergence $y_k^i y_k^m \to y^i y^m$ weakly in $L_2(Q)$ it follows that

$$\lim_{k \to \infty} \int_0^T b\big(y_k(t), y_k(t), \xi_j(t)\big)\, dt = \lim_{k \to \infty} \sum_{i,m=1}^2 \int_Q y_k^i y_k^m D_i \xi_j^m\, dQ =$$

$$\sum_{i,m=1}^2 \int_Q y^i y^m D_i \xi_j^m\, dQ = \int_0^T b\big(y(t), y(t), \xi_j(t)\big)\, dt.$$

Passing to the limit in the equality (11.18), we have

$$\int_0^T \big\langle y'(t), \xi_j(t)\big\rangle\, dt + \nu \int_0^T a\big(y(t), \xi_j(t)\big)\, dt + \int_0^T b\big(y(t), y(t), \xi_j(t)\big)\, dt =$$

$$\int_0^T \big\langle f(t), \xi_j(t)\big\rangle\, dt, \ j = 1, 2, \dots .$$

The function ω is arbitrary here. Then we have

$$\big\langle y'(t), \mu_j(t)\big\rangle + a\big(y(t), \mu_j(t)\big) + b\big(y(t), y(t), \mu_j(t)\big) = \big\langle f(t), \mu_j(t)\big\rangle, \ j = 1, 2, \dots .$$

Now we obtain the equality (11.12) because $\{\mu_j\}$ is the basis of the space W.

Embedding of the space Y in $C(0, T; H)$ is continuous. Then we have the convergence $y_k(0) \to y(0)$ weakly in H. From (11.15), (11.16) the initial condition (11.4) follows. Thus, the function $y \in Y$ satisfies the equalities (11.12), (11.4). Then the problem (11.1)–(11.4) is solvable.

6. Prove the uniqueness of the solution of the considered boundary problem. Suppose there exist solutions y_1 and y_2 of the problem (11.1)–(11.4). Denote $y = y_1 - y_2$. Then we have for all $\eta \in W$

$$\big\langle y'(t), \eta\big\rangle + \nu a\big(y(t), \eta\big) + b\big(y(t), y_1(t), \eta\big) + b\big(y_2(t), y(t), \eta\big) - b\big(y(t), y(t), \eta\big) = 0.$$

Navier–Stokes equations

Determine $\eta = y(t)$. Using the second equality (11.8), we get

$$\langle y'(t), y(t)\rangle + a\big(y(t), y(t)\big) + b\big(y(t), y_1(t), y(t)\big) = 0.$$

Integrating by using the equality $y(0) = 0$, we have

$$\frac{1}{2}\|y(t)\|_H^2 + \nu \int_0^t \|y(\tau)\|_W^2 d\tau = -\int_0^t b\big(y(\tau), y_1(\tau), y(\tau)\big) d\tau. \qquad (11.19)$$

Consider the value

$$\left|\int_0^t b\big(y(\tau), y_1(\tau), y(\tau)\big) d\tau\right| = \left|\sum_{i,j=1}^2 \int_0^t \int_\Omega y^i(\tau) y^j(\tau) D_i y_1^j(\tau) dx d\tau\right| \leq$$

$$c_2 \int_0^t \|y^i(\tau)\|_{[L_4(\Omega)]^2}^2 \|y_1(\tau)\|_W d\tau,$$

where $c_2 > 0$. Using Lemma 11.2, we have

$$\left|\int_0^t b\big(y(\tau), y_1(\tau), y(\tau)\big) d\tau\right| \leq c_3 \int_0^t \|y(\tau)\|_H \|y(\tau)\|_W \|y_1(\tau)\|_W d\tau \leq$$

$$\nu \int_0^t \|y(\tau)\|_W^2 d\tau + c_4 \int_0^t \|y(\tau)\|_H^2 \|y_1(\tau)\|_W^2 d\tau,$$

where $c_3 > 0$, $c_4 > 0$. From (11.19) it follows that

$$\|y(\tau)\|_H^2 \leq 2c_4 \int_0^t \|y(\tau)\|_H^2 \|y_1(\tau)\|_W^2 d\tau.$$

Using the Gronwall lemma, we get $\|y(\tau)\|_H \leq 0$. Then $y_1 = y_2$. This completes the proof of the theorem. $\square$

Consider an optimization control problem for the system described by Navier–Stokes equations.

11.2 Optimization control problems for the evolutional Navier–Stokes equations

Consider the equations

$$y' - \nu \Delta y + \sum_{i=1}^2 y^i D_i y + \nabla q = v + f, \qquad (11.20)$$

420   *Optimization and Differentiation*

$$\operatorname{div} y = 0 \qquad (11.21)$$

in the set Q with boundary conditions

$$y(x,t) = 0, \ (x,t) \in \Sigma, \qquad (11.22)$$

$$y(x,0) = \varphi(x), \ x \in \Omega, \qquad (11.23)$$

where the functions $f \in X'$ and $\varphi \in H$ are known. Choose the control $v = (v^1, v^2)$ from the convex closed subset U of the space $V = L_2(0,T;H)$. From Theorem 11.1 it follows that for all $v \in V$ the problem (11.20)–(11.23) has a unique solution $y = y[v]$ from the space Y. Consider the functional

$$I(v) = \frac{\alpha}{2}\|v\|_V^2 + \frac{1}{2}\|y[v] - y_d\|_X^2,$$

where $\alpha > 0$, $y_d \in X$ is a known function.

Problem 11.1 *Find the control $u \in U$ for the system (11.20)–(11.23) that minimizes the functional I on the set U.*

Lemma 11.3 *The map $y[\cdot] : V \to Y$ for the problem(11.20)–(11.23) is weakly continuous.*

Proof. Suppose the convergence $v_k \to v$ weakly in V. The function $y_k = y[v_k]$ satisfies the equality

$$\langle y_k'(t), \eta \rangle + \nu a\big(y_k(t), \eta\big) + b\big(y_k(t), y_k(t), \eta\big) = \langle v_k(t) + f(t), \eta \rangle \ \forall \eta \in W \quad (11.24)$$

that is the analog of (11.12). Determine $\eta = y_k(t)$; we have

$$\frac{d}{dt}\|y_k(t)\|_H^2 + 2\nu\|y_k(t)\|_W^2 \le$$

$$2\|v_k(t) + f(t)\|_{W'}\|y_k(t)\|_W \le \nu\|y_k(t)\|_W^2 + \nu^{-1}\|v_k(t) + f(t)\|_{W'}^2.$$

Integrating with using (11.23) continuous embedding of the space W' to I, we have

$$\|y_k(t)\|_H^2 + \nu \int_0^t \|y_k(\tau)\|_W^2 d\tau \le \frac{1}{\nu} \int_0^t \Big[\|v_k(\tau)\|_{W'}^2 + \|f(\tau)\|_{W'}^2\Big]d\tau \le$$

$$c_1\|v_k\|_V^2 + \frac{1}{\nu}\|f\|_{L_2(0,T;W')}^2$$

that is the analog of (11.17), where $c_1 > 0$. Then the sequence $\{y_k\}$ is bounded in the space Z.

From the equality (11.24) it follows that

$$\|y_k'(t)\|_{W'}^2 = \sup_{\|\eta\|_W = 1} |\langle y_k'(t), \eta \rangle| \le$$

$$\sup_{\|\eta\|_W=1} \left[\nu \big| a\big(y_k(t),\eta\big) \big| + \big| b\big(y_k(t),y_k(t),\eta\big) \big| + \big| \langle v_k(t)+f(t),\eta \rangle \big| \right]$$

because of the definition of the norm of the adjoint space. We have

$$\big| a\big(y_k(t),\eta\big) \big| \le \|y_k(t)\|_W^2 \|\eta\|_W,$$

$$\big| b\big(y_k(t),y_k(t),\eta\big) \big| = \big| b\big(y_k(t),\eta,y_k(t)\big) \big| \le \|y_k(t)\|_W \|y_k(t)\|_H \|\eta\|_W,$$

$$\big| \langle v_k(t)+f(t),\eta \rangle \big| \le \|v_k(t)+f(t)\|_{W'} \|\eta\|_W.$$

Using the boundedness of the sequence $\{y_k\}$ in the space $L_\infty(0,T;H)$, we get the inequality

$$\|y_k'(t)\|_{W'} \le c_2 \|y_k(t)\|_W + \|v_k(t)+f(t)\|_{W'}.$$

After integration we have

$$\|y_k'\|_{X'}^2 \le 2c_2 \|y_k\|_X^2 + 2\|v_k+f\|_{X'}.$$

The sequence $\{y_k\}$ is bounded in space X, $\{v_k\}$ is bounded in V, and in X' too. Then from the last inequality the boundedness of the sequence $\{y_k'\}$ in the space X' follows. Thus, the sequence $\{y_k\}$ is bounded in Y. After extracting a subsequence we have $y_k \to y$ weakly in Y. Passing to the limit in the equality (11.24) (see passing to the limit in the equality (11.12) of Theorem 11.1), we have $y = y[v]$. $\square$

From Theorem 2.4 by using Lemma 11.3 the result follows.

Theorem 11.2 *Problem* 11.1 *is solvable.*

Try to determine the necessary condition of optimality. Find the derivative of the given functional. Analyze the differentiability of the solution of the boundary problem with respect to the control. Denote by y_0 and y_σ the solutions of the problem (11.20)–(11.23) for the controls u and $u+\sigma h$, where σh is a number, $h \in V$. Then we have

$$\langle \eta_\sigma'[h](t),\zeta \rangle + \nu a\big(\eta_\sigma[h](t),\zeta\big) + b\big(y_\sigma(t),\eta_\sigma[h](t),\zeta\big) +$$

$$b\big(\eta_\sigma[h](t),y_0(t),\zeta\big) = \langle h(t),\zeta \rangle \ \forall \zeta \in W,$$

where $\eta_\sigma[h] = (y_\sigma - y_0)/\sigma$. Multiply it by the function $\omega \in L_2(0,T)$ and integrate the result. We have

$$\int_0^T \langle \eta_\sigma'[h](t),\lambda(t) \rangle dt + \nu \int_0^T a\big(\eta_\sigma[h](t),\lambda(t)\big) dt + \int_0^T b\big(y_\sigma(t),\eta_\sigma[h](t),\lambda(t)\big) dt +$$

$$\int_0^T b\big(\eta_\sigma[h](t),y_0(t),\lambda(t)\big) dt = \int_0^T \langle h(t),\lambda(t) \rangle dt, \qquad (11.25)$$

where $\lambda = \omega \zeta$.

Consider the following boundary problem

$$-p' - \nu\Delta p - \sum_{i=1}^{2} y_\sigma^i D_i p + \sum_{i=1}^{2} p^i Dy_0^i + \nabla r = \mu_Q, \qquad (11.26)$$

$$\operatorname{div} p = 0 \qquad (11.27)$$

in the set Q with the boundary conditions

$$p(x,t) = 0, \ (x,t) \in \Sigma, \qquad (11.28)$$

$$p(x,T) = \mu_\Omega, \ x \in \Omega. \qquad (11.29)$$

We use the following proposition.

Lemma 11.4 *Embedding of the space Z to $L_4(Q)^2$ is continuous, as there exists a constant $c = c(Q) > 0$ such that*

$$\|\psi\|_{L_4(Q)^2}^2 \leq c\|\psi\|_{L_\infty(0,T;H)}\|\psi\|_X \ \forall\psi \in Z.$$

Consider the equations

$$-p_\mu[u]' - \nu\Delta p_\mu[u] - \sum_{i=1}^{2} y^i[u]D_i p_\mu[u] + \sum_{i=1}^{2} p_\mu^i[u]Dy^i[u] + \nabla r\mu[u] = \mu_Q, \quad (11.30)$$

$$\operatorname{div} p_\mu[u] = 0 \qquad (11.31)$$

in the set Q with the boundary conditions

$$p_\mu[u](x,t) = 0, \ (x,t) \in \Sigma, \qquad (11.32)$$

$$p_\mu[u](x,T) = \mu_\Omega, \ x \in \Omega. \qquad (11.33)$$

Lemma 11.5 *For all numbers σ and $h \in V$, $\mu_Q \in X'$, $\mu_\Omega \in H$ the problem (11.26)–(11.29) has a unique solution $p = p_\mu^\sigma[h]$ from the space Y and $r \in D'(Q)$, as if $\sigma \to 0$, then $p_\mu^\sigma \to p_\mu[u]$ weakly in Y.*

Proof. 1. Using the technique of Theorem 11.1, prove the equivalence of the problem (11.26)–(11.29) to the equation

$$-\langle p'(t), \eta\rangle + \nu a\big(p(t), \eta\big) - b\big(y_\sigma(t), p(t), \eta\big) + b\big(\eta, y_0(t), p(t)\big) = \langle \mu_Q(t), \eta\rangle \tag{11.34}$$

for all $\eta \in W$ with condition (11.29). Determine the approximate solution p_k of the problem (11.34), (11.29) by the formula

$$p_k(t) = \sum_{m=1}^{k} \xi_{mk}(t)\mu_m, \ k = 1, 2, \dots, \qquad (11.35)$$

Navier–Stokes equations

where the functions $\xi_{1k}, ..., \xi_{kk}$ satisfy the system of the linear ordinary differential equations

$$-\langle p'_k(t), \mu_j \rangle + \nu a(p_k(t), \mu_j) - b(y_\sigma(t), p_k(t), \mu_j) +$$

$$b(\mu_j, y_0(t), p_k(t)) = \langle \mu_Q(t), \mu_j \rangle, \quad j = 1, ..., k \quad (11.36)$$

with the final condition

$$p_k(T) = p_{0k}, \quad (11.37)$$

where $p_{0k} \to \mu_\Omega$ in I. This problem has a unique solution on the interval $(0, T)$.

2. Determine a priori estimates for this solution. Multiply j-th equality (11.36) by ξ_{jk} and sum the result by j. We get

$$-\frac{d}{dt}\|p_k(t)\|_H^2 + \nu\|p_k(t)\|_W^2 = -b(p_k(t), y_0(t), p_k(t)) = \langle \mu_Q(t), p_k(t) \rangle.$$

We obtain the inequality

$$-\frac{1}{2}\frac{d}{dt}\|p_k(t)\|_H^2 + \nu\|p_k(t)\|_W^2 \leq c_1\|p_k(t)\|_H\|p_k(t)\|_W\|y_0(t)\|_W$$

$$+\|\mu_Q(t)\|_{W'}\|p_k(t)\|_W \leq \frac{\nu}{2}\|p_k(t)\|_W^2 + c_2\|p_k(t)\|_H^2\|y_0(t)\|_W^2 + c_3\|\mu_Q(t)\|_{W'}^2,$$

where $c_1 > 0$, $c_2 > 0$, $c_3 > 0$. Integrating the result by using (11.37), we have

$$\|p_k(t)\|_H^2 \leq \|p_{0k}\|_H^2 + c_2 \int_t^T \|p_k(\tau)\|_H^2\|y_0(\tau)\|_W^2 d\tau + c_3 \int_t^T \|\mu_Q(\tau)\|_{W'}^2 d\tau \leq$$

$$\|\mu_\Omega\|_H^2 + c_3\|\mu_Q\|_{X'}^2 + c_2 \int_t^T \|p_k(\tau)\|_H^2\|y_0(\tau)\|_W^2 d\tau.$$

Using the Gronwall lemma, determine the boundedness of the sequence $\{p_k\}$ in the space $L_\infty(0, T; H)$. From the previous inequality it follows that

$$-\frac{d}{dt}\|p_k(t)\|_H^2 + \nu\|p_k(t)\|_W^2 \leq \frac{\nu}{2}\|p_k(t)\|_W^2 + c_4\|y_0(\tau)\|_W^2 + c_5\|\mu_Q(t)\|_{W'}^2,$$

where $c_4 > 0$, $c_5 > 0$. Then after integration we obtain the boundedness of the sequence $\{p_k\}$ in X.

3. After extracting a subsequence determine the convergence $p_k \to p$ weakly in X. Multiply the equality (11.36) by an arbitrary function $\omega \in C^1[0, T]$ that is equal to zero for $t = 0$ and $t = T$. After integration by using the condition (11.37) we have

$$\int_0^T \langle \xi'_j(t), p_k(t) \rangle dt + \nu \int_0^T a(p_k(t), \xi_j(t)) dt + \int_0^T b(y_\sigma(t), \xi_j(t), p_k(t)) dt +$$

$$\int_0^T b(\xi_j(t), y_0(t), p_k(t)) dt = \int_0^T \langle \mu_Q(t), \xi_j(t) \rangle dt, \quad (11.38)$$

424   *Optimization and Differentiation*

where $\xi_j = \omega\mu_j$. We have the convergence

$$\int\limits_0^T a\big(p_k(t), \xi_j(t)\big)\,dt \to \int\limits_0^T a\big(p(t), \xi_j(t)\big)\,dt.$$

Note the equality

$$\int\limits_0^T b\big(y_\sigma(t), \xi_j(t), p_k(t)\big)\,dt = \sum_{i,m=1}^2 \int\limits_Q y_\sigma^i D_i \xi_j^m p_k^m\,dQ.$$

From $y_\sigma \in X$, $\xi_j \in C(0,T;W)$, and $H^1(\Omega) \subset L_4(\Omega)$ it follows that $y_\sigma^i D_i \xi_j^m \in L_2\big(0,T;L_{4/3}(\Omega)\big)$. Using the convergence $p_k \to p$ weakly in X, we have $p_k^m \to p^m$ weakly in $L_2\big(0,T;L_4(\Omega)\big)$. Therefore, we get the convergence

$$\int\limits_Q y_\sigma^i D_i \xi_j^m p_k^m\,dQ \to \int\limits_Q y_\sigma^i D_i \xi_j^m p^m\,dQ.$$

Then

$$\int\limits_0^T b\big(y_\sigma(t), \xi_j(t), p_k(t)\big)\,dt \to \int\limits_0^T b\big(y_\sigma(t), \xi_j(t), p(t)\big)\,dt.$$

Using the analogical transformations, we obtain the convergence

$$\int\limits_0^T b\big(\xi_j(t), y_0(t), p_k(t)\big)\,dt \to \int\limits_0^T b\big(\xi_j(t), y_0(t), p(t)\big)\,dt.$$

Passing to the limit in the equality (11.38) as $k \to \infty$, we get

$$\int\limits_0^T \langle \xi_j'(t), p(t) \rangle\,dt + \nu \int\limits_0^T a\big(p(t), \xi_j(t)\big)\,dt + \int\limits_0^T b\big(y_\sigma(t), \xi_j(t), p(t)\big)\,dt +$$

$$\int\limits_0^T b\big(\xi_j(t), y_0(t), p(t)\big)\,dt = \int\limits_0^T \langle \mu_Q(t), \xi_j(t) \rangle\,dt, \ j = 1, 2, \dots .$$

The function ω is arbitrary. Hence, for all $j = 1, 2, \dots$ we get

$$\langle p'(t), \mu_j \rangle + \nu a\big(p(t), \mu_j\big) - b\big(y_\sigma(t), \mu_j, p(t)\big) + b\big(\mu_j, y_0(t), p(t)\big) = \langle \mu_Q(t), \mu_j \rangle\,dt.$$

Using the property of the basis, determine that the function p of the space Z satisfies the equality (11.34).

Determine the linear operator $B_\sigma : Z \to X'$ and $B_0 : Z \to X'$ by the equalities

$$\langle B_\sigma \psi, \zeta \rangle = \int_0^T b\big(y_\sigma(t), \zeta(t), \psi(t)\big)dt \ \forall \psi \in Z, \ \zeta \in X,$$

$$\langle B_0 \psi, \zeta \rangle = \int_0^T b\big(\zeta(t), y_0(t), \psi(t)\big)dt \ \forall \psi \in Z, \ \zeta \in X.$$

We have

$$\big\|B_\sigma \psi\big\|_{X'} = \sup_{\|\zeta\|_X = 1} \big|\langle B_\sigma \psi \rangle\big| = \sup_{\|\zeta\|_X = 1} \bigg| \int_0^T b\big(y_\sigma(t), \zeta(t), \psi(t)\big)dt \bigg| \le$$

$$\sup_{\|\zeta\|_X = 1} \sum_{i,j=1}^2 \int_Q \big|y_\sigma^i D_i \zeta^j \psi^j\big| dQ \le \sup_{\|\zeta\|_X = 1} \sum_{i,j=1}^2 \big\|y_\sigma^i \psi^j\big\|_{L_2(Q)} \big\|D_i \zeta^j\big\|_{L_2(Q)}.$$

Using Lemma 11.4 and Hölder inequality, we get

$$\big\|B_\sigma \psi\big\|_{X'} \le c_1 \big\|y_\sigma\big\|_{L_4(Q)^2}^2 \big\|\psi\big\|_{L_4(Q)^2}^2 \le$$

$$c_2 \big\|y_\sigma\big\|_X \big\|y_\sigma\big\|_{L_\infty(0,T;H)} \big\|\psi\big\|_X \big\|\psi\big\|_{L_\infty(0,T;H)},$$

where $c_1 > 0$, $c_2 > 0$. Then the linear operator B_σ is continuous. We determine also the inequality

$$\big\|B_0 \psi\big\|_{X'} \le c_2 \big\|y_0\big\|_X \big\|y_0\big\|_{L_\infty(0,T;H)} \big\|\psi\big\|_X \big\|\psi\big\|_{L_\infty(0,T;H)}.$$

Hence, the linear operator B_0 is continuous too.

Using (11.30), we get the equality

$$p' = \nu Ap + B_\sigma p + B_0 p = \mu_Q.$$

Then we obtain

$$\big\|p'\big\|_{X'} \le \nu \big\|Ap\big\|_{X'} + \big\|B_\sigma p\big\|_{X'} + \big\|B_0 p\big\|_{X'} + \big\|\mu_Q\big\|_{X'}.$$

Thus, we have $p' \in X'$ and $p \in Y$.

5. Suppose there exists two solutions p_1 and p_2 of the given boundary problem. Then its difference p satisfies the equality

$$-\langle p'(t), \eta \rangle + \nu a(p(t), \eta) - b\big(y_\sigma, (p(t), \eta) + b\big(\eta, y_0, p(t)\big) = 0 \ \forall \eta \in W.$$

Determine $\eta = p(t)$; we obtain

$$-\frac{d}{dt}\|p(t)\|_H^2 + \nu\|p(t)\|_W^2 = -b\big(p(t), y_0(t), p(t)\big).$$

426 *Optimization and Differentiation*

Using the standard technique, we get

$$-\frac{d}{dt}\|p(t)\|_H^2 \le c_4 \|p(t)\|_H^2 \|y_0(t)\|_W^2.$$

After integration we obtain

$$\|p(t)\|_H^2 \le c_4 \int_t^T \|p(\tau)\|_H^2 \|y_0(\tau)\|_W^2 d\tau.$$

Using the Gronwall lemma, we have $p = 0$. Therefore, the solution of the problem is unique.

6. Suppose $\sigma \to 0$. Then $y_\sigma \to y_0$ weakly in Y. Therefore, the set $\{y_\sigma\}$ is bounded in the space Y. Denote by $p_\mu^\sigma[h]$ the solution of the problem (11.26)–(11.29). Using the known transformations, determine the inequalities

$$\|p_\mu^\sigma[h](t)\|_H^2 \le \|\mu_\Omega\|_H^2 + c_3 \|\mu_Q\|_{X'}^2 + c_2 \int_t^T \|p(\tau)\|_H^2 \|y_0(\tau)\|_W^2 d\tau,$$

$$\frac{\nu}{2}\|p_\mu^\sigma[h]\|_X^2 \le c_4 \|y_0\|_X^2 + \|\mu_\Omega\|_H^2 + c_5 \|\mu_Q\|_{X'}^2.$$

We have the boundedness of the set $\{p_\mu^\sigma[h]\}$ in Z. Obtain the inequality

$$\|p_\mu^\sigma[h]'\|_{X'} \le \|Ap_\mu^\sigma[h]\|_{X'} + \|B_\sigma p_\mu^\sigma[h]\|_{X'} + \|B_0 p_\mu^\sigma[h]\|_{X'} + \|\mu_Q\|_{X'}.$$

Use the estimates

$$\|B_\sigma p_\mu^\sigma[h]\|_{X'} \le c_2 \pi_\sigma \|y_\sigma\|_{L_\infty(0,T;H)} \|y_\sigma\|_X,$$

$$\|B_0 p_\mu^\sigma[h]\|_{X'} \le c_2 \pi_\sigma \|y_\sigma\|_{L_\infty(0,T;H)} \|y_\sigma\|_X,$$

where

$$\pi_\sigma = \|p_\mu^\sigma[h]\|_{L_\infty(0,T;H)} \|p_\mu^\sigma[h]\|_X.$$

Hence, the set $\{p_\mu^\sigma[h]'\}$ is bounded in the space X', and $\{p_\mu^\sigma[h]\}$ is bounded in Y. After extracting a subsequence we have the convergence $p_\mu^\sigma[h] \to p_\mu$ weakly in Y.

The function $p_\mu^\sigma[h]$ satisfies the equality

$$\int_0^T \langle p_\mu^\sigma[h]'(t), \zeta(t)\rangle dt = \nu \int_0^T a\big(p_\mu^\sigma[h](t), \zeta(t)\big) dt + \int_0^T b\big(y_\sigma(t), \zeta(t), p_\mu^\sigma[h](t)\big) dt +$$

$$\int_0^T b\big(\zeta(t), y_0(t), p_\mu^\sigma[h](t)\big) dt = \int_0^T \langle \mu_Q(t), \zeta(t)\rangle dt \quad \forall \zeta \in X. \tag{11.39}$$

Navier–Stokes equations

427

Determine the convergence to the linear terms

$$\int_0^T \langle p_\mu^\sigma[h]'(t), \zeta(t) \rangle dt \to \int_0^T \langle p_\mu'(t), \zeta(t) \rangle dt,$$

$$\int_0^T a\big(p_\mu^\sigma[h](t), \zeta(t)\big) dt \to \int_0^T a\big(p_\mu(t), \zeta(t)\big) dt,$$

$$\int_0^T b\big(\zeta(t), y_0(t), p_\mu^\sigma[h](t)\big) dt \to \int_0^T b\big(\zeta(t), y_0(t), p_\mu(t)\big) dt.$$

We have the equality

$$\int_0^T b\big(y_\sigma(t), \zeta(t), p_\mu^\sigma[h](t)\big) dt = \sum_{i,j=1}^2 \int_Q y_\sigma^i D_i \zeta^j p_\mu^\sigma[h]^j dQ.$$

Using Lemma 11.4, determine the boundedness of the sets $\{y_\sigma^i\}$ and $\{p_\mu^\sigma[h]^j\}$ in the space $L_4(Q)$. Then $\{y_\sigma^i p_\mu^\sigma[h]^j\}$ is bounded in $L_2(Q)$. Embedding of the space Y to $L_2(0, T; H)$ is compact. Hence, we have $y_\sigma^i \to y_0^i$ a.e. on Q and $p_\mu^\sigma[h]^j \to p_\mu^j$ a.e. on Q. Then $y_\sigma^i p_\mu^\sigma[h]^j \to y_0^i p_\mu^j$ a.e. on Q and weakly in $L_2(Q)$. Determine the convergence

$$\int_Q y_\sigma^i D_i \zeta^j p_\mu^\sigma[h]^j dQ \to \int_Q y_0^i D_i \zeta^j p_\mu^j dQ.$$

Then

$$\int_0^T b\big(y_\sigma(t), \zeta(t), p_\mu^\sigma[h](t)\big) dt \to \int_0^T b\big(y_0(t), \zeta(t), p_\mu(t)\big) dt.$$

Passing to the limit in the equality (11.39), we have

$$\int_0^T \langle p_\mu'(t), \zeta(t) \rangle dt = \nu \int_0^T a\big(p_\mu(t), \zeta(t)\big) dt + \int_0^T b\big(y_0(t), \zeta(t), p_\mu(t)\big) dt +$$

$$\int_0^T b\big(\zeta(t), y_0(t), p_\mu(t)\big) dt = \int_0^T \langle \mu_Q(t), \zeta(t) \rangle dt \quad \forall \zeta \in X.$$

Thus, $p_\mu = p_\mu[u]$. $\square$

Determine the differentiability of the solution of the boundary problem with respect to the control.

428   *Optimization and Differentiation*

Lemma 11.6 *The operator $y[\cdot] : V \to Y$ for the problem (11.20)–(11.23) has $(V, X; V, X^w)$-extended derivative $y'[u]$ at the arbitrary point $u \in V$ as there exists a linear continuous operator $y_T[\cdot] : V \to H$ such that for all $h \in V$ we have the convergence $\{y[u + \sigma h](T) - y[u](T)\}/\sigma \to y_T[u]h$ weakly in H. In addition, for all $h \in V$, $\mu_Q \in X'$, $\mu_\Omega \in H$ the following equality holds:*

$$\langle \mu_Q, y'[u]h \rangle + (\mu_\Omega, y_T[u]h)_H = (h, p_\mu[u])_V. \tag{11.40}$$

Proof. For $\mu_\Omega = 0$ the equality (11.40) determines the linear continuous operator $y[u] : V \to X$. Choose $\lambda = p_\mu^\sigma[h]$ at the equality (11.25). For all $h \in V$, $\mu_Q \in X'$, $\mu_\Omega \in H$ we get

$$\langle \mu_Q, (y[u+\sigma h] - y[u])/\sigma \rangle + (\mu_\Omega, \{y[u+\sigma h](T) - y[u](T)\}/\sigma)_H = (h, p_\mu^\sigma[h])_V.$$

Using (11.40) with $\mu_\Omega = 0$, for all $h \in V$, $\mu_Q \in X'$ we have

$$\langle \mu_Q, (y[u + \sigma h] - y[u])/\sigma - y'[u]h \rangle = (h, p_\mu^\sigma[h] - p_\mu[u])_V.$$

Pass to the limit by using Lemma 11.5. We obtain the convergence $(y[u+\sigma h] - y[u])/\sigma \to y'[u]h$ weakly in X for all $h \in V$. The final proposition of the lemma can be proved as before. $\square$

Lemma 11.7 *The operator $y[\cdot] : V \to X$ for the problem (11.20)–(11.23) is continuous.*

Proof. Consider the controls $v_1, v_2 \in V$ and the corresponding solutions $y_1, y_2 \in Y$ of the problem (11.20)–(11.23). The function $y = y_1 - y_2$ satisfies the following equality (see the proof of the uniqueness of Theorem 11.1):

$$\langle y'(t), \eta \rangle + \nu a\big(y(t), \eta\big) + b\big(y(t), y_1(t), \eta\big) + b\big(y_1(t), y(t), \eta\big) -$$

$$b\big(y(t), y(t), \eta\big) = \langle v(t), \eta \rangle \;\; \forall \eta \in W,$$

where $v = v_1 - v_2$. Choose $\eta = y(t)$. After easy transformations (see the final part of the proof of Theorem 11.1), we get the equality

$$\frac{1}{2}\|y(t)\|_H^2 + \nu \int_0^t \|y(\tau)\|_W^2 d\tau = -\int_0^t b\big(y(\tau), y_1(\tau), y(\tau)\big)d\tau + \int_0^t \langle v(\tau), y(\tau) \rangle d\tau$$

that is the analog of (11.19). The right-hand side of this equality is estimated (see the final part of the proof of Theorem 11.1) by the sum

$$\frac{\nu}{2} \int_0^t \|y(\tau)\|_W^2 d\tau + c_1 \int_0^t \|y(\tau)\|_H^2 \|y_1(\tau)\|_W^2 d\tau + c_2\|v\|_V^2,$$

Navier–Stokes equations

where $c_1 > 0$, $c_2 > 0$. Then we get the inequality

$$\|y(t)\|_H^2 + \nu \int_0^t \|y(\tau)\|_W^2 d\tau \le 2c_1 \int_0^t \|y(\tau)\|_H^2 \|y_1(\tau)\|_W^2 d\tau + 2c_2 \|v\|_V^2.$$

Using Gronwall lemma, we obtain the inequality $\|y(t)\|_H \le c_3 \|v\|_V$, where $c_3 > 0$. Using the previous inequality, we get $\|y\|_X \le c_4 \|v\|_V$, where $c_4 > 0$. This completes the proof of the lemma. $\square$

Now determine the differentiability of the state functional for Problem 11.1. Consider the equations

$$-p' - \nu \Delta p - \sum_{i=1}^2 y^i[u] D_i p + \sum_{i=1}^2 p^i[u] D y^i[u] + \nabla r = \Delta y_d - \Delta y[u], \quad (11.41)$$

$$\operatorname{div} p = 0 \quad (11.42)$$

in the set Q with the boundary conditions

$$p(x,t) = 0, \ (x,t) \in \Sigma, \quad (11.43)$$

$$p(x,T) = 0, \ x \in \Omega. \quad (11.44)$$

Lemma 11.8 *The functional I for Problem 11.1 has Gâteaux derivative $I'(u) = \alpha u + p$ at the arbitrary point $u \in V$.*

Proof. We have the equality

$$\frac{I(u + \sigma h) - I(u)}{\sigma} = \frac{\alpha}{2} \sum_{i=1}^2 \int_Q \left[2u^i h^i + \sigma(h^i)^2\right] dQ +$$

$$\frac{1}{2} \sum_{i,j=1}^2 \int_Q D_j\left(y^i[u + \sigma h] + y^i[u] - 2y_d^i\right) D_j \frac{y^i[u + \sigma h] - y^i[u]}{\sigma} dQ.$$

Passing to the limit as $\sigma \to 0$ by using Lemma 11.6 and Lemma 11.7, we get

$$(I'(u), h)_V = \sum_{i=1}^2 \int_Q \alpha u^i h^i dQ + \sum_{i,j=1}^2 \int_Q D_j\left(y^i[u] - y_d^i\right) D_j\left(y'[u]h\right)^i dQ =$$

$$\sum_{i=1}^2 \int_Q \alpha u^i h^i dQ + \sum_{i=1}^2 \int_Q \left(\Delta y_d^i - \Delta y^i[u]\right)\left(y'[u]h\right)^i dQ \ \forall h \in V.$$

430   *Optimization and Differentiation*

Choosing $\mu_Q = \Delta y_d - \Delta y[u]$, $\mu_\Omega = 0$ at the equality (11.40) and using (11.41), we have the equality

$$\left(I'(u), h\right)_H = \sum_{i=1}^n \int_\Omega (\alpha u^i + p^i) h^i dx \ \ \forall h \in H.$$

This completes the proof of Lemma 11.8. $\square$

The necessary condition minimum of Gâteaux differentiable functional I on the convex subset U of Hilbert space V at the point u is the variational inequality

$$\left(I'(u), v - u\right)_V \ \ \forall v \in V.$$

Using Lemma 11.8, we have the following result.

Theorem 11.3 *If u is the solution of Problem 11.1, then the variational inequality holds:*

$$\left(\alpha u + p, v - u\right)_V \ \ \forall v \in U. \tag{11.45}$$

Thus, we can find the optimal control from the problem that involves the boundary problem (11.20)–(11.23) for $v = u$, the adjoint system (11.41)–(11.44), and the variational inequality (11.45).

Remark 11.3 We could consider optimization control problems with the initial control and the functional that depend upon the state function at the final time (see Chapter 9).

11.3   Stationary Navier–Stokes equations

Now consider a stationary system. Let Ω be an open bounded set of the space $\mathbb{R}^n$ with the boundary Γ, where $2 \leq n \leq 4$. The system described by the boundary problem for the ***stationary Navier–Stokes equations***

$$-\nu\Delta y + \sum_{i=1}^n y D_i y + \nabla q = f, \tag{11.46}$$

$$\operatorname{div} y = 0 \tag{11.47}$$

in the set Ω the boundary condition

$$y(x) = 0 \ x \in \Gamma, \tag{11.48}$$

where $y = \left(y^1, ..., y^n\right)$, $f = \left(f^1, ..., f^n\right)$, $\nu > 0$. Consider the set

$$D = \left\{ y \in \left[D(\Omega)\right]^n \Big| \operatorname{div} y = 0 \right\},$$

Navier–Stokes equations 431

its closure H to the space $\left[L_2(\Omega)\right]^n$, and the closure W to the space $\left[H^1(\Omega)\right]^n$. The spaces H and W are Hilbert, with scalar product

$$(\psi, \eta)_H = \sum_{i=1}^n \int_\Omega \psi^i \eta^i dx \ \forall \psi, \eta \in H,$$

$$(\psi, \eta)_W = \sum_{i,j=1}^n \int_\Omega D_j \psi^i D_j \eta^i dx \ \forall \psi, \eta \in W.$$

Consider again the functionals

$$a(\psi, \eta) = \sum_{i,j=1}^n \int_\Omega D_i \psi^j D_i \eta^j dx \ \forall \psi, \eta \in W,$$

$$b(\psi, \eta, \zeta) = \sum_{i,j=1}^n \int_\Omega \psi^i D_i \eta^j \zeta^j dx \ \forall \psi, \eta, \zeta \in W$$

that satisfy the condition (11.7), (11.8). Note that there exists a constant $\chi > 0$ such that

$$|b(\psi, \eta, \zeta)| \le \chi \|\psi\|_W \|\eta\|_W \|\zeta\|_W \ \forall \psi, \eta, \zeta \in W. \tag{11.49}$$

Theorem 11.4 *For any $f \in W'$ the problem (11.46)–(11.48) has a solution $y \in W$, $q \in D'(\Omega)$ that is unique if.*

$$\nu^2 > \chi \|f\|_{W'}. \tag{11.50}$$

We consider the following result that was used before for the analysis of the nonlinear stationary systems.

Lemma 11.9 *Let P be a continuous map on Euclid space $\mathbb{R}^k$ such that $(P\xi, \xi) > 0$ for any vector $\xi \in \mathbb{R}^k$ with absolute value $r > 0$. Then there exists a vector $\xi \in \mathbb{R}^k$ such that $|\xi| \le r$ and $P\xi = 0$.*

Proof of Theorem 11.4. 1. Consider an element $\eta = (\eta^1, ..., \eta^n)$ from the space W. Suppose $y \in W$, $q \in D'(\Omega)$ satisfy the equalities (11.46)–(11.48). Determine the scalar product with respect to $\mathbb{R}^n$ of the terms at the left-hand side and the right-hand side of the equality (11.1) and η. After integration for all $\eta \in W$ we have

$$\nu \sum_{i,j=1}^n \int_\Omega D_i y^j D_i \eta^j dx + \sum_{i,j=1}^n \int_\Omega y^i D_i y^j D_i \eta^j dx + \sum_{j=1}^n \int_\Omega D_j q \eta^j dx = \sum_{j=1}^n \int_\Omega f^j \eta^j dx.$$

From Green's formula, it follows that

$$\sum_{j=1}^n \int_\Omega D_j q \eta^j dx = - \int_\Omega q \,\mathrm{div}\, \eta dx = 0 \ \forall \eta \in W. \tag{11.51}$$

432   *Optimization and Differentiation*

Using the definition of the functionals a and b, we have

$$\nu a\big(y(t), \eta\big) + b\big(y(t), y(t)\eta\big) = \langle f(t), \eta \rangle \ \ \forall \eta \in W \tag{11.52}$$

that is the analog of (11.12).

Suppose now the function $y \in W$ satisfies the equality (11.52). Then we obtain

$$\sum_{j=1}^{n} \int_{\Omega} \Big(-\nu \Delta y^j + \sum_{j=1}^{n} y^i D_i y^j - f^j \Big) \eta^j dx = 0 \ \ \forall \eta \in W.$$

Denote the value under the brackets by $D_j q$. Using (11.51), we determine the existence of $q \in D'(\Omega)$ such that the equality (11.46) is true. Thus, it is sufficient to prove the existence of the unique function $y \in W$ that satisfies the equality (11.52).

2. Apply the Galerkin method for proving the solvability of the problem. The space W is dense in D. Then there exists a complete set $\{\mu_m\}$ of the space D that is complete in W. Determine the sequence $\{y_k\}$ by the formula

$$y_k = \sum_{m=1}^{k} \xi_{mk} \mu_m, \ k = 1, 2, ..., \tag{11.53}$$

where the numbers $\xi_{1k}, ..., \xi_{kk}$ satisfy the equalities

$$\nu a\big(y_k, \mu_j\big) + b\big(y_k, y_k, \mu_j\big) = \langle f, \mu_j \rangle, \ j = 1, ..., k. \tag{11.54}$$

The problem (11.54) is the system of nonlinear algebraic equations. We prove its solvability by using the technique of the analysis of the nonlinear elliptic equations (see Chapter 4). Determine the operator P on the space $\mathbb{R}^k$ that maps the vector $\xi = \big(\xi_{1k}, ..., \xi_{kk}\big)$ to the vector $P\xi = \big(\eta_1, ..., \eta_k\big)$, where

$$\eta_j = \nu a\big(y_k, \mu_j\big) + b\big(y_k, y_k, \mu_j\big) - \langle f, \mu_j \rangle, \ j = 1, ..., k.$$

Then the system (11.54) is transformed to the operator equation $P\xi = 0$ that is considered by Lemma 11.9. Find the scalar product

$$(P\xi, \xi) = \sum_{j=1}^{k} \eta_j \xi_{kj} = \nu a\big(y_k, y_k\big) + b\big(y_k, y_k, y_k\big) - \langle f, y_k \rangle.$$

Using (11.7), (11.8), we get inequality

$$(P\xi, \xi) = \nu \big\|y_k\big\|_W^2 - \langle f, y_k \rangle \geq \nu \big\|y_k\big\|_W^2 - \big\|y_k\big\|_W \big\|f\big\|_{W'}.$$

Choose the vector ξ with large enough absolute value r such that the value y_k that is determined by the equality (11.53) satisfies the inequality

$$\nu \big\|y_k\big\|_W \geq \big\|f\big\|_{W'}.$$

Navier–Stokes equations

From the previous condition, it follows that $(P\xi, \xi) \geq 0$ for all vector ξ such that $\|\xi\| = r$. Then the equation $P\xi = 0$ and the system (11.54) are solvable by Lemma 11.8. Therefore, the considered function y_k exists in reality.

3. Multiply j-th the equality (11.54) by the number ξ_{jk}. Summing the result, we get

$$\nu a(y_k, y_k) + b(y_k, y_k, y_k) = \langle f, y_k \rangle.$$

Then we obtain the inequality

$$\nu \|y_k\|_W^2 \leq \nu \|y_k\|_W \|f\|_{W'}.$$

Thus, sequence $\{y_k\}$ is bounded in space W.

4. After extracting a subsequence we have the convergence $y_k \to y$ weakly in W. Then $a(y_k, \mu_j) \to a(y, \mu_j)$. Consider the equality

$$b(y_k, y_k, \mu_j) = -b(y_k, \mu_j, y_k) = -\sum_{i,m=1}^{2} \int_\Omega y_k^i y_k^m D_i \mu_j^m dx.$$

Using compact embedding of the space W to I, we have the convergence $y_k^i \to y^i$ strongly in $L_2(\Omega)$. By $D_i \mu_j^m \in C(\overline{\Omega})$, we obtain $y_k^m D_i \mu_j^m \to y^m D_i \mu_j^m$ strongly in $L_2(\Omega)$. Then

$$\sum_{i,m=1}^{2} \int_\Omega y_k^i y_k^m D_i \mu_j^m dx \to \sum_{i,m=1}^{2} \int_\Omega y^i y^m D_i \mu_j^m dx.$$

Passing to the limit in the equality (11.54), we get

$$\nu a(y, \mu_j) + b(y, y, \mu_j) = \langle f, \mu_j \rangle, \quad j = 1, ..., k$$

and the equation (11.46) too.

5. Suppose y_1 and y_2 are solutions of the problem (11.46). Denote $y = y_1 - y_2$. We have

$$\nu a(y, \eta) + b(y, y_1, \eta) + b(y_2, y, \eta) = 0.$$

Choosing $\eta = y$ we get

$$\nu \|y\|_W^2 + b(y, y_1, y) + b(y_2, y, y) = 0.$$

Then we obtain

$$\nu \|y\|_W^2 = -b(y, y_1, y) \leq \chi \|y_1\|_W \|y\|_W \tag{11.55}$$

because of the inequality (11.49).

The function y_1 satisfies the equality (11.52). Choose $\eta = y$. Determine the inequality

$$\nu \|y_1\|_W^2 \leq |\langle f, y_1 \rangle| \leq \|f\|_{W'} \|y_1\|_W.$$

434　　　　　　　　　　*Optimization and Differentiation*

Then we have

$$\|y_1\|_W \le \nu^{-1}\|f\|_{W'}.$$

Using (11.55), we get

$$\left(\nu^2 - \chi\|f\|_{W'}\right)\|y\|_W \le 0.$$

Using the condition (11.50), we determine $y = 0$. Therefore, the solution of the considered problem is unique. $\square$

Now consider an optimization control problem for the system described by the stationary Navier–Stokes equations.

11.4　Optimization control problems for the stationary Navier–Stokes equations

Let Ω be an open bounded n-dimensional set with the boundary Γ, where $2 \le n \le 4$. Consider the system

$$-\nu\Delta y + \sum_{i=1}^{2} y^i D_i y + \nabla q = v + f, \tag{11.56}$$

$$\operatorname{div} y = 0 \tag{11.57}$$

in the set Ω with the boundary condition

$$y(x) = 0, \ x \in \Gamma. \tag{11.58}$$

The vector-function $v = \left(v^1, ..., v^n\right)$ is in control here. This is an element of a bounded convex closed subset U of the space H. From Theorem 11.4 it follows that for all $v \in H$ the problem (11.56)–(11.58) has a solution $y = y[v]$ from the space W that is unique if

$$\nu^2 \ge \chi\|v + f\|_{W'}. \tag{11.59}$$

Consider the functional

$$I(v) = \frac{\alpha}{2}\|v\|_H^2 + \frac{1}{2}\|y[v] - y_d\|_W^2,$$

where $\alpha > 0$, $y_d \in W$ is a known function.

Problem 11.2 *Find the control $u \in U$ for the system (11.56)–(11.58) that minimizes the functional I on the set U.*

Navier–Stokes equations 435

Lemma 11.10 *If $v_k \to v$ weakly in H, then after extracting a subsequence for the solutions of the problem* (11.56)–(11.58) *we have the convergence $y[v_k] \to y[v]$ weakly in W.*

Proof. Suppose the convergence $v_k \to v$ weakly in H. The function $y_k = y[v_k]$ satisfies the equality

$$\nu a(y_k, \eta) + b(y_k, y_k, \eta) = \langle v_k + f, \eta \rangle \ \forall \eta \in W. \tag{11.60}$$

Choose $\eta = y_k$; we have the inequality

$$\nu \|y_k\|_W^2 \le |\langle v_k + f, \eta \rangle| \le \|v_k + f\|_{W'} \|y_k\|_W.$$

Then the sequence $\{y_k\}$ is bounded in the space W. After extracting a subsequence we have $y_k \to y$ weakly in W. Passing to the limit at the equality (11.60) by using the technique of proving Theorem 11.4, we complete the proof of Lemma 11.10. $\square$

Remark 11.4 The results of Lemma 11.9 are true even without the condition (11.59). The boundary problem here can have a non-unique solution. Therefore, we proved a weaker result than weak continuity of control–state mapping.

Now we obtain the existence of the optimal control.

Theorem 11.5 *Problem* 11.2 *is solvable.*

Remark 11.5 We use extracting a subsequence of the minimizing sequence for the proof of the solvability of the optimization problem. Therefore, the possibility of proving the convergence of a subsequence is not the only obstacle for proving of the existence theorem here.

Determine the differentiability of the solution of the boundary problem (11.56)–(11.58) with respect to the control at an arbitrary point $u \in H$. Denote by y_0 and y_σ the solutions of the problem (11.56)–(11.58) for the controls u and $u + \sigma h$, where σ is a number, $h \in H$. Then we have

$$\nu a(\eta_\sigma[h], \zeta) + b(y_\sigma, \eta_\sigma[h], \zeta) + b(\eta_\sigma[h], y_0, \zeta) = \langle h, \zeta \rangle \ \forall \zeta \in W, \tag{11.61}$$

where $\eta_\sigma[h] = (y_\sigma - y_0)/\sigma$.

Consider the following system

$$-\nu \Delta p - \sum_{i=1}^2 y_\sigma^i D_i p + \sum_{i=1}^2 p^i D y_0^i + \nabla r = \mu_Q, \tag{11.62}$$

$$\operatorname{div} p = 0 \tag{11.63}$$

in the set Ω with the boundary condition

$$p(x, t) = 0 \ (x, t) \in \Sigma \tag{11.64}$$

that is the stationary analog of the system (11.26)–(11.29).

We use the following proposition.

Optimization and Differentiation

Lemma 11.11 *For all $u \in H$ under the supposition*

$$\nu > \nu(u) = \sqrt{\chi} \|u + f\|_{W'},$$

we have the inequality

$$\|y[v] - y[u]\|_W \le c\|v - u\|_{W''},$$

where $c > 0$.

Proof. The function $\varphi = y[v] - y[u]$ satisfies the equality

$$\nu a(\varphi, \eta) + b(\varphi, y[u], \eta) + b(y[v], \varphi, \eta) = u - v \ \forall \eta \in W.$$

Choose $\eta = \varphi$. We have

$$\nu\|\varphi\|_W^2 = -b(\varphi, y[u], \varphi) + \langle u - v, \varphi \rangle.$$

Using the condition (11.49), we get the inequality

$$\nu\|\varphi\|_W^2 \le \|\varphi\|_W \|y[u]\|_W + \|u - v\|_{W'} \|\varphi\|_W.$$

Using the estimate

$$\|y[u]\|_W \le \nu^{-1}\|u + f\|_{W''},$$

transform the previous inequality to

$$\left(\nu - \chi\nu^{-1}\|u + f\|_{W'}\right)\|\varphi\|_W^2 \le \|u - v\|_{W'}\|\varphi\|_W.$$

This complete the proof of the lemma if $\nu > \nu(u)$. $\square$

Consider the equations

$$-\nu\Delta p_\mu[u] - \sum_{i=1}^{2} y^i[u]D_i p_\mu[u] + \sum_{i=1}^{2} p_\mu^i[u]Dy^i[u] + \nabla r\mu[u] = \mu, \qquad (11.65)$$

$$\operatorname{div} p_\mu[u] = 0 \qquad (11.66)$$

in the set Ω with the boundary conditions

$$p_\mu[u](x) = 0 \ x \in \Gamma. \qquad (11.67)$$

Lemma 11.12 *Under the inequality $\nu > \nu(u)$ for all number σ, the problem (11.62)–(11.64) has a unique solution $p = p_\mu^\sigma$ from the space W, as we have the convergence $p_\mu^\sigma \to p_\mu[u]$ weakly in W as $\sigma \to 0$.*

Navier–Stokes equations

Proof. 1. The problem (11.62)–(11.64) to the equation

$$\nu a(p, \eta) - b(y_\sigma, p, \eta) + b(\eta, y_0, p) = \langle \mu, \eta \rangle \ \forall \eta \in W. \tag{11.68}$$

Let $\{\mu_m\}$ be a complete set of the space W. Determine an approximate solution of the problem by the formula

$$p_k = \sum_{m=1}^{k} \xi_{mk} \mu_m, \ k = 1, 2, \dots, \tag{11.69}$$

where the functions $\xi_{1k}, \dots, \xi_{kk}$ satisfy the system of the linear algebraic equations

$$\nu a(p_k, \mu_j) - b(y_\sigma, p_k, \mu_j) + b(\mu_j, y_0, p_k) = \langle \mu, \mu_j \rangle, \ j = 1, \dots, k. \tag{11.70}$$

Then we can find the function p_k. Multiply by j-th the equality (11.61) by the number ξ_{jk} and sum the result. We get

$$\nu \|p_k\|_W^2 + b(p_k, y_0, p_k) = \langle \mu, p_k \rangle.$$

Obtain the inequality

$$\nu \|p_k\|_W^2 \le \chi \|p_k\|_W \|y_0\|_W + \|\mu\|_{W'} \|p_k\|_W \le$$

$$\chi \nu^{-1} \|u + f\|_{W'} \|p_k\|_W^2 + \|\mu\|_{W'} \|p_k\|_W.$$

Thus, we have

$$[\nu^2 - \nu(u)^2]\nu^{-1}\|p_k\|_W \le \|\mu\|_{W'}.$$

Using the inequality $\nu > \nu(u)$, we obtain the boundedness of the sequence $\{p_k\}$ in the space W.

After extracting a subsequence, we have the convergence $p_k \to p$ weakly in W. By the linearity of the system (11.70) we can pass to the limit here without any difficulties. We obtain the equality

$$\nu a(p, \mu_j) - b(y_\sigma, p, \mu_j) + b(\mu_j, y_0, p) = \langle \mu, \mu_j \rangle, \ j = 1, \dots.$$

Then the equality (11.68) is true. The uniqueness of the solution of this problem is obvious.

2. Suppose the convergence $\sigma \to 0$. Using Lemma 11.10, we have $y_\sigma \to y_0$ in W. Denote by $p_\mu^\sigma[h]$ the solution of the problem (11.68). Using the known technique, we have the boundedness of the set $\{p_\mu^\sigma[h]\}$ in W. After extracting a subsequence, we obtain the convergence $p_\mu^\sigma[h] \to p_\mu$ weakly in W. Then

$$a(p_\mu^\sigma[h], \eta) \to a(p_\mu, \eta) \ \forall \eta \in W,$$

$$b(p_\mu^\sigma[h], y_0, \eta) \to b(p_\mu, y_0, \eta) \ \forall \eta \in W.$$

438   *Optimization and Differentiation*

We have the equality

$$b\big(y_\sigma, p_\mu^\sigma[h], \eta\big) = -b\big(y_\sigma, \eta, p_\mu^\sigma[h]\big) = -\sum_{i,j=1}^{n} \int_\Omega y_\sigma^i D_i \eta^j p_\mu^\sigma[h]^j dx.$$

Using the convergence $y_\sigma^i \to y_0^i$ strongly in $H^1(\Omega)$ and in $L_4(\Omega)$ and the condition $D_i \eta^j \in L_2(\Omega)$, we have $y_\sigma^i D_i \eta^j \to y_0^i D_i \eta^j$ strongly in $L_{4/3}(\Omega)$. In addition, we have the convergence $p_\mu^\sigma[h]^j \to p_\mu^j$ weakly in $H^1(\Omega)$ and in $L_4(\Omega)$. Then

$$\int_\Omega y_\sigma^i D_i \eta^j p_\mu^\sigma[h]^j dx \to \int_\Omega y_0^i D_i \eta^j p_\mu^j dx.$$

Hence, we get the convergence

$$b\big(y_\sigma, p_\mu^\sigma[h], \eta\big) \to b\big(y_0, p_\mu, \eta\big).$$

Passing to the limit in the equality (11.68) for $p = p_\mu^\sigma[h]$ as $\sigma \to 0$ we prove that the limit function p is the solution of the boundary problem (11.65)–(11.67). $\square$

Prove the differentiability of the solution of the boundary problem with respect to the control.

Lemma 11.13 *The operator* $y[\cdot] : V \to Y$ *for the problem* (11.56)–(11.58) *has* $\big(H, W; H, W^w\big)$*-extended derivative at the arbitrary point* $u \in V$ *under the condition* $\nu > \nu(u)$ *that satisfies the equality*

$$\langle y'[u]h, \mu \rangle = \big(h, p_\mu[u]\big)_H \ \forall h \in H, \mu \in W'. \tag{11.71}$$

Proof. The equality (11.71) determines a linear continuous operator $y'[u] : H \to W$. Choose $\lambda = p_\mu^\sigma[h]$ for the equality (11.61). We get

$$\big\langle (y[u + \sigma h] - y[u])/\sigma, \mu \big\rangle = \big(h, p_\mu^\sigma[h]\big)_H \ \forall h \in H, \mu \in W'.$$

Using (11.71), we have

$$\big\langle (y[u + \sigma h] - y[u])/\sigma - y'[u]h, \mu \big\rangle = \big(h, p_\mu^\sigma[h] - p_\mu[u]\big)_H.$$

Passing to the limit by using Lemma 11.11, for all $h \in H$ we have $(y[u + \sigma h] - y[u])/\sigma \to y'[u]h$ in W. $\square$

Now determine the differentiability of the state functional for Problem 11.2. Consider the equations

$$-\nu \Delta p - \sum_{i=1}^{2} y^i[u] D_i p + \sum_{i=1}^{2} p^i[u] D y^i[u] + \nabla r = \Delta y_d - \Delta y[u], \tag{11.72}$$

$$\operatorname{div} p = 0 \tag{11.73}$$

in the set Ω with the boundary condition

$$p(x) = 0, \ x \in \Gamma. \tag{11.74}$$

Navier–Stokes equations

Lemma 11.14 *The functional I for Problem 11.2 has Gâteaux derivative $I'(u) = \alpha u + p$ at the arbitrary point $u \in V$, if $\nu > \nu(u)$.*

Proof. We have the equality

$$\frac{I(u + \sigma h) - I(u)}{\sigma} = \frac{\alpha}{2} \sum_{i=1}^{n} \int_{\Omega} \left[2u^i h^i + \sigma (h^i)^2 \right] dx +$$

$$\frac{1}{2} \sum_{i,j=1}^{2} \int_{\Omega} D_j \left(y^i[u + \sigma h] + y^i[u] - 2y^i_d \right) D_j \frac{y^i[u + \sigma h] - y^i[u]}{\sigma} dx.$$

Passing to the limit as $\sigma \to 0$ by using Lemma 11.11 and Lemma 11.13, we get

$$\left(I'(u), h \right)_H = \sum_{i=1}^{n} \int_{\Omega} \alpha u^i h^i dx + \sum_{i,j=1}^{n} \int_{\Omega} D_j \left(y^i[u] - y^i_d \right) D_j \left(y'[u]h \right)^i dx =$$

$$\sum_{i=1}^{n} \int_{\Omega} \alpha u^i h^i dx + \sum_{i=1}^{n} \int_{\Omega} \left(\Delta y^i_d - \Delta y^i[u] \right) \left(y'[u]h \right)^i dx \quad \forall h \in H.$$

Choosing $\mu_Q = \Delta y_d - \Delta y[u]$, $\mu_\Omega = 0$ at the equality (11.40) and using (11.41), we have the equality

$$\left(I'(u), h \right)_V = \sum_{i=1}^{2} \int_{Q} (\alpha u^i + p^i) h^i dQ \quad \forall h \in V.$$

This completes the proof of Lemma 11.8. $\square$

Putting the value of the functional derivative in to the standard variational inequality, we obtain the necessary condition of optimality.

Theorem 11.6 *Suppose $\nu > \sup\limits_{v \in U_0}$, where U_0 is the set of all solutions of Problem 11.2. Then the solution u of Problem 11.2 satisfies the variational inequality*

$$\left(\alpha u + p, v - u \right)_H \quad \forall v \in U. \tag{11.75}$$

Thus, we have the system that includes the initial boundary problem (11.56)–(11.58) for $v = u$, the adjoint system (11.72)–(11.74), and the variational inequality (11.75).

11.5 System of Navier–Stokes and heat equations

Let Ω be an open bounded set of the plane with the boundary Γ, $T > 0$, $Q = \Omega \times (0,T)$, $\Sigma = \Gamma \times (0,T)$. Consider the system of the equations

$$y_1' - \nu \Delta y_1 + \sum_{i=1}^{2} y_2^i D_i y_1 = f_1, \qquad (11.76)$$

$$y_2' - \nu \Delta y_2 + \sum_{i=1}^{2} y_2^i D_i y_2 + \nabla q + \gamma y_1 = f_2, \qquad (11.77)$$

$$\operatorname{div} y_2 = 0 \qquad (11.78)$$

in the set Q, where $D_i = \partial/\partial x_i$, $i = 1, 2$, $y_2 = (y_2^1, y_2^2)$, $f_2 = (f_2^1, f_2^2)$, $\nu > 0$, $\gamma = (\gamma^1, \gamma^2)$, $\gamma^1 > 0$, $\gamma^2 > 0$. We have also the boundary conditions

$$y_i(x,t) = 0 \ (x,t) \in \Sigma, \qquad (11.79)$$

$$y_i(x,0) = \varphi_i(x), \ x \in \Omega, \ i = 1, 2, \qquad (11.80)$$

where $\varphi_2 = (\varphi_2^1, \varphi_2^2)$. Denote $y = (y_1, y_2)$, $f = (f_1, f_2)$, $\varphi = (\varphi_1, \varphi_2)$.

Remark 11.6 The equations (11.76)–(11.80) describe the movement of the incompressible viscous fluid with heat exchange. The function u_1 is the temperature here, u_2 is the velocity, and q is the pressure.

Determine again the set

$$D = \left\{ y \in \left[D(\Omega)\right]^2 \middle| \operatorname{div} y = 0 \right\}.$$

Denote by H_2 the closure of the set D to the space $\left[L_2(\Omega)\right]^2$, and W_2 is its closure to $\left[H_0^1(\Omega)\right]^2$. Determine the spaces

$$H_1 = L_2(\Omega), \ W_1 = H_0^1(\Omega), \ H = H_1 \times H_2, \ W = W_1 \times W_2,$$

$$X_i = L_2(0,T;W_i), \ X_i' = L_2(0,T;W_i'), \ i = 1, 2, \ X = X_1 \times X_2, \ X' = X_1 \times X_2,$$

$$Y = \left\{ y \middle| y \in X, \ y' \in X' \right\}, \ Z = L_\infty(0,T;H) \cap X.$$

Consider the functionals

$$a_1(\psi, \eta) = \sum_{i=1}^{2} \int_\Omega D_i \psi D_i \eta \, dx \ \forall \psi, \eta \in W_1,$$

$$a_2(\psi, \eta) = \nu \sum_{i,j=1}^{2} \int_\Omega D_i \psi^j D_i \eta^j \, dx \ \forall \psi, \eta \in W_2,$$

$$b_1(\psi, \eta, \zeta) = \sum_{i=1}^{2} \int_{\Omega} \psi^i D_i \eta \zeta dx \ \forall \psi \in W_2, \ \eta, \zeta \in W_1,$$

$$b_2(\psi, \eta, \zeta) = \sum_{i,j=1}^{2} \int_{\Omega} \psi^i D_i \eta^j \zeta^j dx \ \forall \psi, \eta, \zeta \in W_2.$$

We have as before the equalities

$$a_2(\psi, \psi) = \nu \|\psi\|_{W_2}^2 \ \forall \psi \in W_2, \tag{11.81}$$

$$b_2(\psi, \psi, \eta) = -b_2(\psi, \eta, \psi); \ b_2(\psi, \eta, \eta) = 0 \ \forall \psi, \eta \in W_2. \tag{11.82}$$

We use also its analogs

$$a_1(\psi, \psi) = \|\psi\|_{W_1}^2 \ \forall \psi \in W_1, \tag{11.83}$$

$$b_1(\psi, \psi, \eta) = -b_1(\psi, \eta, \psi); \ b_1(\psi, \eta, \eta) = 0 \ \forall \psi \in W_2, \eta \in W_1. \tag{11.84}$$

We considered before the inequality

$$\left| b_2(\psi, \eta, \zeta) \right|^2 \le c_2 \|\psi\|_{W_2} \|\psi\|_{H_2} \|\eta\|_{W_2} \|\eta\|_{H_2} \|\zeta\|_{W_2}^2 \ \forall \psi, \eta, \zeta \in W_2, \tag{11.85}$$

where $c_2 > 0$. There exists also its analog

$$\left| b_1(\psi, \eta, \zeta) \right|^2 \le c_1 \|\psi\|_{W_2} \|\psi\|_{H_2} \|\eta\|_{W_1} \|\eta\|_{H_1} \|\zeta\|_{W_1}^2 \ \forall \psi \in W_2, \ \eta, \zeta \in W_2. \tag{11.86}$$

where $c_1 > 0$.

Theorem 11.7 *For any $f \in X'$, $\varphi \in H$ the problem (11.76)–(11.80) has a unique solution $y \in Y$, $q \in D'(Q)$.*

Proof. 1. Transform the problem statement. Consider an arbitrary element $\eta = (\eta_1, \eta_2)$ of the space W, where $\eta_2 = (\eta_2^1, \eta_2^2)$. Multiply the equality (11.76) by η_1 and integrate the result. We have

$$\int_{\Omega} y_1' \eta_1 dx + \sum_{i=1}^{2} \int_{\Omega} D_i y_1 D_i \eta_1 dx + \sum_{i=1}^{2} \int_{\Omega} y_2^i D_i y_1 \eta_1 dx = \int_{\Omega} f_1 \eta_1 dx \ \forall \eta_1 \in W_1.$$

Now multiply the first equality (11.71) by η_2^1, and multiply its second equality by η_2^1. After summing and integration we have

$$\sum_{j=1}^{2} \int_{\Omega} (y_2')^j \eta_2^j dx + \nu \sum_{i,j=1}^{2} \int_{\Omega} D_i y_2^j D_i \eta_2^j dx + \sum_{i,j=1}^{2} \int_{\Omega} y_2^i D_i y_2^j \eta_2^j dx +$$

$$\sum_{j=1}^{2} \int_{\Omega} D_j q \eta_2^j dx + \sum_{j=1}^{2} \gamma^j \int_{\Omega} y_1 \eta_2^j dx = \sum_{j=1}^{2} \int_{\Omega} f_2^j \eta_2^j dx \ \forall \eta_2 \in W_2.$$

442 *Optimization and Differentiation*

Using Green's formula, the boundary conditions (11.79), and the equality $\operatorname{div} \eta_2 = 0$, we have

$$\langle y_1'(t), \eta_1 \rangle + a_1(y_1(t), \eta_1) + b_1(y_2(t), y_1(t)\eta_1) = \langle f_1(t), \eta_1 \rangle \ \forall \eta_1 \in W_1, \quad (11.87)$$

$$\langle y_2'(t), \eta_2 \rangle + \nu a_2(y_2(t), \eta_2) + b_2(y_2(t), y_2(t)\eta_2) + \langle \gamma y_1(t), \eta_2 \rangle =$$
$$\langle f_2(t), \eta_2 \rangle \ \forall \eta_2 \in W_2. \quad (11.88)$$

Now the problem (11.76)–(11.80) is transformed to the equations (11.87), (11.88) with the initial condition (11.80). We can prove the equivalence of these problems as before (see Theorem 11.1).

2. There exists the set of positive eigenvalues $\{\lambda_{2j}\}$ for the problem

$$(\eta, \mu)_{W_2} = \lambda(\eta, \mu)_{H_2} \ \forall \eta \in W_2$$

and the sequence $\{\mu_{2j}\} = \{(\mu_{2j}^1, \mu_{2j}^2)\}$ of the eigenfunctions that is the basis of the space W_2. Thus, the set of non-zero functions $\{\mu_{2j}\}$ satisfy the equalities

$$(\eta, \mu_{2j})_{W_2} = \lambda_{2j}(\eta, \mu_{2j})_{H_2} \ \forall \eta \in W_2.$$

We choose also the set of the eigenfunctions $\{\mu_{1j}\}$ for the problem

$$(\eta, \mu)_{W_1} = \lambda(\eta, \mu)_{H_1} \ \forall \eta \in W_1$$

as the basis of the space W_1.

Determine the approximate solution of the considered problem

$$y_k = y_k(t) = \big(y_{1k}(t), y_{2k}(t)\big) = \big(y_{1k}(t); y_{2k}^1(t), y_{2k}^2(t)\big)$$

by the formula

$$y_{ik}(t) = \sum_{m=1}^{k} \xi_{imk}(t)\mu_m, \ k = 1, 2, \dots, \ i = 1, 2. \quad (11.89)$$

The functions ξ_{1mk}, ξ_{2mk} here satisfy the equalities

$$\langle y_{1k}'(t), \mu_{1j} \rangle + a_1(y_{1k}(t), \mu_{1j}) + b_1(y_{2k}(t), y_{1k}(t), \mu_{1j}) = \langle f_1(t), \mu_{1j} \rangle, \quad (11.90)$$

$$\langle y_{2k}'(t), \mu_{2j} \rangle + \nu a_2(y_{2k}(t), \mu_{2j}) + b_2(y_{2k}(t), y_{2k}(t), \mu_{2j}) +$$
$$\langle \gamma y_{1k}(t), \mu_{2j} \rangle = \langle f_2(t), \mu_{2j} \rangle, \ j = 1, \dots, k. \quad (11.91)$$

Consider also the initial conditions

$$y_k(0) = \varphi_k \quad (11.92)$$

such that we have the convergence

$$\varphi_k \to \varphi \ \text{in} \ H. \quad (11.93)$$

Navier–Stokes equations   443

The problem (11.90)–(11.92) has a solution on an interval $(0, T_k)$.

3. Multiply j-th equality (11.87) by $\xi_{1mk}(t)$, multiply j-th equality (11.88) by $\xi_{2mk}(t)$ and sum the results. After summing by j by using the formula (11.89) we get

$$\sum_{i=1}^{2}\left[\langle y_{ik}'(t), y_{ik}(t)\rangle + a_i\big(y_{ik}(t), y_{ik}(t)\big) + b_i\big(y_{2k}(t), y_{ik}(t), y_{ik}(t)\big)\right] =$$

$$\sum_{i=1}^{2}\langle f_i(t), y_{ik}(t)\rangle - \langle \gamma y_{1k}(t), y_{2k}(t)\rangle.$$

Using the conditions (11.83)–(11.86), we have

$$\frac{1}{2}\frac{d}{dt}\big\|y_k(t)\big\|_H^2 + \nu'\big\|y_k(t)\big\|_W^2 \le \big|\langle f(t), y_k(t)\rangle\big| + \sum_{i=1}^{2}\gamma^i\left|\int_\Omega y_{1k}y_{2k}^i dx\right| \le$$

$$\frac{\nu'}{2}\big\|y_k(t)\big\|_W^2 + \frac{1}{2\nu'}\big\|f(t)\big\|_{W'}^2 + \frac{c}{2}\big\|y_k(t)\big\|_H^2,$$

where $\nu' = \min 1, \nu$. We denote by c different positive constants that do not depend upon k. After integration by using (11.92), (11.93), we have

$$\big\|y_k(t)\big\|_H^2 + \nu'\int_0^t\big\|y_k(\tau)\big\|_W^2 d\tau \le \big\|\varphi_k(t)\big\|_H^2 + \frac{1}{\nu'}\int_0^t\big\|f(\tau)\big\|_{W'}^2 d\tau +$$

$$c\int_0^t\big\|y_k(\tau)\big\|_H^2 d\tau \le c + c\int_0^t\big\|y_k(\tau)\big\|_H^2 d\tau.$$

From the inequality

$$\big\|y_k(t)\big\|_H^2 \le c + c\int_0^t\big\|y_k(\tau)\big\|_H^2 d\tau$$

it follows that

$$\big\|y_k(t)\big\|_H^2 \le c$$

because of the Gronwall lemma. Therefore, we can choose $T_k = T$, and the sequence $\{y_k\}$ is bounded in the space Z.

4. Use the property of the special basis for obtaining the additional estimate of the approximate solution. For $i = 1, 2$ determine the operator P_{ik} on the space H_i by the formula

$$P_{ik} = \sum_{m=1}^{k}\big(\psi, \mu_{im}\big)_{H_i}\mu_{im}, \quad k = 1, 2, \dots ,$$

444   *Optimization and Differentiation*

This is the projection operator of the space H_i to the set M_{ik} of the functions that are linear combinations of $\mu_{i1}, ..., \mu_{ik}$, i.e., $P_{ik}\psi = \psi$ for all $\psi \in M_{ik}$. We can prove (see Theorem11.1) that the norm of the operator P_{ik} is not greater than 1. We can consider it as an operator on the space W_i. Then its adjoint operator is determined on the space W_i'. It is equal to P_{ik}.

Determine the operator $B_1 : W_2 \times W_1 \to W_1'$ by the equality

$$\langle B_1(\psi, \eta), \zeta \rangle = b_1(\psi, \eta, \zeta) \ \forall \psi \in W_2, \ \eta, \zeta \in W_1.$$

Multiply j-th equality (11.90) by μ_{1j} and sum by j by using the definition of the operator P_{1k}. Using $y_{1k}'(t) \in M_{1k}$, we get

$$y_{1k}' = -P_{1k}\Delta y_{1k} - P_{1k}B_1(y_{2k}, y_{1k}) + P_{1k}f_1. \tag{11.94}$$

Determine the linear operator $A : W_2 \to W_2'$ and the nonlinear operator $B_2 : W_2 \to W_2'$ by the formulas

$$\langle A\psi, \eta \rangle = a_2(\psi, \eta), \ \langle B_2\psi, \eta \rangle = b_2(\psi, \psi, \zeta) \ \forall \psi, \eta \in W_2.$$

Multiply j-th equality (11.91) by μ_{2j} and sum by j by using the definition of the operator P_{2k}. Using $y_{2k}'(t) \in M_{2k}$, we get

$$y_{2k}' = -P_{2k}Ay_{2k} - P_{2k}B_2y_{2k} + P_{2k}f_2. \tag{11.95}$$

By the boundedness of the sequence $\{y_k\}$ in the space X and the boundedness of the operators $\Delta : W_1 \to W_1'$, $A : W_2 \to W_2'$, $B_1 : W_2 \times W_1 \to W_1'$, $B_2 : W_2 \to W_2'$, and $P_{ik} : W_i \to W_i'$, determine that the sequences $\{P_{1k}\Delta y_{1k}\}$, $\{P_{1k}B_1(y_{1k}, y_{2k})\}$, and $\{P_{1k}f_1\}$ are bounded in the space X_1', and the sequences $\{P_{2k}A_2y_{2k}\}$, $\{P_{2k}B_2y_{2k})\}$, $\{P_{2k}\gamma y_{1k}\}$, and $\{P_{1k}f_1\}$ are bounded in X_2'. From (11.94), (11.95) it follows that the sequences $\{y_{ik}'\}$ are bounded in the spaces X_i', $i = 1, 2$. Therefore, the sequence $\{y_k\}$ is bounded in Y. We can prove (see Theorem 11.1) that the sequence $\{y_{2k}^i y_{2k}^m\}$ is bounded in $L_2(Q)$, where $i, m = 1, 2$. Analogically, the sequence $\{y_{2k}^i y_{1k}\}$ is bounded in $L_2(Q)$.

5. After extracting a subsequence we have the convergence $y_k \to y$ weakly in Y, $y_{2k}^i y_{1k} \to \psi_1^i$ and $y_{2k}^i y_{2k}^m \to \psi_2^{im}$ weakly in $L_2(Q)$. Embedding of the space Y to $L_2(0, T; H)$ is compact by Theorem 8.5. Thus, we get the convergence $y_k \to y$ strongly in $L_2(0, T; H)$ and a.e. on Q. Then $y_{2k}^i y_{1k} \to y_2^i y_1$ and $y_{2k}^i y_{2k}^m \to y_2^i y_2^m$ a.e. on Q, $i, m = 1, 2$. Using the boundedness of the sequence $\{y_{2k}^i y_{2k}^m\}$ and $\{y_{2k}^i y_{1k}\}$ in $L_2(Q)$, we have $y_{2k}^i y_{1k} \to y_2^i y_1$ and $y_{2k}^i y_{2k}^m \to y_2^i y_2^m$ weakly in $L_2(Q)$. Now we can pass to the limit at the equalities (11.90)–(11.92) by using the technique of Theorem 11.1.

Multiply the equality (11.90) by an arbitrary function $\omega \in C^1[0, T]$. After integration we have

$$\int_0^T \langle y_{1k}'(t), \xi_{1j}(t) \rangle dt + \int_0^T a_1\big(y_{1k}(t), \xi_{1j}(t)\big)dt +$$

$$\int_0^T b_1\big(y_{2k(t)}, y_{1k}(t), \xi_{1j}(t)\big)dt = \int_0^T \langle f_1(t), \xi_{1j}(t)\rangle dt, \qquad (11.96)$$

where $\xi_{1j}(t) = \omega(t)\mu_{1j}$, $j = 1, ..., k$. Using the convergence $y'_{1k} \to y'_1$ weakly in X'_1 and $y_{1k} \to y_1$ weakly in X_1, we have

$$\int_0^T \langle y'_{1k}(t), \xi_{1j}(t)\rangle dt \to \int_0^T \langle y'_1(t), \xi_{1j}(t)\rangle dt,$$

$$\int_0^T a_1\big(y_{1k}(t), \xi_{1j}(t)\big)dt \to \int_0^T a_1\big(y_1(t), \xi_{1j}(t)\big)dt.$$

By the first equality (11.84) we obtain

$$\int_0^T b_1\big(y_{2k}(t), y_{1k}(t), \xi_{1j}(t)\big)dt = -\int_0^T b_1\big(y_{2k}(t), \xi_{1j}(t), y_{1k}(t)\big)dt =$$

$$\sum_{i=1}^2 \int_Q y_{2k}^i y_{1k} D_i \xi_{1j} dQ.$$

Using the convergence $y_{2k}^i y_{1k} \to y_2^i y_1$ weakly in $L_2(Q)$, we get

$$\lim_{k\to\infty} \int_0^T b_1\big(y_{2k}(t), y_{1k}(t), \xi_{1j}(t)\big)dt = \lim_{k\to\infty} \sum_{i=1}^2 \int_Q y_{2k}^i y_{1k} D_i \xi_{1j} dQ =$$

$$\sum_{i=1}^2 \int_Q y^{2i} y_1 D_i \xi_{1j} dQ = \int_0^T b_1\big(y_2(t), y_1(t), \xi_{1j}(t)\big)dt.$$

The function ω is arbitrary here. Then we have

$$\langle y'_1(t), \mu_{1j}(t)\rangle + a\big(y_1(t), \mu_{1j}(t)\big) + b_1\big(y_2(t), y_1(t), \mu_{1j}(t)\big) = \langle f_1(t), \mu_{1j}(t)\rangle,$$

$$j = 1, 2, ...$$

Then the equality (11.87) is true.

Analogically, multiply the equality (11.91) by the arbitrary function $\omega \in C^1[0, T]$. After integration we have

$$\int_0^T \langle y'_{2k}(t), \xi_{2j}(t)\rangle dt + \int_0^T a_2\big(y_{2k}(t), \xi_{2j}(t)\big)dt +$$

$$\int_0^T b_2\big(y_{2k(t)}, y_{2k}(t), \xi_{2j}(t)\big)dt + \int_0^T \langle \gamma y_{1k}(t), \xi_{2j}(t)\rangle dt = \int_0^T \langle f_2(t), \xi_{2j}(t)\rangle dt,$$

446   *Optimization and Differentiation*

where $\xi_{2j}(t) = \omega(t)\mu_{2j}$, $j = 1, ..., k$. Using the convergence $y_k \to y$ weakly in Y, we get

$$\int_0^T \langle y'_{2k}(t), \xi_{2j}(t)\rangle dt \to \int_0^T \langle y'_2(t), \xi_{2j}(t)\rangle dt,$$

$$\int_0^T a_2\big(y_{2k}(t), \xi_{2j}(t)\big) dt \to \int_0^T a_2\big(y_2(t), \xi_{2j}(t)\big) dt,$$

$$\int_0^T \langle \gamma y_{1k}(t), \xi_{2j}(t)\rangle dt \to \int_0^T \langle \gamma y_1(t), \xi_{2j}(t)\rangle dt.$$

By the first equality (11.82) we obtain

$$\int_0^T b_2\big(y_{2k}(t), y_{2k}(t), \xi_{2j}(t)\big) dt = - \int_0^T b_2\big(y_{2k}(t), \xi_{2j}(t), y_{2k}(t)\big) dt =$$

$$\sum_{i,m=1}^2 \int_Q y_{2k}^i y_{2k}^m D_i \xi_{2j}^m dQ.$$

Using the convergence $y_{2k}^i y_{2k}^m \to y_2^i y_2^m$ weakly in $L_2(Q)$, we get

$$\lim_{k\to\infty} \int_0^T b_2\big(y_{2k}(t), y_{2k}(t), \xi_{2j}(t)\big) dt = \lim_{k\to\infty} \sum_{i,m=1}^2 \int_Q y_{2k}^i y_{2k}^m D_i \xi_{2j} dQ =$$

$$\sum_{i,m=1}^2 \int_Q y_2^i y_2^m D_i \xi_{2j} dQ = \int_0^T b_2\big(y_2(t), y_2(t), \xi_{2j}(t)\big) dt.$$

The function ω is arbitrary here. Then for all $j = 1, 2, ...$ we have

$$\langle y'_2(t), \mu_{2j}(t)\rangle + a_2\big(y_2(t), \mu_{2j}(t)\big) + b_2\big(y_2(t), y_2(t), \mu_{2j}(t)\big) = \langle f_2(t), \mu_{2j}(t)\rangle.$$

Therefore, the equality (11.88) is true.

By continuous embedding of the space Y to $C(0, T; H)$ we have the convergence $y_k(0) \to y(0)$ weakly in H. From (11.92), (11.93) the initial condition (11.80) follows. Thus, the function $y \in Y$ satisfies the equality (11.87), (11.88), (11.80). Then the boundary problem (11.76)–(11.80) is solvable.

6. Prove the uniqueness of the solution of this problem. Suppose there exist the solutions y and z of the problem (11.76)–(11.80). Then $w = (w_1, w_2)$ satisfies the equalities

$$\langle w'_1(t), \eta_1\rangle + a_1\big(w_1(t), \eta_1\big) + b_1\big(w_2(t), y_1(t), \eta_1\big) + b_1\big(z_2(t), w_1(t), \eta_1\big) = 0,$$

$$\langle w_2'(t), \eta_2 \rangle + a_2(w_2(t), \eta_2) + b_2(w_2(t), y_2(t), \eta_2) + b_2(z_2(t), w_2(t), \eta_2) +$$
$$\langle \gamma w_1(t), \eta_2 \rangle = 0$$

for all $\eta_i \in W_i$, $i = 1, 2$. Determine $\eta_i = y_i(t)$. Using the equalities (11.81)–(11.84), we have

$$\frac{1}{2}\frac{d}{dt}\left\|w_1(t)\right\|_{H_1}^2 + \left\|w_1(t)\right\|_{W_1}^2 = b_1\big(w_2(t), w_1(t), y_1(t)\big) \le$$

$$\left[2\left\|w_2(t)\right\|_{W_2}\left\|w_2(t)\right\|_{H_2}\left\|w_1(t)\right\|_{W_1}\left\|w_1(t)\right\|_{H_1}\right]^{1/2}\left\|y_1(t)\right\|_{H_1},$$

$$\frac{1}{2}\frac{d}{dt}\left\|w_2(t)\right\|_{H_2}^2 + \nu\left\|w_2(t)\right\|_{W_2}^2 = b_2\big(w_2(t), w_2(t), y_2(t)\big) - \langle \gamma w_1(t), w_2 \rangle \le$$

$$\sqrt{2}\left\|w_2(t)\right\|_{W_2}\left\|w_2(t)\right\|_{H_2}\left\|y_2(t)\right\|_{W_2} + \left\|\gamma w_1(t)\right\|_{H_2}\left\|w_2(t)\right\|_{H_2}$$

because of the estimates (11.85), (11.86). Summing the inequalities, we have

$$\frac{1}{2}\frac{d}{dt}\left\|w(t)\right\|_H^2 + \nu'\left\|w(t)\right\|_W^2 \le \sqrt{2}\left\|y_1(t)\right\|_{W_1}\sum_{i=1}^{n}\left\|w_i(t)\right\|_{W_i}\left\|w_i(t)\right\|_{H_i} +$$

$$\sqrt{2}\left\|w_2(t)\right\|_{W_2}\left\|w_2(t)\right\|_{H_2}\left\|y_2(t)\right\|_{W_2} + |\gamma_1 + \gamma_2|\left\|w_1(t)\right\|_{H_1}\left\|w_2(t)\right\|_{H_2} \le$$

$$\frac{\nu'}{2}\left\|w(t)\right\|_W + c\left[\left\|y_1(t)\right\|_{W_1}^2 + \left\|y_2(t)\right\|_{W_2}^2 + |\gamma_1 + \gamma_2|\right]\left\|w(t)\right\|_H^2.$$

After integration we obtain

$$\left\|w(t)\right\|_H^2 + \nu'\int_0^t \left\|w(\tau)\right\|_W^2 d\tau \le$$

$$c\int_0^t \left[\left\|y_1(\tau)\right\|_{W_1}^2 + \left\|y_2(\tau)\right\|_{W_2}^2 + |\gamma_1 + \gamma_2|\right]\left\|w(\tau)\right\|_H^2 d\tau. \qquad (11.97)$$

Using the Gronwall lemma, we get the inequality

$$\left\|w(t)\right\|_H^2 \le 0.$$

Then $y = z$. This completes the proof of the theorem. $\square$

Now consider the optimization control problem for this system.

11.6 Optimization control problems for Navier–Stokes and heat equations

Consider the system, described by the equations

$$y_1' - \nu \Delta y_1 + \sum_{i=1}^{2} y_2^i D_i y_1 = v_1 + f_1, \tag{11.98}$$

$$y_2' - \nu \Delta y_2 + \sum_{i=1}^{2} y_2^i D_i y_2 + \nabla q + \gamma y_1 = v_2 + f_2, \tag{11.99}$$

$$\operatorname{div} y_2 = 0, \tag{11.100}$$

in the set Q with the boundary conditions

$$y_i(x, t) = 0, \ (x, t) \in \Sigma, \tag{11.101}$$

$$y_i(x, 0) = \varphi_i(x), \ x \in \Omega, \ i = 1, 2. \tag{11.102}$$

The control $v = (v_1, v_2)$ belongs to the convex closed subset $U = U_1 \times U_2$ of the space $V = L_2(0, T; H)$. From Theorem 11.7 it follows that for all $v \in V$ the problem (11.98)–(11.102) has a unique solution $y = y[v]$ from the space Y. Consider the functional

$$I(v) = \frac{\alpha}{2} \|v\|_V^2 + \frac{1}{2} \|y[v] - y_d\|_X^2,$$

where $\alpha > 0$, $y_d \in X$ is a known function.

Problem 11.3 *Find the control $u \in U$ for the system (11.98)–(11.102) that minimizes the functional I on the set U.*

Lemma 11.15 *The map $y[\cdot] : V \to Y$ for the problem (11.98)–(11.102) is weakly continuous.*

Proof. Suppose the convergence $v_k \to v$ weakly in V. The function $y_k = y[v_k]$ satisfies for all $\eta_i \in W_i$, $i = 1, 2$ the equalities

$$\langle y_{1k}'(t), \eta_1 \rangle + a_1(y_{1k}(t), \eta_1) + b_1(y_{2k}(t), y_{1k}(t), \eta_1) = \langle v_{1k}(t) + f_1(t), \eta_1 \rangle,$$

$$\langle y_{2k}'(t), \eta_2 \rangle + a_2(y_{2k}(t), \eta_2) + b_2(y_{2k}(t), y_{2k}(t), \eta_2) +$$
$$\langle \gamma y_{1k}(t), \eta_2 \rangle = \langle v_{2k}(t) + f_2(t), \eta_2 \rangle$$

that is the analog of (11.87), (11.88). Choose $\eta_i = y_{ik}(t)$, $i = 1, 2$. Sum these equalities and use the technique of the proof of Theorem 11.7. We obtain the inequality

$$\frac{d}{dt} \|y_k(t)\|_H^2 + 2\nu' \|y_k(t)\|_W^2 \leq \nu' \|y_k(t)\|_W^2 + \nu'^{-1} \|v_k(t) + f(t)\|_{W'}^2 + c\|y_k(t)\|_W^2.$$

Then the sequence $\{y_k\}$ is bounded in Z.

Consider the norms

$$\left\|y'_{1k}(t)\right\|_{W'_1} = \sup_{\|\eta\|_{W_1}=1} \left|\langle y'_{1k}(t), \eta\rangle\right| \leq$$

$$\sup_{\|\eta\|_{W_1}=1} \left[\left|a_1\big(y_{1k}(t), \eta\big)\right| + \left|b_1\big(y_{2k}(t), y_{1k}(t), \eta\big)\right| + \left|\langle v_{1k}(t) + f_1(t), \eta\rangle\right|\right],$$

$$\left\|y'_{2k}(t)\right\|_{W'_2} = \sup_{\|\eta\|_{W_2}=1} \left|\langle y'_{2k}(t), \eta\rangle\right| \leq \sup_{\|\eta\|_{W_2}=1} \left[\left|a_2\big(y_{2k}(t), \eta\big)\right| +\right.$$

$$\left.\left|b_2\big(y_{2k}(t), y_{2k}(t), \eta\big)\right| + \left|\langle \gamma y_{1k}(t), \eta\rangle\right| + \left|\langle v_{2k}(t) + f_2(t), \eta\rangle\right|\right].$$

Then the sequence $\{y'_k\}$ is in the space X' because of the continuity of the functionals a_i and b_i. Hence, the sequence $\{y_k\}$ is bounded in space Y. After extracting a subsequence we have the convergence $y_k \to y$ weakly in Y. Passing to the limit in the previous equalities by using the technique of the proof of Theorem 11.7, we have $y = y[v]$. $\square$

Using Lemma 11.15, we prove the existence of the optimal control.

Theorem 11.8 *Problem 11.3 is solvable.*

Determine the differentiability of the state function with respect to the control. Denote by y_0 and y_σ the solutions of the problem (11.98)–(11.102) for the controls u and $u + \sigma h$, where σh is a number, $h \in V$. Then we have

$$\langle \eta'_{1\sigma}[h](t), \varsigma_1\rangle + a_1\big(\eta_{1\sigma}[h](t), \varsigma_1\big) + b_1\big(y_{2\sigma}(t), \eta_{1\sigma}[h](t), \varsigma_1\big) +$$

$$b\big(\eta_{1\sigma}[h](t), y_{10}(t), \varsigma_1\big) = \langle h_1(t), \varsigma_1\rangle \,\,\forall \varsigma_1 \in W_1,$$

$$\langle \eta'_{2\sigma}[h](t), \varsigma_2\rangle + a_2\big(\eta_{2\sigma}[h](t), \varsigma_2\big) + b_2\big(y_{2\sigma}(t), \eta_{2\sigma}[h](t), \varsigma_2\big) +$$

$$b_2\big(\eta_{2\sigma}[h](t), y_{20}(t), \varsigma_2\big) + \langle \gamma \eta_{1\sigma}[h](t), \varsigma_2\rangle = \langle h_2(t), \varsigma_2\rangle \,\,\forall \varsigma_2 \in W_2,$$

where $\eta_\sigma[h] = (y_\sigma - y_0)/\sigma$. Sum these equalities. Multiplying the result by a function $\omega \in L_2(0, T)$, after the integration we have

$$\sum_{i=1}^{2} \Bigg\{ \int_0^T \langle \eta'_{i\sigma}[h](t), \lambda_i(t)\rangle dt + \int_0^T a_i\big(\eta_{i\sigma}[h](t), \lambda_i(t)\big) dt +$$

$$\int_0^T b_i\big(y_{2\sigma}(t), \eta_{i\sigma[h]}(t), \lambda_i(t)\big) dt + \int_0^T b_i\big(\eta_{2\sigma}[h](t), y_{i0}(t), \lambda_i(t)\big) dt \Bigg\} +$$

$$\int_0^T \langle \gamma \eta_{1\sigma}[h](t), \lambda_2(t)\rangle dt = \sum_{i=1}^{2} \int_0^T \langle h_i(t), \lambda_i(t)\rangle dt, \qquad (11.103)$$

450 *Optimization and Differentiation*

where $\lambda = (\lambda_1, \lambda_2)$, $\lambda_i = \omega\zeta_i$, $i = 1, 2$. Consider the following boundary problem

$$-p_1' - \Delta p_1 - \sum_{i=1}^{2} D_i\big(y_{2\sigma}^i p_1\big) + \sum_{i=1}^{2} \gamma_i p_2^i = \mu_{1Q}, \qquad (11.104)$$

$$-p_2' - \nu\Delta p_2 - \sum_{i=1}^{2} D_i\big(y_{2\sigma}^i p_2\big) + \sum_{i=1}^{2} Dy_{20}^i p_2^i + Dy_{10} p_1 + \nabla r = \mu_{2Q}, \quad (11.105)$$

$$\operatorname{div} p_2 = 0 \qquad (11.106)$$

in the set Q with the boundary conditions

$$p_i(x,t) = 0, \ (x,t) \in \Sigma, \qquad (11.107)$$

$$p_i(x,T) = \mu_{i\Omega}, \ x \in \Omega \ i = 1, 2, \qquad (11.108)$$

where

$$p_2 = (p_2^1, p_2^2), \ \mu_{2Q} = (\mu_{2Q}^1, \mu_{2Q}^2), \ \mu_{2\Omega} = (\mu_{2\Omega}^1, \mu_{2\Omega}^2).$$

Determine the function

$$p = (p_1, p_2), \ \mu_Q = (\mu_{1Q}, \mu_{2Q}), \ \mu_\Omega = (\mu_{1\Omega}, \mu_{2\Omega}).$$

Consider also the equations

$$-p_{1\mu}[u]' - \Delta p_{1\mu}[u] - \sum_{i=1}^{2} D_i\big(y_{20}^i p_{1\mu}[u]\big) + \sum_{i=1}^{2} \gamma_i p_{2\mu}^i[u] = \mu_{1Q}, \qquad (11.109)$$

$$-p_{2\mu}[u]' - \nu\Delta p_{2\mu}[u] - \sum_{i=1}^{2} D_i\big(y_{20}^i p_{2\mu}[u]\big) +$$

$$\sum_{i=1}^{2} Dy_{20}^i p_{2\mu}^i[u] + Dy_{10} p_{2\mu}[u] + \nabla r_\mu[u] = \mu_{2Q}, \qquad (11.110)$$

$$\operatorname{div} p_{2\mu}[u] = 0 \qquad (11.111)$$

in the set Q with the boundary conditions

$$p_\mu[u](x,t) = 0, \ (x,t) \in \Sigma, \qquad (11.112)$$

$$p_\mu[u](x,T) = \mu_\Omega, \ x \in \Omega. \qquad (11.113)$$

Lemma 11.16 *For all numbers σ and $h \in V$, $\mu_Q \in X'$, $\mu_\Omega \in H$ the problem* (11.104)–(11.108) *has a unique solution $p = p_\mu^\sigma[h]$ from the space Y and $r \in D'(Q)$, as if $\sigma \to 0$, then $p_\mu^\sigma \to p_\mu[u]$ weakly in Y.*

Navier–Stokes equations

Proof. 1. The problem (11.104)–(11.108) is equivalent to the equations

$$-\langle p_1'(t), \eta \rangle + a_1(p_1(t), \eta) + b_1(y_{2\sigma}(t), p_1(t), \eta) +$$

$$\langle \gamma \eta, p_2(t) \rangle = \langle \mu_{1Q}(t), \eta \rangle \; \forall \eta \in W_1, \tag{11.114}$$

$$-\langle p_2'(t), \eta \rangle + a_2(p_2(t), \eta) + b_2(y_{2\sigma}(t), \eta, p_2(t)) +$$

$$\sum_{i=1}^{2} \langle \eta, y_{i0} p_i(t) \rangle = \langle \mu_{2Q}(t), \eta \rangle \; \forall \eta \in W_2 \tag{11.115}$$

with condition (11.113). Determine the approximate solution $p_k = (p_{k1}, p_{k2})$ by the formula

$$p_{ik}(t) = \sum_{m=1}^{k} \xi_{imk}(t) \mu_m, \; k = 1, 2, \ldots \, i = 1, 2.$$

The functions ξ_{1mk}, ξ_{2mk} satisfy here the equations

$$-\langle p_{1k}'(t), \mu_{1j} \rangle + a_1(p_{1k}(t), \mu_{1j}) + b_1(y_{2\sigma}(t), \mu_{1j}, p_{1k}(t)) +$$

$$\langle \gamma \mu_{1j}, p_{2k}(t), \rangle = \langle \mu_{Q1}(t), \mu_{1j} \rangle, \tag{11.116}$$

$$-\langle p_{2k}'(t), \mu_{2j} \rangle + a_2(p_{2k}(t), \mu_{2j}) + b_2(y_{2\sigma}(t), \mu_{2j}, p_2(t)) +$$

$$\sum_{i=1}^{2} \langle \mu_{2j}, y_{i0} p_{ik}(t) \rangle = \langle \mu_{Q2}(t), \mu_{2j} \rangle, \; j = 1, \ldots, k \tag{11.117}$$

with the final condition

$$p_k(T) = p_{0k}, \tag{11.118}$$

where $p_{0k} \to \mu_\Omega$ in I. This problem has a unique solution on the interval $(0, T)$.

2. Multiply j-th equality (11.116) by ξ_{1jk}, and multiply (11.117) by ξ_{2jk}. After summing we get inequality

$$-\frac{1}{2}\frac{d}{dt}\big\|p_k(t)\big\|_H^2 + \nu'\big\|p_k(t)\big\|_W^2 \le \sum_{i=1}^{2} \big|b_i(p_{ik}(t), y_{i0}(t), (p_{ik}(t))\big| +$$

$$\big|\langle \gamma p_{1k}(t), p_{2k}(t) \rangle\big| + \big|\langle \mu_Q(t), p_k(t) \rangle\big| \le c \sum_{i=1}^{2} \big\|p_{ik}(t)\big\|_{W_i}\big\|p_{ik}(t)\big\|_{H_i}\big\|y_{i0}(t)\big\|_{W_i} +$$

$$c\big\|p_{1k}(t)\big\|_{H_1}^2\big\|p_{2k}(t)\big\|_{H_2}^2 + \big\|\mu_Q(t)\big\|_{W'}\big\|p_k(t)\big\|_W \le$$

$$\frac{\nu'}{2}\big\|p_k(t)\big\|_W^2 + \big[c\big\|y_0(t)\big\|_W^2 + c\big]\big\|p_k(t)\big\|_H^2 + c\big\|\mu_Q(t)\big\|_{W'}^2.$$

452   *Optimization and Differentiation*

with different positive constant c. Integrating the result by using the condition (11.113), we have

$$\left\|p_k(t)\right\|_H^2 + \nu' \int_t^T \left\|p_k(\tau)\right\|_H^2 d\tau \leq$$

$$\left\|\mu_\Omega\right\|_H^2 + c\left\|\mu_Q\right\|_{X'}^2 + c \int_t^T \left\|p_k(\tau)\right\|_H^2 \left[\left\|y_0(\tau)\right\|_W^2 + 1\right] d\tau.$$

Using the Gronwall lemma, determine the boundedness of the sequence $\{p_k\}$ in the space Z.

3. After extracting a subsequence we have the convergence $p_k \to p$ weakly in X. Multiply the equality (11.116) by an arbitrary function $\omega \in C^1[0,T]$ that is equal to zero for $t = 0$ and $t = T$. After integration we have

$$\int_0^T \langle \xi_j'(t), p_{1k}(t) \rangle dt + \int_0^T a_1\big(p_{1k}(t), \xi_j(t)\big) dt + \int_0^T b_1\big(y_{2\sigma}(t), \xi_j(t), p_{1k}(t)\big) dt +$$

$$\int_0^T \langle \gamma \xi_j(t), p_{2k}(t) \rangle dt = \int_0^T \langle \mu_{1Q}(t), \xi_j(t) \rangle dt,$$

where $\xi_j = \omega \mu_{1j}$. Passing to the limit by using the technique of the proof of Lemma 11.4, we obtain the equality (11.114). Pass to the limit analogically in the equality (11.117). Find the time derivatives from the equalities (11.104), (11.105); we get $p' \in X'$. Therefore, $p \in Y$. The uniqueness of this solution is obvious because of the linearity of the system.

4. Suppose the convergence $\sigma \to 0$. Then $y_\sigma \to y_0$ weakly in Y. Hence, the set $\{y_\sigma\}$ is bounded in the space Y. Using the known technique, we obtain that the solution $p_\mu^\sigma[h] = (p_{1\mu}^\sigma[h], p_{2\mu}^\sigma[h])$ of the problem (11.104)–(11.108) satisfies the inequality

$$\left\|p_\mu^\sigma[h](t)\right\|_H^2 + \nu' \int_t^T \left\|p_\mu^\sigma[h](\tau)\right\|_W^2 d\tau \leq \left\|\mu_\Omega\right\|_H^2 + c\left\|\mu_Q\right\|_{X'}^2 +$$

$$c \int_t^T \left\|p_\mu^\sigma[h](\tau)\right\|_H^2 \left[\left\|y_0(\tau)\right\|_W^2 + 1\right] d\tau.$$

Then the set $\{p_\mu^\sigma[h]\}$ is bounded in the space Z. Find the time derivatives from the equalities (11.104), (11.105). We obtain the boundedness of $\{p_\mu^\sigma[h]'\}$ in the space X'. Then the set $\{p_\mu^\sigma[h]\}$ is bounded in Y. After extracting a subsequence we have the convergence $p_\mu^\sigma[h] \to p_\mu$ weakly in Y, where $p_\mu = (p_{1\mu}, p_{2\mu})$.

$$\text{Navier–Stokes equations} \qquad 453$$

Multiply the equality (11.104) by a function $\zeta \in X_1'$. After integration we have

$$\int_0^T \langle p_{1\mu}^\sigma[h]'(t), \zeta(t) \rangle dt + \int_0^T a_1\big(p_{1\mu}^\sigma[h](t), \zeta(t)\big) dt + \int_0^T b_1\big(y_{2\sigma}(t), \zeta(t), p_{1\mu}^\sigma[h](t)\big) dt +$$

$$\int_0^T \langle p_{2\mu}^\sigma[h](t), \gamma\zeta(t) \rangle dt = \int_0^T \langle \mu_{1Q}(t), \zeta(t) \rangle dt \ \ \forall \zeta \in X_1'. \qquad (11.119)$$

Analogically, from the equality (11.105) it follows that

$$\int_0^T \langle p_{2\mu}^\sigma[h]'(t), \zeta(t) \rangle dt + \int_0^T a_2\big(p_{2\mu}^\sigma[h](t), \zeta(t)\big) dt + \int_0^T b_2\big(y_{2\sigma}(t), \zeta(t), p_{2\mu}^\sigma[h](t)\big) dt +$$

$$\sum_{i=1}^2 \int_0^T b_i\big(\zeta(t), y_{i0}(t), p_{i\mu}^\sigma[h](t)\big) dt = \int_0^T \langle \mu_{2Q}(t), \zeta(t) \rangle dt \ \ \forall \zeta \in X_2'. \qquad (11.120)$$

The convergence for the linear terms of this equality is obvious.

$$\int_0^T \langle p_{i\mu}^\sigma[h]'(t), \zeta(t) \rangle dt \to \int_0^T \langle p_{i\mu}'(t), \zeta(t) \rangle dt,$$

$$\int_0^T a_i\big(p_{i\mu}^\sigma[h](t), \zeta(t)\big) dt \to \int_0^T a_i\big(p_{i\mu}(t), \zeta(t)\big) dt,$$

$$\int_0^T b_i\big(\zeta(t), y_{i0}(t), p_{i\mu}^\sigma[h](t)\big) dt \to \int_0^T b_i\big(\zeta(t), y_{i0}(t), p_{i\mu}(t)\big) dt,$$

$$\int_0^T \langle p_{2\mu}^\sigma[h](t), \gamma\zeta(t) \rangle dt \to \int_0^T \langle p_{2\mu}(t), \gamma\zeta(t) \rangle dt.$$

Consider the equalities

$$\int_0^T b_1\big(y_{2\sigma}(t), \zeta(t), p_{1\mu}^\sigma[h](t)\big) dt = \sum_{i=1}^2 \int_Q y_{2\sigma}^i D_i \zeta p_{1\mu}^\sigma[h] dQ,$$

$$\int_0^T b_2\big(y_{2\sigma}(t), \zeta(t), p_{2\mu}^\sigma[h](t)\big) dt = \sum_{i,j=1}^2 \int_Q y_{2\sigma}^i D_i \zeta^j p_{2\mu}^\sigma[h]^j dQ.$$

454   *Optimization and Differentiation*

By Lemma 11.4 the sets $\{y_{2\sigma}^i\}$, $\{p_{1\mu}^\sigma[h]\}$, and $\{p_{2\mu}^\sigma[h]^j\}$ are bounded in the space $L_4(Q)$. Then we obtain the boundedness of $\{y_{2\sigma}^i p_{1\mu}^\sigma[h]\}$ and $\{y_{2\sigma}^i p_{2\mu}^\sigma[h]^j\}$ in $L_2(Q)$. Using compact embedding of the space Y to $L_2(0,T;H)$, we have $y_{2\sigma}^i \to y_{20}^i$, $p_{1\mu}^\sigma[h] \to p_{1\mu}$, and $p_{2\mu}^\sigma[h]^j \to p_{2\mu}^j$ a.e. on Q. Then $y_{2\sigma}^i p_{1\mu}^\sigma[h] \to y_{20}^i p_{1\mu}$, $y_{2\sigma}^i p_{2\mu}^\sigma[h]^j \to y_{20}^i p_{2\mu}^j$ a.e. on Q, and weakly in $L_2(Q)$. Therefore, we have the convergence

$$\int_Q y_{2\sigma}^i D_i \zeta p_{1\mu}^\sigma[h] dQ \to \int_Q y_{20}^i D_i \zeta p_{1\mu} dQ,$$

$$\int_Q y_{2\sigma}^i D_i \zeta^j p_{2\mu}^\sigma[h]^j dQ \to \int_Q y_{20}^i D_i \zeta^j p_{2\mu}^j dQ.$$

Then

$$\int_0^T b_1\big(y_{2\sigma}(t), \zeta(t), p_{1\mu}^\sigma[h](t)\big) dt \to \int_0^T b_1\big(y_{20}(t), \zeta(t), p_{1\mu}(t)\big) dt,$$

$$\int_0^T b_2\big(y_{2\sigma}(t), \zeta(t), p_{2\mu}^\sigma[h](t)\big) dt \to \int_0^T b_1\big(y_{20}(t), \zeta(t), p_{2\mu}(t)\big) dt.$$

Passing to the limit in the equalities (11.119), (11.120), we prove that P_μ is the solution of the boundary problem (11.109)–(11.113). $\square$

Lemma 11.17 *The operator $y[\cdot] : V \to Y$ for the problem (11.98)–(11.102) has $(V, X; V, X^w)$-extended derivative $y'[u]$ at the arbitrary point $u \in V$ as there exists a linear continuous operator $y_T[\cdot] : V \to H$ such that for all $h \in V$ we have the convergence $\{y[u + \sigma h](T) - y[u](T)\}/\sigma \to y_T[u]h$ weakly in H. Besides, for all $h \in V$, $\mu_Q \in X'$, $\mu_\Omega \in H$ the following equality holds:*

$$\langle \mu_Q, y'[u]h \rangle + \big(\mu_\Omega, y_T[u]h\big)_H = \big(h, p_\mu[u]\big)_V. \tag{11.121}$$

Proof. For $\mu_\Omega = 0$ the equality (11.121) determines the linear continuous operator $y[u] : V \to X$. Choose $\lambda = p_\mu^\sigma[h]$ at the equality (11.103). For all $h \in V$, $\mu_Q \in X'$, $\mu_\Omega \in H$ we get

$$\langle \mu_Q, \big(y[u+\sigma h] - y[u]\big)/\sigma \rangle + \big(\mu_\Omega, \{y[u+\sigma h](T) - y[u](T)\}/\sigma\big)_H = \big(h, p_\mu^\sigma[h]\big)_V.$$

Using (11.121) with $\mu_\Omega = 0$, for all $h \in V$, $\mu_Q \in X'$ we have

$$\langle \mu_Q, \big(y[u + \sigma h] - y[u]\big)/\sigma - y'[u]h \rangle + \big(\mu_\Omega, y_T[u]h\big)_H = \big(h, p_\mu^\sigma[h] - p_\mu[u]\big)_V.$$

Pass to the limit by using Lemma 11.16. We obtain the convergence $(y[u + \sigma h] - y[u])/\sigma \to y'[u]h$ weakly in X for all $h \in V$. The final proposition of the lemma can be proved as before. $\square$

Prove the following result that is an analog of Lemma 11.7.

Navier–Stokes equations 455

Lemma 11.18 *The operator* $y[\cdot] : V \to X$ *for the problem* (11.98)–(11.102) *is continuous.*

Proof. Consider arbitrary controls u and v, and the corresponding solutions y and z of the problem (11.98)–(11.102). Then the difference $w = (w_1, w_2) = y - z$ for all $\eta_1 \in W_1$, $\eta_2 \in W_2$ satisfies the equalities

$$\langle w_1'(t), \eta_1 \rangle + a_1(w_1(t), \eta_1) + b_1(w_2(t), y_1(t), \eta_1) + b_1(z_2(t), w_1(t), \eta_1) = \langle \xi_1(t), \eta_1 \rangle,$$

$$\langle w_2'(t), \eta_2 \rangle + a_2(w_2(t), \eta_2) + b_2(w_2(t), y_2(t), \eta_2) + b_2(z_2(t), w_2(t), \eta_2) +$$

$$\langle \gamma w_1(t), \eta_2 \rangle = \langle \xi_2(t), \eta_2 \rangle,$$

where $\xi = (\xi_1, \xi_2) = u - v$. Determine $\eta_i = y_i(t)$, $i = 1, 2$ and repeat the transformation from the final part of Theorem 11.7. We obtain the inequality

$$\frac{1}{2} \|w(t)\|_H^2 + \nu' \int_0^t \|w(\tau)\|_W^2 d\tau \le$$

$$c \int_0^t \left[\|y_1(\tau)\|_{W_1}^2 + \|y_2(\tau)\|_{W_2}^2 + |\gamma^1| + |\gamma^2| \right] \|w(\tau)\|_H^2 d\tau + \int_0^t \|\xi(\tau)\|_H \|w(\tau)\|_H d\tau \le$$

$$c \int_0^t \left[\|y_1(\tau)\|_{W_1}^2 + \|y_2(\tau)\|_{W_2}^2 + |\gamma^1| + |\gamma^2| + \frac{1}{2c} \right] \|w(\tau)\|_H^2 d\tau + \frac{1}{2} \|\xi\|_V^2$$

that is the analog of (11.97). We complete the proof of the lemma by using the Gronwall lemma. $\square$

Now we obtain the following result that is the obvious analog of Lemma 11.8. Consider the equations

$$-p_1' - \Delta p_1 - \sum_{i=1}^{2} D_i(y_2^i[u]p_1) + \sum_{i=1}^{2} \gamma_i p_2^i = \Delta y_{1d} - \Delta y_1[u], \qquad (11.122)$$

$$-p_2' - \nu \Delta p_2 - \sum_{i=1}^{2} D_i(y_2^i[u]p_2) + \sum_{i=1}^{2} D y_2^i[u]p_2^i + D y_1[u]p_1 + \nabla r = \Delta y_{2d} - \Delta y_2[u],$$

$$(11.123)$$

$$\operatorname{div} p_2 = 0 \qquad (11.124)$$

in the set Q with the boundary conditions

$$p(x, t) = 0, \ (x, t) \in \Sigma, \qquad (11.125)$$

$$p(x, T) = 0, \ x \in \Omega. \qquad (11.126)$$

456 *Optimization and Differentiation*

Lemma 11.19 *The functional I for Problem* 11.3 *has Gâteaux derivative* $I'(u) = \alpha u + p$ *at the arbitrary point* $u \in V$.

Put the value of the functional derivative in to the standard variational inequality. We obtain the necessary condition of optimality.

Theorem 11.9 *If u is the solution of Problem* 11.3, *then the variational inequality holds:*

$$\big(\alpha u + p, v - u\big)_V \ \forall v \in U. \tag{11.127}$$

Now we have the system that includes the boundary problem (11.99)–(11.102) for $v = u$, the adjoint system (11.122)–(11.126), and the variational inequality (11.127) for finding the solution of the considered optimization control problem.

11.7 Comments

The theory of Navier–Stokes equations including the proof of the existence and the uniqueness of its solutions and Lemma 11.1, Lemma 11.2, Lemma 11.4, and the formulas (11.7)–(11.9) is described by O.A. Ladyzhenskaya [292], J.L. Lions [316], and R. Temam [516]. The solvability of the optimization problems for the systems described by the non-stationary second-dimensional Navier–Stokes equations is considered, for example, by J. Baranger [47], V. Barbu [52], T. Bewley, R. Temam, and M. Ziane [68], S. Chaabane, Ferchichi, and K. Kunisch [120], M. Desai and K. Ito [139], O.Yu. Imanuilov [249], S.Ya. Serovajsky [488], and D. Wachsmuth [545]. T. Bewley, R. Temam, and M. Ziane [68] and N. Bilić [71] prove the uniqueness of its solution. V. Barbu [52], N. Bilić [71], M. Desai and K. Ito [139], S.Ya. Serovajsky [488], and S. Sritharan [511] determine the necessary conditions of optimality. D. Wachsmuth [545] obtains the sufficient conditions of optimality. T. Bewley, R. Temam, and M. Ziane [68], N. Bilić [71], M. Desai and K. Ito [139], M. Hintermüller and M. Hinze [243], L.S. Hou and S.S. Ravindran [244], and D. Wachsmuth [545] find the approximate solution of these problems. V. Barbu [52] considers the time optimal problem. J. Haslinger, J. Malek, and J. Stebel [237], A. Henrot and J. Sokolovsky [239] (the review), S. Manservisi [338], and B.A. Ton [526] analyze set control problems for these systems. The optimal control problems for the multi-dimensional Navier–Stokes equations are considered by V. Barbu [52], N. Bilić [71], A.V. Fursikov [194], [195], [197], L.S. Hou and S.S. Ravindran [244], and G. Wang [547], [546]. J.L. Lions [315] and J.P. Raymond [425] analyze the optimization problems for the linearized Navier–Stokes equations.

The solvability of the optimization problems for the stationary Navier–Stokes equations is proved by A.V. Fursikov [197], J.P. Gossez [220], K. Ito and S.S. Ravindran [259], J.C. Reyes and R. Griesse [428], and S.Ya. Serovajsky [488]. E. Casas, M. Mateos, and J.-P. Raymond [14], A.V. Fursikov [197], L.S. Hou and T.P. Svobodny [245], K. Ito and S.S. Ravindran [259], J.C. Reyes and R. Griesse [428], S.Ya. Serovajsky [488], and G. Wang [547] determine the necessary conditions of optimality for these problems. E. Casas, M. Mateos, and J.-P. Raymond [114], J.C. Reyes and R. Griesse [428], and J.C. Reyes and F. Tröltzsch [429] analyze the sufficient conditions of optimality. E. Casas, M. Mateos, and J.-P. Raymond [114], M. Gunzburger, L. Hou, and T. Svobodny [226], [225], and J.C. Reyes and F. Tröltzsch [429] analyze its approximate solutions.

Note also the optimal control problems for the different mathematical physics problems. R. Cuvelier [134] and S.Ya. Serovajsky [488] consider the system of Navier–Stokes equations

Navier–Stokes equations

and heat equation. M. Gunzburger and C. Trenchea [227] and L. Wang [549] analyze magnetohydrodynamics problems. T. Roubiček [434] and Z. Zhao and L. Tian [567] consider *Burgers equation*. H.-C. Lee and O.Yu. Imanuvilov [304] analyze the *Boussinesq equation*. Z. Zhao and L. Tian [567] consider the *Korteweg–de Vries equation*. E. Cances, C. Le Bris, and M. Pilot [106] consider the *Schrödinger equation*. J.Y. Park and J.U. Jeong [396] analyze the *Klein–Gordon equation*. The optimization problems for a Stefan problem are considered by V. Barbu [49], [51], A. Friedman, S. Huang, and J. Yong [189], P. Neittaanmäki and D. Tiba [383], T. Roubiček [434], E. Sachs [442], Ch. Saguez [444], and D. Tiba [519].

Part IV

Addition

Chapter 12

Functors of the differentiation

12.1	Elements of the categories theory	461
12.2	Differentiation functor and its application to the extremum theory	463
12.3	Additional properties of extended derivatives	466
12.4	Extended differentiation functor and its application to the extremum theory	469
12.5	Comments	471

We interpret the operator of the differentiation as a functor. We use this result for obtaining necessary conditions of optimality by using categories theory notions. The classical and extended differentiations are considered.

12.1 Elements of the categories theory

Determine general notions of the categories theory.

Definition 12.1 *The **category** Γ is the pair $(Ob\,\Gamma, Mor\,\Gamma)$ with the class of **objects** $Ob\,\Gamma$ and the class of **morphisms** $Mor\,\Gamma$ that satisfy the following conditions:*
1) for all objects X, Y there exists a set of morphisms $H(X,Y)$;
2) for all morphisms A there exist unique objects X and Y that are called the beginning and the end of the morphism such that $A \in H(X,Y)$;
3) for all morphisms $A \in H(X,Y)$, $B \in H(Y,Z)$ there exists a composition of morphisms $A \circ B \in H(X,Z)$;
4) the composition of morphisms is associative, i.e., the following equality holds $A \circ (B \circ C) = (A \circ B) \circ C$, if these compositions have the sense;
5) for all objects X there exists a unit morphism E_X such that $E_X \circ A = A$, $B \circ E_X = B$, if these compositions have the sense.

Consider examples.

Example 12.1 *Sets category*. The objects of this category are the sets. For all sets X and Y the set of morphisms $H(X,Y)$ is the set of all operators with the domain X and the codomain Y. Hence, the first and the second

461

462   *Optimization and Differentiation*

properties of Definition 12.1 are true. The composition of the morphisms is its superposition. It satisfies the property of associativity. Finally, the unit morphism E_X is the unit operator on the set X.

Remark 12.1 We determine the set $H(X,Y)$, if both objects are non-empty. For all non-empty sets X we determine $(X, \emptyset) = \emptyset$. If $X \neq \emptyset$, then we determine $(\emptyset, X) = \Omega_X$, where $\Omega_X : \emptyset \to X$ maps the set $\emptyset$ to the empty subset of X. Finally, we determine $(\emptyset, \emptyset) = E_\emptyset$, where $E_\emptyset$ is the identity transformation of the empty set to itself.

Example 12.2 *Banach spaces category.* It has Banach spaces as the objects and the linear continuous operators as the morphisms. For all Banach spaces X, Y the set $H(X,Y)$ includes all linear continuous operators from X to Y. The composition of the linear continuous operators is linear continuous, and the unit operator is linear continuous too.

Definition 12.2 *The morphism $A \in H(X,Y)$ of the category Γ is called the* **isomorphism,** *if there exists a morphism $A^{-1} \in H(Y,X)$ that is called the* **inverse morphism** *such that $A \circ A^{-1} = E_X$, $A^{-1} \circ A = E_Y$.*

The inverse morphism of the sets category is the inverse operator, and the bijection is its isomorphism. Consider the Banach spaces category. If a linear continuous operator, i.e., its morphism, is the bijection, then its inverse operator is continuous by the Banach theorem about the inverse operator. Then this is the isomorphism of the considered category.

Definition 12.3 *The category Σ is called the* **subcategory** *of the category Γ, if $Ob\,\Sigma \subset Ob\,\Gamma$, $Mor\,\Sigma \subset Mor\,\Gamma$, as the composition of the morphisms in Σ is equal to the composition of these morphisms in Γ, and the unit morphism of each object in Σ is equal to the unit morphism of this object in Γ.*

The category of Banach spaces is the subcategory of the sets category. It is also the subcategory of the category of linear continuous spaces, where the objects are the linear normalized spaces, and the morphisms are the linear continuous operators. For all category Γ the category with objects of Γ and isomorhpisms of Γ is the subcategory of Γ. For example, the category with all sets as the objects and all bijections as morphisms is the subcategory of the sets category.

Definition 12.4 *Consider categories Γ, Σ and the pair of mapping $F : Ob\,\Gamma \to Ob\,\Sigma$ and $F : Mor\,\Gamma \to Mor\,\Sigma$. This is called the* **functor** *$F : \Gamma \to \Sigma$, if for all objects X of Γ we have $F(E_X) = E_{F(X)}$ and $F(A \circ B) = F(A) \circ F(B)$ for all morphisms A, B of Γ, where the end of A is equal to the start of B. If the previous equality is replaced by $F(A \circ B) = F(B) \circ F(A)$, then F is called the* **cofunctor.**

The transformation that maps the linear normalized space to its completion and each linear continuous operator to mapping of the corresponding completions is the completion functor from the category of the linear normalized spaces to the category of Banach spaces.

Functors of the differentiation 463

Example 12.3 *General cofunctor* of the category of Banach spaces. Determine the map H on the category of Banach spaces. For all objects X of this category determine the set $H(X)$ of all morphisms from X to $\mathbb{R}$ and for all morphisms A with the start X and the end Y determine the transformation $H(A)$ of $H(Y)$ to $H(X)$ such that $H(A)B = A \circ B$ for all $B \in H(Y)$. Thus, each Banach space X that is the object of the category is mapped to the set of all linear continuous functionals on X that is the adjoint space X'. This is a Banach space, i.e., the object of the same category. Then each linear continuous operator A between Banach spaces X and Y that is the morphism of this category is mapped to the linear continuous operator A^* between the adjoint spaces Y' and X' that satisfies the equality

$$\langle p, Ax \rangle = \langle A^*p, y \rangle \ \forall x \in X, \ p \in Y'.$$

This is the adjoint operator. Suppose E_X is the unit operator on the Banach space X. Then $H(E_X)$ is its adjoint operator that is the unit operator on the adjoint space X'. Then we have $H(E_X) = E_{H(X)}$. Let X, Y, Z be Banach spaces and $A : X \to Y$ and $B : Y \to Z$ be linear continuous operators. The composition $A \circ B$ is the linear continuous operator BA. Then $H(A \circ B)$ is the adjoint operator to BA, i.e., the linear continuous operator A^*B^*. This is the superposition of the adjoint operators, i.e., the morphism $H(B) \circ H(A)$. Hence, we obtain the equality $H(A \circ B) = H(B) \circ H(A)$. Thus, H is the cofunctor on the category of Banach spaces that is called the general cofunctor of this category.

Now we prove that the differentiation can be interpreted as a functor, and the necessary conditions of optimality can be transformed by using the categories theory.

12.2 Differentiation functor and its application to the extremum theory

Consider the set of all pairs (X, x), where X is a Banach space, and x is a fixed point of the space X. We have the set of Banach spaces with fixed points. For all two pairs (X, x) and (Y, y) determine an operator $A : X \to Y$ that is determined on a neighborhood of the point x and Fréchet differentiable at this point such that $Ax = y$. Suppose the pairs (Y, y) and (Z, z) have the analogical relation with an operator B. Then the composition BA maps the space X to Z, as $BAx = z$. By the Composite function theorem the operator BA that is determined on the neighborhood of the point x is Fréchet differentiable at this point, as $(BA)'(x) = B'(y)A'(x)$. Then each operator A with the considered properties determines the morphism A_x with the start (X, x) and the end (Y, y), as the operator BA determines the morphism $A_x \circ B_y$. Denote this

464   *Optimization and Differentiation*

category by Γ. Note that we can determine the natural addition on the set $H(X,Y)$.

Consider a morphism A_x of the category Γ such that the operator A is continuously differentiable on neighborhood O of the point x of its domain, as the derivative $A'(x)$ is invertible. By the Inverse function theorem, there exists the inverse operator A^{-1} that is continuously differentiable on a neighborhood O' of the point $y = Ax$, as $(A^{-1})'(\xi) = \{A'[A^{-1}(\xi)]\}^{-1}$ for all $\xi \in O'$. Thus, the inverse operator determines the morphism $(A^{-1})_y$ of the category Γ that is inverse to A_x, i.e., $(A^{-1})_y = (A_x)^{-1}$. Therefore, the set of Banach spaces with fixed points and the considered morphisms is the subcategory of the category Γ, where each morphism is the isomorphism.

Determine the transformation D of Γ to the category Σ Banach spaces. For all objects (X,x) and morphism A_x with the start (X,x) and the end (Y,y) of the category Γ determine $D(X,x) = X$ and $D(A_x) = A'(x)$. The unit morphism $E_{(X,x)}$ of Σ is the unit operator E_X on X. Then $D\big(E_{(X,x)}\big)$ is the derivative of E_X at the point x, i.e., the operator E_X. However, it is the unit morphism of the category Σ for the object $X = D(X,x)$. Then we have the equality $D(A_x) = A'(x)$. Now suppose B_y is a morphism of the category Γ with the start (Y,y). By the Composite function theorem, we have

$$D(A_x \circ B_y) = (BA)'(x) = D(A_x) \circ D(B_y).$$

Therefore, D is the functor from Γ to the category Σ. If the derivative $A'(x)$ is invertible, then from the equality $(A^{-1})'(y) = [A'(y)]^{-1}$ it follows that $D[(A^{-1})_y] = D(A_x)^{-1}$. Thus, we have the following result.

Theorem 12.1 *The differentiation operator is the functor of the category of Banach spaces with fixed points Γ to the category of Banach spaces Σ.*

Apply these results to the extremum theory. Consider the state equation

$$Ay = v, \tag{12.1}$$

where $A : Y \to V$ is a continuously differentiable operator, Y, V are Banach spaces, v is the control, and y is the state function. Suppose for all $v \in V$ the equation (12.1) has a unique solution $y = y[v]$ from the space Y. Determine the functional $I : V \to \mathbb{R}$ by the equality

$$I(v) = J(v) + K(y[v]),$$

where $J : V \to \mathbb{R}$, $K : Y \to \mathbb{R}$ are continuously differentiable functionals.

Problem 12.1 *Find a point $u \in V$ that minimizes the functional I on the space V.*

Functors of the differentiation 465

The necessary condition of extremum for the smooth functional I on all space at the point u is the stationary condition $I'(u)$. This functional determines the morphism I_u. Then the stationary condition is transformed to $D(I_u) = 0$. Suppose u is the solution of the Problem 12.1, and y is the corresponding solution of the equation (12.1). If the derivative $A'(y)$ is invertible, then the operator A^{-1} is continuously differentiable at the point u. Consider the morphisms A_y, J_u, K_y, and $(A^{-1})_u$ of the category Γ. We have $I_u = J_u + (A^{-1})_u \circ K_y$. Using the linearity of the derivative, we find

$$D(I_u) = D(J_u) + D\big[(A^{-1})_u \circ K_y\big] = D(J_u) + D\big[(A^{-1})_u\big] \circ D(K_y).$$

By the form of the functor of the inverse morphism, we get the equality

$$D(I_u) = D(J_u) + \big[D(A_y)\big]^{-1} \circ D(K_y).$$

Consider the general cofunctor of the Banach spaces category that maps the Banach space and the linear continuous operator to its adjoint space and the adjoint operator. Then transform the previous equality

$$D(I_u) = D(J_u) + H\big\{\big[D(A_y)\big]^{-1}\big\}D(K_y) = D(J_u) + \big\{H\big[D(A_y)\big]\big\}^{-1}.$$

Theorem 12.2 *If the derivative of the operator A at the point $y = y[u]$ is invertible, where u is a solution of Problem 12.1, then the following equality holds:*

$$D(J_u) + \big\{H\big[D(A_y)\big]\big\}^{-1} = 0. \tag{12.2}$$

From Theorem 12.2 the standard optimality conditions follow.

Corollary 12.1 *The solution of Problem 12.1 satisfies the equality*

$$J'(u) + p = 0, \tag{12.3}$$

where p is the solution of the equation

$$\big[A'(y)\big]^* p = K'(y). \tag{12.4}$$

Indeed, from the equality (12.2) it follows that

$$J'(u) + \big\{\big[A'(y)\big]^*\big\}^{-1} K'(y) = 0.$$

If the function p is the solution of the equation (12.4), then the last equality can be transformed to (12.2).

The equality (12.3) is the standard optimality condition for the optimization control problem without any constraints (see Chapter 4).

Determine the analogical results by replacing the classic differentiation by the extended one. However, we have the necessity to determine additional properties of the extended derivatives.

12.3 Additional properties of extended derivatives

We determined before the weak form of the extended Gâteaux differentiability of the inverse operator (see Theorem 5.6). Now we consider its extended Fréchet differentiability.

Consider as before Banach spaces Y, Z, an operator $A : Y \to Z$, and points $y_0 \in Y$, $z_0 = Ay_0$. Let us have a Banach space Z_* that is a subspace of Z and its neighborhood O_* of zero. Then the set $O = z_0 + O_*$ is the neighborhood of the point z_0.

Assumption 12.1 *The operator A is invertible on the set O.*

Denote the inverse operator by L. Fix a point $h \in Z_*$. For small enough numbers σ the point $z_\sigma = z_0 + \sigma h$ belongs to the set O. Consider the equality

$$ALz_\sigma = z_0 + \sigma h, \quad ALz_0 = z_0$$

for an arbitrary function $h \in Z_*$ and a number σ that is so small that the following inclusion holds: $z_\sigma \in O$. Then we have

$$ALz_\sigma - ALz_0 = \sigma h.$$

Assumption 12.2 *The operator A is Gâteaux differentiable.*

Using the Lagrange formula, we obtain the equality

$$\langle \lambda, Ay - Ay_0 \rangle = \langle \lambda, A'[y_0 + \delta(\lambda)(y - y_0)] \rangle \quad \forall y \in Y, \lambda \in Z',$$

where $\delta(\lambda) \in (0,1)$. For all $z \in O$, $\lambda \in Z'$ determine the linear continuous operator $G(z, \lambda) : Y \to Z$ by the formula

$$G(z, \lambda)y = A'[y_0 + \delta(\lambda)(Lz - Lz_0)]y \quad \forall y \in Y.$$

Note the equality $G(z_0, \lambda) = A'(y_0)$ for all $\lambda \in Z'$. Then we have

$$\langle \lambda, G(z_\sigma, \lambda)(Lz_\sigma - Lz_0) \rangle = \sigma \langle \lambda, h \rangle \quad \forall \lambda \in Z', h \in Z_*$$

that is the formula (5.20).

For all $z \in O$ consider Banach spaces $Z(z)$ and $Y(z)$ such that embedding of Y, Z_*, and $Z(z)$ to $Y(z)$, $Z(z)$, and Z are continuous and dense correspondingly,

Assumption 12.3 *The operator A is Gâteaux differentiable, as for any $z \in O$, there exists the continuous continuation $\overline{G}(z, \lambda)$ of the operator $G(z, \lambda)$ on the set $Y(z)$ such that its codomain is the subset $Z(z)$.*

Functors of the differentiation 467

Transform the previous equality. We have

$$\left\langle \overline{G}(z_\sigma, \lambda)^* \lambda, (Lz_\sigma - Lz_0)/\sigma \right\rangle = \langle \lambda, h \rangle \ \forall \lambda \in Z(z_\sigma)', h \in Z_* \qquad (12.5)$$

that is the formula (5.21). Consider the equation

$$\left[\overline{G}(z_\sigma, p_\mu[z]) \right]^* p_\mu[z] = \mu. \qquad (12.6)$$

This is the linear equation

$$\left[\overline{A}'(y_0) \right]^* p_\mu = \mu \qquad (12.7)$$

for $z = z_0$, where $\overline{A}'(y_0)$ is a continuation of the operator $A'(y_0)$ on the set $Y(z_0)$. There are the equations (5.22) and (5.23). Consider a Banach space Y_* that is continuous and dense including to the space $Y(z)$ for all $z \in O$.

Assumption 12.4 *For all $z \in O$, $\mu \in Y(z)'$ the equation (12.6) has a unique solution $p_\mu[z] \in Z(z)'$.*

Return to the equality (12.5). Determine $\lambda = p_\mu[z_\sigma]$ for small enough σ. We get

$$\left\langle \mu, [L(z_0 + \sigma h) - Lz_0]/\sigma \right\rangle = \langle p_\mu[z_\sigma], h \rangle \ \forall \mu \in Y(z_\sigma)', h \in Z_* \qquad (12.8)$$

that is the formula (5.24).

In reality we repeat Theorem 5.6 about the extended differentiability of the inverse operator with respect to the weak topology. Particularly, Assumptions 12.1–12.4 are equal to Assumptions 5.1–5.3. However, for obtaining a stronger form of the extended differentiability it is necessary to use a stronger supposition than Assumption 5.4.

Assumption 12.5 *If $\sigma \to 0$, then we have the convergence $p_\mu[z_0 + \sigma h] \to p_\mu$ $*$-weakly in Z'_* uniformly with respect to the closed unit ball M with zero of the space Y'_* as the center for all $h \in Z_*$, where p_μ is a solution of the equation (12.7).*

Assumption 12.6 *There exists a Banach space W such that its embedding to Y_* is continuous and dense, as $p_\mu[z_0 + \sigma h] \to p_\mu$ strongly in W' uniformly with respect to M for all $h \in Z_*$.*

Remark 12.2 Assumption 5.4 is weaker than Assumption 12.5 because we had the convergence for all parameter μ. However, now we use its uniform convergence.

Theorem 12.3 *Under the Assumptions 12.1–12.5, the operator A^{-1} has $\bigl(Z(z_0), Y(z_0); Z_*, Y_*\bigr)$-extended Gâteaux derivative D at the point z_0 that satisfies the equality*

$$\langle \mu, Dh \rangle = \langle p_\mu, h \rangle \ \forall \mu \in Y(z_0)', h \in Z_*. \qquad (12.9)$$

This is $\bigl(Z(z_0), Y(z_0); Z_*, W\bigr)$-extended Fréchet derivative under Assumption 12.6.

468   *Optimization and Differentiation*

Proof. We have the equality

$$\left\langle \mu, \frac{L(z_0 + \sigma h) - Lz_0}{\sigma} - Dh \right\rangle = \langle p_\mu[z_\sigma] - p_\mu, h \rangle \; \forall \mu \in M, h \in Z_*.$$

Then we have the equality

$$\left\| \frac{L(z_0 + \sigma h) - Lz_0}{\sigma} - Dh \right\|_{Y_*} = \sup_{\mu \in M} \left| \langle p_\mu[z_\sigma] - p_\mu, h \rangle \right|.$$

Using Assumption 12.5, we have the convergence $p_\mu[z_\sigma] \to p_\mu$ *-weakly in Z'_* uniformly with respect to M as $\sigma \to 0$. Then we have

$$\frac{A^{-1}(z_0 + \sigma h) - Lz_0}{\sigma} \to Dh \text{ in } Y_* \; \forall h \in Z_*.$$

Thus, D is the extended Gâteaux derivative. Then we have

$$\left\| L(z_0 + h) - Lz_0 - Dh \right\|_{Y_*} = \sup_{\mu \in M} \left| \langle p_\mu[z_0 + h] - p_\mu[z_0], h \rangle \right| \leq$$

$$\sup_{\mu \in M} \left\| p_\mu[z_0 + h] - p_\mu[z_0] \right\|_{W'} \|h\|_W.$$

By Assumption 12.6, we have the convergence $p_\mu[z_0 + h] \to p_\mu[z_0]$ strongly in W' as $h \to 0$. From the last inequality, it follows that D is a Fréchet derivative. This complete the proof of Theorem 12.3. $\square$

Determine an extended analog of the Composite function theorem. Consider Banach spaces $X_*, X_0, X; Y_1, Y^0, Y_*, Y_0, Y; Z_1, Z^0, Z_*$ with continuous embedding $X_* \subset X_0 \subset X; Y_1 \subset Y^0 \subset Y_* \subset Y_0 \subset Y; Z_1 \subset Z^0 \subset Z_*$; operators $A : X \to Y_1; B : Y \to Z_1$; and the points $x \in X, y = Ax$.

Theorem 12.4 *If the operator A has $\left(X_0, Y^0; X_*, Y_*\right)$-extended Gâteaux (correspondingly, Fréchet) derivative D_A at the point x, and the operator B has $\left(Y_0, Z^0; Y_*, Z_*\right)$-extended Fréchet derivative D_B at the point y, then the composition BA has $\left(X_0, Z^0; X_*, Z_*\right)$-extended Gâteaux (correspondingly, Fréchet) derivative D_{BA} at the point x such that $D_{BA} = D_B D_A$.*

Proof. Suppose both operators are Fréchet differentiable. Then we have the equalities

$$A(x + h) = Ax + D_A h + \eta_A(h), \; B(y + g) = By + D_B g + \eta_B(g)$$

with the linear continuous operators $D_A : X_0 \to Y^0$ and $D_B : Y_0 \to Z^0$, as

$$\left\| \eta_A(h) \right\|_{Y_*} = o\left(\|h\|_{X_*} \right), \; \left\| \eta_B(g) \right\|_{Z_*} = o\left(\|g\|_{Y_*} \right).$$

Then we get

$$BA(x + h) = B\left[Ax + D_A h + \eta_A(h)\right] = BAx + D_B\left[D_A h + \eta_A(h)\right] + \eta_B\left[D_A h + \eta_A(h)\right].$$

Determine the value

$$\eta_{BA}(h) = D_B\big[\eta_A(h)\big] + \eta_B(D_A h) + \eta_B\big[\eta_A(h)\big].$$

We have

$$\big\|D_B\big[\eta_A(h)\big]\big\|_{Z_*} \le c_1 \big\|D_B\big[\eta_A(h)\big]\big\|_{Z^0} \le c_1 \|D_B\|\big\|\eta_A(h)\big\|_{Y_0} \le$$

$$c_1 c_2 \|D_B\|\big\|\eta_A(h)\big\|_{Y_*} = o\Big(\|h\|_{X_*}\Big),$$

$$\big\|\eta_B(D_A h)\big\|_{Z_*} = o\Big(\|D_A h\|_{Y_*}\Big),$$

$$\big\|\eta_B\big[\eta_A(h)\big]\big\|_{Z_*} = o\Big(\big\|\eta_A(h)\big\|_{Y_*}\Big) = o\Big(o\Big(\|h\|_{X_*}\Big)\Big) = o\Big(\|h\|_{X_*}\Big),$$

where $c_1 > 0$, $c_2 > 0$. Using the inequality

$$\big\|D_A h\big\|_{Y_*} \le c_3 \big\|D_A h\big\|_{Y^0} \le c_3 \|D_A\|\|h\|_{X_0},$$

where $c_3 > 0$, we obtain

$$\big\|\eta_B(D_A h)\big\|_{Z_*} = o\Big(\|h\|_{X_*}\Big).$$

Then we have

$$\big\|\eta_{BA}(h)\big\|_{Z_*} = o\Big(\|h\|_{X_*}\Big).$$

Therefore, the composition of the operators is Fréchet differentiable. The case of Gâteaux differentiability can be considered by using the analogical transformation and the proof of the standard Composite function theorem.

12.4 Extended differentiation functor and its application to the extremum theory

Determine a category Γ_E. Its objects are the threes (X, x, X_*), where X, X_* are Banach spaces with continuous embedding, and x is a fixed point of X. Consider two threes (X, x, X_*) and (Y, y, Y_*). Suppose there exist Banach spaces X_0, Y^0, Y_1 with continuous embedding $X_* \subset X_0 \subset X$, $Y_1 \subset Y^0 \subset Y_* \subset Y$ and the operator $A : X \to Y_1$ that is determined on a X_*-neighborhood of the point x and $(X_0, Y^0; X_*, Y_*)$-extended Fréchet differentiable at this point such that $Ax = y$. Consider the composition $A_x = \iota(Y_1, Y)A$, where $\iota(Y_1, Y)$ is the embedding operator of the space Y_1 to Y, as the morphism of the category Γ_E for these objects. Determine also the morphism B_y with the start (Y, y, Y_*) and the end (Z, z, Z_*). Then there exist spaces Y_0, Z^0, Z_1 with continuous embedding $Y_* \subset Y_0 \subset Y$,

470   *Optimization and Differentiation*

$Z_1 \subset Z^0 \subset Z_* \subset Z$, and the operator $B : Y \to Z_1$ that is determined
on a Y_*-neighborhood of the point y and $(Y_0, Z^0; Y_*, Z_*)$-extended Fréchet
differentiable at this point such that $By = z$, as $B_y = \iota(Z_1, Z)A$. It is
obvious that the composition BA is determined on a X_*-neighborhood of
the point y and has the codomain in Z_1, as $BAx = z$. By Theorem 12.4,
this is $(X_0, Z^0; X_*, Z_*)$-extended Fréchet differentiable at the point x. Then
$(BA)_z = \iota(Z_1, Z)BA$ is the morphism of the category Γ_E between the objects
(X, x, X_*) and (Z, z, Z_*). It can be interpreted as the composition $A_x \circ B_y$ of
the morphisms A_x and B_y. Therefore, Γ_E is in reality the category. Theorem
12.3 describes its isomorphisms.

Determine the transformation $D_E : \Gamma_E \to \Sigma$. For all objects (X, x, X_*)
determine $D_E(X, x, X_*) = X_*$. For all morphisms A_x with the start (X, x, X_*)
and the end (Y, y, Y_*) suppose $D_E(A_x)$ is the extended Fréchet derivative D_A
of the operator A at the point x that determines the morphism A_x. Consider
the morphism B_y of the category Γ_E with the start (Y, y, Y_*) and the end
(Z, z, Z_*). Then $D_E(B_y) = D_b$, where D_b is the extended Fréchet derivative of
the operator B at the point y that determines the morphism B_y. By Theorem
12.4, we have the equality $D_{BA} = D_B D_A$, where D_{BA} is the extended Fréchet
derivative of the composition BA that determines the morphism $(BA)_z = A_x \circ B_y$. Then we get

$$D_E(A_x \circ B_y) = D_E(A_x) \circ D_E(B_y).$$

Thus, D_E is in reality the functor. We proved the following result.

Theorem 12.5 *The operator of the extended differentiation is the functor
from the category of the threes Γ_E to the category of Banach spaces Σ.*

We can use this result for the analysis of Problem 12.1. Suppose the functionals J and K are Fréchet differentiable on the spaces W and Y_*. Replace
the classical derivatives of Theorem 12.2 by the extended derivative by using
Theorems 12.3 and 12.4. We have the following result.

Theorem 12.6 *If the operator A satisfies the supposition of Theorem 12.2
with respect to the solution u of Problem 12.1, then the following equality
holds:*

$$D_E(J_u) + \left\{ H \left[D_E(A_y) \right] \right\}^{-1} D_E(K_y) = 0,$$

where $y = y[u]$.

Theorem 12.6 uses weaker suppositions than Theorem 12.2. Therefore, it
has a larger class of applications.

Thus, the category theory can interpret the standard and extended differentiations as the functors. Then we obtain another form of the necessary
conditions of optimality.

12.5   Comments

The theory of categories was developed in the forties of the twentieth century by S. Eilenberg and S. MacLane. The basis of the theory of categories is described, for example, by I. Bucur and A. Deleanu [97] and L.A. Scornyakov (Ed.) [449]. V.I. Elkin [163] and A.I. Kuhtenko [289] consider some applications of the theory of categories to the control problems. The interpretation of differentiation, both classic and extended, as functors is proposed by S.Ya. Serovajsky [490], [495].

Bibliography

[1] F. Abergel and R. Temam. Optimality conditions for some non-qualified problems of distributed control. *SIAM J. Control. Optim.*, 27(1):1–12, 1989.

[2] A. Abuladze and R. Klotzler. Distributional control in processes with Hammerstein type integral equations. *Zeits. Anal. Anwen.*, 11(2):269–276, 1992.

[3] R. Acar. Identification of the coefficient in elliptic equations. *SIAM J. Contr. Optim.*, 31(5):1221–1244, 1993.

[4] R. Adams. *Sobolev spaces.* Academic Press, 1975.

[5] S. Agmon. *Lectures on elliptic boundary value problems.* Van Nostrand Reinhold Princeton, New York, 1965.

[6] V.I. Agoshkov. *Methods of optimal control and adjoint equations of mathematical physivs problems.* IVM RAN, Moscow, 2003.

[7] N.I. Ahiezer. *Lectures on the calculus of variations.* GITL, Moscow, 1955.

[8] N.U. Ahmed. Optimal control of generating policies in a power system governed by a second order hyperbolic partial differential equation. *SIAM J. Contr. Optim.*, 15(6):1016–1033, 1977.

[9] N.U. Ahmed. Optimization control of systems governed by a class of nonlinear evolutional equations in a reflexive Banach space. *J. Optim. Theory Appl.*, 28:57–81, 1978.

[10] N.U. Ahmed and K.L. Teo. Necessary conditions for optimality of Cauchy problems for parabolic partial differential systems. *SIAM J. Contr.*, 13(5):981–993, 1975.

[11] N.U. Ahmed and K.L. Teo. On the optimal control systems governed by quasilinear integro partial differential equations of parabolic type. *J. Math. Anal. Appl.*, 59(1):33–59, 1977.

[12] N.U. Ahmed and K.L. Teo. *Optimal control of distributed parameter systems.* North-Holland, Amsterdam, 1981.

473

474   *Bibliography*

[13] S.A. Aisagaliev. *Boundary problems of optimal control.* Kazakh Univ., Almaty, 1999.

[14] V.M. Alekseev, V.M. Tihomirov, and S.V. Fomin. *Optimal control.* Nauka, Moscow, 1979.

[15] J.J. Alibert and J.P. Raymond. Boundary control of semilinear elliptic equations with discontinuous leading coefficients and unbounded control. *Num. Funct. Anal. Optim.*, 18(3–4):235–250, 1997.

[16] W. Alt and U. Mackenroth. Convergence of finite elements approximations of state constrained convex parabolic boundary control problems. *SIAM. J. Control Optim.*, 27(4):718–736, 1989.

[17] N.V. Andreev and V.S. Melnik. Boundary control of nonlinear elliptic systems. *DAN USSR, ser. A*, (8):63–66, 1983.

[18] P. Antosic, J. Mikusinski, and R. Sikorski. *Theory of distributions. The sequential approach.* 1973.

[19] H.A. Antosiewicz. Newton's method and boundary value problem. *J. Comp. Sci.*, 2(2):177–202, 1968.

[20] M. Aoki. *Introduction to optimization techniques.* Macmillan, 1971.

[21] N. Arada and J.P. Raymond. State-constrained relaxed problems for semilinear elliptic equations. *J. Math. Anal. Appl.*, 223(1):248–271, 1998.

[22] N. Arada and J.P. Raymond. Minimax control of parabolic systems with state constraints. *SIAM J. Control.*, 38(1):532–548, 1999.

[23] J.-L. Armand. *Application of the theory of optimal control of distributed-parameter system to structural optimization. PhD Thesis.* Stanford Univ., Stanford, 1971.

[24] V.I. Arnold. *Geometric theory of ordinary differential equations.* Izhevsk. resp. tip., 2000.

[25] A.V. Arutunov. *Extremum conditions. Abnormal and degenerate problems.* Factorial, Moscow, 1997.

[26] A.V. Arutunov. Necessary conditions of extremum and the inverse function theorem without a priori supposition about the normality. *Tr. MIAN*, 236:33–44, 2002.

[27] A.V. Arutunov. Inverse function theorem in the neighborhood of the abnormal point. *Mathematical notes*, 78(4):612–621, 2005.

[28] A. Arutyunov, N. Bobylev, and S. Korovin. One remark to Ekeland's variational principle. *Comput. Math. Appl.*, 34(2–4):267–271, 1997.

Bibliography

[29] E. Asplund. Topics in the theory of convex functions. In *Theory and applications of monotone operators*, pages 1–33. Grisetti, Italy, 1969.

[30] H. Attouch and G. Beer. Stability of the geometric Ekeland variational principle: Convex case. *J. Optim. Theory Appl.*, 81(1):1–19, 1994.

[31] J.P. Aubin and I. Ekeland. *Applied nonlinear analysis*. John Wiley and Sons, New York, 1984.

[32] J.P. Aubin and H. Frankowska. On inverse function theorems for set-valued maps. *J. Math. Pures et Appl.*, 66:71–89, 1987.

[33] A. Auslender. *Optimisation: Méthodes numeriques*. Masson, Paris, 1970.

[34] E. Avakov. Extremum conditions for smooth extremum problems with equality constraints. *J. Comp. Math. and Math. Phys.*, 25(5):680–693, 1985.

[35] E. Avakov and A.V. Arutunov. Inverse function theorem and extremum conditions for the problems with non-closed image. *Math. Sbornik*, 196(9):3–22, 2005.

[36] V.I. Averbuh and O.G. Smolyanov. Differentiation theory in linear topological spaces. *Uspehi Math. Nauk*, 22(6):201–260, 1967.

[37] V.I. Averbuh and O.G. Smolyanov. Different definitions of the derivatives in linear topological spaces. *Uspehi Math. Nauk*, 23(4):67–116, 1968.

[38] B. Baeuzamy. *Introduction to Banach spaces and their geometry*. North-Holland, Amsterdam, 1985.

[39] V.L. Bakke. A maximum principle for an optimal control problem with integral constraints. *J. Optim. Theory Appl.*, 13(1):32–55, 1974.

[40] A.V. Balakrishnan. On a new computing technique in optimal control. *SIAM J. Control*, 6(2):149–173, 1968.

[41] A.V. Balakrishnan. *Introduction to optimization theory in a Hilbert space*. Springer-Verlag, Berlin, New York, 1971.

[42] A.V. Balakrishnan and L.W. Neustadt. *Computing methods in optimization problems*. Academic Press, New York, 1964.

[43] E.J. Balder. An existence problem for the optimal control of certain nonlinear integral equation of Urysohn type. *J. Optim. Theory Appl.*, 42(3):447–463, 1984.

[44] E.J. Balder. Necessary and sufficient conditions for L_1-strong weak semicontinuity of integral functional. *Nonlin. Anal.*, 11:1399–1404, 1987.

Bibliography

[45] N.V. Banichuk. On a variational problem with an unknown boundary definition and forms elastic bodies.

[46] H.T. Banks and C. Wang. Optimal feedback control of infinite-dimensional parabolic evolution systems: approximation techniques. *SIAM J. Contr. Optim.*, 27(6):1182–1219, 1989.

[47] J. Baranger. *Quelque resultats en optimization non convexe.* Thèse Grenoble, 1973.

[48] J. Baranger and R. Temam. Nonconvex optimal problems depending on a parameter. *SIAM J. Contr.*, 13:146–152, 1975.

[49] V. Barbu. Necessary conditions for distributed control problems governed by parabolic variational inequalities. *SIAM J. Contr. Optim.*, 19(1):64–68, 1981.

[50] V. Barbu. Optimal feedback controls for semilinear parabolic equations. *Lect. Notes Math.*, 979:43–70, 1983.

[51] V. Barbu. *Mathematical methods in optimization of differential systems.* Dordrecht, Kluwer, 1994.

[52] V. Barbu. The time optimal control of Navier–Stokes equations. *Systems Control Let.*, 30:93–100, 1997.

[53] V. Barbu and G. Wang. Internal stabilization of semilinear parabolic systems. *J. Math. Anal. Appl.*, 285(2):387–407, 2003.

[54] M. Bardi and S. Bottacin. On the Dirichlet problem for nonlinear degenerate elliptic equations and applications to optimal control. *Rend. Semin. Mat. Univ. e Politecn. Torino*, 56(4):13–39, 1998.

[55] R.G. Bartle. On the openness and inversion of differentiable mappings. *Ann. Acad. Sci. Fenn.*, 257:3–8, 1958.

[56] R. Becker, H. Kapp, and R. Rannacher. Adaptive finite element methods for optimal control of partial differential equations: Basic concept. *SIAM J. Contr. Optim.*, 39(1):113–132, 2000.

[57] R. Bellman. *Dynamic programming.* Princeton University Press, 1957.

[58] R. Bellman and R. Kalaba. *Dynamic programming and modern control theory.* Academic Press, 1965.

[59] A. Belmiloudi. Bilinear minimax control problems for a class of parabolic systems with applications to control of nuclear reactors. *J. Math. Anal. Appl.*, 327(1):620–642, 2007.

Bibliography 477

[60] M.L. Bennati. Un theorema di esistenza per controlli ottimi di sistemi definiti da eqazioni integrali di Urysohn. *Publ. Ist. Mat. Univ. Genova*, 224:1–16, 1978.

[61] A. Bensoussan and P. Kenneth. Sur l'analogie entre les méthods de regularisation et de penalization. *RIRO*, 13, 1969.

[62] M. Bergounioux. A penalization method for optimal control of elliptic problems with state constraints. *SIAM J. Contr. Optim.*, 30(2):305–323, 1992.

[63] M. Bergounioux. Augmented Lagrangian method for distributed optimal control problems with state constraints. *J. Optim. Theory Appl.*, 78(3):493–521, 1993.

[64] M. Bergounioux and F. Tröltzsch. Optimality conditions and generalized bang-bang principle for a state-constrained semilinear parabolic problem. *Num. Funct. Anal. Optim.*, 17(5–6):517–536, 1996.

[65] L.D. Berkovitz. A penalty function proof of the maximum principle. *Appl. Math. Optimiz.*, 2(4):291–303, 1976.

[66] Ya.M. Bershchansky. Conjugation of singular and non-singular optimal control plots. *Autom. and Telemech.*, (3):5–11, 1979.

[67] O.M. Besov, V.P. Ilin, and S.M. Nikolsky. *Integral representations of functions and embedding theorems*. Nauka, Moscow, 1975.

[68] T. Bewley, R. Temam, and M. Ziane. Existence and uniqueness of optimal control to the Navier–Stokes equations. *C.R. Acad. Sci. Paris, Ser. I*, 330(11):1007–1011, 2001.

[69] M.F. Bidaut. Existence theorems for usual and approximate solutions of optimizations problems. *J. Optim. Theory and Appl.*, 15(4):397–411, 1975.

[70] M.F. Bidaut. Un problème de contrôle optimal à function coût en norm L_1. *C. R. Acad. Sci. Paris*, 281:A273–A276, 1975.

[71] N. Bilić. Approximation of optimal distributed control problem for Navier–Stokes equations. *Numer. Meth. Approx. Theory. Novi Sad*, 4–6:177–185, 1985.

[72] A.V. Bitsadze. *Boundary problems for second order elliptic equations*. Nauka, Moscow, 1966.

[73] V.I. Blagodatskikh. Sufficient conditions for optimality. *Differ. Equat.*, 9(3):416–422, 1973.

[74] V.I. Blagodatskikh. Sufficient conditions for optimality in problems with state constraints. *Appl. Math. Optim.*, 7(2):149–157, 1981.

478   Bibliography

[75] G.A. Bliss. *Lecture on the calculus of variations.* University of Chicago Press, Chicago, 1949.

[76] I. Bock and J. Lovisek. An optimal control problem for an elliptic variational inequality. *Math. Slov.*, 33(1):23–28, 1983.

[77] V.G. Boltyansky. Sufficient conditions for optimality. *DAN SSSR*, 140(5):994–997, 1961.

[78] V.G. Boltyansky. *Mathematical methods of optimal control.* Nauka, Moscow, 1969.

[79] V.G. Boltyansky. *Optimal control of discrete systems.* Nauka, Moscow, 1973.

[80] V.G. Boltyansky. *Theory of tents as the device for solving extremal problems.* VNIISI, Moscow, 1985.

[81] V.G. Boltyansky and P.S. Soltan. Combinatorial geometry and convexity classes. *Uspehi Math. Nauk*, 33(1):3–42, 1978.

[82] O. Bolza. *Lectures on the calculus of variations.* New York, 1960.

[83] J.F. Bonnans. Second-order analysis for control constrained optimal control problems of semilinear elliptic systems. *Appl. Math. Optimiz.*, 38(3):303–325, 1998.

[84] J.F. Bonnans and E. Casas. An extension of Pontryagin's principle for state-constrained optimal control of semilinear elliptic equations and variational inequalities. *SIAM J. Contr. Optim.*, 33(1):274–298, 1995.

[85] Yu.G. Borisovich and V.V. Obuhovsky. On an optimization problem for control systems of parabolic type. *Tr. MIAN*, 211:95–101, 1995.

[86] Yu.G. Borisovich, V.G. Zvyagin, and Yu.I. Sapronov. Nonlinear Fredholm maps and Leray – Schauder theory. *Uspehi Math. Nauk*, 27(4):3–54, 1977.

[87] N. Bourbaki. *Groupes et algebres de Lie.* Hermann, Paris, 1971,1972.

[88] V.N. Brandin. Sufficient optimality conditions in nonlinear systems with constraints on the control. *Izv. AN SSSR, thech. kib.*, (6):164–166, 1977.

[89] A.S. Bratus. Condition of extremum for eigenvalues of elliptic boundary-value problems. *J. Optim. Theory Appl.*, 68(3):423–436, 1991.

[90] H. Bremermann. *Distribution, complex variables, and Fourier transforms.* Addison-Wesley, Reading, MA, 1965.

[91] H. Brésis. *Analyse fonctionnelle. Théorie et applications.* Dunod, Paris, 1999.

Bibliography

[92] R. Brockett. Functional expansions and higher order necessary conditions in optimal control. *Lect. Not. Ec. Math. Syst.*, 131:111–121, 1976.

[93] M. Brokate. Necessary optimality conditions for the control of semi-linear hyperbolic boundary value problems. *SIAM J. Contr. Optim.*, 25(5):1353–1369, 1987.

[94] J. C. Bruch, Jr. S. Adali, J.M. Sloss, and I.S. Sadek. Maximum principle for the optimal control of a hyperbolic equation in one space dimension, part 2: application. *J. Optim. Theory Appl.*, 87(2):287–300, 1995.

[95] A.E. Bryson and Ho Yu-Chi. *Applied optimal control.* Hemisphere, Washington, DC, 1975.

[96] R. Buckdahn and J. Li. Stochastic differential games and viscosity solutions of Hamilton-Jacobi-Bellman-Isaacs equations. *SIAM J. Contr. Optim.*, 47(1):444–475, 2008.

[97] I. Bucur and A. Deleanu. *Introduction to the theory of categories and functors.* John Wiley and Sons Ltd., London, 1968.

[98] A.L. Bukhgeim. *Introduction to the theory of inverse problems.* Utrecht, Boston, 1999.

[99] P. Burgmeier and H. Jasinski. Über die Auwendung des Algorithmus von Krylov und Černous'ko and Problems der Optimalen Steurung von Systemen mit verteilten Parametern. *Wiss. Z. TH Leuna Mereseburg*, 21(3/4):470–476, 1981.

[100] F. Burk. *A garden of Integrals.* MAA, California State University, Chico, 2007.

[101] A.G. Butkovsky. *The optimal control theory of distributed parameter systems.* Nauka, Moscow, 1965.

[102] A.G. Butkovsky. *Control methods of distributed parameter systems.* Nauka, Moscow, 1975.

[103] T. Butler and A.V. Martin. On a method of Courant for minimizing functionals. *Math. And Physics*, 41:291–299, 1962.

[104] G. Buttazzo, G. Mariano, and S. Hildebrandt. *One-dimensional variational problems.* Oxford University Press, New York, 1998.

[105] A. Cañada, J.L. Gámez, and J.A. Montero. Study of an optimal control problem for diffusive nonlinear elliptic equations of logistic type. *SIAM J. Contr. Optim.*, 36(4):1171–1189, 1998.

[106] E. Cances, C. Le Bris, and M. Pilot. Contrôle optimal bilinéaire d'un equation de Schrödinger. *C.R. Acad. Sci. Paris, ser. I*, 330(7):567–571, 2000.

480 *Bibliography*

[107] R. Cannarsa and O. Cârjua. On the Bellman equation for the minimum time problem in infinite dimensions. *SIAM J. Control*, 43(2):532–548, 2004.

[108] R. Cannarsa and G. Da Prato. Nonlinearity of control with infinite horizon for distributed parameter systems and stationary Hamilton - Jacobi equations. *SIAM J. Contr. Optim.*, 27(4):861–875, 1989.

[109] R. Cannarsa and M.E. Tessitore. Infinite-dimensional Hamilton-Jacobi equations and Dirichlet boundary control problems of parabolic type. *SIAM J. Contr. Optim.*, 34(1):1831–1847, 1996.

[110] E. Casas. Boundary control of semilinear elliptic equations with point-wise state constraints. *SIAM J. Contr. Optim.*, 33(4):993–1006, 1993.

[111] E. Casas. Pontryagin's principle for state-constrained boundary control problems of semilinear parabolic equations. *SIAM. J. Contr. Optim.*, 35(4):1297–1327, 1997.

[112] E. Casas and L.A. Fernández. Dealing with integral state constraints in boundary control problems of quasilinear elliptic equations. *SIAM J. Contr. Optim.*, 33(2):568–589, 1995.

[113] E. Casas and M. Mateos. Second order optimality conditions for semilinear elliptic control problems with finitely many state constraints. *SIAM J. Contr. Optim.*, 40(5):1431–1454, 2002.

[114] E. Casas, M. Mateos, and J.-P. Raymond. Error estimates for the numerical approximation of a distributed control problem for the steady-state Navier–Stokes equations. *SIAM J. Contr. Optim.*, 46(2):952–982, 2007.

[115] E. Casas and F. Tröltzsch. Second-order necessary and sufficient optimality conditions for optimization problems and applications to control theory. *SIAM J. Optim.*, 13(2):406–431, 2002.

[116] L. Cesari. An existence theorem in problems of optimal control. *SIAM J. Contr.*, A3(1):7–22, 1965.

[117] L. Cesari. Closure, lower closure, and semicontinuity theorems in optimal control. *SIAM J. Contr. Optim.*, 9(2):287–315, 1971.

[118] L. Cesari. An existence theorem without convexity conditions. *SIAM J. Control*, 12(2):319–331, 1974.

[119] L. Cesari. *Optimization. Theory and applications. Problems with ordinary differential equations.* Springer-Verlag, New York, 1983.

[120] S. Chaabane, Ferchichi, and K. Kunisch. Optimal distributed-control of vortices in Navier–Stokes flows. *C. R. Acad. Sci. Paris, Ser. I*, 341(12):147–152, 2005.

Bibliography 481

[121] A.K. Chaudhuri. Dynamic programming, maximum principle and minimum time control. *Intern. J. Control.*, 1(1):13–19, 1965.

[122] S. Chaumont. Uniqueness to elliptic and parabolic Hamilton–Jacobi Bellman equations with non-smooth boundary. *C. R. Acad. Sci. Paris, Ser. I*, 339(8):555–560, 2004.

[123] Q. Chen, D. Chu, and R.C.E. Tan. Optimal control of obstacle for quasilinear elliptic variational bilateral problems. *SIAM J. Contr. Optim.*, 44(3):1067–1080, 2005.

[124] A. Cheng and K. Morris. Well-posedness of boundary control systems. *SIAM J. Contr. Optim.*, 42(4):1244–1265, 2003.

[125] F.L. Chernousko and N.V. Banichuk. *Variational problems of mechanics and control. Numerical methods.* Nauka, Moscow, 1973.

[126] F.L. Chernousko and V.V. Kolmanovsky. Numerical and approximate methods of optimization. In *Math. Analysis. The results of science and technique*, volume 14, pages 101–166. Nauka, Moscow, 1977.

[127] K.-S. Chou, G.-X. Li, and C. Qu. A note on optimal systems for the heat equation. *J. Math. Anal. Appl.*, 261(2):741–751, 2001.

[128] F.H. Clarke. *Optimization and nonsmooth analysis.* John Wiley and Sons, New York, 1988.

[129] J.F. Colombeau. *Elementary introduction to new generalized functions.* North-Holland, Amsterdam, 1985.

[130] R. Courant. Variational methods for the solution of problems of equilibrium and vibrations. *Bull. Amer. Math. Soc.*, 49:1–23, 1943.

[131] D. Cowles. An existence theorem for optimal problem involving integral equation. *SIAM J. Control*, 11(4):595–606, 1973.

[132] M. Cristea. Local inversion theorems without assuming continuous differentiability. *J. Math. Anal. Appl.*, 143:259–263, 1989.

[133] J. Cullum. Penalty function and continuous in optimal control problems. *Bul. Inst. Polit. Iasi.*, 15(1–2):1/93–1/98, 1969.

[134] R. Cuvelier. Resolution numerique d'un problème de contrôle d'un couplage des équations de Navier–Stokes et celle de la chaleur. *Calcolo*, 15(4):349–379, 1978.

[135] G. Da Prato and A. Ichikawa. Optimal control for integrodifferential equations of parabolic type. *SIAM J. Contr. Optim.*, 31(5):1167–1182, 1993.

482 Bibliography

[136] B. Dacorogna. Weak continuity and weak lower semicontinuity of nonlinear functionals. *Lect. Notes Math.*, 922, 1982.

[137] V.F. Demyanov and A.M. Rubinov. *Basis of nonsmooth analysis and quasidifferential calculus.* Nauka, Moscow, 1990.

[138] D. Dentcheva and S. Helbig. On variational principles, level sets, well-posedness, and ε-solutions in vector optimization. *J. Optim. Theory Appl.*, 89(2):325–329, 1996.

[139] M. Desai and K. Ito. Optimal controls of Navier–Stokes equations. *SIAM J. Optim. Contr.*, 32(5):1428–1446, 1994.

[140] J. Dieudonné. *Foundation of modern analysis.* Academic Press, 1960.

[141] V.V. Dikussar and A.A. Milutin. *Qualitative and numerical methods in the maximum principle.* Nauka, Moscow, 1989.

[142] A.V. Dmitruk, A.A. Milutin, and N.P. Osmolovsky. Lusternik theorem and extremum theory. *Uspehi Math. Nauk*, 35(6):11–46, 1980.

[143] A. Domokos. Implicit function theorems and variational inequalities. *Stud. Univ. Babes-Bolyai Math.*, 39(4):29–36, 1994.

[144] A. Domokos. Equivalence between implicit function theorems. *Stud. Univ. Babes-Bolyai Math.*, 41(4):41–46, 1996.

[145] L. Doyen. Inverse function theorems and shape optimization. *SIAM J. Optim. Contr.*, 32(6):1621–1642, 1994.

[146] J. Dronion and J.P. Raymond. Optimal pointwise control of semilinear parabolic equations. *Nonlinear Anal. Theory, Meth. Appl.*, 39(2):135–156, 2000.

[147] A.D. Dubovitsky and A.A. Milutin. *Necessary conditions of weak extremum.* Nauka, Moscow, 1971.

[148] E. Dubuc and J. Zilber. Inverse function theorems for Banach spaces in a topos. *Cah. topol. et geom. differ. categor.*, 41(3):207–224, 2000.

[149] G. Duvaut and J.L. Lions. *Les inéquations en mécanique et en physique.* Dunod, Paris, 1972.

[150] M.T. Dzhenaliev. Boundary value problems and optimal control problems for linear loaded hyperbolic equations. *Differential equations*, 28(2):232–241, 1992.

[151] M. Edelstein. Farthest points of sets in uniformly convex Banach spaces. *Isr. J. Math.*, 4(3):171–176, 1966.

Bibliography

483

[152] M. Edelstein. On nearest points of sets in uniformly convex Banach spaces. *J. London Math. Soc.*, 43:375–377, 1968.

[153] J. Eells. A setting for global analysis. *Bull. Amer. Math. Soc.*, 72:751–807, 1966.

[154] A.I. Egorov. *Optimal control of heat and diffusion processes*. Nauka, Moscow, 1978.

[155] A.I. Egorov. *Fundamentals of control theory*. Physmathlit, Moscow, 2004.

[156] Yu.V. Egorov. Necessary optimality conditions in a Banach space. *Math. Sbornik*, 64(1):79–101, 1964.

[157] Yu.V. Egorov. On the generalized functions theory. *Uspehi Math. Nauk*, 45(5):3–40, 1990.

[158] I. Ekeland. Sur le contrôle optimale de systèmes gouvernés par des équations elliptic. *J. Funct. Anal.*, 9(1):1–62, 1972.

[159] I. Ekeland. On the variational principle. *J. Math. Anal. Appl.*, 47(2):324–353, 1974.

[160] I. Ekeland. Legendre duality in nonconvex optimization and calculus of variations. *SIAM J. Control Optim.*, 15(6):905–934, 1977.

[161] I. Ekeland. The ε-variational principle. *Lect. Notes Math.*, 1446:1–15, 1990.

[162] I. Ekeland and R. Temam. *Convex analysis and variational problems*. North-Holland Publish. Comp., Amsterdam, Oxford; American Elsevier Publish. Comp., New York, 1976.

[163] V.I. Elkin. On categories and the foundations of the theory of nonlinear control dynamical systems. *Differ. Equat.*, 38(11):1467–1482, 2002.

[164] E. Elsgolts. *Differential equations and calculus of variations*. Nauka, Moscow, 1969.

[165] G. Emmanuele and A. Villani. A linear hyperbolic system and an optimal control problem. *J. Optim. Theory Appl.*, 44(2):213–229, 1984.

[166] H.W. Engl. Discrepancy principles for Tikhonov regularization of ill-posed problems leading to optimal convergence rates. *J. Optim. Theory Appl.*, 52(2):209–215, 1987.

[167] S.H. Farag and M.H. Farag. On a optimal control problem for a quasi-linear parabolic equation. *Appl. Math.*, 27(2):239–250, 2000.

484   Bibliography

[168] S. Farlow. *Partial differential equations for scientists and engineers.* Dover, New York, 1993.

[169] H.O. Fattorini. Nonlinear infinite dimensional optimal control problems with constraints and unbounded control sets. *Rend. Ist. Mat. Univ. Trieste*, 28:121–146, 1996.

[170] H.O. Fattorini. The maximum principle for control systems described by linear parabolic equations. *J. Math. Anal. Appl.*, 259(2):630–651, 2001.

[171] H.O. Fattorini and T. Murphy. Optimal problems for nonlinear parabolic boundary control systems. *SIAM J. Contr. Optim.*, 32(2):1577–1596, 1994.

[172] R.P. Fedorenko. The ill-posedness of optimal control problems and regularization of numerical solutions. *Differ. Equat.*, 15(4):1043–1047, 1975.

[173] R.P. Fedorenko. *Approximate solution of the optimal control problems.* Nauka, Moscow, 1978.

[174] W. Fenchel. *Convex cones, sets and functions, mimeographed lecture notes.* Princeton University, Princeton, 1951.

[175] M. Fečkan. Note on a local invertibility. *Math. Slov.*, 46(2–3):285–289, 1996.

[176] G. Fichera. Problemi elastostatici con vincoli unilaterali: il problema di Signorini non ambigue condizioni al contorno. *Mem. Acad. Naz. Lincei*, 7:91–140, 1964.

[177] A.F. Filippov. On some issues of optimal control theory. *Vest. MGU, Ser. Math. Mech. Astr. Phys. Shim.*, (2):25–32, 1959.

[178] W.H. Fleming and R.W. Rishell. *Deterministic and stochastic optimal control.* Springer-Verlag, Berlin, 1975.

[179] R. Fletcher. *Optimization.* Academic Press, 1969.

[180] F. Flores-Bazán. Existence theorems for a class of nonconvex problems in the calculus of variations. *J. Optim. Theory Appl.*, 78(1):31–48, 1993.

[181] H. Frankowska. The maximum principle for a differential inclusion problem. *Lect. Notes. Contr. Inf. Sci.*, 62:517–531, 1984.

[182] H. Frankowska. Theoremes d'application ouverte et de function inverse. *C.R.Acad. Sci. Paris., ser. I.*, 305:773–776, 1987.

[183] H. Frankowska. High order inverse function theorems. *Ann. Inst. Poincare*, (6):283–303, 1989.

Bibliography 485

[184] G. Fraser-Andrews. A multiple-shooting technique for optimal control. *J. Optim. Theory Appl.*, 102(2):299–311, 1999.

[185] F.G. Friedlander. *Introduction to the theory of distributions*. Cambridge University Press, Cambridge, 1998.

[186] A. Friedmann. *Partial differential equations*. Holt, Rinehart and Winston, New York, 1969.

[187] A. Friedmann. *Foundations of modern analysis*. Holt, New York, 1970.

[188] A. Friedmann. Nonlinear optimal control problems for parabolic equations. *SIAM J. Contr. Optim.*, 22(5):805–816, 1984.

[189] A. Friedmann, S. Huang, and J. Yong. Optimal periodic control for the two-phase Stefan problem. *SIAM J. Contr. Optim.*, 26(1):23–41, 1988.

[190] A. Frölicher and W. Bucher. *Calculus in vector spaces without norm*. Springer-Verlag, Berlin, 1966.

[191] J. Fu. A generalization of Ekeland's variational principle. *Acta Anal. Funct. Appl.*, 2(1):43–45, 2000.

[192] N. Fujii. Necessary conditions for a domain optimization problem in elliptic boundary value problems. *SIAM J. Contr. Optim.*, 24(3):346–360, 1986.

[193] N. Fujii. Lower semicontinuity in domain optimization problems. *J. Optim. Theory Appl.*, 59(2):407–422, 1988.

[194] A.V. Fursikov. Control problems and theorems concerning the unique solvability of the boundary value problem for three-dimensional Navier–Stokes and Euler. *Math. Sbornik*, 115(2):281–306, 1981.

[195] A.V. Fursikov. Properties of solutions of some extremal problems connected with the Navier–Stokes equations. *Math. Sbornik*, 118(3):323–349, 1982.

[196] A.V. Fursikov. Lagrange principle for problems of optimal control of ill-posed or single distributed systems. *J. Math. Pure et Appl.*, 71:139–195, 1992.

[197] A.V. Fursikov. *Optimal control of distributed systems. Theory and applications*. Amer. Math. Soc., Providence, 1999.

[198] R. Gabasov and F.M. Kirillova. *Qualitative theory of optimal processes*. Nauka, Moscow, 1971.

[199] R. Gabasov and F.M. Kirillova. *Singular optimal control*. Nauka, Moscow, 1973.

486   *Bibliography*

[200] H. Gajewski, K. Gröger, and K. Zacharias. *Nichtlineare Operatorgleichungen und Operatordifferentiagleichungen.* Academie Verlag, Berlin, 1974.

[201] I.M. Gali. Optimal control of systems governed by elliptic operators of infinite order. *Lect. Notes Math.*, 962:263–275, 1982.

[202] R.V. Gamkrelidze. *Optimal control basics.* Tbilisi Univ., 1977.

[203] H. Gao and N.H. Pavel. Optimal control problems for a class of semilinear multi-solution elliptic equations. *J. Optim. Theory Appl.*, 118(2):353–380, 2003.

[204] V.S. Gavrilov and M.I. Sumin. Parametric optimization of nonlinear Goursat – Darboux systems problem with states constraints. *J. Comp. Math. and Math. Phys.*, 44(6):1002–1022, 2004.

[205] I.M. Gelfand and S.V. Fomin. *Variational calculus.* Physmathgiz, Moscow, 1961.

[206] I.M. Gelfand and G.E. Shilov. *Generalized functions I,II.* Physmathgiz, Moscow, 1958.

[207] B.D. Gelman. Generalized implicit function theorem. *Funct. Anal. Appl.*, (3):28–35, 2001.

[208] I.D. Georgieva. A necessary optimality condition for Mosco-differentiable functions. *Dokl. Bolg. AN*, 55(2):13–16, 2002.

[209] J. Gil de Lamadrid. *Topology of mapping in locally convex topological vector spaces, their differentiation and integration, and application to gradient mapping. Thesis.* Univ. Michigan, 1955.

[210] D. Gilbarg and N.S. Trudinger. *Elliptic partial differential equations of second order.* Springer-Verlag, Berlin, 1977.

[211] J.R. Giles and S. Sciffer. On weak Hadamard differentiability of covex function on Banach spaces. *Bull. Austral. Math. Soc.*, 54(1):155–166, 1996.

[212] P.E. Gill, W. Murrey, and M.H. Wright. *Practical optimization.* Academic Press, London, 1981.

[213] A. Gleason. *Introduction to abstract analysis.* Addison-Wesley, Reading, MA, 1966.

[214] M. Goebel. Optimal control of coefficient in linear elliptic equations. *Math. Oper. Stad. Optim.*, 12(4):525–533, 1981.

Bibliography 487

[215] M. Goebel and U. Raitums. Optimality conditions for systems governed by a two point boundary value problem. *Optimization*, 20(5):671–685, 1989.

[216] I.K. Gogodze. Necessary optimality conditions in elliptic control problems with integral constraints. *Soob. AN GruzSSR*, 81(1):17, 1976.

[217] C.J. Goh and X.Q. Yang. Nonlinear Lagrangian theory for nonconvex optimization. *J. Optim. Theory Appl.*, 109(1):99–121, 2001.

[218] H. Goldberg and F. Tröltzsch. Second-order sufficient optimality conditions for a class of nonlinear parabolic boundary control problems. *SIAM J. Contr. Optim.*, 31(1):1007–1025, 1993.

[219] L. Gong and P. Fiu. Optimal control of nonsmooth system governed by quasilinear elliptic equations. *Int. J. Math. Mech. Sci.*, 20(2):339–346, 1997.

[220] J.P. Gossez. Existence of optimal control for some nonlinear processes. *J. Optim. Theory Appl.*, (3):89–97, 1969.

[221] K.D. Graham and D.L. Russell. Boundary value control of the wave equation in a spherical region. *SIAM J. Control*, 13(1):174–196, 1975.

[222] L.M. Graves. On the problem of Lagrange.

[223] L.M. Graves. Implicit functions and differential equations in general analysis. *Trans. Amer. Math. Soc.*, 29:514–552, 1927.

[224] B. Grunbaum. *Studies in combinatorial geometry and the theory of convex bodies*. Nauka, Moscow, 1971.

[225] M. Gunzburger, L. Hou, and T. Svobodny. Analysis and finite element approximation of optimal control problems for the stationary Navier–Stokes equations with distributed and Neumann controls. *Math. Comp*, 57:123–151, 1991.

[226] M. Gunzburger, L. Hou, and T. Svobodny. Analysis and finite element approximation of optimal control problems for the stationary Navier–Stokes equations with Dirichlet control. *RAIRO Model. Math. Anal. Numer.*, 25(6):711–748, 1991.

[227] M. Gunzburger and C. Trenchea. Analysis of an optimal control problem for the three-dimensional coupled modified Navier–Stokes and Maxwell equations. *J. Math. Anal. Appl.*, 333(1):295–310, 2007.

[228] X. Guo and J. Ma. Lagrange multiplier theorem in Banach space. *J. Nanjing Univ. Math. Biquarterly*, 21(2):189–193, 2004.

[229] V.I. Gurman. *The principle of extension in control problems*. Nauka, Moscow, 1975.

488   *Bibliography*

[230] C.-W. Ha. On a Theorem of Lyusternik. *J. Math. Anal. Appl.*, 110:227–229, 1985.

[231] J. Ha and S. Nakagiri. Optimal control problem for nonlinear hyperbolic distributed parameter systems with damping terms. *Funct. Equat.*, 47(1):1–23, 2004.

[232] J. Ha, S. Nakagiri, and H. Tanabe. Fréchet differentiability of solution mappings for semilinear second order evolution equations. *J. Math. Anal. Appl.*, 346(2):374–383, 2008.

[233] T.X.D. Ha. Some variants of the Ekeland variational principle for a set-valued map. *J. Optim. Theory Appl.*, 124(1):187–206, 2005.

[234] W. Hackbusch. Fast solution of elliptic control problems. *J. Optim. Theory Optim.*, 31(2):565–581, 1980.

[235] H. Halkin. Mathematical foundations of systems optimization. In *Topics in Optimization. Mathematical Foundations of System Optimization*, pages 197–262. 1967.

[236] P. Halmos. *A Hilbert space problem book*. Springer-Verlag, New York, 1982.

[237] J. Haslinger, J. Malek, and J. Stebel. Optimization in problems governed by Generalized Navier–Stokes equations. *Contr. And. Cybern.*, 34(1):283–303, 2005.

[238] J.W. He and R. Glowinski. Neumann control of unstable parabolic systems: Numerical approach. *J. Optim. Theory Appl.*, 96(1):1–55, 1998.

[239] A. Henrot and J. Sokolovsky. Mathematical analysis in shape optimization. *Contr. And. Cybern.*, 31(1):37–57, 2005.

[240] D. Henry. *Geometric theory of semilinear parabolic equations*. Springer-Verlag, Berlin, 1981.

[241] M.R. Hestens. *Calculus of variations and optimal control theory*. John Wiley and Sons, New York, 1966.

[242] S. Hiltunen. Implicit functions from locally convex spaces to Banach spaces. *Stud. Math.*, 134(3):235–250, 1999.

[243] M. Hintermüller and M. Hinze. A SQP-semismooth Newton-type algorithm applied to control of the instationary Navier–Stokes system subject to control constraints. *SIAM J. Optim.*, 16(4):1177–1200, 2006.

[244] L.S. Hou and S.S. Ravindran. A penalized Neumann control problem for solving an optimal Dirichlet control problem for the Navier–Stokes equations. *SIAM J. Contr. Optim.*, 36(5):1795–1814, 1998.

Bibliography 489

[245] L.S. Hou and T.P. Svobodny. Optimization problems for the Navier–Stokes equations with regular boundary controls. *J. Math. Anal. Appl.*, 177(2):342–367, 1993.

[246] X.X. Huang. Pointwise well-posedness of perturbed vector optimization problems in a vector-valued variational principle. *J. Optim. Theory Appl.*, 108(1):671–684, 2001.

[247] V. Hutson and J.S. Pym. *Applications of functional analysis and operator theory.* Academic Press, New York, London, 1980.

[248] W. Huyer and A. Neumaier. A new exact penalty function. *SIAM J. Optim.*, 13(4):1141–1158, 2003.

[249] O.Yu. Imanuilov. On some optimal control problems associated with the Navier–Stokes system. *Proc. Petrovsky Semin.*, 15:108–127, 1991.

[250] A.D. Ioffe. An existence theorem for a general Bolza problem. *SIAM J. Contr. Optim.*, 14(3):458–466, 1976.

[251] A.D. Ioffe. On lower semicontinuity of integral functionals. II. *SIAM J. Contr. Optim.*, 15(6):991–1000, 1977.

[252] A.D. Ioffe and V.N. Tihomirov. Extension of variational problems. *Trudy MMM*, 18:187–246, 1968.

[253] A.D. Ioffe and V.N. Tihomirov. *Theory of extremum problems.* Nauka, Moscow, 1974.

[254] A.D. Ioffe and V.N. Tihomirov. Some remarks on variational principles. *Math. Notes*, 61(2):305–311, 1997.

[255] A.D. Ioffe and A.J. Zaslavski. Variational principles and well-posedness in optimization and calculus of variations. *SIAM J. Contr. Optim.*, 43(4):566–581, 2000.

[256] K. Iosida. *Functional analysis.* Springer-Verlag, Berlin, New York, 1980.

[257] V.K. Isaev and Sonin V.V. A modification of Newton's method of numerical solution of boundary problems. *J. Comp. Math. and Math. Phys.*, 3(6), 1963.

[258] A.D. Iskenderov and R.K. Tagiev. Optimization problem with control in the coefficients for parabolic equations. *Differential Equations*, 19(8):1324–1334, 1983.

[259] K. Ito and S.S. Ravindran. Optimal control of terminally convected fluid flows. *SIAM J. Sci. Comput.*, 19(6):1847–1869, 1998.

[260] V.I. Ivanenko and V.S. Melnik. *Variational methods in control problems for distributed parameter systems.* Naukova Dumka, Kiev, 1989.

490   *Bibliography*

[261] N.V. Ivanov. Approximation of smooth manifolds by real algebraic sets. *Uspehi Math. Nauk*, 37(1):3–52, 1982.

[262] W. Johnson, J. Lindenstrauss, D. Priess, and G. Schechtman. Almost Frechet differentiability of Lipschitz mapping between infinite-dimensional Banach space. *Proc. London Math. Soc.*, 84(3):711–745, 2002.

[263] M.C. Joshi and R.K. Bose. *Some topics in nonlinear functional analysis.* Wiley, New York, 1984.

[264] A. Just. Optimal control of an object described by the two-dimensional equation of heat conduction. *Zes. Nauk. Plodz. Mat.*, 14:73–92, 1982.

[265] S.I. Kabanihin. *Inverse and ill-posed problems.* Sib. nauch. izd., 2009.

[266] R. Kachurovsky. On monotone operators and convex functionals. *Uspehi Math. Nauk*, (4):213–215, 1960.

[267] L.V. Kantorovich and G.P. Akilov. *Functional analysis.* Nauka, Moscow, 1977.

[268] H.H. Keller. Differenzierbarkeit in topologischen vektoräumen. *Comment. Math. Helv.*, 38(4):308–320, 1964.

[269] H.J. Kelley, E.R. Kopp, and H.G. Moyer. *Successive approximation technique for trajectory optimization.* Proc. Symp. Vehicle Syst. Opt., New York, 1961.

[270] D. Kinderlehrer and G. Stampacchia. *An introduction to variational inequalities and their applications.* Academic Press, New York, 1980.

[271] U. Knöpp. Explicit solutions to the optimal boundary control problem of a vibrating string. *Int. J. Control*, 34(1):95–110, 1981.

[272] G. Knowles. Some remarks on infinite-dimensional nonlinear control without convexity. *SIAM J. Contr. Optim.*, 15(5):830–840, 1977.

[273] G. Knowles. Time-optimal control of parabolic systems with boundary conditions involving time delays. *J. Optim. Theory Appl.*, 25(4):563–574, 1978.

[274] G. Knowles. Finite element approximation of parabolic time optimal control problems. *SIAM J. Contr. Optim.*, 20(3):414–427, 1982.

[275] T. Kobayashi. Some remarks on controllability for distributed parameter systems. *SIAM J. Contr.*, 16(5):733–742, 1978.

[276] A.N. Kolmogorov and S.V. Fomin. *Elements of the functions theory and functional analysis.* Nauka, Moscow, 1988.

Bibliography

[277] M. Kong, G. Wu, and Z. Shen. On the inverse function theorem. *J. Nanjing Univ. Math. Biquarterly*, 20(1):58–63, 2002.

[278] M.M. Kostreva and A.L. Ward. Linear formulation of a distributed boundary control problem. *J. Optim. Theory Optim.*, 103(2):385–399, 1999.

[279] W. Kotarsky. Optimal control of a system governed by a parabolic equation with an infinite number of variables and time delay. *J. Optim. Theory Appl.*, 63(1):57–67, 1990.

[280] A. Kowalewsky. Optimal control of distributed hyperbolic systems with boundary condition involving a time lag. *Arch. Autom. Telemech.*, 33(4):537–545, 1988.

[281] V. Krabs. Boundary control of the higher-dimensional wave equation. *Lect. Notes Econ. Math. Sci.*, 117:183–189, 1976.

[282] M.A. Krasnoselsky and P.P. Zabreiko. *Geometric methods of nonlinear analysis*. Nauka, Moscow, 1975.

[283] S.G. Krein et al. *Functional analysis*. Nauka, Moscow, 1972.

[284] Yu.P. Krivenkov. Sufficient optimality conditions for the problem with second order differential equation of elliptic type with phase constraints. *Differ. Equat.*, (1):85–99, 1975.

[285] V.F. Krotov. *Basics of optimal control theory*. Vysh. Shkola, Moscow, 1990.

[286] V.F. Krotov, V.Z. Bukreev, and V.I. Gurman. *New methods of the calculus of variations in the flight dynamics*. Mashinostroenie, Moscow, 1969.

[287] V.F. Krotov and V.I. Gurman. *Methods and problems of optimal control*. Nauka, Moscow, 1973.

[288] I.A. Krylov and F.L. Chernousko. The algorithm of the method of successive approximations for optimal control problems. *J. Comp. Math. and Math. Phys.*, (12):14–34, 1972.

[289] A.I. Kuhtenko. On the theory of categories and topoi in control problems. In *Difficult control systems*, pages 4–15. Birkhäuser, Basel, 1989.

[290] K. Kunisch and L.W. White. Identifiability under approximation for an elliptic boundary value problem. *SIAM J. Contr. Optim.*, 25(2):279–297, 1987.

[291] S.S. Kutateladze. *Basics of functional analysis*. Inst. Math. SO RAN, Novosibirsk, 2001.

492   *Bibliography*

[292] O.A. Ladyzhenskaya. *Mathematical theory of viscous incompressible fluid.* Physmathgiz, Moscow, 1961.

[293] O.A. Ladyzhenskaya, V.A. Solonnikov, and N.N. Uraltseva. *Linear and quasi-linear parabolic equations.* Nauka, Moscow, 1967.

[294] O.A. Ladyzhenskaya and N.N. Uraltseva. *Linear and quasilinear elliptic equations.* Nauka, Moscow, 1973.

[295] C. Lanczos. *The variational principles of mechanics.* Courier Corp., 1970.

[296] S. Lang. *Introduction to differentiable manifolds.* John Wiley and Sons New York, London, 1962.

[297] I. Lasiecka. Ritz–Galerkin approximation of the time optimal boundary control problem for parabolic systems with Dirichlet boundary conditions. *SIAM J. Contr. Optim.*, 22(3):477–500, 1984.

[298] I. Lasiecka and K. Malanowski. On discrete-time Ritz–Galerkin approximation on control constrained optimal control problem for parabolic equation. *Lect. Not. Contr. Inf. Sci.*, 6:334–342, 1978.

[299] I. Lasiecka and R. Triggiani. *Deterministic control theory for partial differential equations. Vol. 1.* Cambridge Univ. Press, Boston, 1998.

[300] M.A. Lavrentiev and L.A. Lusternik. *Course of variational calculus.* ONTI, Moscow, Leningrad, 1950.

[301] G. Le Bourg. Perturbed optimal problems in Banach spaces. *Bull. Soc. Math. France Mem.*, 60:95–111, 1979.

[302] U. Ledzewicz and A. Nowakowsky. Optimality conditions for control problem governed by abstract semilinear differential equations in complex Banach spaces. *J. Appl. Anal.*, 1(1):67–91, 1997.

[303] U. Ledzewicz and H. Schättler. An extended maximum principle. *Nonlinear Anal. Theory, Meth. Appl.*, 29(2):159–183, 1997.

[304] H.-C. Lee and O.Yu. Imanuvilov. Analysis of optimal control problems for the 2-D stationary Boussinesq equations. *J. Math. Anal. Appl.*, 242(1):191–211, 2000.

[305] S. Lefschetz. *Differential equations: Geometric theory.* Interscience Publ., Inc., New York, London 1957.

[306] L. Lei and G. Wang. Optimal control of semilinear parabolic equations with k-approximate periodic solutions. *SIAM J. Contr. Optim.*, 46(5):1754–1778, 2007.

Bibliography 493

[307] G. Leitmann. *Calculus of variations and optimal control.* Plenum Publ. Corp., New York, 1981.

[308] B. Lemaire, C.O.A. Salem, and J.P. Revalski. Well-posedness by perturbations of variational problems. *J. Contr. Theory Appl.*, 115(2):346–368, 2002.

[309] J. Leray. *Hyperbolic equations.* Princeton Lect. Note, 1954.

[310] L. Levaggi. Sliding modes in Banach space. *Prepr. Dip. Mat. Univ. Genova*, 410:1–21, 2000.

[311] E.S. Levitin, A.A. Milutin, and N.P. Osmolovsky. Conditions of high order for a local minimum in problems with constraints. *Uspehi Math. Nauk*, 28(6):85–148, 1978.

[312] R. Li, W. Liu, H. Ma, and T. Tang. Adaptive finite element approximation for distributed elliptic optimal control problems. *SIAM J. Contr. Optim.*, 42(4):1244–1265, 2003.

[313] X. Li and J. Yong. *Optimal control theory for infinite dimensional systems.* Birkhauser, Boston, 1995.

[314] J.L. Lions. Sur le contrôle optimale des systèmes d'écrits par des équations aux derivées partielles lineaires. *C. R. Acad. Sci. Paris*, 263:713–715, 1966.

[315] J.L. Lions. *Contrôle optimal de systèmes gouvernés par des équations aux dérivées partielles.* Dunod, Gauthier-Villars, Paris, 1968.

[316] J.L. Lions. *Quelques méthods de résolution des problèmes aux limites non linéaires.* Dunod, Gauthier-Villars, Paris, 1969.

[317] J.L. Lions. *Contrôle de systèmes distribués singuliers.* Gauthier-Villars, Paris, 1983.

[318] J.L. Lions. Contrôle de Pareto de systèmes distribués. Le cas stationaire. *C. R. Acad. Sci. Paris, ser. 1*, 302(6):223–227, 1986.

[319] J.L. Lions and E. Magenes. *Problèmes aux limites non homogenes et applications.* Dunod, Paris, 1968.

[320] V.G. Litvinov. Optimal control of coefficients in elliptical systems. *Differ. Equat.*, (6):1036–1047, 1982.

[321] V.G. Litvinov. *Optimization in elliptic boundary value problems.* Nauka, Moscow, 1987.

[322] W.B. Liu, P. Neittaanmäki, and D. Tiba. Existence for shape optimization problems in arbitrary dimension. *SIAM J. Optim. Contr.*, 41(5):1440–1454, 2003.

494 *Bibliography*

[323] Ya.B. Lopatinsky. *Introduction to the modern theory of partial differential equations.* Naukova Dumka, Kiev, 1980.

[324] K. Lorenz. An implicit function theorem in locally convex spaces. *Math. Nachr.*, 129:91–101, 1986.

[325] A. Lorenzi. *An introduction to identification problems via functional analysis.* De Gruyter, Utrecht, Boston, 2001.

[326] P. Loridan. Necessary conditions for ε-optimality. *Math. Progr. Study*, 19:140–152, 1982.

[327] H.W. Lou. Existence of optimal control for semilinear elliptic equations without Cesari-type conditions. *ANZIAM Journal*, 45(1):115–131, 2003.

[328] H.W. Lou. Maximum principle of optimal control for degenerate quasilinear elliptic equations. *SIAM. J. Control Optim.*, 42(1):1–23, 2003.

[329] H.W. Lou. Existence of optimal controls in the absence of Cesari-type conditions for semilinear elliptic and parabolic systems. *J. Optim. Theory Appl.*, 125(2):367–391, 2005.

[330] F.V. Lubyshev and M.E. Fairuzov. Approximation and regularization of optimal control problems for quasilinear elliptic equations. *J. Comp. Math. and Math. Phys.*, 41(8):1148–1164, 2001.

[331] R. Lucchetti. Stability in optimization. *SERDICA Bulg. Math. Publ.*, 15:34–48, 1989.

[332] R. Lucchetti and F. Patrone. Hadamard and Tyhonov well-posedness of a certain class of convex functions. *J. Math. Anal. Appl.*, 88:204–215, 1982.

[333] K.A. Lurie. *Optimal control in mathematical physics.* Nauka, Moscow, 1975.

[334] L.A. Lusternik. On conditional extremum of functionals. *Math. Sbornik*, (3):390–401, 1934.

[335] A.M. Lyapunov. *Collected works. Vol. 2.* AN SSSR, Moscow, 1956.

[336] U. Mackenroth. On some elliptic control problems with state constraints. *Optimization*, 17:595–607, 1986.

[337] V.I. Maksimov. A boundary control problem for a nonlinear parabolic equation. *Differential Equations*, 39(11):1626–1632, 2003.

[338] S. Manservisi. Some shape optimal control computations for Navier–Stokes flows. In *Control and Estim. Dist. Param. Syst. Conf.*, pages 231–248. Birkhausep, 2003.

Bibliography

[339] P. Marcellini and C. Sbordone. Relaxation of non-convex variational problems. *Atti. Acad. Naz. Lincei Cl. Sci. Fis. Mat. Nat.*, 63(5):341–344, 1977.

[340] L. Markus. A brief history of control. In *Differ. Equat. Syst. Contr. Sci.*, pages 11–25. New York, 1994.

[341] E. Martinez. Reduction in optimal control theory. *Repts. Math. Phys.*, 53(1):79–90, 2004.

[342] F. Masiero. Stochastic optimal control problems and parabolic equations in Banach spaces. *SIAM J. Contr. Optim.*, 47(1):251–300, 2008.

[343] A.S. Matveev and V.A. Yakubovich. *The abstract theory of optimal control*. Sankt Peterb. Univ., 1994.

[344] H. Maurer and H. Mittelmann. Optimization techniques for solving elliptic control problems with control and state constraints. Pt. 1. Boundary control. *Comput. Optimiz. Appl.*, 16(1):29–55, 2000.

[345] R. McGill. Optimal control, inequality state constraints and the generalized Newton - Raphson algorithm. *J. Soc. Ind. Appl. Math.*, A3(2):291–298, 1965.

[346] E.J. McShane. Generalized curves. *Duke Math. J.*, (6):513–536, 1940.

[347] N.G. Medhin. Penalty technique for a min-max control problem. *J. Optim. Theory Appl.*, 52(1):81–95, 1987.

[348] N.G. Medhin. Optimal processes governed by integral equation with unilateral constraints. *J. Math. Anal. Appl.*, 129:269–283, 1988.

[349] D. Meidner and B. Vexler. A priori error estimates for space-time finite element discretization of parabolic optimal control problems. Part II: Problems with control constraints. *SIAM J. Contr. Optim.*, 47(3):1301–1329, 2008.

[350] V.S. Melnik. The method of monotone operators in the theory of optimal control with constraints. *DAN YSSR*, (7):71–73, 1984.

[351] C. Meyer, P. Philip, and F. Tröltzsch. Optimal control of a semilinear PDE with nonlocal radiation interface conditions. *SIAM J. Contr. Optim.*, 45(2):699–721, 2006.

[352] C. Meyer and A. Rösch. Superconvergence properties of optimal control problems. *SIAM J. Contr. Optim.*, 43(3):970–985, 2004.

[353] P. Michel. Necessary conditions for optimality of elliptic systems with positivity constraints on the state. *SIAM J. Control Optim.*, 18(1):91–97, 1980.

496   *Bibliography*

[354] F. Mignot and J.P. Puel. Contrôle optimal d'un systèmes, gouvernée par une unéquation variationelle parabolique. *C. R. Acad. Sci. Paris*, 297(12):277–280, 1984.

[355] S.G. Mihlin. *Variational methods in mathematical physics*. Nauka, Moscow, 1970.

[356] S.G. Mihlin. *Linear partial differential equations*. VSh, Moscow, 1977.

[357] J. Milnor. *Morse theory*. Princeton University Press, 1963.

[358] J. Milnor. Analytic proof of the hairy ball theorem and the Brouwer fixed point theorem. *Am. Math. Mon.*, 85:521–524, 1978.

[359] A.A. Milutin. *The maximum principle in the general problem of optimal control*. Phyzmathlit, Moscow, 2001.

[360] A.A. Milutin. On amplification of the conditions of Clarke and Smirnov for convex value inclusions. *Math. Sbornik*, 194(2):87–116, 2003.

[361] G.J. Minti. Monotone (non linear) operators in Hilbert space. *Duke Math. J.*, 29:341–346, 1962.

[362] C. Miranda. *Partial differential equations of elliptic type*. Springer, 1970.

[363] S. Mizohata. *The theory of partial differential equations*. Cambridge, 1973.

[364] N.N. Moiseev. Dynamic programming methods in optimal control theory. *J. Comp. Math. and Math. Phys.*, 4(3):485–494, 1964.

[365] N.N. Moiseev. *Numerical methods in the theory of optimal systems*. Nauka, Moscow, 1971.

[366] B.S. Mordukhovich. *Approximation methods in problems of optimization and control*. Nauka, Moscow, 1988.

[367] B.S. Mordukhovich and J.V. Outrata. On second-order subdifferentials and their applications. *SIAM J. Optim.*, 12(1):139–169, 2001.

[368] B.S. Mordukhovich and J.-P. Raymond. Neumann boundary control of hyperbolic equations with pointwise state constraints. *SIAM J. Contr. Optim.*, 43(4):1354–1372, 2005.

[369] B.S. Mordukhovich and B. Wang. Necessary suboptimality and optimality conditions via variational principles. *SIAM J. Contr. Optim.*, 21(2):623–640, 2002.

[370] J.J. Moreau. *Fonctionnelles covexes*. Collèges de France, Paris, 1967.

Bibliography 497

[371] S.F. Morozov and V.I. Plotnikov. On necessary and sufficient conditions for the continuity and semicontinuity of the functionals of variational calculation. *Math. Sbornik*, (3):265–280, 1962.

[372] S.F. Morozov and V.I. Sumin. Necessary optimality conditions for the control of stationary transfer processes. *Izv. Vuzov. Math.*, (10):46–56, 1974.

[373] F. Murat. Thèoremes de non existènce pour des problèmes de contrôle dans les coefficients. *C. R. Acad. Sci. Paris*, (5):A395–A398, 1972.

[374] F. Murat. Contre-examples pour divers problèmes ou le contrôle intervient dans les coefficients. *Ann. di Math. Pura et Appl.*, 112:46–68, 1977.

[375] F. Murat and L. Tartar. On the control of coefficients in partial equations. In *Topics Math. Model. Compos. Mater.*, pages 1–8. Boston, 1997.

[376] A. Myslinski and J. Sokolowski. Nondifferentiable optimization problems for elliptic systems. *SIAM J. Contr. Optim.*, 23(4):632–648, 1985.

[377] S. Nababan and E.S. Noussair. Theory for optimal control of quasilinear parabolic differential equations. *J. Math. Anal. Appl.*, 94:222–236, 1983.

[378] S. Nababan and K.L. Teo. Necessary conditions for optimal controls for systems governed by parabolic partial delay-differential equations in divergence form with first boundary conditions. *J. Optim. Theory Appl.*, 37(4):565–613, 1982.

[379] M. Nacinovich and R. Schianchi. Semicontinuity of a functional depending on curvature. *Calc. Var.*, 15(2):203–214, 2002.

[380] J. Nash. The analyticity of solutions of problems implicit function with analytic data. *Uspehi Math. Nauk*, 26(4):216–226, 1971.

[381] A. Nedic and D.P. Bertsekas. Incremental subgradient methods for nondifferentiable optimization. *SIAM J. Optim.*, 12(1):109–138, 2001.

[382] Z. Nehari. Sufficient conditions in the calculus of variations and in the theory of optimal control. *Proc. Amer. Math. Soc.*, 39(3):535–539, 1973.

[383] P. Neittaanmäki and D. Tiba. On the finite element approximation of the boundary control for two-phase Stefan problem. *Lect. Not. Contr. Inf. Sci.*, 62:356–370, 1984.

[384] P. Neittaanmäki and D. Tiba. *Optimal control of nonlinear parabolic systems. Theory, algorithms, and applications.* Marcel Dekker, New York, 1994.

498   Bibliography

[385] L.W. Neustadt. An abstract variational theory with application to a broad class of optimization problems. *SIAM J. Control*, (4):505–527, 1966.

[386] L.W. Neustadt. *Optimization*. Princeton University Press, Princeton, 1976.

[387] L. Nirenberg. *Topics in nonlinear functional analysis*. Lecture Notes, New York University, 1974.

[388] N.N. Novozhenov and V.I. Plotnikov. Generalized Lagrange multiplier rule for distributed systems with phase constraints. *Differ. Equat.*, 18(4):584–592, 1982.

[389] C. Olech. Existence theory and optimal control theory. *Topol. Funct. Anal., Vienna*, 1:291–328, 1976.

[390] J.V. Outrata and W. Römisch. On optimality conditions for some non-smooth optimization problems over L_p spaces. *J. Optim. Theory Appl.*, 126(2):411–438, 2005.

[391] Z. Páles and V. Zeidan. Optimal control problems with set-valued control and state constraints. *SIAM J. Optim.*, 14(2):334–358, 2003.

[392] L.P. Pan and J. Yong. Optimal control for quasilinear retarded parabolic systems.

[393] N. Papageorgiou. On the optimal control of strongly nonlinear evolution equations. *J. Math. Anal. Appl.*, 164:83–103, 1992.

[394] N. Papageorgiou. Optimal control of nonlinear evolution equations with nonmonotone nonlinearities. *J. Optim. Theory Appl.*, 77(3):643–660, 1993.

[395] N.S. Papageorgiou. On the variational stability of a class of nonlinear parabolic equations. *Z. Anal. Anwend.*, 15(1):245–262, 1996.

[396] J.Y. Park and J.U. Jeong. Optimal control of damped Klein–Gordon equations with state constraints. *J. Math. Anal. Appl.*, 334(1):11–27, 2007.

[397] F. Patrone. On the optimal control of variational inequalities. *Pubbl. Ist. Mat. Univ. Genova*, 146:1–15, 1976.

[398] J. Peetre. Espaces d'interpolation at theorems de Sobolev. *Ann. Inst. Fourier*, 16:279–317, 1966.

[399] J.P. Penot. Well-behavior, well-posedness and nonsmooth analysis. In *4th Int. Conf. Math. Meth. Oper. Res. and 6 Workshop Well-posed and Stab. Optim. Probl. Pliska*, pages 141–190, 1998.

Bibliography 499

[400] H.J. Pesch and R. Bulirsch. The maximum principle, Bellman's equation, and Carathéodory's work. *J. Optim. Theory Appl.*, 80(2):199–225, 1994.

[401] D. Peterson and J.H. Zalkind. A review of direct sufficient conditions in optimal control theory. *Int. J. Control*, 28(4):589–610, 1978.

[402] L.V. Petuhov and V.A. Troitsky. Variational optimization problem for hyperbolic equations with boundary controls. *Prikl. Math. Mech.*, 39(2):260–270, 1975.

[403] V.I. Plotnikov. On the convergence of finite-dimensional approximations in the problem the optimal heating of massive bodies of arbitrary shape. *J. Comp. Math. and Math. Phys.*, 8(1):136–157, 1968.

[404] V.I. Plotnikov. Necessary and sufficient conditions of optimality conditions and uniqueness of optimizing functions for control systems of general type. *Izv. AN SSSR, Ser. Math.*, 36(3):652–679, 1972.

[405] N.I. Pogodaev. On solutions of the Goursat–Darboux System with boundary controls and distributed controls. *Differential Equations*, 43(8):1142–1152, 2007.

[406] E. Polak. *Computational methods in optimization: A unified approach.* Academic Press, New York, 1975.

[407] B.T. Polyak. Semicontinuity of integral functionals and existence theorems on extremal problems. *Mathem. Sbornik*, 78(1):65–84, 1969.

[408] L.S. Pontryagin, V.G. Boltyansky, R.V. Gamkrelidze, and E.F. Mishchenko. *The mathematical theory of optimal processes.* Nauka, Moscow, 1974.

[409] M.M. Postnikov. *Introduction to Morse theory.* Nauka, Moscow, 1971.

[410] M.M. Potapov. Difference approximation and regularization of optimal control problems for Goursat – Darboux systems. *Vestn. MGU, VMK*, (2):17–26, 1978.

[411] E.G. Poznyak and E.V. Shikin. *Differential geometry: First look.* Moscow Univ., 1990.

[412] D. Priess. Gataux differentiable Lipschitz functions need not be Frechet differentiable on a resident subset. *Rend. Circ. Mat. Palermo*, 31:217–222, 1982.

[413] A.J. Pritchard and M.J.E. Mayhew. Feedback from discrete points for distributed parameter systems. *Intern. J. Control*, 14(4):619–630, 1971.

[414] B.N. Pshenichny. *Necessary optimality conditions.* Nauka, Moscow, 1969.

500   *Bibliography*

[415] B.N. Pshenichny. *Convex analysis and variational problems*. Nauka, Moscow, 1980.

[416] B.N. Pshenichny. Implicit function theorems for multi-valued mapping. In *Nonsmooth Optimization*, pages 371–391. New York, 1989.

[417] C. Pucci. Un problema variazionale per I coefficienti di equazioni differenziali di tipo ellittico.

[418] C. Pucci. Operatori ellittico estremanti. *Ann. Mat. Pura ed Appl.*, 72:141–170, 1966.

[419] J.H. Qiu. Ekeland's variational principle in Frechet spaces and the density of extremal points. *Stud. Math.*, 168(1):81–94, 2005.

[420] R. Radulescu and M. Radulescu. Local inversion theorems without assuming continuous differentiability. *J. Math. Anal. Appl.*, 138:581–590, 1989.

[421] U. Raitums. Necessary conditions for an extremum in optimal control problems for nonlinear elliptic equations. *Sib. Math. J.*, 23(1):144–152, 1982.

[422] U. Raitums. *Optimal control problems for elliptic equations*. Zinatne, Riga, 1989.

[423] P.K. Rashevsky. *Geometric theory of partial differential equations*. URSS, Moscow, 2003.

[424] J.P. Raymond. Existence and uniqueness results for minimization problems with nonconvex functionals.

[425] J.P. Raymond. Feedback boundary stabilization of the two-dimensional Navier–Stokes equations. *SIAM J. Contr. Optim.*, 45(3):790–828, 2006.

[426] M. Reed and B. Simon. *Methods of modern mathematical physics. I. Functional analysis*. Academic Press, New York, London, 1972.

[427] J.P. Revalski and N.V. Zhivkov. Well-posed constrained optimization problems in metric spaces. *J. Optim. Theory Appl.*, 76(1):145–163, 1993.

[428] J.C. Reyes and R. Griesse. State-constrained optimal control of the three-dimensional stationary Navier–Stokes equations. *J. Math. Anal. Appl.*, 343(1):257–272, 2008.

[429] J.C. Reyes and F. Tröltzsch. Optimal control of the stationary Navier–Stokes equations with mixed control–state constraints. *SIAM J. Contr. Optim.*, 46(2):604–629, 2007.

[430] J.I. Richards and H.K. Joun. *Theory of distributions: A non technical introduction*. Cambridge University Press, Cambridge, 1990.

Bibliography 501

[431] S.M. Robinson. Stability theory for the systems of inequalities. Part 2. Differentiable nonlinear systems. *SIAM J. Num. Anal.*, 13(4):497–513, 1976.

[432] R.T. Rockafellar. *Convex analysis*. Princeton Mathematical Series, Princeton, 1970.

[433] A. Rösch and F. Tröltzsch. Existence of regular Lagrange multipliers for a nonlinear elliptic optimal control problem with pointwise control–state constraints. *SIAM J. Control Optim.*, 45(2):548–564, 2006.

[434] T. Roubiček. *Relaxation on optimization theory and variational calculus*. W. de Guyter, Berlin, New York, 1997.

[435] T. Roubiček. Optimal control of nonlinear Fredholm integral equation. *J. Optim. Theory Appl.*, 97(3):707–729, 1998.

[436] E. Roxin. The existence of optimal control. *Mich. Math. J.*, 9(2):109–112, 1962.

[437] L.I. Rozonoer. The maximum principle of Pontryagin in the theory of optimal systems. *Autom. and Telem.*, 20(10–12):1320–1334, 1441–1458, 1561–1578, 1959.

[438] J.E. Rubio. Generalized curves and extremal points. *SIAM J. Control*, 13(1):28–48, 1975.

[439] W. Rudin. *Functional analysis*. McGraw-Hill Science, 1991.

[440] B. Russak. An indirect sufficient proof for problem with bounded state variables. *Pacif. J. Math.*, 62(1):219–253, 1976.

[441] T.L. Saaty and J. Bram. *Nonlinear mathematics*. McGraw-Hill, New York, 1964.

[442] E. Sachs. A parabolic control problem with a bounded condition of the Stefan - Boltzman type. *Z. Angew. Math. Mech.*, 58(10):443–450, 1978.

[443] A. Sage and P. Chaudhuri. Gradient and quasigradient linear computing technic for distributed parameter systems. *Int. J. Control*, 6(1):81–98, 1967.

[444] C. Saguez. Contrôle optimal d'un systèmes, gouvernée par une unéquation variationelle parabolique. Observation du domaine de contact. *C. R. Acad. Sci. Paris*, 287:957–959, 1979.

[445] M.E. Salukvadze and M.T. Tcutcunava. An optimal control problem for systems described by partial differential equations of hyperbolic type with delay. *J. Optim. Theory Appl.*, 52(2):311–322, 1987.

502 Bibliography

[446] W. Schmidt. Necessary conditions for discrete integral processes in Banach space. *Beitr. Anal.*, 16:137–145, 1981.

[447] L. Schwartz. *Analyse mathématique. I, II.* Hermann, Paris, 1967.

[448] L. Schwartz. *Théorie des distributions.* Hermann, Paris, 1967.

[449] L.A. Scornyakov et al. *General algebra. Vol. 2.* Nauka, Moskow, 1991.

[450] J. Séa. *Optimisation. Théorie et algorithmes.* Dunod, Paris, 1971.

[451] J. Sebastiãoe e Silva. Le calcul differentiel et integral dans les éspaces localment convexes, reels on complexes. *Atti. Acc. Dei Lincei Rend.*, 20(6):743–750, 1956.

[452] T. Seidman and H.X. Zhou. Existence and uniqueness of optimal control for a parabolic equations quasilinear. *SIAM J. Contr. Optim.*, 20(6):749–762, 1982.

[453] T. Seidman and H.X. Zhou. Existence and uniqueness of optimal control for a quasilinear parabolic equations. *SIAM J. Contr. Optim.*, 20(6):747–762, 1982.

[454] G. Seifert and W. Trelfall. *Calculus of variations in the large.* Izhevsk Univ., 2000.

[455] S. Serovajsky. An optimal control problem for an elliptic system with power nonlinearity. *Sib. Math. J.*, 25(1):120–125, 1984.

[456] S. Serovajsky. The regularization method of Tikhonov in the problem of optimal control for nonlinear parabolic system. *Sib. Math. J.*, 30(1):212–215, 1989.

[457] S. Serovajsky. Linearizability of infinite dimensional control systems. *Izv. Vuzov. Math.*, (12):71–80, 1990.

[458] S. Serovajsky. Approximate optimal conditions for a system described by a nonlinear parabolic equation. *Izv. Vuzov. Math.*, (11):52–60, 1991.

[459] S. Serovajsky. Extended differentiability of an implicit function theorem in spaces without norm. *Izv. Vuzov. Math.*, (12):55–63, 1991.

[460] S. Serovajsky. Necessary and sufficient conditions of optimality for a system described by a nonlinear elliptic equation. *Sib. Math. J.*, 32(3):141–150, 1991.

[461] S. Serovajsky. Optimal control in nonlinear stationary system with a non-monotonic operators. *Differ. Equat.*, 28(9):1579–1587, 1992.

[462] S. Serovajsky. Pareto optimality for systems described by nonlinear equations of parabolic type. *Izv. Vuzov. Math.*, (11):55–64, 1992.

Bibliography

[463] S. Serovajsky. Quasilinearization method in terms of the irreversibility of the operator derivative. *Izv. AN KazSSR, Ser. Phys. Math.*, (1):69–72, 1992.

[464] S. Serovajsky. The method of regularization in the problem of optimal control of nonlinear hyperbolic system. *Differ. Equat.*, 28(12):2188–2190, 1992.

[465] S. Serovajsky. The stability in the linear approximation in infinite dimensional systems. *Izv. Vuzov. Math.*, (8):57–64, 1992.

[466] S. Serovajsky. Differentiation of inverse functions in spaces without norm. *Functional analysis and applications*, 27(4):84–87, 1993.

[467] S. Serovajsky. Inverse function theorem and extended differentiability in Banach spaces. *Izv. Vuzov. Math.*, (8):39–49, 1995.

[468] S. Serovajsky. Necessary optimality conditions in the case of nondifferentiability of control–state mapping. *Differ. Equat.*, 31(6):1055–1059, 1995.

[469] S. Serovajsky. Optimization in a nonlinear parabolic system with a control in the coefficients. *Rus. Acad. Sci. Sb. Math.*, 81(2):533–543, 1995.

[470] S. Serovajsky. Principles of extended differentiability. *Izv. NAN HK, Ser. Phys. Math.*, (1):44–48, 1995.

[471] S. Serovajsky. Extremum problems on differentiable submanifolds of Banach spaces. *Izv. Vuzov. Math.*, (5):83–86, 1996.

[472] S. Serovajsky. Gradient methods in an optimal control problems of nonlinear elliptic system. *Sib. Math. J.*, 37(5):1154–1166, 1996.

[473] S. Serovajsky. Optimal control of a nonlinear singular system with state constraints. *Mathematical Notes*, 60(4):383–388, 1996.

[474] S. Serovajsky. Extendedly differentiable manifolds. *Izv. Vuzov. Math.*, (1):55–65, 1997.

[475] S. Serovajsky. Approximate solution of optimization problems for infinite-dimensional singular systems. *Sib. Math. J.*, 39(3):660–673, 2003.

[476] S. Serovajsky. Approximate solution of singular optimization problems. *Mathematical Notes*, 74(5):685–694, 2003.

[477] S. Serovajsky. Lower completion and extension of extremum problems. *Izv. Vuzov. Math.*, (5):30–41, 2003.

504   Bibliography

[478] S. Serovajsky. Optimal control for elliptic equations with nonsmooth nonlinearity. *Differ. Equat.*, 39(10):1420–1424, 2003.

[479] S. Serovajsky. Sequential extension in the coefficients control problems for elliptic type equations. *Journal of Inverse and Ill-Posed Problems*, 11(5):523–536, 2003.

[480] S. Serovajsky. An approximate solution of the optimal control problem for elliptic singular equation with a nonsmooth nonlinearity. *Izv. Vuzov. Math.*, (1):80–86, 2004.

[481] S. Serovajsky. Approximate penalty method in optimal control problems for nonsmooth singular systems. *Mathematical Notes*, 76(5):834–843, 2004.

[482] S. Serovajsky. *Counterexamples in the optimal control theory.* Brill Academic Press, Netherlands, Utrecht-Boston, 2004.

[483] S. Serovajsky. A control problem in coefficients with extended derivative with respect to the convex set. *Izv. Vuzov. Math.*, (12):46–55, 2005.

[484] S. Serovajsky. Calculation of functional gradients and extended differentiation of operators. *Journal of Inverse and Ill-Posed Problems*, 13(4):383–396, 2005.

[485] S. Serovajsky. Optimal control for singular equation with nonsmooth nonlinearity. *Journal of Inverse and Ill-Posed Problems*, 14(6):621–631, 2006.

[486] S. Serovajsky. *Optimization and differentiation. Minimization of functionals. Stationary systems.* Print-S, 2006.

[487] S. Serovajsky. Optimal control of a singular evolutional equation with nonsmooth operator and fixed terminal state. *Differential equations*, 43(2):259–266, 2007.

[488] S. Serovajsky. *Optimization and differentiation. Evolutional systems.* Kazakh univ., 2009.

[489] S. Serovajsky. Optimization for nonlinear hyperbolic equations without the uniqueness theorem for a solution of the boundary-value problem. *Russian Mathematics (Izv. VUZ)*, 53(1):76–83, 2009.

[490] S. Serovajsky. Differentiation of operators and the extremum conditions with category interpretation. *Izv. Vuzov. Math.*, (2):66–77, 2010.

[491] S. Serovajsky. Necessary conditions of optimality for nonlinear stationary systems without the differentiability of control-state mapping. *Izv. Vuzov. Math.*, (6):32–47, 2010.

Bibliography

[492] S. Serovajsky. Approximation methods for optimal control problems of nonlinear infinite dimensional systems. *Mathematical notes*, 94(4):567–582, 2013.

[493] S. Serovajsky. Optimal control for nonlinear evolutional systems without the differentiability of control–state mapping. *Mathematical Notes*, 93(4):586–603, 2013.

[494] S. Serovajsky. Optimal control for systems described by hyperbolic equation with strong nonlinearity. *Journal of Applied Analysis and Computation*, 3(2):183–195, 2013.

[495] S. Serovajsky. Differentiation functor and its application in the optimization control theory. In *Fourier Analysis. Trends in Mathematics*, pages 335–347. Springer Int. Publ., Switzerland, 2014.

[496] S. Serovajsky. State-constrained optimal control of nonlinear elliptic variational inequalities. *Optimization Letters*, 8(7):2041–2051, 2014.

[497] S. Serovajsky. Optimization control problems for systems described by elliptic variational inequalities with state constraints. In *Methods of Fourier Analysis and Approximation Theory. Applied and Numerical Harmonic Analysis*, pages 211–224. Springer International Publishing (Birkäuser), Switzerland, 2016.

[498] A. Shapiro. On concepts of directional differentiability. *J. Optim. Theory Appl.*, 66(6):477–488, 1990.

[499] R.E. Showalter. *Monotone operators in Banach space and nonlinear differential equations*. Amer. Math. Soc., Providence, 1997.

[500] M. Sibony. Some numeric techniques for optimal control governed by partial differential equations. In *Num. Meth. Optim. Basel*, pages 123–136. 1973.

[501] M. Sidar. An iterative algorithm for optimal control problem. *Int. J. Nonlin. Mech.*, 9(1):1–16, 1968.

[502] C. Simionescu. Optimal control problem in Sobolev spaces with weight. *SIAM J. Control Optim.*, 14(1):137–143, 1976.

[503] T.K. Sirazetdinov. *Optimization of distributed parameter systems*. Nauka, Moscow, 1977.

[504] T. Slawing. Shape optimization for semi-linear elliptic equations based on a embedding domain method. *Appl. Math. and Optim.*, 49(2):183–199, 2004.

[505] J.M. Sloss, I.S. Sadek, Jr.J.C. Bruch, and S. Adali. Maximum principle for the optimal control of a hyperbolic equation in one space dimension, part 1: theory. *J. Optim. Theory Appl.*, 87(1):33–45, 1995.

506   Bibliography

[506] E.R. Smolyakov. *Unknown pages of the history of optimal control*. URSS, Moscow, 2002.

[507] S.L. Sobolev. *Sur les equations aux dérivées partielles hyperboliques non-linéaires*. Edizione Cremoneze, Roma, 1961.

[508] S.L. Sobolev. *Some applications of functional analysis in mathematical physics*. Nauka, Moscow, 1988.

[509] J. Sokolowsky. On parametric optimal control for a class of linear and quasilinear equations of parabolic type. *Contr. Cybern.*, 4(1):19–38, 1975.

[510] H.A. Spang. A review of minimization techniques for non linear function. *SIAM Review*, 4(4):343–365, 1962.

[511] S. Sritharan. Dynamic programming of the Navier–Stokes equations. *Systems Control Lett.*, 16(4):299–307, 1991.

[512] A.M. Steinberg and H.L. Stalford. On existence of optimal control. *J. Optim. Theory and Appl.*, 11(3):266–273, 1973.

[513] M.I. Sumin. Suboptimal control of semilinear elliptic equations with phase constraints and boundary control. *Differ. Equat.*, 37(3):260–275, 2001.

[514] M.B. Suryanarayana. Existence theorems for problems concerning by hyperbolic partial equations. *J. Optimal Theory Appl.*, 15(4):361–392, 1975.

[515] L. Tartar. Problèmes de contrôle des coefficients dans les équations aux derivées partielles. *Lect. Notes Econom. Math. Syst.*, 107:420–426, 1977.

[516] R. Temam. *Navier–Stokes equations*. North-Holland Publ. Comp., Amsterdam, 1979.

[517] K.L. Teo. Optimal control of systems governed by time-delayed, second-order, linear, parabolic partial differential equations with a first boundary condition. *J. Optim. Theory Appl.*, 29(3):437–481, 1979.

[518] D. Tiba. Optimal control of nonsmooth distributed parameter systems.

[519] D. Tiba. Boundary control for a Stefan problem. In *Optimal control in partial differential equations*, pages 229–242. Birkhäuser, Basel, 1984.

[520] D. Tiba. Optimal control for second order semilinear hyperbolic equations. *Contr. Theory Adv. Techn.*, 3(1):33–46, 1987.

[521] D. Tiba. New results in space optimization. *Math. Repts.*, 5(4):389–398, 2003.

Bibliography

[522] V.M. Tihomirov. *Lagrange principle and optimal control problems.* MGU, Moscow, 1982.

[523] A.N. Tihonov. On the methods of regularization of optimal control problems. *DAN SSSR*, 162(4):763–765, 1965.

[524] A.N. Tihonov and V.Ya. Arsenin. *Methods for solving of ill-posed problems.* Nauka, Moscow, 1986.

[525] A.N. Tihonov and F.P. Vasiliev. Methods for solving of ill-posed extremum problems. In *Mat. Mod. Num. Met.*, volume 3, pages 297–342. Banach Centr. Publ., Warzawa, 1978.

[526] B.A. Ton. Optimal shape control problem for the Navier–Stokes equations. *SIAM J. Optim. Contr.*, 41(6):1733–1747, 2003.

[527] L. Tonelli. Un teorema di calcolo delle variazioni. *Rend. Acad. Naz. Lincei*, 15:417–423, 1932.

[528] J.S. Treiman. The linear nonconvex generalized gradient and Lagrange multipliers. *SIAM J. Optim.*, 5(3):670–680, 1995.

[529] C. Trenchea. Optimal control of an elliptic equation under periodic conditions. *Mem. Sec. Sti. Ser. 4, Acad. Rom.*, 25:23–35, 2002.

[530] V.A. Trenogin. *Functional analysis.* Nauka, Moscow, 1980.

[531] F. Tröltzsch. *Optimality conditions for parabolic control problems and applications.* Teubner Texte, Leipzig, 1984.

[532] F. Tröltzsch. On the Lagrange–Newton–SQP method for the optimal control of semilinear parabolic equations. *SIAM J. Contr. Optim.*, 38(1):294–312, 1999.

[533] C. Trombetti. On the lower semicontinuity and relaxation properties of a certain classes of variational integrals. *Rend. Accad. Naz. Sci. XL Mem. Mat.*, 21(1):25–51, 1997.

[534] C. Ursescu. Continuability, differentiability and calculus adjacent and tangent sets. *Mathematica*, 42(1):97–112, 2000.

[535] M.M. Vainberg. *Variational method and method of monotone operators.* Nauka, Moscow, 1972.

[536] E.M. Vaisbord. An optimal control problem for systems with distributed parameters. *Izv. AN SSSR, Thech. Kib.*, (6):167–172, 1966.

[537] F.P. Vasiliev. On the regularization of ill-posed problems of minimization on sets that are given approximately. *J. Comp. Math. and Math. Phys.*, 20(1):38 40, 1980.

508   Bibliography

[538] F.P. Vasiliev. *Methods for solving extremum problems*. Nauka, Moscow, 1981.

[539] F.P. Vasiliev. *Optimization methods. I,II*. MCNMO, Moscow, 2011.

[540] O.V. Vasiliev. *Optimization methods in functional spaces*. Irkutsk Univ., 1979.

[541] R.T. Vescan. On Pareto minima and optimal points for injective functionals. *Bull. Math. Soc. Sci. Math. RSR*, 19(3–4):401–407, 1975.

[542] V.S. Vladimirov. *Generalized functions in mathematical physics*. Springer, 1979.

[543] V.S. Vladimirov. *Equations of mathematical physics*. Nauka, Moscow, 1988.

[544] M.D. Voisie. First-order necessary conditions of optimality for strongly monotone nonlinear control problems. *J. Optim. Theory Appl.*, 116(2):421–436, 2003.

[545] D. Wachsmuth. Analysis of the SQP-Method for optimal control problems governed by the nonstationary Navier–Stokes equations based on L_p-theory. *SIAM J. Contr. Optim.*, 46(2):1133–1153, 2007.

[546] G. Wang. Optimal controls of 3-dimensional Navier–Stokes equations with state constraints. *J. Optim. Control Theory*, 41(2):583–606, 2002.

[547] G. Wang. Pontryagin's maximum principle for optimal control of the stationary Navier–Stokes equations. *Nonlinear Anal.*, 52(8):1853–1866, 2003.

[548] G. Wang and C. Liu. Maximum principle for state–constrained control of semilinear parabolic differential equation. *J. Optim. Theory Appl.*, 115(1):183–209, 2002.

[549] L. Wang. Optimal control of magnetohydrodynamic equations with state constraint. *J. Optim. Theory Appl.*, 122(3):599–626, 2004.

[550] P.K.C. Wang. Optimal control of parabolic systems with boundary conditions involving time delays.

[551] D.E. Ward and G.M. Lee. On relations between vector optimization problems and vector variational inequalities. *J. Optim. Theory Appl.*, 113(3):583–596, 2002.

[552] J. Warga. Relaxed variational problem. *J. Math. Ann. Appl.*, (4):111–127, 1962.

[553] J. Warga. *Optimal control of differential and functional equations*. Academic Press, New York, 1972.

Bibliography 509

[554] L. Wolfersdorf. Optimal control problem with elliptic differential equations systems of first order in the plan. In *Compl. Anal. Meth. Trends Appl.*, pages 270–279. Berlin, 19.

[555] L. Wolfersdorf. On some optimal control problem for linear elliptic systems in the plan. *Beitr. Anal.*, 17:95–98, 1981.

[556] Z.S. Wu and K.L. Teo. A conditional gradient method for an optimal control problem involving a class of nonlinear second order hyperbolic partial differential equation. *J. Math. Anal. Appl.*, 91:376–397, 1983.

[557] Z.S. Wu and K.L. Teo. Optimal control problem involving second boundary value problem of parabolic type. *SIAM J. Contr. Optim.*, 21(5):729–757, 1983.

[558] V.A. Yakubovich. On the abstract theory of optimal control. *Sib. Math. J.*, 18(3):685–707, 19.

[559] V.A. Yakubovich. Sufficient optimality conditions in optimal control problems. *Vest. LGU*, (1):53–58, 1984.

[560] J.J. Ye. Nondifferentiable multiplier rules for optimization and bilevel optimization problems. *SIAM J. Optim.*, 15(1):252–274, 2004.

[561] B.P. Yeo. A modified quasilinearization algorithm for the computation of optimal singular control. *Int. J. Control*, 32(4):723–730, 1980.

[562] J. Yong. Existence theory of optimal controls for distributed parameter systems. *Kodai Math. J.*, 15:193–220, 1992.

[563] L. Young. *Lectures on the calculus of variations and optimal control theory*. W.B. Saunders Co., Philadelphia, 1969.

[564] C. Yu and J. Yu. Unified approach to Hadamard well-posedness. *Acta Annual Funct. Appl.*, 4(4):311–316, 2002.

[565] J. Zabkzyk. *Mathematical control theory: An introduction*. Birkhauser, Boston, 1992.

[566] E. Zarantonello. *Contributions to nonlinear functional analysis*. Academic Press, New York, 1973.

[567] Z. Zhao and L. Tian. Optimal control of sufficient nonlinear KdV Burgers equation. *J. Jiangsu Univ. Ntur. Sci.*, 26(2):140–143, 2005.

[568] X.P. Zhou. Weak lower semicontinuity of a functional with any order. *J. Math. Anal. Appl.*, 221(1):217–237, 1998.

[569] V. Zizler. On some extremal problem in Banach spaces. *Math. Scand.*, 32:214–229, 1973.

510 *Bibliography*

[570] L. Zlaf. *Calculus of variations and integral equations.* Nauka, Moscow, 1970.

[571] T. Zolezzi. Necessary conditions for optimal controls of elliptic or parabolic problems. *SIAM J. Control,* 10(4):594–607, 1972.

[572] T. Zolezzi. A characterization of well-posed optimal control systems. *SIAM J. Control Optim.,* 19(5):604–616, 1981.

[573] T. Zolezzi. Well-posedness and the Lavrentiev phenomenon. *SIAM J. Control Optim.,* 30(4):787–799, 1992.

[574] T. Zolezzi. Extended well-posedness of optimization problems. *J. Optim. Theory Appl.,* 91(1):257–266, 1996.

[575] G. Zoutendijk. Non linear programming: A numerical survey. *SIAM Control,* 4(1):194–210, 1966.

Index

δ-function, 296

calculus of variations, 11
  on the large, 50
category, 461
  of Banach spaces, 462
  of sets, 461
closure, 22
coefficient
  Fourier, 61
cofunctor, 462
  general, 463
condition
  Lusternik, 181
  of extremum
    first order, 11
    high order, 11
    necessary, 6
    sufficient, 6
  of optimality
    degenerate, 288
    necessary, 7
    sufficient, 7, 54, 306
  stationary
    for the function, 5
    for the functional, 31
constraint
  local, 107
continuation, 207
control, 92
  initial, 354
  singular, 115, 288
control-state mapping, 93
convergence, 151

derivative
  Clarke, 84

conormal, 120
extended
  Fréchet, 193
  Gâteaux, 193
Fréchet, 78
Gâteaux
  of the functional, 27
generalized, 36
of operator
  Fréchet, 131
  Gâteaux, 130
of the abstract function, 264
partial, 244
with respect to the direction, 26
with respect to the subspace, 42
diffeomorphism, 136
dimension, 20
distribution, 36
  on the interval, 267
divergence, 412

eigenfunction, 413
eigenvalue, 413
elements
  linear dependent, 20
embedding
  compact, 151
  continuous, 25
  dense, 25
equality
  Parceval, 59
equation
  adjoint, 91
  Bellman, 301
  Boussinesq, 457
  Burgers, 457
  continuity, 412

512 *Index*

elliptic
  linear, 98
  nonlinear, 152
Euler, 15
evolutional
  first order, 271
  second order, 366
Hamilton–Jacobi, 299
heat, 283
  first boundary problem, 283
Klein–Gordon, 457
Korteweg–de Vries, 457
linearized, 244
Poisson, 39, 98
Schrödinger, 457
state, 92
wave, 380
equations
  motion, 412
  Navier–Stokes, 411
    stationary, 430
estimate
  a priori, 99
example
  Fréchet, 133
extremal, 15

form
  quadratic, 30
formula
  Green, 120
  Lagrange, 131
  of integration by parts, 269
function
  abstract, 264
    differentiable, 264
    essentially bounded, 267
    simple, 265
  abstract Bochner
    integrable, 265
    measurable, 265
  Bellman, 304
  Caratheodory, 132
  essentially bounded, 34
  implicit, 243, 244

functional
  ∗-weakly lower semicontinuous,
    63
  Bellman, 304
  convex, 18
  cost, 93
  Fréchet differentiable, 78
  Gâteaux differentiable, 27
  Lagrange, 111
  lower semicontinuous, 62
  proper, 64
  strictly convex, 19
  strictly uniformly convex, 69
  subdifferentiable, 75
  weakly lower semicontinuous, 62
functor, 462
  of differentiation, 464
  of extended differentiation, 470

gradient, 28

Hamiltonian, 301
homeomorphism, 182

inequality
  Hölder, 35
  variational, 46
integral
  Bochner, 265
  Dirichlet, 38, 56
intersection
  of spaces, 150
isomorphism, 462
  canonic, 26

Jacobian, 131

lemma
  Euler–Lagrange, 13
  Gronwall, 366
  lemma, 62
  of variations calculus, general,
    13
limit, 21
  lower, 62

maximum

Index 513

absolute, 6, 32
local, 6, 32
strict, 6
strict local, 32
method
  Bellman, 303
  extension, 83
  Faedo–Galerkin, 271
  Galerkin, 100
  gradient, 81
  Krotov, 82
  Lagrange multipliers, 110
  Newton–Raphson, 127
  penalty, 116
  projection gradient, 81
  successive approximation, 127, 282
  successive approximations, 95
  Tihonov regularization, 70, 289
minimum
  absolute, 6, 32
  local, 6, 32
  strict, 6
  strict local, 32
  strong, 15
  weak, 15
morphism, 461
  inverse, 462
Morse theory, 50
multi-index, 33
multiplier
  Lagrange, 111

norm, 20
  of the operator, 23

object
  of category, 461
operator
  ∗-weakly continuous, 92
  adjoint, 24
  affine, 20
  bounded, 23
  coercive, 144
  continuous, 22

continuous at a point, 22
continuously differentiable, 134
control, 92
differentiable
  extended with respect to the convex set, 228
  with respect to the convex set, 215
embedding, 25
extendedly differentiable, 193
extendedly Gâteaux differentiable, 193
Fréchet differentiable, 131
Gâteaux differentiable, 130
implicit, 243
Laplace, 38
linear, 20
monotone, 144
Nemytsky, 132
power, 133, 185
self-adjoint, 25, 105
state, 92
strictly monotone, 144
trace, 119
weakly continuous, 91

point
  Lebesgue, 34
  limit, 22
  normal, 181
  of inflection, 10
  stationary, 5
principle
  Ekeland, 73
  Lagrange, 112
  maximum, 115
  optimality of Bellman, 303
problem
  boundary, 15
  Dirichlet, 39, 98
    nonhomogeneous, 121
  extremum
    Hadamard ill-posed, 71
    Hadamard well-posed, 71
    Tihonov ill-posed, 68

514 *Index*

Tihonov well-posed, 68
    identification, 221
    inverse, 221
    Lagrange, 11, 55
    Neumann, 199
    optimal control, 93
    spectrum, 413
product
    of spaces, 90
    scalar, 25
projector, 47
property
    topological, 62

sequence
    bounded, 22
    convergent, 21
        *-weakly, 61
        strongly, 21
        uniformly, 22
        weakly, 60
    fundamental, 23
    minimizing, 58
series, 91
    convergent, 91
set
    *-weak closed, 62
    bounded, 22
    closed, 22
    complete, 91
    convex, 18
    dense, 22
    of admissible controls, 93
    open, 22
    weak closed, 62
sets
    diffeomorphic, 136
solution
    classical, 102
    generalized, 102
    global, 270
    local, 270
    of extremum problem
        approximate, 68, 73
    smooth, 121

space
    $D'(\Omega)$, 36
    n-dimensional, 20
    adjoint, 23, 149
    Banach, 23
    complete, 23
    control, 92
    finite dimensional, 20
    functional
        $C(\bar{\Omega})$, 33
        $C^m(0,T;W)$, 265
        $C^m(\bar{\Omega})$, 33
        $D(\Omega)$, 35
        $L_\infty(0,T;W)$, 267
        $L_\infty(\Omega)$, 34
        $L_p(0,T;W)$, 266
        $L_p(\Omega)$, 33
    Hilbert, 25
    infinite dimensional, 20
    linear, 17
    linear normalized, 21
    linear topological, 22
    reflexive, 24
    self-adjoint, 24
    separable, 90
    Sobolev
        $H^m(\Omega)$, 36
        $H_0^m(\Omega)$, 36
        $H^s(\Omega)$, 119
        $H^s(\mathbb{R}^n)$, 118
        $H^{-1/2}(\Gamma)$, 119
        $H^{1/2}(\Gamma)$, 119
        $W_p^m(\Omega)$, 36
        $\dot{W}_p^m(\Omega)$, 36
    state, 92
    value, 92
    vector, 17
state, 92
subcategory, 462
subdifferential, 75
subgradient, 75
subspace, 18
subvariety
    extended differentiable, 210
sum

of series, 91

of spaces, 150

support, 35

system

bilinear, 216

control, 92

strongly nonlinear, 140

weakly nonlinear, 140

theorem

Banach inverse operator, 91

Banach–Alaoglu, 62

Banach–Steinhaus, 145

Bolzano–Weierstrass, 58

composite function, 133

implicit function, 244

inverse function, 135

Krasnoselsky, 132

Lebesgue, 34

Lusternik, 52

Peano, 270

Riesz, 25

Sobolev embedding, 151

trace, 119

Weierstrass, 82

topology

∗-weak, 62

strong, 62

weak, 62

Trace, 119

transformation

Fourier, 117

inverse, 118

uniqueness

of minimum, 66

variation

needle, 48

of the function, 12

of the functional, 13

variety

affine, 18